Metaheuristics in Water, Geotechnical and Transport Engineering

Metaheuristics in Water, Geotechnical and Transport Engineering

Edited by

Xin-She Yang
Mathematics and Scientific Computing
National Physical Laboratory
Teddington, Middlesex, TW11 0LW, UK

Amir Hossein Gandomi
Department of Civil Engineering
The University of Akron
Akron, OH 44325, USA

Siamak Talatahari
Marand Faculty of Engineering
University of Tabriz
Tabriz, Iran

Amir Hossein Alavi
School of Civil Engineering
Iran University of Science and Technology
Tehran, Iran

AMSTERDAM • BOSTON • HEIDELBERG • LONDON • NEW YORK • OXFORD
PARIS • SAN DIEGO • SAN FRANCISCO • SINGAPORE • SYDNEY • TOKYO

ELSEVIER

Elsevier
32 Jamestown Road, London NW1 7BY
225 Wyman Street, Waltham, MA 02451, USA

First edition 2012

Notices
Knowledge and best practice in this field are constantly changing. As new research and experience broaden our understanding, changes in research methods, professional practices, or medical treatment may become necessary.

Practitioners and researchers must always rely on their own experience and knowledge in evaluating and using any information, methods, compounds, or experiments described herein. In using such information or methods they should be mindful of their own safety and the safety of others, including parties for whom they have a professional responsibility.

To the fullest extent of the law, neither the Publisher nor the authors, contributors, or editors, assume any liability for any injury and/or damage to persons or property as a matter of products liability, negligence or otherwise, or from any use or operation of any methods, products, instructions, or ideas contained in the material herein.

British Library Cataloguing-in-Publication Data
A catalogue record for this book is available from the British Library

Library of Congress Cataloging-in-Publication Data
A catalog record for this book is available from the Library of Congress

ISBN: 978-0-323-28260-4

For information on all Elsevier publications visit our website at store.elsevier.com

This book has been manufactured using Print On Demand technology. Each copy is produced to order and is limited to black ink. The online version of this book will show color figures where appropriate.

Working together to grow
libraries in developing countries

www.elsevier.com | www.bookaid.org | www.sabre.org

ELSEVIER   BOOK AID
            International   Sabre Foundation

Contents

Part Three   Transport Engineering

List of contributors

A. Ahangar-Asr Computational Geomechanics Group, College of Engineering, Mathematics and Physical Sciences, University of Exeter, UK

Amir Hossein Alavi School of Civil Engineering, Iran University of Science and Technology, Tehran, Iran

M. Tamer Ayvaz Department of Civil Engineering, Pamukkale University, Denizli, Turkey

H. Md Azamathulla River Engineering and Urban Drainage Research Centre, Universiti Sains Malaysia, Penang, Malaysia

Jafar Bolouri Bazaz Civil Engineering Department, Ferdowsi University of Mashhad, Mashhad, Iran

Jessica Andrea Carballido Departamento de Ciencias e Ingeniería de la Computación, Universidad Nacional del Sur, Blanca, Argentina

Y.M. Cheng Department of Civil and Structural Engineering, Hong Kong Polytechnic University, Kowloon, Hong Kong, People's Republic of China

C.G. Chua Jalan Putra Permai 8G, Taman Equine, Seri Kembangan, Selangor, Malaysia

Sarat Kumar Das Department of Civil Engineering, National Institute of Technology, Rourkela, Odisha, India

Alper Elçi Department of Environmental Engineering, Dokuz Eylül University, Izmir, Turkey

A. Faramarzi Department of Civil Engineering, School of Engineering, University of Greenwich, UK

Mariano Frutos Departamento de Ingeniería, Universidad Nacional del Sur, Blanca, Argentina

Amir Hossein Gandomi Department of Civil Engineering, The University of Akron, Akron, OH

Z.W. Geem Department of Energy and Information Technology, Gachon University, Seongnam, South Korea

Anthony T.C. Goh School of Civil and Environmental Engineering, Nanyang Technological University, Singapore

Yousef Hassanzadeh Department of Civil Engineering, University of Tabriz, Tabriz, Iran

Sharad Kumar Jain Water Resources Development and Management Department, IIT Roorkee, Roorkee, India

A.A. Javadi Computational Geomechanics Group, College of Engineering, Mathematics and Physical Sciences, University of Exeter, UK

Manoj K. Jha Center for Advanced Transportation and Infrastructure Engineering Research, Department of Civil Engineering, Morgan State University, Baltimore, MD

Nikos Kallioras Department of Transportation Planning and Engineering, National Technical University of Athens, Athens, Greece

Matthew G. Karlaftis Department of Transportation Planning and Engineering, National Technical University of Athens, Athens, Greece

Khewal Bhupendra Kesur School of Statistics and Actuarial Science, University of the Witwatersrand, Johannesburg, South Africa

Nikos D. Lagaros Institute of Structural Analysis and Seismic Research, National Technical University of Athens, Athens, Greece

Ali Mollahasani Department of Civil, Environmental, and Material Engineering, University of Bologna, Bologna, Italy

N. Mottaghifard Computational Geomechanics Group, College of Engineering, Mathematics and Physical Sciences, University of Exeter, UK

Fereidoon Moghadas Nejad Department of Civil and Environmental Engineering, Amirkabir University of Technology, Tehran, Iran

Ali Nikjoofar Faculty of Civil Engineering, University of Tabriz, Tabriz, Iran

Ana Carolina Olivera Departamento de Ciencias e Ingeniería de la Computación, Universidad Nacional del Sur, Blanca, Argentina

Muthiah Perumal Department of Hydrology, IIT Roorkee, Roorkee, India

George Piliounis Institute of Structural Analysis and Seismic Research, National Technical University of Athens, Athens, Greece

Deepti Rani Department of Hydrology, IIT Roorkee, Roorkee, India

Pijush Samui Centre for Disaster Mitigation and Management, VIT University, Vellore, Tamil Nadu, India

Mohamed A. Shahin Department of Civil Engineering, Curtin University, Perth, WA, Australia

Vijay P. Singh Department of Biological and Agricultural Engineering and Department of Civil and Environmental Engineering, Texas A&M University, College Station, TX

Dinesh Kumar Srivastava Department of Hydrology, IIT Roorkee, Roorkee, India

Siamak Talatahari Marand Faculty of Engineering, University of Tabriz, Tabriz, Iran

A. Vasan Department of Civil Engineering, Birla Institute of Technology and Science, Pilani—Hyderabad Campus, Andhra Pradesh, India

Xin-She Yang Centre for Mathematics and Scientific Computing, National Physical Laboratory, Teddington, UK

Mahdi Zarghami Faculty of Civil Engineering, University of Tabriz, Tabriz, Iran

Hamzeh Zakeri Department of Civil and Environmental Engineering, Amirkabir University of Technology, Tehran, Iran

1 Optimization and Metaheuristic Algorithms in Engineering

Xin-She Yang

Centre for Mathematics and Scientific Computing, National Physical Laboratory, Teddington, UK

1.1 Introduction

Optimization is everywhere, and thus it is an important paradigm with a wide range of applications. In almost all applications in engineering and industry, we are trying to optimize something—whether to minimize the cost and energy consumption or to maximize profit, output, performance, and efficiency. In reality, resources, time, and money are always limited; consequently, optimization is far more important in practice (Yang, 2010b; Yang and Koziel, 2011). The optimal use of available resources of any sort requires a paradigm shift in scientific thinking because most real-world applications have far more complicated factors and parameters to affect how the system behaves.

Contemporary engineering design is heavily based on computer simulations, which introduces additional difficulties to optimization. Growing demand for accuracy and ever-increasing complexity of structures and systems results in the simulation process being more and more time consuming. In many engineering fields, the evaluation of a single design can take as long as several days or even weeks. Any method that can speed up the simulation time and optimization process can thus save time and money.

For any optimization problem, the integrated components of the optimization process are the optimization algorithm, an efficient numerical simulator, and a realistic representation of the physical processes that we wish to model and optimize. This is often a time-consuming process, and in many cases, the computational costs are usually very high. Once we have a good model, the overall computation costs are determined by the optimization algorithms used for searching and the numerical solver used for simulation.

Search algorithms are the tools and techniques used to achieve optimality of the problem of interest. This search for optimality is complicated further by the fact that uncertainty is almost always present in the real world. Therefore, we seek not only the optimal design but also the robust design in engineering and industry.

Metaheuristics in Water, Geotechnical and Transport Engineering. DOI: http://dx.doi.org/10.1016/B978-0-12-398296-4.00001-5

Optimal solutions, which are not robust enough, are not practical in reality. Suboptimal solutions or good robust solutions are often the choice in such cases.

Simulations are often the most time-consuming part. In many applications, an optimization process often involves evaluating objective function many times (with often thousands, hundreds of thousands, and even millions of configurations). Such evaluations often involve the use of extensive computational tools such as a computational fluid dynamics simulator or a finite element solver. Therefore, efficient optimization with an efficient solver is extremely important.

Optimization problems can be formulated in many ways. For example, the commonly used method of least squares is a special case of maximum-likelihood formulations. By far, the best-known formulation is to write a nonlinear optimization problem as

$$\text{minimize } f_i(x), \quad i = 1, 2, \ldots, M \tag{1.1}$$

subject to the constraints

$$h_j(x) = 0, \quad j = 1, 2, \ldots, J \tag{1.2}$$

and

$$g_k(x) \leq 0, \quad k = 1, 2, \ldots, K \tag{1.3}$$

where f_i, h_j, and g_k are general nonlinear functions. Here, the design vector $x = (x_1, x_2, \ldots, x_n)$ can be continuous, discrete, or mixed in n-dimensional space. The functions f_i are called *objective* or *cost functions*, and when $M > 1$, the optimization is multiobjective or multicriteria (Sawaragi et al., 1985; Yang, 2010b). It is possible to combine different objectives into a single objective, though multiobjective optimization can give far more information and insight into the problem. It is worth pointing out here that we write the problem as a minimization problem, but it can also be written as a maximization by simply replacing $f_i(x)$ by $-f_i(x)$.

When all functions are nonlinear, we are dealing with nonlinear constrained problems. In some special cases when f_i, h_j, g_k are linear, the problem becomes linear, and we can use widely linear programming techniques such as the simplex method. When some design variables can take only discrete values (often integers), while other variables are real and continuous, the problem is of mixed type, which is often difficult to solve, especially for large-scale optimization.

A very special class of optimization is the convex optimization, which has guaranteed global optimality. Any optimal solution is also the global optimum, and most importantly, there are efficient algorithms of polynomial time to solve such problems (Conn et al., 2009). These efficient algorithms, such as the interior-point methods (Karmarkar, 1984), are widely used and have been implemented in many software packages.

1.2 Three Issues in Optimization

There are three main issues in the simulation-driven optimization and modeling, and they are the efficiency of an algorithm, the efficiency and accuracy of a numerical simulator, and the assignment of the right algorithms to the right problem. Despite their importance, there are no satisfactory rules or guidelines for such issues. Obviously, we try to use the most efficient algorithms available, but the actual efficiency of an algorithm depends on many factors such as the inner working of an algorithm, the information needed (such as objective functions and their derivatives), and implementation details. The efficiency of a solver is even more complicated, depending on the actual numerical methods used and the complexity of the problem of interest. As for choosing the right algorithms for the right problems, there are many empirical observations, but no agreed guidelines. In fact, there is no universally efficient algorithms for all types of problems. Therefore, the choice depends on many factors and is sometimes subject to the personal preferences of researchers and decision makers.

1.2.1 Efficiency of an Algorithm

An efficient optimizer is very important to ensure the optimal solutions are reachable. The essence of an optimizer is a search or optimization algorithm implemented correctly so as to carry out the desired search (though not necessarily efficient). It can be integrated and linked with other modeling components. There are many optimization algorithms in the literature, and no single algorithm is suitable for all problems, as dictated by the No Free Lunch Theorems (Wolpert and Macready, 1997).

Optimization algorithms can be classified in many ways, depending on the focus or the characteristics that we are trying to compare. Algorithms can be classified as gradient-based (or derivative-based) and gradient-free (or derivative-free). The classic methods of steepest descent and the Gauss—Newton methods are gradient based, as they use the derivative information in the algorithm, while the Nelder—Mead downhill simplex method (Nelder and Mead, 1965) is a derivative-free method because it uses only the values of the objective, not any derivatives.

Algorithms can also be classified as deterministic or stochastic. If an algorithm works in a mechanically deterministic manner without any random nature, it is called *deterministic*. For such an algorithm, it will reach the same final solution if we start with the same initial point. The hill-climbing and downhill simplex methods are good examples of deterministic algorithms. On the other hand, if there is some randomness in the algorithm, the algorithm will usually reach a different point every time it is run, even starting with the same initial point. Genetic algorithms and hill climbing with a random restart are good examples of stochastic algorithms.

Analyzing stochastic algorithms in more detail, we can single out the type of randomness that a particular algorithm is employing. For example, the simplest and yet often very efficient method is to introduce a random starting point for a deterministic algorithm. The well-known hill-climbing method with random restart is a good example. This simple strategy is both efficient in most cases and easy to implement in practice. A more elaborate way to introduce randomness to an algorithm is to use randomness inside different components of an algorithm, and in this case, we often call such algorithm *heuristic* or, more often, *metaheuristic* (Talbi, 2009; Yang, 2008, 2010b). A very good example is the popular genetic algorithms, which use randomness for crossover and mutation in terms of a crossover probability and a mutation rate. Here, *heuristic* means to search by trial and error, while *metaheuristic* is a higher level of heuristics. However, modern literature tends to refer to all new stochastic algorithms as metaheuristic. In this book, we will use *metaheuristic* to mean either. It is worth pointing out that metaheuristic algorithms are a hot research topic, and new algorithms appear almost yearly (Yang, 2008, 2010b).

From the mobility point of view, algorithms can be classified as local or global. Local search algorithms typically converge toward a local optimum, not necessarily (often not) the global optimum, and such algorithms are often deterministic and have no ability of escaping local optima. Simple hill climbing is an example. On the other hand, we always try to find the global optimum for a given problem, and if this global optimality is robust, it is often the best, though it is not always possible to find such global optimality. For global optimization, local search algorithms are not suitable. We have to use a global search algorithm. Modern metaheuristic algorithms in most cases are intended for global optimization, though the process is not always successful or efficient. A simple strategy such as hill climbing with random restart may change a local search algorithm into a global search. In essence, randomization is an efficient component for global search algorithms. In this chapter, we will provide a brief review of most metaheuristic optimization algorithms.

Straightforward optimization of a given objective function is not always practical. In particular, if the objective function comes from a computer simulation, it may be computationally expensive, noisy, or nondifferentiable. In such cases, so-called surrogate-based optimization algorithms may be useful where the direct optimization of the function of interest is replaced by iterative updating and reoptimization of its model—i.e., a surrogate. The surrogate model is typically constructed from the sampled data of the original objective function; however, it is supposed to be cheap, smooth, easy to optimize, and yet reasonably accurate so that it can produce a good prediction of the function's optimum. Multifidelity or variable-fidelity optimization is a special case of surrogate-based optimization, where the surrogate is constructed from the low-fidelity model (or models) of the system of interest (Koziel and Yang, 2011). Using variable-fidelity optimization is particularly useful, as the reduction of the computational cost of the optimization process is of primary importance.

Whatever the classification of an algorithm is, we have to make the right choice to use an algorithm correctly, and sometimes using a proper combination of algorithms may achieve far better results.

1.2.2 The Right Algorithms?

From the optimization point of view, the choice of the right optimizer or algorithm for a given problem is crucially important. The algorithm chosen for an optimization task will largely depend on the type of the problem, the nature of an algorithm, the desired quality of solutions, the available computing resource, time limit, availability of the algorithm implementation, and the expertise of the decision makers (Yang, 2010b; Yang and Koziel, 2011).

The nature of an algorithm often determines if it is suitable for a particular type of problem. For example, gradient-based algorithms such as hill climbing are not suitable for an optimization problem with a discontinuous objective. Conversely, the type of problem we are trying to solve also determines the algorithms we may choose. If the objective function of an optimization problem at hand is highly nonlinear and multimodal, classic algorithms such as hill climbing and downhill simplex are not suitable, as they are local search algorithms. In this case, global optimizers, such as particle swarm optimization and cuckoo search, are most suitable (Yang, 2010a; Yang and Deb, 2010).

Obviously, the choice is also affected by the desired solution quality and available computing resources. Because computing resources are limited in most applications, we have to obtain good solutions (if not necessary the best) in a reasonable and practical time. Therefore, we have to balance resource availability with solution quality. We cannot achieve solutions with guaranteed quality, though we strive to obtain the best-quality solutions that we possibly can. If time is the main constraint, we can use some greedy methods, or hill climbing with a few random restarts.

Sometimes, even with the best possible intentions, the availability of an algorithm and the expertise of the decision makers are the ultimate defining factors for choosing an algorithm. Even though some algorithms are better for the given problem at hand, we may not have that algorithm implemented in our system or we do not have such access, which limits our choice. For example, Newton's method, hill-climbing, Nelder–Mead downhill simplex, trust-region methods (Conn et al., 2009), and interior-point methods are implemented in many software packages, which may also increase their popularity in applications. In practice, even with the best possible algorithms and well-crafted implementation, we still may fail to get the desired solutions. This is the nature of nonlinear global optimization, as most of such problems are Non-deterministic polynomial-time hard (NP-hard), and no efficient (in the polynomial sense) solutions exist for a given problem. Thus, the challenges of research in computational optimization and applications are to find the right algorithms most suitable for a given problem so as to obtain good solutions (perhaps also the best solutions globally), in a reasonable timescale with a limited amount of resources. We aim to do this in an efficient, optimal way.

1.2.3 Efficiency of a Numerical Solver

To solve an optimization problem, the most computationally extensive part is probably the evaluation of the design objective to see if a proposed solution is feasible and/or if it is optimal. Typically, we have to carry out these evaluations many times, often thousands, hundreds of thousands, and even millions of times (Yang, 2008, 2010b). Things become even more challenging computationally, when each evaluation task takes a long time to complete using some black-box simulators. If this simulator is a finite element or computational fluid dynamics solver, the running time of each evaluation can take from a few minutes to a few hours or even weeks. Therefore, any approach to save computational time either by reducing the number of evaluations or by increasing the simulator's efficiency will save time and money. In general, a simulator can be a simple function subroutine, a multiphysics solver, or an external black-box evaluator.

The main way to reduce the number of objective evaluations is to use an efficient algorithm, so that only a small number of such evaluations are needed. In most cases, this is not possible. We have to use some approximation techniques to estimate the objectives, or to construct an approximation model to predict the solver's outputs without actually using the solver. Another way is to replace the original objective function by its lower-fidelity model, e.g., obtained from a computer simulation based on coarsely discretized structure of interest. The low-fidelity model is faster, but not as accurate as the original one, and therefore it has to be corrected. Special techniques have to be applied to use an approximation or corrected low-fidelity model in the optimization process so that the optimal design can be obtained at a low computational cost (Koziel and Yang, 2011).

1.3 Metaheuristics

Metaheuristic algorithms are often nature-inspired, and they are now among the most widely used algorithms for optimization. They have many advantages over conventional algorithms, as we can see from many case studies presented in later chapters in this book. There are a few recent books that are solely dedicated to metaheuristic algorithms (Talbi, 2009; Yang, 2008, 2010a,b). Metaheuristic algorithms are very diverse, including genetic algorithms, simulated annealing, differential evolution (DE), ant and bee algorithms, particle swarm optimization, harmony search, firefly algorithm, cuckoo search, and others. Here, we will introduce some of these algorithms briefly.

1.3.1 Ant Algorithms

Ant algorithms, especially the ant colony optimization (Dorigo and Stütle, 2004), mimic the foraging behavior of social ants. Primarily, ants use pheromones as a chemical messenger, and the pheromone concentration can also be considered as the indicator of quality solutions to a problem of interest. As the solution is often linked with the pheromone concentration, the search algorithms often produce routes and paths marked by the higher pheromone concentrations, and therefore, ant-based algorithms are particularly suitable for discrete optimization problems.

The movement of an ant is controlled by pheromones that will evaporate over time. Without such time-dependent evaporation, ant algorithms will lead to premature convergence to the (often wrong) solutions. With proper pheromone evaporation, they usually behave very well.

There are two important issues here: the probability of choosing a route and the evaporation rate of the pheromones. There are a few ways of solving these problems, although this is still an area of active research. For a network routing problem, the probability of ants at a particular node i to choose the route from node i to node j is given by

$$p_{ij} = \frac{\phi_{ij}^{\alpha} d_{ij}^{\beta}}{\sum_{i,j=1}^{n} \phi_{ij}^{\alpha} d_{ij}^{\beta}} \qquad (1.4)$$

where $\alpha > 0$ and $\beta > 0$ are the influence parameters, and their typical values are $\alpha \approx \beta \approx 2$. Here, ϕ_{ij} is the pheromone concentration on the route between i and j and d_{ij}, the desirability of the same route. Some *a priori* knowledge about the route, such as the distance s_{ij}, is often used so that $d_{ij} \propto 1/s_{ij}$, which implies that shorter routes will be selected due to their shorter traveling time; and thus the pheromone concentrations on these routes are higher. This is because the traveling time is shorter, and thus the less amount of the pheromone has been evaporated during this period.

1.3.2 Bee Algorithms

Bee-inspired algorithms are more diverse—a few use pheromones, but most do not. Almost all bee algorithms are inspired by the foraging behavior of honeybees in nature. Interesting characteristics, such as waggle dancing, polarization, and nectar maximization, are often used to simulate the allocation of the foraging bees along flower patches, and thus in different regions of the search space. For a more comprehensive review, see Yang (2010a) and Parpinelli and Lope (2011).

Different variants of bee algorithms use slightly different characteristics of the behavior of bees. For example, in the honeybee-based algorithms, forager bees are allocated to different food sources (or flower patches) so as to maximize the total nectar intake (Karaboga, 2005; Nakrani and Tovey, 2004; Pham et al., 2006; Yang, 2005). In the virtual bee algorithm (VBA), pheromone concentrations can be linked with the objective functions more directly (Yang, 2005). The artificial bee colony (ABC) optimization algorithm was first developed by Karaboga (2005). In the ABC algorithm, the bees in a colony are divided into three groups: employed bees (forager bees), onlooker bees (observer bees), and scouts. Unlike the honeybee algorithm, which has only two groups of bees (forager bees and observer bees), bees in ABC are more specialized (Afshar et al., 2007; Karaboga, 2005).

Similar to the ant-based algorithms, bee algorithms are very flexible in dealing with discrete optimization problems. Combinatorial optimization, such as routing and optimal paths, has been solved by ant and bee algorithms. In principle, they can solve both continuous optimization and discrete optimization problems; however, they should not be the first choice for continuous problems.

1.3.3   The Bat Algorithm

The bat algorithm is a relatively new metaheuristic (Yang, 2010c). Microbats use a type of sonar called *echolocation* to detect prey, avoid obstacles, and locate their roosting crevices in the dark, and the bat algorithm was inspired by this echolocation behavior. These bats emit a very loud sound pulse and listen for the echo that bounces back from the surrounding objects. Their pulses vary in properties and can be correlated with their hunting strategies, depending on the species. Most bats use short, frequency-modulated signals to sweep through about an octave, while others more often use constant-frequency signals for echolocation. Their signal bandwidth varies depending on the species and often increased by using more harmonics.

The bat algorithm uses three idealized rules: (1) all bats use echolocation to sense distance, and they also "know" the difference between food/prey and background barriers in some unknown way; (2) a bat flies randomly with a velocity v_i at position x_i with a fixed frequency range $[f_{min}, f_{max}]$, varying its emission rate $r \in [0,1]$ and loudness A_0 to search for prey, depending on the proximity of their target; (3) although the loudness can vary in many ways, we assume that it varies from a large (positive) A_0 to a minimum constant value A_{min}. These rules can be translated into the following formulas:

$$f_i = f_{min} + (f_{max} - f_{min})\varepsilon, \quad v_i^{t+1} = v_i^t + (x_i^t - x^*)f_i, \quad x_i^{t+1} = x_i^t + v_i^t \qquad (1.5)$$

where ε is a random number drawn from a uniform distribution and x^* is the current best solution found so far during iterations. The loudness and pulse rate can vary with iteration t in the following way:

$$A_i^{t+1} = \alpha A_i^t, \quad r_i^t = r_i^0[1 - \exp(-\beta t)] \qquad (1.6)$$

Here, α and β are constants. In fact, α is similar to the cooling factor of a cooling schedule in the simulated annealing, which will be discussed next. In the simplest case, we can use $\alpha = \beta$, and we have, in fact, used $\alpha = \beta = 0.9$ in most simulations.

The bat algorithm has been extended to the multiobjective bat algorithm (MOBA) by Yang (2011a), and preliminary results suggested that it is very efficient (Yang and Gandomi, 2012).

1.3.4   Simulated Annealing

Simulated annealing is among the first metaheuristic algorithms (Kirkpatrick et al., 1983). It was essentially an extension of the traditional Metropolis—Hastings algorithm but applied in a different context. The basic idea of the simulated annealing algorithm is to use random search in terms of a Markov chain, which not only accepts changes that improve the objective function but also keeps some changes that are not ideal.

In a minimization problem, for example, any better moves or changes that decrease the value of the objective function f will be accepted; however, some changes that increase f will also be accepted with a probability P. This probability P, also called the transition probability, is determined by

$$P = \exp\left[-\frac{\Delta E}{k_B T}\right] \tag{1.7}$$

where k_B is Boltzmann's constant, T is the temperature for controlling the annealing process, and ΔE is the change of the energy level. This transition probability is based on the Boltzmann distribution in statistical mechanics.

The simplest way to link ΔE with the change of the objective function Δf is to use $\Delta E = \gamma \Delta f$, where γ is a real constant. For simplicity without losing generality, we can use $k_B = 1$ and $\gamma = 1$. Thus, the probability P simply becomes

$$P(\Delta f, T) = e^{-\Delta f/T} \tag{1.8}$$

Whether or not a change is accepted, a random number r is often used as a threshold. Thus, if $P > r$, the move is accepted.

Here, the choice of the right initial temperature is crucial. For a given change Δf, if T is too high ($T \to \infty$), then $P \to 1$, which means almost all the changes will be accepted. If T is too low ($T \to 0$), then any $\Delta f > 0$ (worse solutions) will rarely be accepted as $P \to 0$, and thus the diversity of the solution is limited, but any improvement Δf will almost always be accepted. In fact, the special case $T \to 0$ corresponds to the classical hill-climbing method because only better solutions are accepted, and the system is essentially climbing or descending a hill. So, a proper temperature range is very important.

Another important issue is how to control the annealing or cooling process so that the system cools gradually from a higher temperature, ultimately freezing to a global minimum state. There are many ways of controlling the cooling rate or the decrease of the temperature. Geometric cooling schedules are often widely used, which essentially decrease the temperature by a cooling factor $0 < \alpha < 1$, so that T is replaced by αT or

$$T(t) = T_0 \alpha^t, \quad t = 1, 2, \ldots, t_f \tag{1.9}$$

where t_f is the maximum number of iterations. The advantage of this method is that $T \to 0$ when $t \to \infty$, and thus, there is no need to specify the maximum number of iterations if a tolerance or accuracy is prescribed.

1.3.5 Genetic Algorithms

Genetic algorithms are a class of algorithms based on the abstraction of Darwin's evolution of biological systems, pioneered by Holland and his collaborators in the 1960s and 1970s (Holland, 1975). Holland was probably the first to use genetic operators such as the crossover and recombination, mutation, and selection in the study of adaptive and artificial systems. Three main components or genetic operators in genetic algorithms are crossover, mutation, and selection of the fittest. Each solution is encoded in a string (often binary or decimal) called *chromosome*. The crossover of two parent strings produce offsprings (new solutions) by swapping part or genes of the chromosomes. Crossover has a higher probability, typically 0.8–0.95. On the other hand, mutation is performed by flipping some digits of a string, which generates new solutions. This mutation probability is typically low, from 0.001 to 0.05. New solutions generated in each generation will be evaluated by their fitness, which is linked to the objective function of the optimization problem. The new solutions are selected according to their fitness—i.e., selection of the fittest. Sometimes, to make sure that the best solutions remain in the population, the best solutions are passed onto the next generation without much change, a process called *elitism*.

Genetic algorithms have been applied to almost all areas of optimization, design, and applications. There are hundreds of good books and thousands of research articles. There are many variants and hybridization with other algorithms, and interested readers can refer to more advanced literature such as Goldberg (1989).

1.3.6 Differential Evolution

DE was developed by Storn and Price (Storn, 1996; Storn and Price, 1997). It is a vector-based evolutionary algorithm that can be considered as a further development in genetic algorithms. As with genetic algorithms, design parameters in a d-dimensional search space are represented as vectors, and various genetic operators are operated over their bits of strings. However, unlike genetic algorithms, DE carries out operations over each component (or each dimension of the solution). Almost everything is done in terms of vectors. For a d-dimensional optimization problem with d parameters, a population of n solution vectors are initially generated, we have x_i where $i = 1,2,\ldots, n$. For each solution x_i at any generation t, we use the conventional notation:

$$x_i^t = (x_{1,i}^t, x_{2,i}^t, \ldots, x_{d,i}^t) \tag{1.10}$$

which consists of d components in the d-dimensional space. This vector can be considered as chromosomes or genomes.

DE consists of three main steps: mutation, crossover, and selection. Mutation is carried out by the mutation scheme. For each vector x_i at any time or generation t, we first randomly choose three distinct vectors x_p, x_q, and x_r at t, and then generate a so-called donor vector by the mutation scheme

$$v_i^{t+1} = x_p^t + F(x_q^t - x_r^t) \tag{1.11}$$

where $F \in [0,2]$ is a parameter, often referred to as the *differential weight*. This requires that the minimum population size is $n \geq 4$. In principle, $F \in [0,2]$, but in practice, a scheme with $F \in [0,1]$ is more efficient and stable.

The crossover is controlled by a crossover probability $C_r \in [0,1]$, and actual crossover can be carried out in two ways: binomial and exponential. Selection is essentially the same as that used in genetic algorithms. The goal is to select the fittest, and for the minimization problem, the minimum objective value. Therefore, we have

$$x_i^{t+1} = \begin{cases} u_i^{t+1} & \text{if} \quad f(u_i^{t+1}) \leq f(x_i^t) \\ x_i^t & \text{otherwise} \end{cases} \tag{1.12}$$

Most studies have focused on the choice of F, C_r, and n, as well as the modification of Eq. (1.11). In fact, when generating mutation vectors, we can use many different ways of formulating Eq. (1.11), and this leads to various schemes with the naming convention: DE/*x*/*y*/*z*, where x is the mutation scheme (rand or best), y is the number of difference vectors, and z is the crossover scheme (binomial or exponential). The basic DE/Rand/1/Bin scheme is given in Eq. (1.11). Following a similar strategy, we can design various schemes. In fact, more than 10 different schemes have been formulated in the literature (Price et al., 2005).

1.3.7 Particle Swarm Optimization

Particle swarm optimization (PSO) was based on swarm behavior in nature, such as fish and bird schooling (Kennedy and Eberhart, 1995). Since then, PSO has generated much wider interest and forms an exciting, ever-expanding research subject called *swarm intelligence*. This algorithm searches the space of an objective function by adjusting the trajectories of individual agents, called *particles*, as the piecewise paths formed by positional vectors in a quasi-stochastic manner.

The movement of a swarming particle consists of two major components: a stochastic component and a deterministic component. Each particle is attracted to the position of the current global best g^* and its own best location x_i^* in history, while at the same time, it has a tendency to move randomly. Let x_i and v_i be the position vector and velocity for particle i, respectively. The new velocity vector is determined by the following formula:

$$v_i^{t+1} = v_i^t + \alpha \varepsilon_1 [g^* - x_i^t] + \beta \varepsilon_2 [x_i^* - x_i^t] \tag{1.13}$$

where ε_1 and ε_2 are two random vectors, with each entry taking a value between 0 and 1. The Hadamard product of two matrices $(u \odot v)$ is defined as the entrywise product, i.e., $[u \odot v]_{ij} = u_{ij} v_{ij}$. The parameters α and β are the learning parameters or acceleration constants, which can typically be taken as, for example, $\alpha \approx \beta \approx 2$.

The initial locations of all particles should distribute relatively uniformly so that they can sample over most regions, which is especially important for multimodal problems. The initial velocity of a particle can be taken as zero, i.e., $v_i^{t=0} = 0$. The new position can then be updated by

$$x_i^{t+1} = x_i^t + v_i^{t+1} \qquad (1.14)$$

Although v_i can be any value, it is usually located in some range $[0, v_{\max}]$.

There are many variants that extend the standard PSO algorithm (Kennedy et al., 2001; Yang, 2008, 2010b), and the most noticeable improvement is probably to use inertia function $\theta(t)$ so that v_i^t is replaced by $\theta(t)v_i^t$:

$$v_i^{t+1} = \theta v_i^t + \alpha \varepsilon_1 \odot [g^* - x_i^t] + \beta \varepsilon_2 \odot [x_i^* - x_i^t] \qquad (1.15)$$

where θ takes the value between 0 and 1. In the simplest case, the inertia function can be taken as a constant, typically $\theta \approx 0.5 - 0.9$. This is equivalent to introducing a virtual mass to stabilize the motion of the particles, and thus the algorithm is expected to converge more quickly.

1.3.8 Harmony Search

Harmony search (HS) is a music-inspired algorithm (Geem et al., 2001), which can be explained in more detail with the aid of the discussion of a musician's improvisation process. When a musician is improvising, he or she has three possible choices: (1) play any famous piece of music (a series of pitches in harmony) exactly from his or her memory; (2) play something similar to a known piece (thus adjusting the pitch slightly); or (3) compose new or random notes. If we formalize these three options for optimization, we have three corresponding components: usage of harmony memory, pitch adjusting, and randomization.

The usage of harmony memory is important, as it is similar to choose the best-fitting individuals in the genetic algorithms. This will ensure that the best harmonies will be carried over to the new harmony memory. An important step is pitch adjustment, which can be considered a local random walk. If x_{old} is the current solution (or pitch), then the new solution (pitch) x_{new} is generated by

$$x_{new} = x_{old} + b_p(2\varepsilon - 1) \qquad (1.16)$$

where ε is a random number drawn from a uniform distribution $[0,1]$. Here, b_p is the bandwidth, which controls the local range of pitch adjustment. In fact, we can see that the pitch adjustment (Eq. (1.16)) is a random walk.

Pitch adjustment is similar to the mutation operator in genetic algorithms. Although adjusting pitch has a similar role, it is limited to certain local pitch adjustment, and thus, it corresponds to a local search. The use of randomization can drive the system further to explore various regions with high solution diversity so as to find the global optimality.

1.3.9 Firefly Algorithm

The firefly algorithm (FA), first developed Yang (2008, 2009), was based on the flashing patterns and behavior of fireflies. In essence, FA uses the following three idealized rules:

1. Fireflies are unisexual, so one firefly will be attracted to other fireflies regardless of their sex.
2. Their attractiveness is proportional to their brightness, and both decrease as their distance increases. Thus, for any two flashing fireflies, the less brighter one will move toward the brighter one. If a particular firefly does not find a brighter one, it will move randomly.
3. The brightness of a firefly is determined by the landscape of the objective function.

As a firefly's attractiveness is proportional to the light intensity seen by adjacent fireflies, we can now define the variation of attractiveness β with distance r by

$$\beta = \beta_0 \, e^{-\gamma r^2} \tag{1.17}$$

where β_0 is the attractiveness at $r = 0$.

The movement of a firefly i that is attracted to another more attractive (brighter) firefly j is determined by

$$x_i^{t+1} = x_i^t + \beta_0 e^{-\gamma r_{ij}^2}(x_j^t - x_i^t) + \alpha \varepsilon_i^t \tag{1.18}$$

where the second term is based on the attraction. The third term is randomized, with α being the randomization parameter and ε_i^t is a vector of random numbers drawn from a Gaussian distribution or uniform distribution at time t. If $\beta_0 = 0$, it becomes a simple random walk. Furthermore, the randomization ε_i^t can easily be extended to other distributions such as Lévy flights.

The Lévy flight essentially provides a random walk whose random step length is drawn from a Lévy distribution:

$$L(s, \lambda) = s^{-(1+\lambda)}, \quad 0 < \lambda \le 2 \tag{1.19}$$

which has an infinite variance with an infinite mean. Here the steps essentially form a random walk process with a power-law step-length distribution with a heavy tail. Some of the new solutions should be generated by a Lévy walk around the best solution obtained so far, which will speed up the local search (Pavlyukevich, 2007).

A demo version of FA implementation, without Lévy flights, can be found at the Mathworks file exchange web site.[1] FA has attracted much attention

---

[1] http://www.mathworks.com/matlabcentral/fileexchange/29693-firefly-algorithm.

(Apostolopoulos and Vlachos, 2011; Gandomi et al., 2011; Sayadi et al., 2010). A discrete version of FA can efficiently solve NP-hard scheduling problems (Sayadi et al., 2010), while a detailed analysis has demonstrated the efficiency of FA over a wide range of test problems, including multiobjective load dispatch problems (Apostolopoulos and Vlachos, 2011). A chaos-enhanced FA with a basic method for automatic parameter tuning has also been developed (Yang, 2011b).

1.3.10   Cuckoo Search

Cuckoo search (CS) is one of the latest nature-inspired metaheuristic algorithms developed by Yang and Deb (2009). CS is based on the brood parasitism of some cuckoo species. In addition, this algorithm is enhanced by the so-called Lévy flights (Pavlyukevich, 2007), rather than by simple isotropic random walks. Recent studies show that CS is potentially far more efficient than the PSO and genetic algorithms (Yang and Deb, 2010).

Cuckoos are fascinating birds, not only because of the beautiful sounds they can make but also because of their aggressive reproduction strategy. Some species such as the *ani* and *Guira* cuckoos lay their eggs in communal nests, though they may remove others' eggs to increase the hatching probability of their own. Quite a number of species engage in the obligate brood parasitism by laying their eggs in the nests of other host birds (often other species).

For simplicity in describing the standard CS, we now use the following three idealized rules:

1. Each cuckoo lays one egg at a time and dumps it in a randomly chosen nest.
2. The best nests with high-quality eggs will be carried over to the next generation.
3. The number of available host nests is fixed, and the probability that an egg laid by a cuckoo is discovered by the host bird is $p_a \in [0,1]$. In such a case, the host bird can either get rid of the egg or abandon the nest and build a completely new nest.

As a further approximation, this last assumption can be approximated by stating that a fraction p_a of the n host nests are replaced by new nests (with new random solutions).

For a maximization problem, the quality or fitness of a solution can simply be proportional to the value of the objective function. Other forms of fitness can be defined in a similar way to the fitness function in genetic algorithms.

For the implementation point of view, we can use the following simple representations that each egg in a nest represents a solution, and each cuckoo can lay only one egg (thus representing one solution), the aim being to use the new and potentially better solutions (cuckoos) to replace less good solutions in the nests. Obviously, this algorithm can be extended to the more complicated case, where each nest has multiple eggs representing a set of solutions. For this discussion, we will use the simplest approach, where each nest has only a single egg. In this case, there is no distinction between egg, nest, and cuckoo: each nest corresponds to one egg, which also represents one cuckoo.

Figure 1.1 Pseudocode of
the CS.

Objective function $f(x)$, $x = (x_1,...,x_d)^T$
Generate initial population of n host nests x_i
while (t < MaxGeneration) or (stop criterion)
  Get a cuckoo randomly/generate a solution by Lévy flights
    and then evaluate its quality/fitness F_i
  Choose a nest among n (say, j) randomly
  if $(F_i > F_j)$,
    Replace j by the new solution
  end
  A fraction (p_a) of worse nests are abandoned
    and new ones/solutions are built/generated
  Keep best solutions (or nests with quality solutions)
  Rank the solutions and find the current best
end while

Based on these three rules, the basic steps of the CS can be summarized as the pseudocode shown in Figure 1.1.

This algorithm uses a balanced combination of a local random walk and the global explorative random walk, controlled by a switching parameter p_a. The local random walk can be written as

$$x_i^{t+1} = x_i^t + \alpha s \otimes H(p_a - \varepsilon) \otimes (x_j^t - x_k^t) \qquad (1.20)$$

where x_j^t and x_k^t are two different solutions selected by random permutation, $H(u)$ is a Heaviside function, ε is a random number drawn from a uniform distribution, and s is the step size. On the other hand, the global random walk is carried out using Lévy flights:

$$x_i^{t+1} = x_i^t + \alpha L(s, \lambda) \qquad (1.21)$$

where

$$L(s, \lambda) = \frac{\lambda \Gamma(\lambda) \sin(\pi \lambda / 2)}{\pi} \frac{1}{s^{1+\lambda}}, \qquad s \gg s_0 > 0 \qquad (1.22)$$

Here, $\alpha > 0$ is the step size scaling factor, which should be related to the scales of the problem of interest. In most cases, we can use $\alpha = O(L/10)$, where L is the characteristic scale of the problem of interest, while in some cases, $\alpha = O(L/100)$ can be more effective and avoid the need to fly too far. Equation (1.22) is essentially the stochastic equation for a random walk. In general, a random walk is a Markov chain whose next status/location only depends on the current location (the first term in Eq. (1.22)) and the transition probability (the second term). However, a substantial fraction of the new solutions should be generated by far-field randomization and whose locations should be far enough from the current best solution to make sure that the system will not be trapped in a local optimum (Yang and Deb, 2010).

The pseudocode given here is sequential; however, vectors should be used from an implementation point of view, as vectors are more efficient than loops. A Matlab implementation is given by Yang and can be downloaded.[2] CS is very efficient in solving engineering optimization problems (Gandomi et al., 2011).

1.3.11   Other Algorithms

There are many other metaheuristic algorithms that are equally popular and powerful, including Tabu search (Glover and Laguna, 1997), artificial immune system (Farmer et al., 1986), and others (Koziel and Yang, 2011; Yang, 2010a,b).

The efficiency of metaheuristic algorithms can be attributed to the fact that they imitate the best features in nature, especially the selection of the fittest in biological systems that have evolved by natural selection over millions of years.

Two important characteristics of metaheuristics are intensification and diversification (Blum and Roli, 2003). Intensification intends to search locally and more intensively, while diversification makes sure the algorithm explores the search space globally (and hopefully also efficiently). A fine balance between these two components is very important to the overall efficiency and performance of an algorithm. Too little exploration and too much exploitation could cause the system to be trapped in local optima, which makes it very difficult or even impossible to find the global optimum. On the other hand, if there is too much exploration but too little exploitation, it may be difficult for the system to converge, which would slow down the overall search performance. A proper balance itself is an optimization problem, and one of the main tasks of designing new algorithms is to find an optimal balance concerning this optimality and/or trade-off.

Furthermore, just exploitation and exploration are not enough. During the search, we have to use a proper mechanism or criterion to select the best solutions. The most common criterion is to use the Survival of the Fittest, i.e., to keep updating the solution with the best one found so far. In addition, a certain elitism is often used, which ensures that the best or fittest solutions are not lost and are passed onto the next generations.

1.4   Artificial Neural Networks

As we will see, artificial neural networks are in essence optimization algorithms, working in different contexts (Yang, 2010a).

[2] www.mathworks.com/matlabcentral/fileexchange/29809-cuckoo-search-cs-algorithm.

1.4.1 Artificial Neurons

The basic mathematical model of an artificial neuron was first proposed by W. McCulloch and W. Pitts in 1943, and this fundamental model is referred to as the McCulloch–Pitts model. Other models and neural networks are based on it. An artificial neuron with n inputs or impulses and an output y_k will be activated if the signal strength reaches a certain threshold θ. Each input has a corresponding weight w_i. The output of this neuron is given by

$$y_l = \Phi\left(\sum_{i=1}^{n} w_i u_i\right) \tag{1.23}$$

where the weighted sum $\xi = \sum_{i=1}^{n} w_i u_i$ is the total signal strength, and Φ is the so-called activation function, which can be taken as a step function. That is, we have

$$\Phi(\xi) = \begin{cases} 1 & \text{if} \quad \xi \geq \theta \\ 0 & \text{if} \quad \xi < \theta \end{cases} \tag{1.24}$$

We can see that the output is only activated to a nonzero value if the overall signal strength is greater than the threshold θ.

The step function has discontinuity; sometimes, it is easier to use a nonlinear, smooth function called a Sigmoid function:

$$S(\xi) = \frac{1}{1 + e^{-\xi}} \tag{1.25}$$

which approaches 1 as $U \rightarrow \infty$ and becomes 0 as $U \rightarrow -\infty$. An interesting property of this function is

$$S'(\xi) = S(\xi)[1 - S(\xi)] \tag{1.26}$$

1.4.2 Neural Networks

A single neuron can perform only a simple task—it is either on or off. Complex functions can be designed and performed using a network of interconnecting neurons or perceptrons. The structure of a network can be complicated, and one of the most widely used is to arrange them in a layered structure, with an input layer, an output layer, and one or more hidden layers (Figure 1.2). The connection strength between two neurons is represented by its corresponding weight. Some artificial neural networks (ANNs) can perform complex tasks and can simulate complex mathematical models, even if there is no explicit functional form mathematically. Neural networks have been developed over the last few decades and applied in almost all areas of science and engineering.

The construction of a neural network involves the estimation of the suitable weights of a network system with some training/known data sets. The task

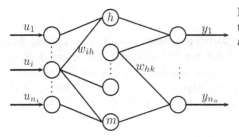

Figure 1.2 Schematic representation of a three-layer neural network with n_i inputs, m hidden nodes, and n_o outputs.

of the training is to find the suitable weights w_{ij} such that the neural networks not only can best-fit the known data but also can predict outputs for new inputs. A good artificial neural network should be able to minimize both errors simultaneously—the fitting/learning errors and the prediction errors.

The errors can be defined as the difference between the calculated (or predicated) output o_k and real output y_k for all output neurons in the least-square sense:

$$E = \frac{1}{2} \sum_{k=1}^{n_o} (o_k - y_k)^2 \tag{1.27}$$

Here, the output o_k is a function of inputs/activations and weights. In order to minimize this error, we can use the standard minimization techniques to find the solutions of the weights.

A simple and yet efficient technique is the steepest descent method. For any initial random weights, the weight increment for w_{hk} is

$$\Delta w_{hk} = -\eta \frac{\partial E}{\partial w_{hk}} = -\eta \frac{\partial E}{\partial o_k} \frac{\partial o_k}{\partial w_{hk}} \tag{1.28}$$

where η is the learning rate. Typically, we can choose $\eta = 1$.

From

$$S_k = \sum_{h=1}^{m} w_{hk} o_h, \quad k = 1, 2, \dots, n_o \tag{1.29}$$

and

$$o_k = f(S_k) = \frac{1}{1 + e^{-S_k}} \tag{1.30}$$

we have

$$f' = f(1 - f) \tag{1.31}$$

$$\frac{\partial o_k}{\partial w_{hk}} = \frac{\partial o_k}{\partial S_k}\frac{\partial S_k}{\partial w_{hk}} = o_k(1-o_k)o_h \qquad (1.32)$$

and

$$\frac{\partial E}{\partial o_k} = (o_k - y_k) \qquad (1.33)$$

Therefore, we have

$$\Delta w_{hk} = -\eta\delta_k o_h, \quad \delta_k = o_k(1-o_k)(o_k-y_k) \qquad (1.34)$$

1.4.3 The Back Propagation Algorithm

There are many ways of calculating weights by supervised learning. One of the simplest and widely used methods is to use the back propagation algorithm for training neural networks, often called back propagation neural networks (BPNNs).

The basic idea is to start from the output layer and propagate backward to estimate and update the weights. From any initial random weighting matrices w_{ih} (for connecting the input nodes to the hidden layer) and w_{hk} (for connecting the hidden layer to the output nodes), we can calculate the outputs of the hidden layer o_h:

$$o_h = \frac{1}{1 + \exp\left[-\sum_{i=1}^{n_i} w_{ih}u_i\right]}, \quad h = 1, 2, \ldots, m \qquad (1.35)$$

and the outputs for the output nodes:

$$o_k = \frac{1}{1 + \exp\left[-\sum_{h=1}^{m} w_{hk}o_h\right]}, \quad k = 1, 2, \ldots, n_o \qquad (1.36)$$

The errors for the output nodes are given by

$$\delta_k = o_k(1-o_k)(y_k - o_k), \quad k = 1, 2, \ldots, n_o \qquad (1.37)$$

where $y_k(k = 1, 2, \ldots, n_o)$ are the data (real outputs) for the inputs $u_i(i = 1, 2, \ldots, n_i)$. Similarly, the errors for the hidden nodes can be written as

$$\delta_h = o_h(1-o_h)\sum_{k=1}^{n_o} w_{hk}\delta_k, \quad h = 1, 2, \ldots, m \qquad (1.38)$$

The updating formulas for weights at iteration t are

$$w_{hk}^{t+1} = w_{hk}^t + \eta\delta_k o_h \qquad (1.39)$$

and

$$w_{ih}^{t+1} = w_{ih}^t + \eta \delta_h u_i \tag{1.40}$$

where $0 < \eta \leq 1$ is the learning rate.

Here, we can see that the weight increments are

$$\Delta w_{ih} = \eta \delta_h u_i \tag{1.41}$$

with similar updating formulas for w_{hk}. An improved version is to use the so-called weight momentum α to increase the learning efficiency:

$$\Delta w_{ih} = \eta \delta_h u_i + \alpha w_{ih}(\tau - 1) \tag{1.42}$$

where τ is an extra parameter. There are many good software packages for ANNs, and there are dozens of good books fully dedicated to implementation. ANNs have been very useful in solving problems in civil engineering (Alavi and Gandomi, 2011a,b; Gandomi and Alavi, 2011).

1.5 Genetic Programming

Genetic programming is a systematic method of using evolutionary algorithms to produce computer programs in a Darwinian manner. Fogel was probably one of the pioneers in primitive genetic programming (Fogel et al., 1966), as he first used evolutionary algorithms to study finite-state automata. However, the true formulation of modern genetic programming was introduced and pioneered by Koza (1992), and the publication of his book *Genetic Programming: On the Programming of Computers by Means of Natural Selection* was a major milestone.

In essence, genetic programming intends to evolve computer programs in an iterative manner by chromosome representations, often in terms of tree structures where each node corresponds a mathematical operator and end nodes represent operands. Evolution is carried out by genetic operators such as crossover, mutation, and selection of the fittest. In the tree-structured representation, crossover often takes the form of subtree exchange crossover, while mutation may take the form of subtree replacement mutation.

According to Koza (1992), there are three stages in the process: preparatory steps, a genetic programming engine, and a new computer program. The genetic programming engine has preparatory steps as inputs and a computer program as its output. First, we have to specify a set of primitive ingredients such as the function set and terminal set. For example, if we wish a computer program to be able to design an electronic circuit, we have to specify the basic components such as transistors, capacitors, and resistors, and their basic functions. Then we have to produce a fitness measure (such as time, cost, stability, and performance) to define what solutions are better than others by that measure. In addition, we have to produce some initialization of algorithm-dependent parameters, such as population size and number of generations, and the termination criteria, which essentially controls when the evolution should stop.

Though computationally expansive, genetic programming has aleady produced human-competitive novel results in many areas such as electronic design, game playing, quantum computing, and invention generation. New invention often requires illogical steps in producing new ideas, and this can often be mimicked as a randomization process in evolutionary algorithms. As pointed out by Koza et al. (2003), genetic programming is a systematic method for getting computers to solve a problem automatically, starting from a high-level statement outlining what needs to be done, which virtually turns a computer into an "automated invention machine." Obviously, that is the ultimate aim of genetic programming.

For applications in engineering, readers can use more specialized literature (Alavi and Gandomi, 2011a,b; Gandomi and Alavi, 2012a,b). There is an extensive literature concerning genetic programming; interested readers can refer to works such as Koza (1992) and Langdon (1998).

References

Afshar, A., Haddad, O.B., Marino, M.A., Adams, B.J., 2007. Honey-bee mating optimization (HBMO) algorithm for optimal reservoir operation. J. Franklin Inst. 344, 452–462.

Alavi, A.H., Gandomi, A.H., 2011a. Prediction of principal ground-motion parameters using a hybrid method coupling artificial neural networks and simualted annealing. Comput. Struct. 89 (23–24), 2176–2194.

Alavi, A.H., Gandomi, A.H., 2011b. A robust data mining approach for formulation of geotechnical engineering systems. Eng. Comput. 28 (3), 242–274.

Apostolopoulos, T., Vlachos, A., 2011. Application of the firefly algorithm for solving the economic emissions load dispatch problem. Int. J. Combinatorics. 2011, Article ID 523806. <http://www.hindawi.com/journals/ijct/2011/523806.html>. Accessed: 15 March 2012.

Blum, C., Roli, A., 2003. Metaheuristics in combinatorial optimization: overview and conceptual comparison. ACM Comput. Surv. 35, 268–308.

Conn, A.R., Schneinberg, K., Vicente, L.N., 2009. Introduction to Derivative-Free Optimization, MPS-SIAM Series on Optimization. SIAM, Philadelphia, PA.

Dorigo, M., Stütle, T., 2004. Ant Colony Optimization. MIT Press, Cambridge, MA, USA.

Farmer, J.D., Packard, N., Perelson, A., 1986. The immune system, adapation and machine learning. Physica D. 2, 187–204.

Fogel, L.J., Owens, A.J., Walsh, M.J., 1966. Artificial Intelligence Through Simulated Evolution. John Wiley & Sons, New York, NY.

Gandomi, A.H., Alavi, A.H., 2011. Applications of computation intelligence in behaviour simulation of concrete maerials. In: Yang, X.S., Koziel, S. (Eds.), Computational Optimization and Applications in Engineering and Industry. Studies in Computational Intelligence, vol. 359. Springer, Heidelberg, Germany, pp. 221–243.

Gandomi, A.H., Alavi, A.H., 2012a. A new multi-gene genetic programming approach to nonlinear system modeling. Part I: Materials and structural engineering. Neural Comput. Appl. 21 (1), 171–187.

Gandomi, A.H., Alavi, A.H., 2012b. A new multi-gene genetic programming approach to nonlinear system modeling. Part II: Geotechnical and earthquake engineering. Neural Comput. Appl. 21 (1), 189–201.

Gandomi, A.H., Yang, X.S., Alavi, A.H., 2011. Mixed variable structural optimization using firefly algorithm. Comput. Struct. 89 (23–24), 2325–2336.

Gandomi, A.H., Yang, X.S., Alavi, A.H., 2011. Cuckoo search algorithm: a metaheuristic approach to solve structural optimization problems. Engineering With Computers. 27, 1–19. doi: 10.1007/s00366-011-0241-y.

Geem, Z.W., Kim, J.H., Loganathan, G.V., 2001. A new heuristic optimization: harmony search. Simulation. 76, 60–68.

Glover, F., Laguna, M., 1997. Tabu Search. Kluwer Academic Publishers, Boston, MA.

Goldberg, D.E., 1989. Genetic Algorithms in Search, Optimization and Machine Learning. Addison Wesley, Reading, MA.

Holland, J., 1975. Adaptation in Natural and Artificial Systems. University of Michigan Press, Ann Arbor, MI.

Karaboga, D., 2005. An idea based on honey bee swarm for numerical optimization. Technical Report TR06. Erciyes University, Turkey.

Karmarkar, N., 1984. A new polynomial-time algorithm for linear programming. Combinatorica. 4 (4), 373–395.

Kennedy, J., Eberhart, R.C., 1995. Particle swarm optimization. In: Proceedings of IEEE International Conference on Neural Networks. Piscataway, NJ, pp. 1942–1948.

Kennedy, J., Eberhart, R.C., Shi, Y., 2001. Swarm Intelligence. Morgan Kaufmann Publishers, San Francisco.

Kirkpatrick, S., Gelatt, C.D., Vecchi, M.P., 1983. Optimization by simulated annealing. Science. 220 (4598), 671–680.

Koza, J.R., 1992. Genetic Programming: On the Programming of Computers by Means of Natural Selection. MIT Press, Cambridge, MA, USA.

Koza, J.R., Keane, M.A., Streeter, M.J., Yu, J., Lanza, G., 2003. Genetic Computing IV: Routine Human-Competitive Machine Intelligence. Kluwer Academic Publishers, Norwell, MA, USA.

Koziel, S., Yang, X.S., 2011. Computational Optimization, Methods and Algorithms. Springer, Germany.

Langdon, W.B., 1998. Genetic Programming + Data Structures = Automatic Programming!. Kluwer Academic Publishers, Norwell, MA, USA.

Nakrani, S., Tovey, C., 2004. On honey bees and dynamic server allocation in internet hosting centers. Adapt. Behav. 12 (3–4), 223–240.

Nelder, J.A., Mead, R., 1965. A simplex method for function optimization. Comput. J. 7, 308–313.

Parpinelli, R.S., Lopes, H.S., 2011. New inspirations in swarm intelligence: a survey. Int. J. Bio-Inspired Comput. 3, 1–16.

Pavlyukevich, I., 2007. Lévy flights, non-local search and simulated annealing. J. Comput. Phys. 226, 1830–1844.

Pham, D.T., Ghanbarzadeh, A., Koc, E., Otri, S., Rahim, S., Zaidi, M., 2006. The bees algorithm: a novel tool for complex optimisation problems. In: Proceedings of IPROMS 2006 Conference, pp. 454–461.

Price, K., Storn, R., Lampinen, J., 2005. Differential Evolution: A Practical Approach to Global Optimization. Springer, Heidelberg, Germany.

Sawaragi, Y., Nakayama, H., Tanino, T., 1985. Theory of Multiobjective Optimisation. Academic Press, Orlando, FL, USA.

Sayadi, M.K., Ramezanian, R., Ghaffari-Nasab, N., 2010. A discrete firefly meta-heuristic with local search for makespan minimization in permutation flow shop scheduling problems. Int. J. Ind. Eng. Comput. 1, 1–10.

Storn, R., 1996. On the usage of differential evolution for function optimization. In: Biennial Conference of the North American Fuzzy Information Processing Society (NAFIPS), pp. 519–523.

Storn, R., Price, K., 1997. Differential evolution—a simple and efficient heuristic for global optimization over continuous spaces. J. Global Optim. 11, 341–359.

Talbi, E.G., 2009. Metaheuristics: From Design to Implementation. John Wiley & Sons, Hoboken, NJ, USA.

Wolpert, D.H., Macready, W.G., 1997. No free lunch theorems for optimization. IEEE Trans. Evol. Comput. 1, 67–82.

Yang, X.S., 2005. Engineering optimization via nature-inspired virtual bee algorithms. In: Mira, J. and Alvarez, J.R. (Eds.), Artificial Intelligence and Knowledge Engineering Applications: A Bioinspired Approach. Lecture Notes in Computer Science, vol. 3562. Springer, Berlin/Heidelberg, pp. 317–323.

Yang, X.S., 2008. Nature-Inspired Metaheuristic Algorithms. first ed. Luniver Press, Frome.

Yang, X.S., 2009. Firefly algorithms for multimodal optimization. In: Watanabe, O., Zeugmann, T. (Eds.), 5th Symposium on Stochastic Algorithms, Foundation and Applications (SAGA 2009). LNCS, vol. 5792, Sapporo, Japan, pp. 169–178.

Yang, X.S., 2010a. Nature-Inspired Metaheuristic Algoirthms. second ed. Luniver Press, Frome.

Yang, X.S., 2010b. Engineering Optimization: An Introduction with Metaheuristic Applications. John Wiley & Sons, Hoboken, NJ, USA.

Yang, X.S., 2010c. A new metaheuristic bat-inspired algorithm, In: Gonzalez, J.R., Pelta, D.A., Cruz, C., Terrazas G., Krasnogor N. (Eds.), Nature-Inspired Cooperative Strategies for Optimization (NICSO 2010). Studies in Computational Intelligence, vol. 284. Springer, pp. 65–74.

Yang, X.S., 2011a. Bat algorithm for multi-objective optimisation. Int. J. Bio-Inspired Comput. 3 (5), 267–274.

Yang, X.S., 2011b. Chaos-enhanced firefly algorithm with automatic parameter tuning. Int. J. Swarm Intell. Res. 2 (4), 1–11.

Yang, X.S., Deb, S., 2009. Cuckoo search via Lévy flights. In: Proceedings of World Congress on Nature and Biologically Inspired Computing (NaBic 2009). IEEE Publications, USA, pp. 210–214.

Yang, X.S., Deb, S., 2010. Engineering optimization by cuckoo search. Int. J. Math. Model. Numer. Optim. 1 (4), 330–343.

Yang, X.S., Gandomi, A.H., 2012. Bat algorithm: a novel approach for global engineering optimization. Eng. Comput. 29 (5), 1–18.

Yang, X.S., Koziel, S., 2011. Computational Optimization and Applications in Engineering and Industry. Springer, Germany.

Part One

Water Resources

2 A Review on Application of Soft Computing Methods in Water Resources Engineering

H. Md Azamathulla

River Engineering and Urban Drainage Research Centre (REDAC), Universiti Sains Malaysia, Engineering Campus, Seri Ampangan, 14300 Nibong Tebal, Pulau Pinang, Malaysia

2.1 Introduction

When solving problems in water resources and hydraulic engineering, there are basically two approaches: knowledge-based and data-driven (Deo, 2011). Nowadays, engineering practices suffer from unpredictable problems that need more serious attention and care. The introduction of soft computing techniques may fulfill the need and demands of solving engineering problems. Soft computing was initiated by Zadeh (1981) in order to construct the new generation of artificial intelligence, which was called *computational intelligence.* Multidisciplinary approaches were designed to model mathematically the problems associated with the complexity and uncertainty of logical systems. Soft computing includes the concepts and techniques used to solve or overcome the difficulties in the real world, especially in the engineering sciences (Alavi and Gandomi, 2011a,b; Gandomi and Alavi, 2011, 2012).

The basic method of soft computing consists of fuzzy logic (FL), neural networks (NNs), and genetic algorithms. These soft computing methods have been applied to many real-world problems, especially in water resources engineering. The focus of these applications is more on optimization in predicting the scour phenomenon that occurs at the laid pipelines under seas or rivers, bridge piers, abutments, culverts, and spillways. Soft computing applications are also applied to the prediction of sediment load in hydraulics and the stage−discharge curve in hydrology.

In many cases, good results have been achieved by combining different soft computing methods. The hybrid system is now growing, as are most of the neuro-fuzzy systems in which NN techniques are used for calibration and induction. Deo (2011) stated that problem solving in water engineering has seen four stages: analytical equations, empirical/experimental methods, numerical methods, and finally data-driven approaches.

Metaheuristics in Water, Geotechnical and Transport Engineering. DOI: http://dx.doi.org/10.1016/B978-0-12-398296-4.00002-7

The author and his research associates have used radial basis function (RBF), adaptive neuro-fuzzy inference system (ANFIS), gene-expression programming (GEP), and linear genetic programming (LGP) for a variety of purposes: prediction of spillway scour depth, bridge pier scour depth, culvert scour depth, dispersion coefficients, and sediment transport in hydraulics. The application in water resources includes the development of the stage—discharge curve for the Pahang River in Malaysia. Details of these studies can be seen in the list of resulting publications shown on the web portal http://redac.eng.usm.my/html/publication.htm. This chapter focuses on reviewing some major applications of soft computing methods in water resources engineering.

2.2 Soft Computing Techniques

2.2.1 Neural Networks

A neural network (NN) represents an interconnection of neurons, each of which basically carries out the task of combining the input, determining its strength by comparing the combination with a bias (or, alternatively, passing it through a nonlinear transfer function), and firing out the result in proportion to such a strength. The known input—output patterns are first used to train a network, and strengths of interconnections (or weights) and bias values are accordingly fixed. A supervised type of training involves feeding input—output examples until the network develops its generalization capability, while an unsupervised training would involve classification of the input into clusters by some rule; the former type of learning is more common. During such training, the network output is compared to the desired or actual/target one and the error or the difference that results is processed through a mathematical algorithm (Azmathullah et al., 2005).

Most of the previous works on NN applications to water resources have included the feed-forward type of the architecture, where there are no backward connections (Figure 2.1), trained using the feed-forward back propagation (FFBP) configuration (Azamathulla et al., 2008). The RBF network (Figure 2.2) is also similar to this, in that it has three layers of neurons: input, hidden, and output. However, it uses only one hidden layer, each neuron in which operates as per the Gaussian transfer function, as against the sigmoid function of the common FFBP. Further, while training of the latter is fully supervised (where both input—output examples are required), the training of the former is fragmented, wherein unsupervised learning of the input information first classifies it into clusters, which in turn are used to yield the output after a supervised learning. This local tuning could be not only more efficient but also more satisfactory in modeling nonlinear data than the common FFBP (Azamathulla et al., 2008).

ANFIS, on the other hand, is a hybrid scheme that uses the learning capability of the artificial neural network (ANN) to derive the fuzzy *if—then* rules with appropriate membership functions worked out from the training pairs leading finally to the inference (Jang and Gulley, 1995; Tay and Zhang, 1999). The difference

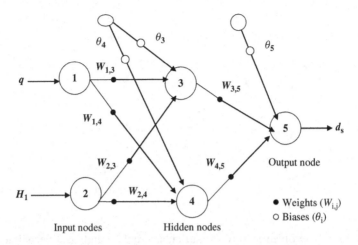

Figure 2.1 FFBP architecture.

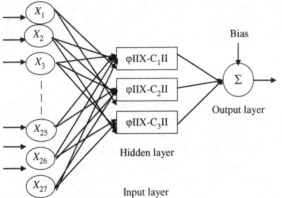

Figure 2.2 RBF neural network architecture.

between the common NN and the ANFIS is that while the former captures the underlying dependency in the form of the trained connection weights, the latter does so by establishing the fuzzy language rules. The input in ANFIS (Figure 2.3) is first converted into fuzzy membership functions, which are combined together, and after following an averaging process, used to obtain the output membership functions and finally the desired output.

2.2.2 Gene-Expression Programming

GEP, which is an extension of genetic programming (GP) (Koza, 1992), is a search technique that involves computer programs (e.g., mathematical expressions, decision trees, polynomial constructs, and logical expressions). GEP computer programs are all encoded in linear chromosomes, which are then expressed as or translated into expression trees (ETs). ETs are sophisticated computer programs

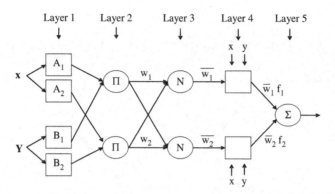

Figure 2.3 ANFIS network architecture.

that have usually evolved to solve a particular problem and are selected according to their fitness at solving that problem.

GEP is a full-fledged genotype and phenotype system, with the genotype totally separated from the phenotype, whereas in GP, genotype and phenotype are mixed together in a simple replicator system. As a result, the full-fledged genotype and phenotype system of GEP surpasses the old GP system by a factor of 100–60,000 (Ferreira, 2001).

Initially, the chromosomes of each individual in the population are generated randomly. Then, the chromosomes are expressed, and each individual is evaluated based on a fitness function and selected to reproduce with modification, leaving progeny with new traits. The individuals in this new generation are in turn subjected to several developmental processes, such as expression of the genomes, confrontation of the selection environment, and reproduction with modification. These processes are repeated for a predefined number of generations, or until a solution is achieved (Ferreira, 2001). The functionality of each genetic operator included in the GEP system has been explained by Guven and Aytek (2009) and Azamathulla and Ahmad (2012).

2.2.3 Linear Genetic Programming

LGP, which is an extension of conventional tree-based GP, evolves developing sequences of instructions from an imperative programming language (C or C++) or from a machine language. The name *linear* refers to the structure of the (imperative) program representation; it does not reflect functional genetic programs that are restricted to only a linear list of nodes. On the contrary, genetic programs normally represent highly nonlinear solutions (Brameier, 2004). The main differences between LGP and conventional, tree-based GP are the graph-based data flow that results from a multiple usage of indexed variable (register) contents and the existence of structurally ineffective code (introns) (Brameier, 2004; Brameier and Banzhaf, 2001). This concept was expanded to the Automatic Induction of Machine code by Genetic Programming (AIMGP) technique, in which the solutions are directly computed as binary machine codes and executed without using an interpreter. In this way, the computer program can evolve very quickly (Brameier and Banzhaf, 2001).

GEP/LGP can be applied in two different ways. One of them manipulates computer programs, while the other one operates on equations. The commercial software gene and DTREG generates the GEP programs (Sherrod, 2008).

2.3 Implementation of Soft Computing Techniques

The applications in hydraulics include spillway scour, bridge pier/abutment scour, river pipeline scour, and sediment bed load predictions. The readers are referred to Azmathullah et al. (2005, 2006) and Azmathulla et al. (2008, 2010, 2011) for hydraulic applications, and to Azamathulla et al. (2011) for river stage−discharge relationship.

2.3.1 Soft Computing Techniques for Spillway Scour

Spillways are provided as an integral part of a dam or as an auxiliary structure that is separate from the main dam to release surplus floodwater which is in excess of the storage space in the reservoir, as provided in the operation plan, and must be passed downstream. Thus, spillways work as safety valves for a dam and the adjoining countryside. The discharge intensity is the most predominant factor in determining the depth of scour. The scour depth also varies with the height of the falling jet. A great height would cause high kinetic energy of the falling jet (Damle et al., 1966). The angle of penetration of a ski-jump jet into the pool would also be important from the perspective of geometry and depth of scour hole. As such, it would be safer to assume that the energy dissipation due to air entrainment is considerably less for the buckets located near the tail water levels.

Referring to Figure 2.4, the equilibrium depth of scour (d_s), measured from the tail water surface, can be written as a function of discharge per width or unit discharge of spillway (q), total head (H_1), radius of the bucket (R), lip angle of the bucket (ϕ), tail water depth (d_w), mean sediment size (d_{50}), acceleration due to gravity (g), densities of water and sediment, ρ_w, and ρ_s:

$$d_s = f(q, H_1, R, \phi, d_w, d_{50}, g, \rho_w, \rho_s) \qquad (2.1)$$

In the present study, the standard deviation of fragmented bed material σ_g was not considered (Table 2.1). The maximum width of scour hole (w_s) and the distance of maximum scour depth from spillway bucket lip (ℓ_s) can be written in a similar form:

$$w_s = f(q, H_1, R, \phi, d_w, d_{50}, g, \rho_w, \rho_s) \qquad (2.2)$$

$$\ell_s = f(q, H_1, R, \phi, d_w, d_{50}, g, \rho_w, \rho_s) \qquad (2.3)$$

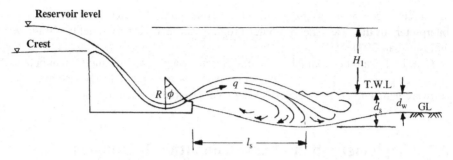

Figure 2.4 Scour below flip bucket spillway.

Table 2.1 Range of Experimental Data

No.	Parameter	Unit	Range
1	Discharge intensity, q	m³/s/m	0.0089–0.3810
2	Total head, H_1	m	0.2791–1.7962
3	Bucket radius, R	m	0.1000–0.6096
4	Lip angle, ϕ	radian	0.1740–0.7800
5	Tail water depth, d_w	m	0.0286–0.2650
6	Bed material size, d_{50}	m	0.0020–0.0080
7	Depth of scour, d_s	m	0.0512–0.5500
8	Distance of maximum scour from bucket lip, ℓ_s	m	0.4200–2.2400
9	Width of scour hole, w_s	m	0.6000–2.1400

By using the Buckingham π theorem, nondimensional equations in functional form were obtained and then nonlinear regression equations were developed. The same inputs were used to develop the ANN/GEP models.

The width prediction is less satisfactory, indicating that the lateral dissipation of energy is too uncertain to predict accurately. Although the percentage error and root mean square error (RMSE) involved in depth prediction are small, the correlation coefficient is low, and absolute deviation is high, indicating that overall, the prediction may be viewed with skepticism. Hence, the author applied soft computing techniques to predict scour below spillways. Among 95 pairs of scour data, 80% of input–output patterns were chosen randomly until the best training performance appeared and was used for network training, while the remaining data were used for testing or validating the trained FFBP, FFC, RBF, ANFIS, and GEP models (Azamathulla, 2011; Azamathulla and Zakaria, 2011) (Table 2.2).

The results were compared with the regression equation formulas and NN schemes. It was found that the GEP models (Eqs. (2.4)–(2.6); Azamathulla and Zakaria, 2011)) are highly satisfactory, as seen in Table 2.2, for depth of scour

Table 2.2 Comparison of Network Yielded and True Values

Parameter	Correlation Coefficient (R)	Average Error (AE)	Average Absolute Deviation (δ)	RMSE
GEP				
d_s/d_w	0.970	−6.680	12.664	0.583
ℓ_s/d_w	0.988	−2.054	3.703	1.629
w_s/d_w	0.992	−2.336	11.187	1.777
ANFIS				
d_s/d_w	0.976	−0.09882	12.50924	0.56921
ℓ_s/d_w	0.936	4.849	14.116	3.515679
w_s/d_w	0.965	4.6170	18.340	3.995247
FFBP (Azamathulla et al., 2005)				
d_s/d_w	0.970	−6.680	13.845	0.579655
ℓ_s/d_w	0.989	−2.876	3.725	0.720417
w_s/d_w	0.990	−2.336	9.111	1.672423
Feed-forward cascade correlation (FFCC; Azamathulla et al. 2005)				
d_s/d_w	0.949	−16.738	19.109	0.841427
ℓ_s/d_w	0.972	−5.286	10.138	1.740402
w_s/d_w	0.965	−3.625	18.1756	3.053031
RBF (Azamathulla et al. 2005)				
d_s/d_w	0.967	2.390	15.13	0.662571
ℓ_s/d_w	0.943	8.801	11.91661	2.722683
w_s/d_w	0.974	7.663	12.97874	2.624881
Regression equations (Azamathulla et al. 2005)				
d_s/d_w	0.842	−1.427	22.790	1.343503
ℓ_s/d_w	0.929	3.900	13.550	3.569314
w_s/d_w	0.883	−19.570	20.150	5.612486

downstream of the flip bucket spillway, width, and location of maximum scour from the bucket lip:

$$\frac{d_s}{d_w} = \left[\frac{d_{50} + q}{\phi} - (H_1 - q - 1.199)\left(5.616\frac{d_{50}}{\phi} \right) \right] \\ \times \left[0.309\phi(R + (d_{50} + 0.185)(H_1 d_{50}))^{-0.5} \right] \tag{2.4}$$

$$\frac{w_s}{d_w} = \left[\frac{q - 0.006^{d_{50}} + 1.168^{R+H_1}}{2.336\phi} \right]^{0.5} \left[\frac{7.521 d_{50} + 3.955 H_1 - 2q}{0.428\phi^{-1}} + 15.24\phi \right]^{0.5} \tag{2.5}$$

$$\frac{\ell_s}{d_w} = \left[\frac{e^{\phi}}{R}(H_1 - d_{50} + 0.495 + 2.878(q)^{-1}) \right] \left[R(q + H_1 - 9.948 d_{50})^{0.5}(2H_1 + q + \phi) \right] \tag{2.6}$$

Table 2.3 Comparison of Network Results with Observed Scour Depths

No.	Method	R	AE	δ	RMSE
1	NN (FFBP-based)	0.92	−8.89	13.27	10.75
2	NN (FFCC-based)	0.90	−33.56	24.12	12.44
3	RBF	0.91	10.02	19.93	8.04
4	ANFIS	0.95	13.64	12.09	7.44
5	Martins (1975)	0.69	21.15	28.52	13.92
6	Veronese (1937)	0.73	−18.5	22.57	11.40
7	Wu (1973)	0.73	16.94	27.60	11.35
8	Incyth (1982)	0.73	21.96	26.76	12.24
9	Azamathulla et al., 2008	0.78	−24.91	26.89	11.69

Network Based on Prototype Data

Another study to estimate the scour based on past prototype measurements rather than against the above-described case of scale model observations was also conducted. A survey of available publications reporting such observations was done. This indicated that only three types of information (namely, scour depth below tail water level d_s, discharge intensity q, and head drop H_1) are uniformly reported in all references. An NN with two input nodes for q and H_1 and one output node for the scour depth d_s were developed (see Figure 2.1). In total, there were 91 input−output pairs formed out of the published data.

RBF and ANFIS yielded more or less similar predictions, with ANFIS producing middle-ranged estimates in a slightly better way than RBF, which is understandable considering the similarities in the data processing with these methods. The results were that adequate training is really necessary while applying the NN; otherwise, one may not get better results than traditional regression (Table 2.3). In other words, the scour data are more amenable to fuzzy if−then rules rather than crisp-value processing in the RBF or FFBP network. ANFIS ensures a localized functioning of the transfer function against the globalized one of a general FFBP. This results in a smaller number of values participating in the mapping process, which in turn requires limited data for training (as against the FFBP). This could also be another reason for the more acceptable performance of ANFIS in the present case. Thus, the research showed that traditional equation-based methods of predicting the design scour downstream of a ski-jump bucket could better be replaced by the NN and similar soft computing schemes. Within the different networks employed, the relatively advanced ANFIS could produce more satisfactory results (Azamathulla et al., 2008). Recently, Guven and Azamathulla (2012) showed that the results of GEP (Eqs. (2.7) and (2.8)) in predicting spillway scour rivaled those of ANN (Table 2.4):

$$d_s = \log(H_1 - q - 92.857) + \left(\frac{q}{H_1} + 9.688H_1\right)^{0.5} + \log\left(\frac{q^{0.5} - 10^{-7.578}}{\log(q)}\right) \qquad (2.7)$$

Table 2.4 Statistical Performance of GEP, GP, and Other Formulas for the Testing Set

Model	RMSE (m)	δ	R
GEP1 (Eq. (2.7))	1.596	0.109	0.917
GEP2 (Eq. (2.8))	1.998	0.210	0.867
GP (Azamathulla et al., 2010)	2.347	0.377	0.842
Veronese (1937)	5.853	0.394	0.816
Wu (1973)	4.415	0.887	0.819
Martins (1975)	2.649	0.277	0.763

$$d_s/H_1 = -0.362(10^{F_1} - 2.734)^{-1} + (F_1^2 \, e^{0.86_1/F_1})^{0.5} + (0.895 F_1^{0.5} - 0.024) \quad (2.8)$$

For future studies on obtaining the pattern of scour, it is suggested to consider its location with respect to the bucket lip and the rock quality designation (RQD) for prototype data. The preliminary studies by the author indicated a very good prediction ability. As such, it is concluded that GEP is more efficient in predicting scour parameters downstream of the flip bucket to other NN schemes.

2.3.2 Soft Computing Techniques for Submerged Pipeline Scour Depth

Scour underneath the pipeline may expose a section of the pipe, causing it to become unsupported. If the free span of the pipe is long enough, the pipe may experience resonant flow-induced oscillations, leading to settlement and potentially structural failure (Azamathulla and Zakaria, 2011).

Details of the river pipeline studies can be seen in the list of resulting publications in the REDAC web portal http://redac.eng.usm.my/html/publication.htm.

The performance of the GP model was compared to that of the RBF NN and conventional regression-based equations (Azamathulla and Zakaria, 2011). The comparison showed that the GP models have a better ability to predict the scour depth with higher R values. Further, Azamathulla et al. (2011) applied the LGP technique to predict the scour depth below the pipeline and compared it with ANFIS. It was concluded that the LGP models have a better ability to predict the river pipeline scour.

Development of the Stage–Discharge Curve for the Pahang River

The conventional models that predict stage–discharge relationships are the stage-rating curve (SRC) and regression (REG). Azamathulla et al. (2011) developed the following explicit formulation of the GEP for discharge as a function of stage (S):

$$Q = 9.84S^2 - 64.391S - 4033.296 \quad (2.9)$$

Furthermore, GEP was used as an alternative tool for modeling the stage–discharge relationship for the Pahang River. The overall results produced coefficient of determination (R^2) values very close to 1, suggesting very little discrepancy between observed and predicted discharges. Besides, the RMSE values remained at a very low level, also confirming GEP as an effective tool to be used for forecasting and the estimation of daily discharge data in flood events.

2.3.3 Soft Computing Techniques to Predict Total Bed Material Load

The rising demand of river sand has led to a mushrooming of river sand mining activities, which jeopardizes the natural and health of the river, as well as causing environmental problems. Generally, the conventional approaches used in most modeling efforts begin with an assumption of empirical and analytical equations. Although much research on the total bed load transport has been recorded throughout the last few decades based on conventional approaches, they still have constrained the wider application of theoretical models. Alternatively, various kinds of soft computing techniques have been introduced and applied in water engineering problems since the last two decades (Azamathulla et al., 2011; Nagy et al., 2002; Yang et al., 2009).

ANNs were developed for sediment data sets, the network input–output. The ANN model was able to predict total load transport successfully in a wide variety of fluvial environments, including both sand and gravel rivers. Moreover, the ANN prediction of mean total load was in almost perfect agreement with the measured mean total load. The high value of the coefficient of determination $(R^2 = 0.958)$ implies that the ANN model provides an excellent fit for the measured data.

A few soft computing techniques have been developed to evaluate and predict the total bed material load; i.e., NNs, ANFIS, and GEP. The study area and data used for the application cover the six sites at each of the three rivers (i.e., Kurau, Langat, and Muda) that have different levels of sand mining activities. Table 2.5 shows the range of field data from the three rivers.

The GEP model was able to predict total load transport successfully in a wide variety of fluvial environments, including both sand and gravel rivers. Also, the GEP estimation of mean total load was in almost perfect agreement with the measured mean total load. The high value of the coefficient of determination $(R^2 = 0.97)$, mean square error (MSE $= 0.057$) that the GP model provides an excellent fit for the measured data. These results suggest that the proposed GEP model is a robust total load predictor. This study demonstrates a successful application of the GEP modeling concept to total bed material load transport. From this study, it can be concluded that only eight parameters are required to predict total bed material load, which is in agreement with previous works (Yang et al., 2009). The value of the GEP approach is that the nonlinear function need not be the same for all fluvial environments (Zakaria et al., 2010).

GEP for Sediment Load

The important parameters that affect the total bed material load during implementation of GEP for sediment data sets can be given in the form $T_j = f(Q, V, B, Y_0, R,$

Table 2.5 Range of Field Data from the Kurau, Langat, and Muda Rivers

Parameters	Study Area		
	Kurau River	**Langat River**	**Muda River**
Flow discharge, Q (m³/s)	0.63–28.94	2.75–120.76	2.59–343.71
Flow velocity, V (m/s)	0.27–1.12	0.23–1.01	0.14–1.45
Water-surface width, B (m)	6.30–26.00	16.4–37.6	9.0–90.0
Flow depth, Y_0 (m)	0.36–1.91	0.64–5.77	0.73–6.90
Cross-sectional area of flow, A (m²)	1.43–33.45	8.17–153.57	5.12–278.34
Hydraulic radius, R (m)	0.177–1.349	0.45–3.68	0.55–3.90
Channel slope, S_0	0.00050–0.00210	0.00065–0.00185	0.00008–0.000235
Bed load, T_b (kg/s)	0.080–0.488	0.027–0.363	0–0.191
Suspended load, T_t (kg/s)	0.001–2.660	0.2860–99.351	0.024–15.614
Total bed material load, T_j (kg/s)	0.089–2.970	0.525–33.398	0.099–15.644
Median sediment size, d_{50} (mm)	0.41–1.90	0.31–3.00	0.29–2.10
Manning's coefficient , n	0.014–0.066	0.034–0.195	0.21–0.108

S_0, W_s, d_{50}) (Zakaria et al., 2010). Initially, the training set is selected from the whole data and the rest is used as the testing set. The implementation of the GEP includes five major steps: (i) select the fitness function, (ii) In this study, select the set of functions which consists of four basic arithmetic operators $(+, -, \times, /)$ and some basic mathematical functions ($\sqrt{}$, x^2, x^3, power) are utilized. (iii) choose the chromosomal architecture which is the length of the head and number of genes, (iv) to choose the linking function, and (v) choose the set of genetic operators that cause variation and their rates. In the present study, the first step is to choose the fitness function. For this problem, the fitness, f_i, of an individual program, i, is measured by

$$f_i = \sum_{j=1}^{C_t} (M - |C_{(i,j)} - T_j|) \tag{2.10}$$

where M is the range of selection, $C_{(i,j)}$ is the value returned by the individual chromosome i for fitness case j (out of C_t fitness cases), and T_j is the target value for fitness case j. If $|C_{(i,j)} - T_j|$ (the precision) is less than or equal to 0.01, then the precision is equal to zero, and $f_i = f_{max} = C_t M$. In this case, $M = 100$ was used; therefore, $f_{max} = 1000$. The advantage of this kind of fitness function is that the system can find the optimal solution by itself. Then, the length of head ($l_h = 10$) and two genes per chromosome were employed, addition and multiplication were used as the linking functions and a combination of all genetic operators (mutation, transportation, and crossover) was used in GEP (Table 2.6).

Table 2.6 Parameters of the Optimized GEP Model

Parameter	Definition	Value
p_1	Function set	$+, -, \times, /, \sqrt{}$, power, sin, cos, tan
p_2	Mutation rate (%)	40
p_3	Inversion rate (%)	30
p_4	One-point and two-point recombination rate (%)	30, 30
p_5	Gene recombination rate	95
p_6	Gene transposition rate	0.1

Table 2.7 Error Measures for Different Sediment Models for Comparison of Traditional Predictors and the GEP Model for Total Bed Material Load

Model	R^2	MSE
GEP	0.97	0.057
Yang (1972)	0.722	10.376
Engelund and Hansen (1967)	0.623	12.735

The best of generation individual, chromosome 10, had fitness 470 for this GEP model of sediment transport. The formulations of GEP for total bed material load, as a function of Q, V, B, Y_0, R, S_0, W_s, d_{50}, were obtained as

$$T_j = \left[\left(-0.39 \times RY_0 \times \sqrt{S_0}\right)/(-0.72 + S_0)\right] + \left(R + e^{\sin(QVR)}\right) \\ + \tan^{-1}\left(-0.16 \times RB\right) + R\sqrt{Q}) + (d_{50} - 3.39) \times d_{50}^3 \times S_0 + \tan^{-1}(V) \\ \times e^V - \log(6.93 - Y_0) \times ((Ws \times B)/(-2.075))$$

$$(2.11)$$

The calibration of the GEP model was performed based on 214 input-target pairs of collected data. Among the 214 data sets, 64 (30%) were reserved for validation, and the remaining 150 sets for the calibration purpose and remaining were used for testing or validating the GEP model. The performance of the GEP model was compared with the traditional total bed material load equations of Yang (1972) and Engelund and Hansen (1967). Table 2.7 presents a comparison of R^2 and MSE of the predicted total bed material load of the different models, and it can be concluded that for all the data sets, the GEP model give either better or comparable results. Overall, particularly for field measurements, the GEP models give better estimations than the existing models. The GEP model produced the

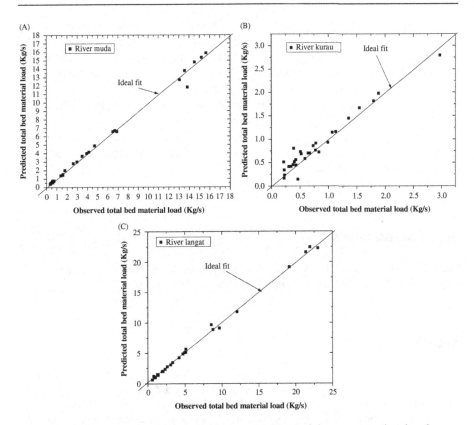

Figure 2.5 Observed and estimated total bed material load of the unseen testing data for three rivers: (A) Muda, (B) Kurau, (C) Langat.

least errors ($R^2 = 0.97$ and MSE $= 0.057$). Figure 2.5A−C shows the observed and estimated total bed material load of the unseen testing data.

2.4   Conclusion

In this chapter, the major applications of different soft computing techiques (namely RBF, ANFIS, GEP, and LGP) in water resources engineering are reviewed. The models developed using these techniques have shown great performance compared to the conventional approcahes. The obtained results support the use of these intelligent techniques for the prediction of hydraulic and hydrologic variables. The application of soft computing techniques in water resources engineering provides the possibility of overcoming the complexity and uncertainty of the existing problems associated with good and acceptable results.

It is worth mentioning that many investigators have recently compared the performance of GP with traditional statistical methods, as well as NN. Deo (2011) concluded in his keynote speech, "GP can automatically select input and tell which are more important." Based on the author's experience, there is good potential to exploit the full potential of the modern soft computing methods in the future. However, the relative advantages of most of the soft computing methods over traditional methods are that they do not assume any dependency between input and output beforehand—they learn directly from example rather than rules, and they are more tolerant toward data errors by virtue of their distributed processing and better adaptability to new data.

Acknowledgments

The writer is grateful to Professor M.C. Deo, IIT Bombay, Mumbai, India, and ASCE/ELSEVIER/IAHR for giving permission to reproduce figures and text.

References

Alavi, A.H., Gandomi, A.H., 2011a. A robust data mining approach for formulation of geotechnical engineering systems. Eng. Comput. 28 (3), 242–274.

Alavi, A.H., Gandomi, A.H., 2011b. Prediction of principal ground-motion parameters using a hybrid method coupling artificial neural networks and simulated annealing. Comput. Struct. 89 (23–24), 2176–2194.

Azamathulla, H.Md., Ahmad, Z., 2012. Gene-expression programming for transverse mixing coefficient. J. Hydrol. 434–435, 142–148.

Azamathulla, H.Md., Zakaria, N.A., 2011. Prediction of scour below submerged pipeline crossing a river using ANN. Water Sci. Tech. 63 (10), 2225–2230.

Azmathullah, H.Md., Deo, M.C., Deolalikar, P.B., 2005. Neural networks for estimation of scour downstream of a ski-jump bucket. J. Hydraul. Eng. 131 (10), 898–908.

Azmathullah, H.Md., Deo, M.C., Deolalikar, P.B., 2006. Estimation of scour below spillways using neural networks. J. Hydraul. Res. Int. Assoc. Hydraul. Eng. Res. 44 (1), 61–69.

Azamathulla, H.Md., Deo, M.C., Deolalikar, P.B., 2008. Alternative neural networks to estimate the scour below spillways. Adv. Eng. Software. 39 (8), 689–698.

Azamathulla, H.Md., Ghani, A.A., Zakaria, N A, Guven, A., 2010. Genetic programming to predict bridge pier scour. ASCE J. Hydraul. Eng. 136 (3), 165–169.

Azamathulla, H.Md., Guven, A., Demir, Y.K., 2011. Linear genetic programming to scour below submerged pipeline. Ocean Eng. 38, 995–1000.

Brameier, M., 2004. On linear genetic programming. Ph.D. thesis, University of Dortmund.

Brameier, M., Banzhaf, W., 2001. A comparison of linear genetic programming and neural networks in medical data mining. IEEE Trans. Evol. Comput. 5, 17–26.

Damle, P.M., Venkatraman, C.P., Desai, S.C., 1966. Evaluation of scour below ski-jump buckets of spillways. In: Proceedings of the CWPRS Golden Jubilee Symposium, Poona, India, Vol. I, 154–163.

Deo, M.C., 2011. Application of data driven methods in hydrology and hydraulics. In: Key note speech, International Conference on Managing Rivers in the 21st Century, Penang, Malaysia, 6–9 December 2011.

Engelund, F., Hansen, E., 1967. A Monograph on Sediment Transport in Alluvial Streams. Teknisk Forlag, Copenhagen, Denmark.

Ferreira, C., 2001. Gene expression programming: a new adaptive algorithm for solving problems. Complex Syst. 13 (2), 87–129.

Gandomi, A.H., Alavi, A.H., 2011. Multi-Stage genetic programming: a new strategy to nonlinear system modeling. Inform. Sci. 181 (23), 5227–5239.

Gandomi, A.H., Alavi, A.H., 2012. A new multi-gene genetic programming approach to nonlinear system modeling. Part II: geotechnical and earthquake engineering problems. Neural Comput. Appl. 21 (1), 189–201.

Guven, A., Aytek, A., 2009. A new approach for stage–discharge relationship: Gene-Expression Programming. J. Hydrolog. Eng. 14 (8), 812–820.

Guven, A., Azamathulla, H.Md., 2012. Gene-expression programming for flip bucket spillway scour. Water Sci. Tech. 65 (11), 1982–1987.

Incyth Lha, 1982. Estudio sobre modelo del aliviadero de la Presa Casa de Piedra, Informe Final. DOH-044-03-82, Ezeiza, Argentina.

Jang, J.S.R., Gulley, N., 1995. Fuzzy Logic Tool Box User's Guide. The Mathworks, Inc.

Koza, J.R., 1992. Genetic Programming: On the Programming of Computers by Means of Natural Selection. A Bradford Book, MIT Press, Cambridge, MA.

Martins, R.B.F., 1975. Scouring of rocky river beds by free jet spillways. Int. Water Power Dam Constr. 27 (4), 152–153.

Nagy, H.M., Watanabe, K., Hirano, M., 2002. Prediction of sediment load concentration in rivers using artificial neural network model. J. Hydraulic Eng. 128 (6), 588–595.

Sherrod, P.H., 2008. DTREG predictive modeling software, http://www.dtreg.com/.

Tay, J.H., Zhang, X., 1999. Neural fuzzy modelling of anaerobic biological waste water treatment systems. J. Environ. Eng. 125 (12), 1149–1159.

Veronese, A., 1937. Erosioni de Fondo a Valle di uno Scarico. Annali dei Lavori Publicci. 75 (9), 717–726, Italy.

Wu, C.M., 1973. Scour at downstream end of dams in Taiwan. In: International Symposium on River Mechanics, Bangkok, Thailand, vol. I, A 13, pp. 1–6.

Yang, C.T., 1972. Unit stream power and sediment transport. J. Hydraul. Eng. 98 (10), 1805–1826.

Yang, C.T, Reza, M, Aalami, M.T., 2009. Evaluation of total load sediment transport using AAN. Int. J. Sediment Res. 24 (3), 274–286.

Zadeh, L.A., 1981. What Is Soft Computing, Soft Computing. Springer-Verlag, Germany.

Zakaria, N.A, Azamathulla, H.Md., Chang, C.K., Ab Ghani, A., 2010. Gene-expression programming for total bed material load estimation—a case study. Sci. Total Environ. 408 (21), 5078–5085.

3 Genetic Algorithms and Their Applications to Water Resources Systems

Deepti Rani[1], Sharad Kumar Jain[2], Dinesh Kumar Srivastava[1] and Muthiah Perumal[1]

[1]Department of Hydrology, IIT Roorkee, Roorkee, India, [2]Water Resources Development and Management Department, IIT Roorkee, Roorkee, India

3.1 Introduction

According to Mitchell (1999), "In the 1950s and the 1960s several computer scientists independently studied evolutionary systems with the idea that evolution could be used as an optimization tool for engineering problems. The idea was to evolve a population of candidate solutions to a given problem, using operators inspired by natural genetic variation and natural selection." All these techniques are collectively referred to as *evolutionary computation (EC) techniques.* EC techniques, also known as heuristic search methods, mostly involve nature-inspired metaheuristic optimization algorithms such as evolutionary algorithms (EAs), comprising genetic algorithms (GAs); evolutionary programming, evolution strategy, and genetic programming; swarm intelligence, comprising ant colony optimization and particle swarm optimization; simulated annealing; and tabu search (Rani and Moreira, 2010).

GAs are a particular class of EA based on the mechanics of natural selection and natural genetics (Goldberg, 1989). GA uses techniques inspired by evolutionary biology such as inheritance, mutation, selection, and crossover. The method was invented by John Holland (1975) and was later popularized by one of his students, David Goldberg, who solved a difficult problem involving the control of gas-pipeline transmission for his dissertation. His book (Goldberg, 1989) provides GA methodology using both mathematical and computational aspects. He was the first to develop a theoretical basis for GAs through the schema theorem. The work of De Jong (1975) showed the usefulness of the GA for function optimization and made the first concerted effort to find optimized GA parameters. Unlike conventional optimization search methods based on gradients, GAs work on a population of possible solutions, attempting to find

Metaheuristics in Water, Geotechnical and Transport Engineering. DOI: http://dx.doi.org/10.1016/B978-0-12-398296-4.00003-9

a solution set that either maximizes or minimizes the value of a function of those solution values (Loucks and van Beek, 2005).

Like other optimization algorithms, a GA starts by defining decision variables and objective function. It terminates like other optimization algorithms too, by testing for convergence. Nevertheless, it is very different than the others with regard to the steps involved in the process. GAs are typically implemented as a computer simulation. GAs have a main generational process cycle. The GA process begins with a population of chromosomes, which is the set of possible solutions for the decision variables of an optimization problem, and moves toward achieving better solutions through evolution. The decision variables are encoded as binary or real-valued strings (genes) for a given search space. A chromosome is the set of these substrings (genes). The evolution starts from a population of completely random chromosomes and occurs in generations. In each generation, the fitness of the whole population is evaluated, and multiple chromosomes are stochastically selected from the current population (based on their fitness) and modified using genetic operators such as crossover and mutation to form a new population. The new population is then used in the next iteration (generation) of the algorithm (Davis, 1991). Population size depends on the nature of the problem, but typically there are hundreds or thousands of possible solutions. Traditionally, the population is generated randomly, covering the entire range of possible solutions (the search space). This algorithm is repeated sequentially until the desired stopping criterion is achieved.

Advantageous features of GAs in solving large-scale, nonlinear optimization problems are that they can be used with continuous or discrete parameters, require no simplifying assumptions about the problem, and, unlike gradient methods, they do not require computation of derivative information during the optimization (Haupt and Haupt, 2004). Davis (1991) has identified three main advantages of GAs in optimization: "First, they generally find nearly global optima in complex spaces. This is important because the search spaces for our problems are highly multimodal, a property that leads hill-climbing algorithms to get stuck in local optima. Second, genetic algorithms do not require any form of smoothness, that is, they can handle nonlinearity and discontinuity and third, considering their ability to find global optima, genetic algorithms are fast, especially when tuned to the domain on which they are operating." Another advantage of GAs is their inherently parallel nature, i.e., the evaluation of individuals within a population can be conducted simultaneously, as in nature.

Most of the early works in GAs came in the fields of computer science and artificial intelligence. More recently, interest has extended to essentially all branches of science, engineering, economy, and research and development, where search and optimization are of interest. The widespread interest in GAs appears to be due to the success in solving many difficult optimization problems. Today, many applications of GAs in different fields can be found in literature. GAs have been applied to many real-life optimization problems by several researchers. Goldberg and Kuo (1987) developed a study for pipeline optimization by making use of GAs. Soh and Yang (1996) used GAs in combination with fuzzy logic for structural-shaped optimization problems. Feng et al. (1997) applied GAs to the problem of cost-time trade-offs in construction projects. Halhal et al. (1997) applied GAs to a network rehabilitation problem having multiple objectives. A methodology based on GAs has been developed by Li and Love (1998) for optimizing the layout of construction-site-level facilities. Wang and Zheng (2002) studied a job shop scheduling problem with a modified GA. Wei et al. (2005) employed GAs in their research, with the aim of optimization of truss size and shaping with frequency constraints. In water resources, GAs have been applied in many fields, for example, rainfall-runoff modeling (Wang, 1991), water supply network design (Dandy and Engelhardt, 2001; Simpson et al., 1994), and groundwater management problems (Cieniawski et al., 1995; McKinney and Lin, 1994; Ritzel et al., 1994). Davidson and Goulter (1995) used GAs to optimize the layout of rectilinear-branched distribution (natural gas/water) systems.

Theoretical aspects of GAs are already available in many textbooks, and this chapter does not aim to discuss them. It is intended here to give a simple presentation that can be helpful in understanding the basic GA procedure, and one can apply the GA to solve problems related to water resource development and management. The references cited within the text and provided at the end of this chapter should be able to guide the readers to more advanced topics in GAs. Overall, this chapter will provide enough material for anyone curious about GAs and their applications in water resources.

This chapter is organized as follows. Section 3.2 provides an overview of GA and the individual steps involved in a typical GA process. This is followed by Section 3.3 giving a review of applications of GAs in water resource problems, followed by an example of a reservoir operation problem and its solution, describing the steps involved in the GA procedure.

3.2   Genetic Algorithms

There are many publications that give excellent introductions to GAs: see, for example, Holland (1975), Goldberg (1989), Davis (1991), Michalewicz (1999), Mitchell (1999), Deb (2003), and Haupt and Haupt (2004). A GA is a mix of principles behind natural evolution in biology and artificial intelligence in computer science. Therefore, GA terminology uses both natural and artificial terms.

As stated earlier, GAs search for the optimum solution from one set of possible solutions that is an array of decision-variable values. This set of possible solutions is called a *population*. There are several populations in a GA run, and each of these populations is called a *generation*. Generally, at each new generation, better solutions (i.e., decision-variable values) that are closer to the optimum solution as compared to the previous generation are created. In the GA context, the set of possible solutions (array of decision-variable values) is defined as a *chromosome*, while each decision-variable value present in the chromosome is formed by genes.

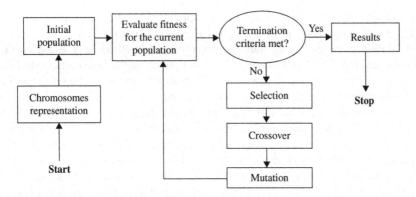

Figure 3.1 Overall GA process.

Population size is the number of chromosomes present in a population. The GA process is briefly described below, and the overall GA process is shown in Figure 3.1.

At the start of the GA optimization, the user has to define the GA operator, such as type of chromosome representation, population size, selection process, types of crossover and mutation, and crossover and mutation probabilities. The initial population is generated according to the selected chromosomal representation at random or using *a priori* knowledge of the search space. For example, given the upper and lower bounds for each decision variable, the chromosomes are created randomly so as to remain within their upper and lower limits. The initial population provides the set of all possible solutions for the first generation, according to the user-defined decision-variable ranges, which have been created randomly. The objective function is used to evaluate each chromosome in the population. Each chromosome in the population has an assigned fitness value, which is used to select the chromosomes from the current population. This process is known as *selection*. Genetic operators, such as crossover and mutation, are performed on the selected chromosomes to create a new set of chromosomes that make the population for the next generation. This algorithm is repeated sequentially until the stopping criterion is achieved. The stopping criterion of a GA is governed either by the number of generations or by the rate of change in the objective function value. Fitness values are expected to improve, indicating the creation of better individuals in new generations. Several generations are considered in the GA process until the user-defined termination criteria is reached.

3.2.1 GA Operators

The GA operators, namely chromosome representation, population size, selection type, and crossover and mutation, control the process of GAs. These operators play an important role in the efficiency of GA optimization in reaching the optimum solution. One of the challenging aspects of using GAs is to choose the optimum GA operator set for the relevant problem.

Representation of Chromosomes

Physical parameters in the search space constituting the phenotypes are encoded into genotypes. The genotype of an individual is the chromosome, and the potential solution to a problem corresponding to the chromosome is the phenotype. In GAs, genetic operators are applied to the genotype to generate better solutions until the optimum is obtained. Then the individual (genotype) representing the optimum solution is decoded to phenotypes. Chromosome representation or encoding is a process of representing the decision-variable values in GAs such that the computer can interact with these values. The decision variables, or phenotypes, in the GA are obtained by applying some mapping from the chromosome representation into the search space. Coding in GA is defined by the type of gene expression, which may be expressed using binary, gray, integers, or real coding. In general, a chromosome (genotype) is presented as

$$(x_1, x_2, \ldots, x_n) \text{ such that } x_1 \in X_1, \ x_2 \in X_2, \ldots, x_n \in X_n \tag{3.1}$$

where, $x_1, x_2, \ldots, x_n$ are bits, integers, real numbers or a mixture of these, and X_1, $X_2, \ldots, X_n$ are the respective search spaces for $x_1, x_2, \ldots, x_n$.

In principle, any character set and coding scheme can be used for chromosome representation. However, the initial GA work of Holland (1975) was done with binary representation, as it was computationally easy. Furthermore, the binary character set can yield the largest number of possible solutions for any given parameter representation, thereby giving more information to guide the genetic search. The GA operators work directly on this representation of the chromosomes to get the optimal solution.

The conventional GA operations and theory were developed on the basis of binary coding, which was used in many applications (Goldberg, 1989). The use of real-valued genes in GAs is claimed by Wright (1991) to offer a number of advantages in numerical function optimization over binary coding. Binary coding and real coding differ mainly in how the crossover and mutation operators are performed in the GA process. There has been growing interest in real-value coding for GAs. In real-value coding, each chromosome is coded as an array of real numbers, with the same length for the decision variable. Gray coding is another type of bit string coding, which uses adjacent variable values where the code occurs as only one binary digit. It was developed to overcome a problem called "Hamming Cliffs," which exists in binary coding and has been used in a number of studies in the water resources field (Dandy et al., 1996).

Binary Coding

The most commonly used representation of chromosomes in the GA is binary coding by using binary numbers 0 and 1. In this coding, each decision variable in

String 1 String 2 **Figure 3.2** Formation of chromosome.
Chromosome   10001   01111

the parameter set is encoded as a binary string, and these are concatenated to form a chromosome. The length of the binary substring (i.e., number of bits) for a variable depends on the size of the search space and the number of decimal places required for accuracy of the decoded variable values (Michalewicz, 1999). If each decision variable is given a string of length L, and there are n such variables, then the chromosome will have a total string length of nL. For example, let there be two decision variables, x_1 and x_2, and let the string length be 5 for each variable. Then the chromosome length is 10, as shown in Figure 3.2.

The search space is divided into 2^L intervals, each having a width equal to $(x_{i,max} - x_{i,min})/2^L$ for a binary string of length L, where $x_{i,max}$ is the upper bound of the decision variable, and $x_{i,min}$ is the lower bound of the decision variable:

$$d = (x_{i,max} - x_{i,min})/2^L \text{ defines the solution accuracy} \tag{3.2}$$

The binary numbers have a base of 2 and use only two characters, 0 and 1. A binary string, therefore, is decoded using Eq. (3.3):

$$N = a_n 2^n + a_{n-1} 2^{n-1} + \cdots + a_1 2^1 + a_0 2^0 \tag{3.3}$$

where

a_i is either 0 or 1 (ith bit in the string),
2^n represents the power of 2 of digit a_i,
n is the number of bits in binary-coded decision variable (i.e., $L-1$),
N is the decoded integer value of the binary string,

and the corresponding actual value of the variables is obtained using Eq. (3.4):

$$x_i = x_{i,min} + \frac{x_{i,max} - x_{i,min}}{2^L - 1} N \tag{3.4}$$

For this example, let the search space for decision variables x_1 and x_2 range from 0 to 5 and 1 to 10, respectively. For the chromosome shown in Figure 3.2, the decoded value for the substrings and the corresponding value of the decision variables will be as shown in Table 3.1.

Using string length $L = 5$, the entire search space for decision variable x_1 can be divided into 31 intervals of 0.16 width each, as shown in Table 3.2. The solution accuracy may be increased by increasing the length of the string. The lower and upper bounds of the real-value search space (i.e., 0 and 5) can be mapped into binary numbers using Eq. (3.2), and all the other intermediate values (i.e., 0–31) can also be easily expressed in binary numbers using Eq. (3.3). The entire search

Table 3.1 Coding and Decoding in GAs

	Decision Variables	
	x_1	x_2
Chromosomes represented as binary strings assuming string length, $L = 5$	10001	01111
Decoded integer value	$1.2^4 + 0.2^3 + 0.2^2 + 0.2^1 + 0.2^0 = 17$	$0.2^4 + 1.2^3 + 1.2^2 + 1.2^1 + 1.2^0 = 15$
Corresponding value of decision variable with solution accuracy $(x_{max} - x_{min})/(2^L - 1)$	2.74	4.84

space for x_1 in binary encoding and decoded real values are given in Table 3.2. Different ranges and accuracies can be considered in GAs through different binary substring lengths for different decision variables. All GA operators are performed on binary strings and once GA optimization is completed, the binary strings can be decoded into real values.

Gray Coding

Gray coding is an ordering of binary character sets such that all adjacent numerical numbers differ by only one bit, whereas in binary coding, adjacent numbers may differ in many bit positions. Gray coding representation has the property that any two points next to each other in the search space differ by one bit only (Haupt and Haupt, 2004). In other words, an increase of one step in the value of the decision variable corresponds to a change of only a single bit. The advantage of gray coding is that random bit flips in mutation are likely to make small changes and therefore result in a smooth mapping between the real search space and the encoded strings. To convert binary coding to gray coding, truth table conversion, as shown in Table 3.3, is followed.

When converting from binary to gray, the first bit of the binary code remains as it is, and the remaining bits follow the truth table conversion, two bits taken sequentially at a time, giving the next bit in gray coding. An example of representation of binary and gray coding of numeric numbers of 1−31 is shown in Table 3.4.

The number of bit positions that differ in two adjacent bit strings of equal length is known as *Hamming distance*. For example, the *Hamming distance* between 01111 and 10000 is 5, since all bit positions differ, and require alteration of 5 bits when converting the number 15 to 16 in binary representation. The Hamming distance associated with certain strings, such as 01111 and 10000, poses difficulty in transition to a neighboring solution in real space, as it requires the alteration of many bits. In gray coding, this distance between any two adjacent binary strings is always 1. Caruana and Schaffer (1988) reported that gray coding can eliminate

Table 3.2 Binary and Real Value Search Space for Decision
Variable x_1

Binary encoding	Decoded value	Corresponding real value of x_1, in the search space
00000	0	0.00
00001	1	0.16
00010	2	0.32
00011	3	0.48
00100	4	0.65
00101	5	0.81
00110	6	0.97
00111	7	1.13
01000	8	1.29
01001	9	1.45
01010	10	1.61
01011	11	1.77
01100	12	1.94
01101	13	2.10
01110	14	2.26
01111	15	2.42
10000	16	2.58
10001	17	2.74
10010	18	2.90
10011	19	3.06
10100	20	3.23
10101	21	3.39
10110	22	3.55
10111	23	3.71
11000	24	3.87
11001	25	4.03
11010	26	4.19
11011	27	4.35
11100	28	4.52
11101	29	4.68
11110	30	4.84
11111	31	5.00

Table 3.3 Truth Table Conversions (B_1 and B_2 are
adjacent bits in a binary string)

B_1	1	0
B_2		
1	0	1
0	1	0

Table 3.4 Representations of Integer Numbers in Binary and Gray Coding

Integers	Binary Coding	Gray Coding
0	00000	00000
1	00001	00001
2	00010	00011
3	00011	00010
4	00100	00110
5	00101	00111
6	00110	00101
7	00111	00100
8	01000	01100
9	01001	01101
10	01010	01111
11	01011	01110
12	01100	01010
13	01101	01011
14	01110	01001
15	01111	01000
16	10000	11000
17	10001	11001
18	10010	11011
19	10011	11010
20	10100	11110
21	10101	11111
22	10110	11101
23	10111	11100
24	11000	10100
25	11001	10101
26	11010	10111
27	11011	10110
28	11100	10010
29	11101	10011
30	11110	10001
31	11111	10000

the hidden bias in binary coding and that the large Hamming distances in the binary representation could result in the search process being deceived or unable to locate the global optimum efficiently. Gray coding has been preferred by several researchers while using GAs in water resource applications (Wardlaw and Sharif, 1999).

Real-Value Coding
For problems with a large number of decision variables, having large search spaces, and requiring a higher degree of precision, binary-coded GAs have performed

poorly (Michalewicz, 1999). Wright (1991) claims that the use of real-valued genes in GAs overcomes a number of drawbacks of binary coding. In real coding, each variable is represented as a vector of real numbers with the same length as that of the solution vector. Efficiency of the GA is increased because genotype into phenotype conversion is not required. In addition, less memory is required because efficient floating-point internal computer representations can be used directly; there is no loss in precision due to formation of discreteness to binary or other values; and there is greater freedom to use different genetic operators. Nonetheless, real coding is more applicable and it seems to fit continuous optimization problems better than binary coding. Eshelman and Schaffer (1993) suggested choosing any of these coding mechanisms, whichever is most suitable for the fitness function. Other authors, such as Michalewicz (1999), justify the use of real coding, showing their advantages with respect to the efficiency and precision reached compared to the binary one. Real coding has been the preferred choice for variable representation in most of the applications found in water resources using GA.

Another form of real number representation is integer coding. In integer coding, the chromosomes are composed of integer values rather than real numbers. The only difference between real coding and integer coding is in the operation of mutation.

Population Size

The population size is the number of chromosomes in the population. The size of a population depends on the nature of the problem, but typically a population contains hundreds or thousands of possible solutions. Traditionally, the population is generated randomly, covering the entire search space. Given upper and lower bounds for each chromosome (decision variable), chromosomes are created randomly so as to remain within the given limits. The principle is to maintain a population of chromosomes, which represents candidate solutions to the problem that evolve over time through a process of competition and controlled variation. Each chromosome in the population has an assigned fitness to determine which chromosomes are used to form new ones in the competition process, which is called *selection*. The new ones are created using genetic operators such as cross-over and mutation.

Larger population sizes increase the amount of variation present in the population but require more fitness evaluations (Goldberg, 1989). Therefore, when the population size is too large, users tend to reduce the number of generations in order to reduce the computing effort, since the computing effort depends on the multiple of population size and number of generations. Reduction in the number of generations reduces the overall solution quality. On the other hand, a small population size can cause the GAs to converge prematurely to a suboptimal solution. Goldberg (1989) reported that a population size ranging from 30 to 200 was the general choice of many GA researchers. Furthermore, Goldberg pointed out that the population size was both application dependent and related to string length. For longer chromosomes and challenging optimization problems, larger population sizes were needed to maintain diversity because it allowed better exploration.

Selection

Selection is the survival of the fittest within the GA. The selection process determines which chromosomes are preferred for generating the next population, according to their fitness values in the current population. The key notion in selection is to give a higher priority or preference to better individuals. During each generation, a proportion of the existing population is selected to breed a new generation; therefore, the selection operator is also known as the *reproduction operator*. All chromosomes in the population, or in a proportion of the existing population, can undergo the selection process using a selection method. This percentage is known as the *generation gap*, which is defined by the user as an input in GAs. The selection process emphasizes to copy the chromosomes with better fitness for the next generation than those with lower fitness values. This may lose population diversity or the variation present in the population and could lead to a premature convergence. Therefore, the method used in the selection process should be able to maintain the balance between selection pressure and population diversity. There are several selection techniques available for GA optimization. Proportional selection, rank selection, and tournament selection (Goldberg and Deb, 1991) are among the most commonly used selection methods. These are briefly discussed below.

Proportional Selection Method

The proportional selection method selects chromosomes for reproduction of the next generation with a probability proportional to the fitness of the chromosomes. In this method, the probability (P) of selecting a chromosome for reproduction can be expressed as

$$P = \frac{\mathrm{ft}_i}{\sum_{i=1}^{N} \mathrm{ft}_i} \tag{3.5}$$

where ft_i is the fitness value of the ith chromosome in the current population of size N, and $\sum_{i=1}^{N} \mathrm{ft}_i$ is the total fitness, which is the sum of fitness values of all chromosomes in the current population.

This method provides noninteger copies of chromosomes for reproduction. Therefore, various methods have been suggested to select the integer number of copies of selected chromosomes for the next generation, including Monte Carlo, roulette wheel, and stochastic universal selection. The roulette wheel selection method is discussed next.

Roulette Wheel Selection The most common selection method is roulette wheel selection. Goldberg (1989) reported that it is also the simplest method. The basic implementation of the roulette wheel selection method assigns each chromosome a "slice" of the wheel, with the size of the slice proportional to the fitness value of the chromosome. In other words, the fitter a member is, the bigger slice of the wheel it gets. To select a chromosome for selection, the roulette wheel is "spun," and the chromosome corresponding to the slice at the point where the wheel stops is grabbed as the one to survive in the offspring generation.

The main steps for the roulette wheel selection algorithm may be generalized as follows:

1. The fitness of each chromosome, ft_i, and their sum $\sum_{i=1}^{N} ft_i$ are calculated, where the population size is N.
2. A real random number, rand ( ), within the range [0,1] is generated, and s is set to be equal to the multiplication of this random number by the sum of the fitness values, $s = \text{rand}(\cdot) \times \sum_{i=1}^{N} ft_i$.
3. A minimal k is determined such that $s \leq \sum_{i=1}^{k} ft_i$, and the kth chromosome is selected for the next generation.
4. Steps 2 and 3 are repeated until the number of selected chromosomes becomes equal to the population size, N.

Tournament Selection

Another selection technique is tournament selection, where randomly selected pairs of chromosomes "fight" to become parents in the mating pool through their fitness function value (Goldberg, 1989). Tournament selection runs a "tournament" among two or more chromosomes chosen at random from the population, and selects the winner in accordance with their fitness values, such that the one with the best fitness is selected for crossover. This process is continued until the required number of chromosomes is selected for the next generation. Selection pressure can be easily adjusted by changing the tournament size. If the tournament size is larger, weak chromosomes have less chance to be selected. In general, in tournament selection, N chromosomes are selected at random and the fittest is selected. The most common type of tournament selection is binary tournament selection, where just two chromosomes are selected.

Rank Selection

In the rank-selection approach, each population is sorted in order of fitness, assigning a numerical rank to each chromosome based on fitness, and the chromosomes are selected based on this ranking rather than the fitness value using the proportionate selection operator. The advantage of this method is that it can prevent very fit chromosomes from gaining dominance early at the expense of less fit ones, thereby increasing the population's genetic diversity (Goldberg and Deb, 1991).

Roulette wheel selection, tournament selection, and rank selection are considered to be the most common and popular selection techniques and have been used frequently in many studies. However, there are many other selection techniques, namely, elitist selection, generational selection, steady-state selection, and hierarchical selection. These techniques may be used independently or in combination. Brief introduction of those selection techniques are given next. A detailed review of selection techniques used in GAs is presented by Shivraj and Ravichandran (2011).

Elitist Selection

The fittest chromosomes from each generation are selected for the next generation, a process known as *elitism*. Most GAs do not use pure elitism, but instead use a modified form where a single chromosome or a few of the best chromosomes from each generation are copied into the next generation. Elitism can be combined with any other selection technique.

Generational Selection
The offspring of the chromosomes selected from each generation become the entire next generation. No chromosomes are retained between generations.

Steady-State Selection
The offspring of the chromosomes selected from each generation go back into the previous generation and replaces some of the less fit members. This process helps to keep some chromosomes between generations.

Hierarchical Selection
Chromosomes go through multiple rounds of selection each generation. Lower-level evaluations are faster and less discriminating, while those that survive to higher levels are evaluated more rigorously. The advantage of this method is that it reduces overall computation time by using faster, less selective evaluation to weed out the majority of chromosomes that show little or no promise, and subjecting only those who survive this initial test to more rigorous and more computationally expensive fitness evaluation.

Crossover

The crossover operator is used to create new chromosomes for the next generation by combining randomly two selected chromosomes from the current generation. Crossover helps to transfer the information between successful candidates— chromosomes can benefit from what others have learned, and schemata can be mixed and combined, with the potential to produce an offspring that has the strengths of both its parents and the weaknesses of neither. However, some algorithms use an elitist selection strategy, which ensures that the fittest chromosome from one generation is propagated into the next generation without any disturbance. The crossover rate is the probability that crossover reproduction will be performed and is an input to GAs. For example, a crossover rate of 0.9 means that 90% of the population is undergoing the crossover operation. A higher crossover rate encourages better mixing of the chromosomes.

There are several crossover methods available for reproducing the next generation. In general, crossover methods can be classified into two groups depending on the chromosomes representation (i.e., binary coding or real-value coding). A number of crossover methods are discussed by Herrera et al. (1998) for binary coding and real coding. The choice of crossover method primarily depends on the application.

Crossover Operators for Binary Coding
In bit string coding, crossover is performed by simply swapping bits between the crossover points. Different types of bit string crossover methods (Davis, 1991; Goldberg, 1989) are discussed next.

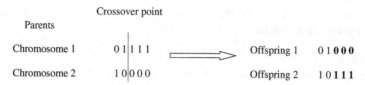

Figure 3.3 Single-point crossover.

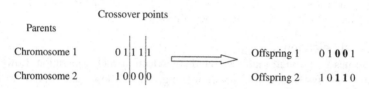

Figure 3.4 Multipoint crossover.

Single-Point Crossover Two parent chromosomes are combined randomly at a randomly selected crossover point somewhere along the length of the chromosome, and the sections on either side are swapped. For example, consider the following two chromosomes, each having 6 binary bits. After crossover, the new chromosomes (i.e., referred as offspring or children) are created as follows if the randomly chosen crossover point is 2 (Figure 3.3).

Multipoint Crossover In multipoint crossover, the number of crossover points are chosen at random, with no duplicates, and sorted in ascending order. Then, the bits between successive crossover points are exchanged between the two parents to produce two new chromosomes. The section between the first bit and the first crossover point is not exchanged between chromosomes. For example, consider the same example of two chromosomes used in a single crossover. If the randomly chosen crossover points are 2 and 4, the new chromosomes are created as shown in Figure 3.4.

The two-point crossover is a subset of the multipoint crossover. The disruptive nature of multipoint crossover appears to encourage the exploration of the search space, rather than favoring the convergence to highly fit chromosomes early in the search, thus making the search more robust.

Uniform Crossover Single-point and multipoint crossover define crossover points between the first and last bit of two chromosomes to exchange the bits between them. Uniform crossover generalizes this scheme to make every bit position a potential crossover point. In uniform crossover, one offspring is constructed by choosing every bit with a probability P from either parent, as shown next using the same example, by exchanging bits at the first, third, and fifth position between the parents (Figure 3.5).

Parents

| Chromosome 1 | 0 1 1 1 1 |
| Chromosome 2 | 1 0 0 0 0 |

Offspring 1 **1 1 0 1 0**

Offspring 2 **0 0 1 0 1**

Figure 3.5 Uniform crossover.

Crossover Operators for Real Coding

In real coding, crossover is simply performed by swapping real values of the genes between the crossover points. Different types of real-value crossover methods have been used. Assume that $x_1 = (x_1^1, x_2^1, \ldots, x_n^1)$ and $x_2 = (x_1^2, x_2^2, \ldots, x_n^2)$ are the two chromosomes selected for crossover operation from the current population. Different crossover operators that can be used in real-coded GAs are discussed next.

Two-Point Crossover Two points of crossover $i,j \in (1, 2, \ldots, n-1)$ are randomly selected, provided that $i < j$ and the segments of the parent, defined by them, are exchanged for generating two offspring (Eshelman et al., 1989), y_1 and y_2, such that:

$$y_1 = (x_1^1, x_2^1, \ldots, x_i^2, x_{i+1}^2, \ldots, x_j^2, x_{j+1}^1, \ldots, x_n^1) \tag{3.6}$$

$$y_2 = (x_1^2, x_2^2, \ldots, x_i^1, x_{i+1}^1, \ldots, x_j^1, x_{j+1}^2, \ldots, x_n^2) \tag{3.7}$$

Random Crossover Two offspring are created:

$$y_1 = (y_1^1, y_2^1, \ldots, y_n^1) \quad \text{and} \quad y_2 = (y_1^2, y_2^2, \ldots, y_n^2)$$

The value of each gene in the offspring is determined by the random uniform choice of the values of this gene in the parents:

$$y_i^k = \begin{cases} x_i^1 & \text{if } u = 0 \\ x_i^2 & \text{if } u = 1 \end{cases}, \quad k = 1, 2 \tag{3.8}$$

where u is a random number that can have a value 0 or 1 (Syswerda, 1989).

Arithmetic Crossover Two offspring, $y_1 = (y_1^1, y_2^1, \ldots, y_n^1)$ and $y_2 = (y_1^2, y_2^2, \ldots, y_n^2)$, are produced, such that

$$y_i^1 = \lambda \cdot x_i^1 + (1 - \lambda) \cdot x_i^2 \tag{3.9}$$

$$y_i^2 = \lambda \cdot x_i^2 + (1 - \lambda) \cdot x_i^1 \tag{3.10}$$

where $\lambda \in [0, 1]$.

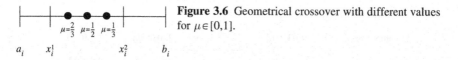

Figure 3.6 Geometrical crossover with different values for $\mu \in [0,1]$.

Geometrical Crossover Two offspring, $y_1 = (y_1^1, y_2^1, \ldots, y_n^1)$ and $y_2 = (y_1^2, y_2^2, y_n^2)$, are created, where

$$y_i^1 = x_i^{1\mu} \cdot x_i^{2(1-\mu)} \tag{3.11}$$

$$y_i^2 = x_i^{2\mu} \cdot x_i^{1(1-\mu)} \tag{3.12}$$

where $\mu \in [0,1]$.

Geometric crossover in shown in Figure 3.6 (Michalewicz, 1999).

BLX-α Crossover Two offspring, $y_1 = (y_1^1, y_2^1, \ldots, y_n^1)$ and $y_2 = (y_1^2, y_2^2, \ldots, y_n^2)$ are generated. where, y_i^k is a randomly, uniformly chosen number from the interval $[X_{min} - I\alpha, X_{max} + I\alpha]$ and X_{max}, X_{min}, and I are defined as shown here:

$$X_{max} = \max\{x_i^1, x_i^2\} \tag{3.13}$$

$$X_{min} = \min\{x_i^1, x_i^2\} \tag{3.14}$$

$$I = X_{max} - X_{min} \tag{3.15}$$

Generally, BLX-α crossover gives the best results. And it is observed that the higher value of α results in better solutions. As α increases the exploration level increases, since the relaxed exploitation zones are spread over exploration zones, thereby increasing the diversity levels in the population (Herrera et al., 1998).

For detailed descriptions of these and other crossover operators (e.g., Fuzzy, SBX (Simulated binary crossover), UNDX (Unimodal normally distributed crossover), and simplex crossover), real-coding readers are referred to Deb (2003) and Herrera et al. (1998).

Mutation

One further operator in GA is the mutation operator, which works on the level of chromosome genes by randomly altering a gene value (Deb, 2003). Mutation introduces innovation into the population by randomly modifying the chromosomes. The operation is designed to prevent GA from premature termination, since it prevents the population from becoming saturated with chromosomes that look alike. Usually considered as a background operator, the role of the mutation operator is often seen as guaranteeing that the probability of searching any given chromosome will never be zero. In GAs, mutation is randomly applied with low probability and modifies elements in the chromosomes. Large mutation rates increase the probability of destroying good chromosomes but prevent premature convergence. The mutation rate determines the probability that mutation will occur. For example, if the population size is 200, string length is 10, and mutation rate is 0.005, then only 10-bit positions will mutate in the whole population (i.e., $200 \times 10 \times 0.005 = 10$).

Similar to crossover techniques, mutation methods can be classified according to the binary coding or real-value coding of the GA.

Mutation for Binary and Gray Coding
In binary and gray coding systems, a chromosome mutation is performed at randomly chosen genes by flipping bit 0 to 1 and vice versa (Goldberg, 1989; Holland, 1975).

Mutation for Real Coding
Mutation in real-coded GA is performed, either by disarranging the gene values or by randomly selecting the new values. For example, let $x = (x_1, x_2, \ldots, x_n)$ be a chromosome and x_i be a gene to be mutated. Then a random number x_i' may be chosen from a given search space of x_i and will replace x_i. Mutation for integer coding is performed analogous to real-value coding, except that after mutation, the value for that gene is rounded to the nearest integer.

A detailed discussion of other mutation operators may be found in related textbooks and publications, for instance, see Herrera et al. (1998) and Deb (2003).

3.3   Review of GA Applications to Water Resource Problems

GAs can successfully deal with a wide range of problem areas. Briefly, the reasons for this success, according to Goldberg (1994), are "(1) GAs can solve hard problems quickly and reliably, (2) GAs are easy to interface to existing simulations and models, (3) GAs are extensible, and (4) GAs are easy to hybridize. All these reasons may be summed up in only one statement: GAs are *robust*." GAs are more powerful in difficult environments where the search space usually is large, discontinuous, complex, and poorly understood. They are not guaranteed to find the global optimum solution to a problem, but they are generally efficient at finding acceptable solutions to many real-life problems.

Goldberg (1989) gives a comprehensive review of GA applications before 1989. During the recent years, GA applications have grown enormously in many fields. GAs have been the most commonly applied nature-inspired metaheuristic algorithms in the water resource planning and management literature. Reviews of their application in different fields of water resources are reported in Nicklow et al. (2010), Rani and Moreira (2010), Labadie (2004), and Cunha (2002).

In this section, applications of GAs in the field of water resources have been classified in different groups.

3.3.1 Water Distribution Systems and Pump Scheduling Problems

Over the past two decades, considerable investment has been made in developing and applying GAs to improve the design and performance of water distribution systems. Interestingly, one of the earlier applications of GAs in water engineering was the optimization of pump schedules for a serial liquid pipeline (Goldberg and Kuo, 1987). Since then, there has been increasing interest in the application of GAs to a wide variety of water distribution system problems, such as calibration of water distribution models, optimal system design, and operation and pump scheduling.

Simpson et al. (1994) were the first to use GAs for water distribution systems. They applied and compared a GA solution to enumeration and to nonlinear programming. Vairavamoorthy and Ali (2005) presented a GA for the least-cost pipe network design problem that discards the regions of the search space where impractical or infeasible solutions are likely to exist, therefore improving search efficiency.

Dandy and Engelhardt (2001) demonstrated the use of a GA to find a near-optimal pipe replacement schedule so as to minimize the present value of capital, repair, and damage costs. Mackle et al. (1995) were among the first to apply a binary GA to pump scheduling problems by minimizing energy costs, subject to reservoir filling and emptying constraints. Subsequently, Savic et al. (1997) developed a multiobjective GA (MOGA) approach to determine pump scheduling. To reduce the excessive run times required by the GA, van Zyl et al. (2004) developed a hybrid optimization approach, in which they combined a steady-state GA with the Hooke and Jeeves hill-climbing method. Rao and Salomons (2007) developed a process based on the combined use of an artificial neural network (ANN) for predicting the consequences of different pump and valve control settings and a GA for selecting the best combination of those settings. The methodology has successfully been demonstrated on the distribution systems of Valencia (Spain) and Haifa (Israel). Munavalli and Mohan-Kumar (2003) and Prasad et al. (2004) used GA for optimal scheduling of multiple chlorine sources.

Besides the above-mentioned papers, many other applications of MOGAs have appeared in the water distribution system literature (Savic and Walter, 1997). Prasad and Park (2004) and Vamvakeridou-Lyroudia et al. (2005) employed the MOGA approach for optimal design of water distribution networks.

3.3.2 Sewer System Design Optimization

The optimal design of a sewer network aims to minimize construction costs while ensuring adequate system performance under specified design criteria. GAs have been the most popular and successful optimization techniques for the design of sewer systems (Afshar et al., 2006; Farmani et al., 2006). Hybrid GAs and MOGAs are becoming attractive in this field of study as well. Farmani et al. (2006) and Guo et al. (2006) employed local search techniques to seed an NSGA II (Non-dominated sorting genetic algorithm II) in the design of sewer networks.

3.3.3 Water Quality and Waste Management

GAs have been applied successfully in the design and operation of water and wastewater treatment plants and to other water quality management problems. GA was applied by Suggala and Bhattacharya (2003) to identify process parameters to remove organics from wastewater cost-effectively to meet pollutant removal standards. For operation of a domestic wastewater treatment plant, Chen et al. (2003) investigated the use of a GA to identify real-time control strategies, such as pH and nutrient levels, electricity consumption, and effluent flow rates, for meeting cost goals and effluent standards. Chen and Chang (1998) introduced a GA to solve a nonlinear fuzzy multiobjective programming model, considering biochemical oxygen demand and dissolved oxygen as water quality parameters, where the water quality calculation was based on the Streeter–Phelps equation. Burn and Yulianti (2001) explored waste load allocation problems using GAs. Yandamuri et al. (2006) similarly proposed optimal waste load allocation models for rivers using NSGA II. Kerachian and Karamouz (2005) extended some of the classical waste load allocation models for river water quality management for determining the monthly treatment or removal fraction to evaporation ponds. The high dimensionality of the problem (large number of decision variables) was handled by using a sequential dynamic GA.

3.3.4 Watershed Planning and Management

Yeh and Labadie (1997) introduced the application of GAs to watershed planning and presented a multiobjective watershed-level planning of stormwater detention systems using MOGAs to generate nondominated solutions for the system cost and detention effect for a watershed-level detention system. Harrell and Ranjithan (2003) applied a GA-based methodology to identify detention pond designs and land-use allocations within subbasins to manage water quality at a watershed scale. Combined use of GA and simulation models can be seen in many watershed management studies. Muleta and Nicklow (2005) linked a GA with the Soil and Water Assessment Tool to identify land-use patterns to meet water quality and cost objectives. Perez-Pedini et al. (2005) combined a distributed hydrologic model with GA for an urban watershed to determine the optimal location of infiltration-based best management practices for stormwater management.

3.3.5 Groundwater System Optimization

Groundwater optimization problems include groundwater remediation design, monitoring network design, groundwater and coastal aquifer management, parameter estimation, and source identification. Cunha (2002) and Mayer et al. (2002) presented reviews of design optimization problems that apply traditional and heuristic solution approaches to solving groundwater flow and contaminant transport processes and remediation problems, while Qin et al. (2009) also reviewed both simulation and optimization approaches used in groundwater systems.

Rogers et al. (1995) was the first to apply a GA to a field-scale remediation problem by using an ANN in place of numerical groundwater flow and the contaminant transport simulation model. Yan and Minsker (2006) proposed an adaptive neural network GA that incorporates an ANN as an approximation model that is adaptively and automatically trained within a GA, to reduce the computational requirement of groundwater remediation problems. Zheng and Wang (2002) applied a GA with response functions to solve a field-scale remediation problem at the Massachusetts Military Reservation that included 500,000 nodes in the simulation model and a 30-year planning horizon. Espinoza et al. (2005) proposed the self-adaptive hybrid GA and demonstrated its ability to reduce computation cost for groundwater remediation problems. Sinha and Minsker (2007) proposed multiscale island injection GAs, which includes multiple population functions at different spatial scales, to reduce the computational time to solve a field-scale pump-and-treat remediation optimization problem.

Many studies have considered parameter uncertainty in solving groundwater remediation optimization problems. Smalley et al. (2000) applied a noisy GA to bioremediation design, with health risk included in the formulation. Wu et al. (2006) compared a Monte Carlo simple GA (SGA) with a noisy GA to solve a sampling network problem with uncertainty in the hydraulic conductivity. Hu et al. (2007) presented an application of two-objective optimization of an *in situ* bioremediation system for a hypothetical site under uncertainty. Singh and Minsker (2008) developed a probabilistic MOGA, which combines a method similar to the noisy GA, with an additional archiving step with the NSGA II, and applied it to two pump-and-treat problems—a hypothetical and a field-scale case study.

A number of works have proposed GA approaches to groundwater monitoring network design (Chadalavada and Datta, 2008).

3.3.6 Parameter Identification

Parameter identification can be defined as a generalized term that denotes any practice, including field or experimental work, to identify parameters for a model. The parameter identification problem for most hydrologic applications is ill-posed, multimodal, nonlinear, and nonconvex (Yeh, 1986). Wang (1991) was among the first to apply the "simple" GA to the calibration problem (Nicklow et al., 2010). Subsequently, many other studies have applied GAs and their variants to watershed calibration. Zechman and Ranjithan (2007) developed a combined GA and genetic programming methodology to address the difficulties associated with models used for parameter estimation.

Tsai et al. (2003) and Mahinthakumar and Sayeed (2005) presented a similar global–local optimization approach, where a GA was used as a global optimizer to provide approximately optimal solutions that were fed in local optimization approaches. Other applications of GAs for groundwater calibration can be found in Wang and Zheng (1998).

NSGA II (Deb et al., 2000) and its variants have been widely used for multiobjective parameter identification in watershed modeling (Khu and Madsen, 2005;

Tang et al., 2006). NSGA II, used for watershed calibration, has employed the Pareto ranking scheme (used in Goldberg (1989)) to deal with multiple objectives.

3.3.7 Optimization of Reservoir System Operation

Optimization of reservoir operations involves allocation of resources, development of stream flow regulation strategies, formulating operating rules, and making real-time release decisions. A reservoir regulation plan, which is also referred to as an *operating procedure* or a *release policy*, is a group of rules quantifying the amount of water to be stored, released, or withdrawn from a reservoir or system of reservoirs under various conditions.

In a multireservoir system, input to a reservoir includes natural inflows, including all other inflows from surface runoffs, streams, and undammed rivers and all releases from adjacent upstream reservoirs on the same river or its tributaries. The output from a reservoir may be through diversion (for irrigation or other uses), spillways (for flood management), release to maintain ecological flow required in the river, and penstocks (to generate power). Also, some water is lost due to evaporation from the water surface and seepage into the ground.

A typical reservoir operation optimization model deals with constraints such as the continuity equation, maximum and minimum storage in the reservoirs, maximum and minimum releases from the reservoirs, and some case-specific obligations. The most commonly accepted objectives are the optimality of the water supply for irrigation, industrial and domestic use, hydropower generation, water quality improvement, recreation, fish and wildlife enhancement, flood control, and navigation.

The reservoir operation rule is commonly defined by a function in which the release of water from a reservoir for the given time interval is computed by using the values of current reservoir storage and current and expected demands and inflows. Generally speaking, the optimization problem takes the following form.

Maximization or minimization of the objective function, subject to the following constraints:

- The continuity equation is satisfied.
- Storage is within the upper and lower bounds.
- Releases are within the upper and lower bounds.
- Final storages are satisfied.

Several approaches have been developed for the optimization of reservoir operations, defining reservoir operating rules, and many different techniques have been studied with regard to this optimization problem. Numerous optimization models have been proposed and reviewed by many scientists (Labadie, 2004).

Esat and Hall (1994) applied a GA to the four-reservoir problem. The objective of this problem was to maximize the benefits from power generation and irrigation water supply, having constraints on both storage and release from the reservoirs. They concluded that GAs have a significant potential in reservoir operation optimization, and GAs are superior over standard dynamic programming (SDP) techniques in many aspects.

Oliveira and Loucks (1997) used a GA to evaluate operating rules for multireservoir systems and indicated that optimum reservoir operating policies can be determined by means of GAs. Wardlaw and Sharif (1999) evaluated GA formulation to different reservoir operation problems, along with a range of sensitivity analysis using different combinations of chromosome representation (binary, gray, and real coding) and crossover and mutation probabilities. Further, they applied GAs to the optimization of multireservoir systems (Sharif and Wardlaw, 2000), and the results were found to be comparable with DDDP (Discrete differential dynamic programming). Jothiprakash and Shanthi (2006) developed a GA model to derive optimal operational strategies for a single reservoir and concluded that GA can be a good alternative for real-time operation. It is important to highlight here that most researchers have agreed that GA could be a potential alternative to SDP.

A number of researchers have come to advocate that a real-coded (or floating-point) GA has a definite advantage over a binary-coded GA (Michalewicz, 1999). In real-value coding, there is no discretization of decision-variable space. Attempts have been made by many researchers to compare the performance of both GA approaches in the context of reservoir systems optimization. Chang et al. (2005) and Jian-Xia et al. (2005) compared the two approaches and found that real-coded GAs were more efficient and faster than binary-coded GAs. Chen and Chang (2007) proposed a real-coded, hypercubic-distributed GA (HDGA). Application of this method to a multireservoir system in northern Taiwan showed that HDGAs can provide much better performance than conventional GAs.

To reduce the computational requirements of the GA, it has been applied in combination with other optimization methods. Cai et al. (2001) presented a combined genetic algorithm—linear programming (GA—LP) strategy to solve the large nonlinear reservoir systems optimization model. GA was used to linearize the original problem in each time period, which is later solved sequentially using LP. The hybrid GA—LP approach was able to find good approximate solutions to the nonlinear models. In view of the computational advantages of combined GA—LP strategies to deal with nonlinearities, Reis et al. (2006) proposed and evaluated a stochastic hybrid GA—LP approach to the operation of reservoir systems, which admits a variety of future inflow variability through a treelike structure of synthetically generated inflows.

Huang et al. (2002) presented a GA-based SDP model to cope with the dimensionality problem of a multiple-reservoir system. A combination of GA and DDDP was proposed by Tospornsampan et al. (2005) for the irrigation reservoir operation problem. The main advantage of the hybrid approach is to save computational resources for optimizing parameters. Also, the good solutions obtained from the GA are used as the initial policy for DDDP, therefore reducing the probability of DDDP to trap in the local optima. Kuo et al. (2006) used a hybrid-neural GA for water quality management of the Feitsui Reservoir in Taiwan.

Ganji et al. (2007) developed a modified version of the SGA, for application to a reservoir operation problem. The SGA reduces the overall run time compared to the SGA through dynamically updating the length of chromosomes. Karamouz et al. (2007) solved a similar problem using a GA-K nearest neighborhood

(GA-KNN) based optimization model. In this methodology, the lengths of chromosomes are increased based on the results of a K-nearest neighborhood (KNN) forecasting model. Nagesh Kumar et al. (2006) developed a GA model for obtaining an optimal reservoir operating policy, but focusing on optimal crop water allocations from an irrigation reservoir in the state of Karnataka, India. The objective of the study was to maximize the relative yield from a specified cropping pattern. Kerachian and Karamouz (2006, 2007) used an algorithm combining a water quality simulation model and a stochastic conflict resolution GA-based optimization technique for determining optimal reservoir operation rules. Zahraie et al. (2008) solved a similar problem using a GA KNN-based optimization model. The KNN method is a nonparametric regression methodology that uses the similarity between observations of predictors and K similar sets of historical observations to obtain the best estimate for a dependent variable. K vectors of the past observations are obtained based on the minimum Euclidean norm from the present condition among all candidates.

Recently, Hinçal et al. (2011) applied GAs to a multireservoir system operation to maximize the energy production in the system by using two different approaches: the conventional (monthly) approach and the real-time approach. Comparison of the results revealed that the energy amounts optimized by using the conventional approach were higher than the energy produced in a real-time operation. However, by using the real-time approach, a close approximation to the real operational data had been achieved.

3.4 The GA Process for a Reservoir Operation Problem

The purpose of optimal reservoir operation is to obtain a policy to specify how water in a reservoir is regulated to satisfy the desired objectives. The optimal operating policy serves to reap the maximum benefit from the reservoir system satisfying the system demands. Here, we assume that the operating policy is composed of a decision variable, which is the release from the reservoir at each time period. The benefit is the return from release of water, and the benefit function is supposed to be given for each time period. Figure 3.7 shows a single reservoir system and the variables associated with a reservoir operation problem.

I_t + Pp$_t$ (Inflow + precipitation)

Figure 3.7 Variables associated with a reservoir operation problem.

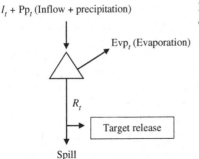

Optimization aims to find the optimum combination of releases that will maximize the return for the system. There are upper and lower limits for releases and storages. These limitations form the constraints of the problem. Another constraint of the problem is that the continuity equation is to be satisfied for each time period. In general, a reservoir operation optimization problem may be expressed as follows:

The objective function is:

Maximization of net benefit

$$g_t(R_t) = \text{Max} \sum\nolimits_{t=1}^{N} [\text{NB}_t(R_t)] \tag{3.16}$$

where $\text{NB}_t(R_t)$ is the benefit function, which is a function of the release at time period t. R_t is the release for period t.

The objective function is subject to:

The continuity equation being satisfied, which is stated as:

$$S_{t+1} = S_t + I_t - R_t + \text{Pp}_t - \text{Evp}_t \quad \forall t = 1, \dots, N \tag{3.17}$$

where S_t, I_t, and R_t are the storage, inflow, and releases for the given reservoir at time period t, and N is the time horizon for the problem under consideration. Pp_t and Evp_t are precipitation over reservoir surface and evaporation from reservoir surface during time period t, respectively.

Limits on storage impose constraints are of the form,

$$S_{\min} \leq S_t \leq S_{\max} \quad \forall t = 1, \dots, N \tag{3.18}$$

which ensures that storage (S_t) will be within specified minimum and maximum values.

Limits on release are as follows:

$$R_{\min} \leq R_t \leq R_{\max} \quad \forall t = 1, \dots, N \tag{3.19}$$

and release (R_t) should be within specified minimum and maximum ranges.

Releases are the decision variables in the problem. Decision variables exist in the composition of the chromosomes of the population in the GA. Constraints of releases are identified during the generation of the initial population, and as a matter of fact, they are satisfied. The continuity equation is readily satisfied since the storages are computed by using the continuity equation given in Eq. (3.17).

Other constraints are embedded into the objective function as a penalty function (Chang et al., 2010; Hinçal et al., 2011). Thus, the constrained optimization problem is converted to an unconstrained optimization problem. The reason why a constraint problem is transformed into an unconstrained problem is to be able to handle the problem by means of the GA. This is done as follows:

If $S_t > S_{\max}$, then the penalty term $\sum_{t=1}^{N} \{c_1 (S_{\max} - S_t)^2\}$ is introduced in an objective function, i.e., Eq. (3.16).

If $S_t < S_{\min}$, then the penalty term $\sum_{t=1}^{N} \{c_2(S_{\min} - S_t)^2\}$ is introduced in an objective function, i.e., Eq. (3.16), where the deviations from maximum and minimum storage are penalized by the square of deviation from constraints. Constants c_1, and c_2 are defined as the weight of the penalty term in order for them to be in the order of the benefit terms. The optimization problem, the objective function, and constraints of which are given above are adapted into the GA. Forthcoming steps will show how GAs are used to solve this problem.

3.4.1 Generation of Initial Population

A chromosome representing search space will be

$$C_j = \{R_1, R_2, \ldots, R_t, \ldots, R_N\} \tag{3.20}$$

Each gene within the chromosome represents a release made from a reservoir at a specific time period and can take up any value between the upper and lower bounds of releases. Since the decision variables are releases (R_t), and the maximum and minimum releases are known for the reservoir, the number of chromosomes generated within these upper and lower limits represent the entire search space for the problem. The population may be generated using binary or real coding. In real coding, randomly generated numbers within the upper and lower limits of the releases will constitute chromosomes of the population. The number of chromosomes generated will depend on the assumed size of the population (population size $j = 1, \ldots, M$).

3.4.2 Calculation of State Variables

After the generation of the initial population, which is composed of chromosomes containing releases (decision variables), calculation of storages (state variables) comes next. Storage for every gene of the individuals is computed making use of continuity Eq. (3.17), which is the equality constraint of the problem. Usage of Eq. (3.17) in calculation of storage ensures that the continuity equation is satisfied for every gene created. The inequality constraints ensure that the storages remain within their limits and are satisfied by incorporating the related penalty terms into the objective function.

3.4.3 Calculation of Fitness Values

In the next step, the fitness value for each chromosome is calculated. The fitness assigned to each gene has direct influence on the eligibility for each chromosome to live in the next generation. Penalty terms originating from violation of the constraints will make sure that the chromosomes violating the storage constraints will not be selected in the next population.

3.4.4   GA Operators

GA operators and selection, crossover, and mutation operators are implemented on the population to get the best solution, as already discussed in Section 3.3. The choice of selection technique depends on the nature of the problem, and various techniques may be applied and compared to choose the best solution for a particular problem.

Crossover operation is performed using a predefined crossover probability. Crossover probability leads to deciding whether to put the parent chromosomes under the process of crossover. A random number is generated and compared with the crossover probability to specify whether to apply the crossover operators. The selected chromosomes undergo crossover operation only if the random number generated is greater than the probability of crossover. There is no rule to define crossover probability, and usually a sensitivity analysis is carried out to get the best value for crossover probability for a particular problem. The crossover operator chosen also depends on the problem, and different crossover techniques may be compared to select the best one for the problem chosen.

Mutation is randomly applied with low probability, typically in the range 0.001 and 0.02, to modify the genes of some chromosomes. The role of mutation is often seen as a safety net to recover good genetic material that may be lost during selection and crossover operations. The mutation operator has been constructed to alter the gene randomly with consideration to the predefined probability of mutation. If the random number generated is greater than the probability of mutation, the gene is reproduced using a suitable mutation operator; otherwise, it remains the same.

3.4.5   Example: A Four-Time-Period Reservoir Operation Problem

To illustrate the main features of GAs, let us consider a reservoir operation problem. The reservoir has an active storage capacity of 20 Million Cubic Meter (MCM). The active storage volume, S_t, in the reservoir can vary from 0 to 20. Let R_t be the release or discharge from the reservoir in time period t. Each variable is expressed as a volume unit for the period, $t = 1, 2, 3, 4$. In these time periods, the inflows to the reservoir are $I_t = 14, 12, 6$, and 8, respectively. The net benefit function for each time period for unit release from the reservoir is defined by $f_t = 11.5 + 1.5R_t^2$. Suppose that only integer solutions are to be considered and the maximum release from the reservoir cannot exceed 9, which is fixed as the target demand for each time period. What is the optimal release R_t for each time period? Evaporation losses and precipitation may be ignored.

Solution

Here, the objective of the problem is to maximize the net benefit from the released water; therefore, the overall objective function may be written as

$$\text{Max} \sum_{t=1}^{4} (11.5 + 1.5R_t^2) - c(R_t - 9)^2 \tag{3.21}$$

where the first term defines the benefit from releases and the second term minimizes the deviation from target demand. In this problem, c is a coefficient to be chosen such that the objective function remains positive, and the value of c is supposed to be 0.1.

The following are the constraints for this function:

The continuity equation:

$$S_{t+1} = S_t + I_t - R_t \quad \forall t \tag{3.22}$$

Limits on storages are as follows:

$$0 \le S_t \le 20 \quad \forall t \tag{3.23}$$

Limits on releases are as follows:

$$0 \le R_t \le 9 \quad \forall t \tag{3.24}$$

Since GA cannot explicitly handle constraints, these are taken care of by penalty functions. If $S_t \ge 20$, then penalty term $\sum_{t=1}^{4}\{c_1(20-S_t)^2\}$ will be introduced in the objective function, and if $S_t \le 0$, then penalty term $\sum_{t=1}^{4}\{c_1(0-S_t)^2\}$ will be introduced instead.

Each individual solution set contains the values of all the decision variables whose best values are being sought. For example, if there are four decision variables x_1, x_2, x_3, and x_4 to be obtained, these variables are arranged into a string or chromosome, $x_1x_2x_3x_4$. If each decision variable is expressed using three digits, then the chromosome 005021050279 would represent $x_1 = 5$, $x_2 = 21$, $x_3 = 50$, and $x_4 = 279$.

Pairs of chromosomes from two parents join together and produce offspring, who in turn inherit some of the genes of the parents. Altered genes may result in improved values of the objective function. These genes will tend to survive from generation to generation, while those that are inferior will tend to die.

A population of possible feasible solutions is generated randomly. Each chromosome contains the values of all the decision variables whose best values are being sought. In this example, we are using numbers to the base 10; therefore, a sample chromosome 8376 will represent the releases $R_1 = 8$, $R_2 = 3$, $R_3 = 7$, and $R_4 = 6$. Another chromosome representing the solution to this problem, chosen randomly, would be 2769. These two chromosomes, each containing four genes, can pair up and have two children. Population size is a GA parameter—that is, the number of solutions being considered. To show the iterations for this example, we assume a population of 10 individuals. However, the best values of GA parameters are usually decided by trial and error.

The GA process begins with the random generation of an initial population of feasible solutions, proceeds with selection, random crossover, and mutation operations, and then randomly generates the new population for the next iteration. This process repeats itself with the new population and continues until there is no significant improvement in the best solution found.

For this example, the process includes the following steps:

1. Initial population is randomly generated, which is a set of solutions/chromosomes having randomly generated decision-variable values within the range 0–9. The release cannot exceed 9, therefore satisfying the release constraint.
2. The corresponding storages in the reservoir are calculated using the continuity Eq. (3.22), assuming that the storage at the beginning of operation is 0. This ensures that the continuity equation is satisfied. If the value of storage for any chromosome is bigger than 20 or less than 0, then the penalty terms are added in the objective function. A negative value of penalty coefficients is considered here, which will decrease the value of objective function and the chromosome will not appear in the next generation. This will prevent the violation of storage constraints.
3. The initial population undergoes the selection operation, and best decision variables are selected using the roulette wheel selection method, as discussed in section "Selection" earlier in this chapter.
4. The selected chromosomes are paired to determine whether a crossover is to be performed on each pair, assuming that the probability of a crossover is 60% ($P_c = 0.6$). If a crossover is to occur, we find the crossing site randomly, by creating a random number between 1 and 3. Note that not all five pairs will undergo crossover operation. With 60% probability, in iteration 1, it was seen that only the first, second, and fifth pairs (shown in bold in Table 3.5) will crossover at randomly chosen site 3. The single-point crossover operation is used in this example.
5. Next, determine if any chromosome in the resulting intermediate population is to be mutated. For mutation, we assume the probability of mutation ($P_m = 0.05$) for each gene. For this example, we assume that mutation increases the value of the number by 1, or if the original number is 9, mutation does not change it to 10; rather, it keeps it as it is. With this probability ($10 \times 4 \times 0.05 = 2$), two chromosomes will undergo mutation randomly. In iteration 1, chromosomes 3 and 6 are randomly chosen for the mutation operation, and for these chromosomes, the genes to be mutated are also chosen randomly by generating a random number between 1 and 4. The mutated genes are shown in bold and italics for iteration 1 in Table 3.5.
6. The last step creates a new population, and steps 2–5 are repeated for a predefined number of generations or until the best solution is obtained.

These steps are performed for two iterations (see Table 3.5). The solution found in the second iteration increases the sum of the fitness value from 2949.2 to 3747.2. This process can continue till the process has converged to the best solution it can find. The whole process may be repeated for different probabilities of crossover and mutation to find out the optimal parameters for the GA process for this problem.

3.5 Conclusions

The GA has become a popular tool for researchers to solve a wide variety of water resource management problems. This chapter has presented a brief review of the theory of GAs and their applications to reservoir operation, groundwater management, water quality, parameter estimation, and other problems related to water resource management. GA has its own advantages and limitations when applied to

Table 3.5 GA Iteration for Reservoir Operation Example

Index for Chromosomes	Initial Population $(R_1 R_2 R_3 R_4)$	Storages S_1 S_2 S_3 S_4	Fitness (ft_i)	Probability of Selection (P)	R and s ( )	s	Cumulative Fitness	Selected Index	Selected Chromosome	Crossover $(P_c = 0.5)$	Mutation $(P_m = 0.05)$
Iteration 1											
1	2695	12 18 15 18	257.6	0.09	0.35	1039.33	257.6	4	8279	**8279**	8279
2	3595	11 18 15 18	249.2	0.08	0.41	1199.69	506.8	4	8279	**8279**	8279
3	9849	5 9 11 10	406.4	0.14	0.56	1653.02	913.2	6	3497	**3499**	3599
4	8279	6 16 15 14	337.6	0.11	0.93	2734.76	1250.8	10	6799	**6797**	6797
5	6538	8 15 18 18	240.8	0.08	0.20	580.40	1491.6	3	9849	9849	9849
6	3497	11 19 16 17	272	0.09	0.01	20.63	1763.6	1	2695	2695	**3695**
7	2676	12 18 17 19	226.4	0.08	0.92	2720.55	1990	10	6799	6799	6799
8	5882	9 13 11 17	274.8	0.09	0.04	121.27	2264.8	1	2695	2695	2695
9	2687	12 18 16 17	269.2	0.09	0.62	1837.25	2534	7	2676	**2679**	2679
10	6799	8 13 10 9	415.2	0.14	0.97	2868.57	2949.2	10	6799	**6796**	6796
Total			$\sum_{i=1}^{N} ft_i = 2949.2$								
Iteration 2											
1	8279	6 16 15 14	337.6	0.10	0.15	502.67	337.6	2	8279	**8299**	8299
2	8279	6 16 15 14	337.6	0.10	0.66	2203.79	675.2	7	6799	**6779**	6779
3	3599	11 18 15 14	334.8	0.10	0.62	2077.63	1010	7	6799	**6779**	6779
4	6797	8 13 10 11	366.8	0.11	0.87	2910.91	1376.8	9	2679	**2699**	2699

(Continued)

Table 3.5 (Continued)

Index for Chromosomes	Initial Population ($R_1 R_2 R_3 R_4$)	S_1	S_2	S_3	S_4	Fitness (ft_i)	Probability of Selection (P)	R and s ( )	s	Cumulative Fitness	Selected Index	Selected Chromosome	Crossover ($P_c = 0.5$)	Mutation ($P_m = 0.05$)
5	9849	5	9	11	10 10	406.4	0.12	0.63	2135.69	1783.2	7	6799	**6749**	6749
6	3695	11	17	14	17	266.4	0.08	0.43	1457.26	2049.6	5	9849	**9899**	9899
7	6799	8	13	10	9	415.2	0.12	0.40	1354.91	2464.8	4	6797	6797	6797
8	2695	12	18	15	18	257.6	0.08	0.29	970.45	2722.4	3	3599	3599	3699
9	2679	12	18	17	16	294.8	0.09	0.52	1747.39	3017.2	5	9849	9849	9849
10	6796	8	13	10	12	346.8	0.10	0.09	311.00	3364	1	8279	8279	8279
Total						$\sum_{i=1}^{N} ft_i = 3364$								

Iteration 3

Index for Chromosomes	Initial Population ($R_1 R_2 R_3 R_4$)	S_1	S_2	S_3	S_4	Fitness (ft_i)
1	8299	6	12	5	5	386
2	6779	8	7	7	5	366.8
3	6779	8	7	7	5	366.8
4	2699	12	8	5	5	343.2
5	6749	8	7	10	5	315.2
6	9899	5	6	5	5	506.4
7	6797	8	7	5	7	366.8
8	3699	11	8	5	5	352
9	9849	5	6	10	5	406.4
10	8279	6	12	7	5	337.6
Total						$\sum_{i=1}^{N} ft_i = 3747.2$

these complex problems, and researchers continue to modify the algorithm itself or combine the use of the algorithm with other techniques. The description of the GA procedure, along with the illustrative example given at the end of the chapter, will be helpful for understanding the basics of the algorithm and its application to a water resource problem.

References

Afshar, M.H., Afshar, A., Mariño, M.A., Darbandi, A.A.S., 2006. Hydrograph-based storm sewer design optimisation by genetic algorithm. Can. J. Civ. Eng. 33 (3), 319−325.

Burn, D.H., Yulianti, J.S., 2001. Waste-load allocation using genetic algorithms. J. Water Resour. Planning Manage. 127 (2), 121−129.

Cai, X.M., McKinney, D.C., Lasdon, L.S., 2001. Solving nonlinear water management models using a combined genetic algorithm and linear programming approach. Adv. Water Resour. 24 (6), 667−676.

Caruana, R.A., Schaffer, J.D., 1988. Representation and hidden bias: gray vs. binary coding for genetic algorithms. In: Laird J. (Ed.), Proceedings of Fifth International Conference on Machine Learning. Morgan Kaufmann, Los Altos, CA, pp. 153−161.

Chadalavada, S., Datta, B., 2008. Dynamic optimal monitoring network design for transient transport of pollutants in groundwater aquifers. Water Resour. Manage. 22 (6), 651−670.

Chang, F.-J., Chen, L., Chang, L.-C., 2005. Optimizing the reservoir operating rule curves by genetic algorithms. Hydrolog. Process. 19 (11), 2277−2289.

Chang, L.-C., Chang, F.-J., Wang, K.-W., Dai, S.-Y., 2010. Constrained genetic algorithms for optimizing multi-use reservoir operation. J. Hydrol. 390, 66−74.

Chen, H.W., Chang, N.B., 1998. Water pollution control in the river basin by genetic algorithm-based fuzzy multi-objective programming modeling. Water Sci. Tech. 37 (8), 55−63.

Chen, L., Chang, F.J., 2007. Applying a real-coded multi-population genetic algorithm to multi-reservoir operation. Hydrolog. Process. 21 (5), 688−698.

Chen, W.C., Chang, N.B., Chen, J.C., 2003. Rough set-based hybrid fuzzy-neural controller design for industrial wastewater treatment. Water Res. 37 (1), 95−107.

Cieniawski, S.E., Eheart, J.W., Ranjithan, S., 1995. Using genetic algorithms to solve a multiobjective groundwater monitoring problem. Water Resour. Res. 31 (2), 399−409.

Cunha, M.D., 2002. Groundwater cleanup: the optimization perspective a literature review. Eng. Optim. 34 (6), 689−702.

Dandy, G.C., Engelhardt, M., 2001. Optimal scheduling of water pipe replacement using genetic algorithms. J. Water Resour. Planning Manage. 127 (4), 214−223.

Dandy, G.C., Simpson, A.R., Murphy, L.J., 1996. An improved genetic algorithm for pipe network optimization. Water Resour. Res. 32 (2), 449−458.

Davidson, J.W., Goulter, I.C., 1995. Evolution program for design of rectilinear branched networks. J. Comput. Civ. Eng. 9 (2), 112−121.

Davis, L., 1991. Handbook of Genetic Algorithms. Van Nostrand Reinhold, New York, NY.

De Jong, K.A., 1975. Analysis of the behavior of a class of genetic adaptive systems. Ph.D. Dissertation. University of Michigan, Ann Arbor.

Deb, K., 2003. Multi-Objective Optimization Using Evolutionary Algorithms. John Wiley & Sons, Singapore.

Deb, K., Agrawal, S., Pratap, A., Meyarivan, T., 2000. A fast elitist non-dominated sorting genetic algorithm for multi-objective optimization: NSGA-II. In: Schoenauer M., Deb K., Rudolph G., Yao X., Lutton E., Merelo J.J., Schwefel H.-P. (Eds.), Parallel Problem Solving From Nature, vol. VI (PPSN-VI). Springer, Berlin/Heidelberg, pp. 849–858.

Esat, V., Hall, M.J., 1994. Water resources system optimization using genetic algorithms. In: Verwey A., Minns A.W., Babovic V., Maksimovic C. (Eds.), Proceedings of First International Conference on Hydroinformatics. Balkema Rotterdam, The Netherlands, pp. 225–231.

Eshelman, L.J., Schaffer, J.D., 1993. Real coded genetic algorithms and interval schemata. In: Whitley, L.D. (Ed.), Foundation of Genetic Algorithms 2. Morgan Kaufmann, San Mateo, CA, pp. 187–202.

Eshelman, L.J., Caruana, A., Schaffer, J.D., 1989. Biases in the crossover landscape. In: Schaffer, J.D. (Ed.), Proceedings of the Third International Conference on Genetic Algorithms. Morgan Kaufmann, San Mateo, CA, pp. 86–91.

Espinoza, F., Minsker, B., Goldberg, D.E., 2005. Adaptive hybrid genetic algorithm for groundwater remediation design. J. Water Resour. Planning Manage. 131 (1), 14–24.

Farmani, R., Savic, D.A., Walters, G.A., 2006. A hybrid technique for optimisation of branched urban water systems. In: Gourbesville P., Cunge J., Guinot V., Liong S.-Y. (Eds.), Proceedings of Seventh Hydroinformatics Conference, vol. 1. Research Publishing, Chennai, India, pp. 985–992.

Feng, C.W., Liu, L.A., Burns, S.A., 1997. Using genetic algorithms to solve construction time-cost trade-off problems. J. Comput. Civ. Eng. 11 (3), 184–189.

Ganji, A., Karamouz, M., Khalili, D., 2007. Development of stochastic conflict resolution models for reservoir operation. II. The value of players' information availability and cooperative behavior. Adv. Water Resour. 30, 528–542.

Goldberg, D.E., 1989. Genetic Algorithms in Search, Optimization, and Machine Learning. Addison-Wesley, Reading, MA.

Goldberg, D.E., Deb, K., 1991. A comparative analysis of selection schemes used in genetic algorithms. In: Rawlins G.J.E. (Ed.), Foundations of Genetic Algorithms. Morgan Kaufman, San Mateo, CA, pp. 69–93.

Goldberg, D.E., 1994. Genetic and evolutionary algorithms come of age. Communication of the Association for Computing Machinery (ACM). 37 (3), 113–119.

Goldberg, D.E., Kuo, C.H., 1987. Genetic algorithms in pipeline optimization. J. Comput. Civ. Eng. 1 (2), 128–141.

Guo, Y., Walters, G.A., Khu, S.T., Keedwell, E.C., 2006. Optimal design of sewer networks using hybrid cellular automata and genetic algorithm. In: Proceedings of Fifth IWA World Water Congress. IWA Publication, London.

Halhal, D., Walters, G.A., Ouazar, D., Savic, D.A., 1997. Water network rehabilitation with structured messy genetic algorithm. J. Water Resour. Planning Manage. 123 (3), 137–146.

Harrell, L.J., Ranjithan, S., 2003. Integrated detention pond design and land use planning for watershed management. J. Water Resour. Planning Manage. 129 (2), 98–106.

Haupt, R.L., Haupt, S.E., 2004. Practical Genetic Algorithms. John Wiley & Sons, Hoboken, NJ.

Herrera, F., Lozano, M., Verdegay, J.L., 1998. Tackling realcoded genetic algorithms: operators and tools for behavioural analysis. Artif. Intell. Rev. 12, 265–319.

Hınçal, O., Altan-Sakarya, A.B., Ger, A.M., 2011. Optimization of multireservoir systems by genetic algorithm. Water Resour. Manage. 25, 1465–1487.

Holland, J.H., 1975. Adaptation in Natural and Artificial Systems. University of Michigan Press, Ann Arbor, MI.

Hu, Z.Y., Chan, C.W., Huang, G.H., 2007. Multi-objective optimization for process control of the *in-situ* bioremediation system under uncertainty. Eng. Appl. Artif. Intell. 20 (2), 225–237.

Huang, W.C., Yuan, L.C., Lee, C.M., 2002. Linking genetic algorithms with stochastic dynamic programming to the long-term operation of a multireservoir system. Water Resour. Res. 38 (12), 9.

Jian-Xia, C., Qiang, H., Yi-min, W., 2005. Genetic algorithms for optimal reservoir dispatching. Water Resour. Manage. 19 (4), 321–331.

Jothiprakash, V., Shanthi, G., 2006. Single reservoir operating policies using genetic algorithm. Water Resour. Manage. 20 (6), 917–929.

Karamouz, M., Mojahedi, A., Ahmadi, A., 2007. Economic assessment of operational policies of inter-basin water transfer. Water Resour. Res. 3 (2), 86–101.

Kerachian, R., Karamouz, M., 2005. Waste-load allocation for seasonal river water quality management: application of sequential dynamic genetic algorithms. J. Sci. Iran. 12 (2), 117–130.

Kerachian, R., Karamouz, M., 2006. Optimal reservoir operation considering the water quality issues: a stochastic conflict resolution approach. Water Resour. Res. 42, W12401.

Kerachian, R., Karamouz, M., 2007. A stochastic conflict resolution model for water quality management in reservoir–river system. Adv. Water Resour. 30 (4), 866–882.

Khu, S.-T., Madsen, H., 2005. Multiobjective calibration with Pareto preference ordering: an application to rainfall-runoff model calibration. Water Resour. Res. 41, 1367–1376.

Kuo, J.-T., Wang, Y.Y., Lung, W.S., 2006. A hybrid neural-genetic algorithm for reservoir water quality management. Water Res. 40 (7), 1367–1376.

Labadie, J.W., 2004. Optimal operation of multireservoir systems: state-of the-art-review. J. Water Resour. Planning Manage. 130 (2), 93–111.

Li, H., Love, P.E.D., 1998. Site-level facilities layout using genetic algorithms. J. Comput. Civ. Eng. 12 (4), 227–231.

Loucks, D.P., van Beek, E., 2005. Water Resources Systems Planning and Management: An Introduction to Methods, Models and Applications. UNESCO, Paris.

Mackle, G., Savic, D.A., Walters, G.A., 1995. Application of genetic algorithms to pump scheduling for water supply. In: Proceedings of Genetic Algorithms in Engineering Systems: Innovations and Applications. GALESIA '95, IEE, London, pp. 400–405.

Mahinthakumar, G., Sayeed, M., 2005. Hybrid genetic algorithm: local search methods for solving groundwater source identification inverse problems. J. Water Resour. Planning Manage. 131 (1), 45–57.

Mayer, A.S., Kelley, C.T., Miller, C.T., 2002. Optimal design for problems involving flow and transport phenomena in saturated subsurface systems. Adv. Water Resour. 25 (8–12), 1233–1256.

McKinney, D.C., Lin, M.D., 1994. Genetic algorithm solution of groundwater management models. Water Resour. Res. 30 (6), 1897–1906.

Michalewicz, Z., 1999. Genetic Algorithms + Data Structures = Evolution Programs. third rev. and extended ed. Springer, New York, NY.

Mitchell, M., 1999. An Introduction to Genetic Algorithms. MIT Press, Cambridge, MA.

Muleta, M.K., Nicklow, J.W., 2005. Decision support for watershed management using evolutionary algorithms. J. Water Resour. Planning Manage. 131 (1), 35–44.

Munavalli, G.R., Mohan, M.S., 2003. Optimal scheduling of multiple chlorine sources in water distribution systems. J. Water Resour. Planning Manage. 129 (6), 493–504.

Nagesh Kumar, D., Srinivasa Raju, K., Ashok, B., 2006. Optimal reservoir operation for irrigation of multiple crops using genetic algorithms. J. Irrigat. Drain. Eng. 132 (2), 123–129.

Nicklow, J., Reed, P., Savic, D., Dessalegne, T., Harrell, L., Chan-Hilton, A., et al., 2010. State of the art for genetic algorithms and beyond in water resources planning and management. J. Water Resour. Planning Manage. 136 (4), 412–432.

Oliveira, R., Loucks, D.P., 1997. Operating rules for multireservoir systems. Water Resour. Res. 33 (4), 839–852.

Perez-Pedini, C., Limbrunner, J.F., Vogel, R.M., 2005. Optimal location of infiltration-based best management practices for storm water management. J. Water Resour. Planning Manage. 131 (6), 441–448.

Prasad, T.D., Park, N.S., 2004. Multiobjective genetic algorithms for design of water distribution networks. J. Water Resour. Planning Manage. 130 (1), 73–82.

Prasad, T.D., Walters, G.A., Savic, D.A., 2004. Booster disinfection of water supply networks: a multi-objective approach. J. Water Resour. Planning Manage. 130 (5), 367–376.

Qin, X.S., Huang, G.H., He, L., 2009. Simulation and optimization technologies for petroleum waste management and remediation process control. J. Environ. Manage. 90 (1), 54–76.

Rani, D., Moreira, M.M., 2010. Simulation–optimization modeling: a survey and potential application in reservoir systems operation. Water Resour. Manage. 24 (6), 1107–1138.

Rao, Z., Salomons, E., 2007. Development of a real-time, near optimal control process for water-distribution networks. J. Hydroinform. 9 (1), 25–37.

Reis, L., Bessler, F., Walters, G., Savic, D., 2006. Water supply reservoir operation by combined genetic algorithm–linear programming (GA–LP) approach. Water Resour. Manage. 20 (2), 227–255.

Ritzel, B.J., Eheart, J.W., Ranjithan, S., 1994. Using genetic algorithms to solve a multiple-objective groundwater pollution containment-problem. Water Resour. Res. 30 (5), 1589–1603.

Rogers, L.L., Dowla, F.U., Johnson, V.M., 1995. Optimal fieldscale groundwater remediation using neural networks and the genetic algorithm. Environ. Sci. Tech. 29 (5), 1145–1155.

Savic, D., Walters, G., 1997. Genetic algorithms for least cost design of water distribution networks. J. Water Resour. Planning Manage. 123 (2), 67–77.

Savic, D.A., Walters, G.A., Schwab, M., 1997. Multiobjective genetic algorithms for pump scheduling in water supply. In: Corne, D., Shapiro, J.L. (Eds.), AISB '97, Lecture Notes in Computer Science, 1305. Springer, Berlin, pp. 227–236.

Sharif, M., Wardlaw, R., 2000. Multireservoir systems optimization using genetic algorithms: case study. J. Comput. Civ. Eng. 14 (4), 255–263.

Shivraj, R., Ravichandran, T., 2011. A review of selection methods in genetic algorithm. Int. J. Eng. Sci. Technol. 3 (5), 3792–3797.

Simpson, A.R., Dandy, G.C., Murphy, L.J., 1994. Genetic algorithms compared to other techniques for pipe optimization. J. Water Resour. Planning Manage. 120 (4), 423–443.

Singh, A., Minsker, B.S., 2008. Uncertainty-based multiobjective optimization of groundwater remediation design. Water Resour. Res. 44, W02404.

Sinha, E., Minsker, B.S., 2007. Multiscale island injection genetic algorithms for groundwater remediation. Adv. Water Resour. 30 (9), 1933–1942.

Smalley, J.B., Minsker, B.S., Goldberg, D.E., 2000. Risk-based *in situ* bioremediation design using a noisy genetic algorithm. Water Resour. Res. 36 (10), 3043–3052.

Soh, C.K., Yang, J.P., 1996. Fuzzy controlled genetic algorithm search for shape optimization. J. Comput. Civ. Eng. 10 (2), 143–150.

Suggala, S.V., Bhattacharya, P.K., 2003. Real coded genetic algorithm for optimization of pervaporation process parameters for removal of volatile organics from water. Ind. Eng. Chem. Res. 42 (13), 3118–3128.

Syswerda, G., 1989. Uniform crossover in genetic algorithms. In: Schaffer, J.D. (Ed.), Proceedings of the Third International Conference on Genetic Algorithms. Morgan Kaufmann, San Mateo, CA, pp. 2–9.

Tang, Y., Reed, P., Wagener, T., 2006. How efficient and effective are evolutionary multiobjective algorithms at hydrologic model calibration? Hydrol. Earth Syst. Sci. 10, 289–307.

Tospornsampan, J., Kita, I., Ishii, M., Kitamura, Y., 2005. Optimization of a multiple reservoir system operation using a combination of genetic algorithm and discrete differential dynamic programming: a case study in Mae Klong system, Thailand. Paddy Water Environ. 3, 29–38.

Tsai, F.T.C., Sun, N.Z., Yeh, W.W.G., 2003. A combinatorial optimization scheme for parameter structure identification in groundwater modeling. Ground Water. 41 (2), 156–169.

Vairavamoorthy, K., Ali, M., 2005. Pipe index vector: a method to improve genetic-algorithm-based pipe optimization. J. Hydraul. Eng. 131 (12), 1117–1125.

Vamvakeridou-Lyroudia, L.S., Walters, G.A., Savic, D.A., 2005. Fuzzy multiobjective optimization of water distribution networks. J. Water Resour. Planning Manage. 131 (6), 467–476.

van Zyl, J., Savic, D.A., Walters, G.A., 2004. Operational optimization of water distribution systems using a hybrid genetic algorithm method. J. Water Resour. Planning Manage. 130 (2), 160–170.

Wang, Q.J., 1991. The genetic algorithm and its application to calibrating conceptual rainfall-runoff models. Water Resour. Res. 27 (9), 2467–2471.

Wang, L., Zheng, D.Z., 2002. A modified genetic algorithm for job shop scheduling. Int. J. Adv. Manufact. Technol. 20 (1), 72–76.

Wang, M., Zheng, C., 1998. Groundwater management optimization using genetic algorithms and simulated annealing: formulation and comparison. J. Am. Water Resour. Assoc. 34 (3), 519–530.

Wardlaw, R., Sharif, M., 1999. Evaluation of genetic algorithms for optimal reservoir system operation. J. Water Resour. Planning Manage. 125 (1), 25–33.

Wei, L.Y., Zhao, M., Wu, G.M., Meng, G., 2005. Truss optimization on shape and sizing with frequency constraints based on genetic algorithm. Comput. Mech. 35 (5), 361–368.

Wright, A., 1991. Genetic algorithms for real parameter optimization. Foundation of Genetic Algorithms. Morgan Kaufmann, San Mateo, CA, pp. 205–218.

Wu, J., Zheng, C., Chien, C., Zheng, L., 2006. A comparative study of Monte Carlo simple genetic algorithm and noisy genetic algorithm for cost-effective sampling network design under uncertainty. Adv. Water Resour. 29, 899–911.

Yan, S., Minsker, B., 2006. Optimal groundwater remediation design using an adaptive neural network genetic algorithm. Water Resour. Res. 42, W05407.

Yandamuri, S.R.M., Srinivasan, K., Bhallamudi, S.M., 2006. Multiobjective optimal waste load allocation models for rivers using nondominated sorting genetic algorithm-II. J. Water Resour. Planning Manage. 132 (3), 133–143.

Yeh, W.W.-G., 1986. Review of parameter identification procedures in groundwater hydrology: the inverse problem. Water Resour. Res. 22 (2), 95–108.

Yeh, C.-H., Labadie, J.W., 1997. Multiobjective watershed-level planning of storm-water detention basins. J. Water Resour. Planning Manage. 123 (6), 336—343.

Zahraie, B., Kerachian, R., Malekmohammadi, B., 2008. Reservoir operation optimization using adaptive varying chromosome length genetic algorithm. Water Int. 33 (3), 380—391.

Zechman, E.M., Ranjithan, S., 2007. Evolutionary computation based approach for model error correction and calibration. Adv. Water Resour. 30 (5), 1360—1370.

Zheng, C.M., Wang, P.P., 2002. A field demonstration of the simulation optimization approach for remediation system design. Ground Water. 40 (3), 258—266.

4 Application of the Hybrid HS—Solver Algorithm to the Solution of Groundwater Management Problems

Mustafa Tamer Ayvaz[1] and Alper Elçi[2]

[1]Department of Civil Engineering, Pamukkale University, Denizli, Turkey,
[2]Department of Environmental Engineering, Dokuz Eylül University, Izmir, Turkey

4.1 Introduction

Mathematical simulation models are essential tools to analyze and manage groundwater systems. These models are based on analytical or numerical solutions of partial differential equations (PDEs), which represent groundwater flow for any given set of initial and boundary conditions. Analytical solutions are possible only for simplified cases where the solution space has a fairly regular shape and aquifer properties are homogeneous. However, these simplifications may not always be valid. Therefore, numerical solutions are used. Many numerical solutions have been proposed to solve the governing PDEs. Among them, the finite difference (FD) and finite element methods are perhaps the most widely used techniques, and many commercial and public domain software packages have been developed to solve groundwater modeling problems by implementing one of these methods.

Although mathematical simulation models are widely used in groundwater modeling, they are not capable of solving management problems per se. Therefore, they are usually integrated with optimization algorithms to determine the best management strategies for utilizing groundwater resources efficiently. It is notable that in groundwater modeling, mathematical models are integrated with optimization algorithms using the so-called response matrix or embedding approach. In the response matrix approach, an external mathematical simulation model is developed to compute groundwater hydraulic head values at available observation locations for given unit changes in pumping or injection rates. This approach is based on the principles of superposition and requires a linear relationship between the groundwater system and the given pumping or injection rates. After establishing mathematical relationships between rates and groundwater heads, these relationships can

Metaheuristics in Water, Geotechnical and Transport Engineering. DOI: http://dx.doi.org/10.1016/B978-0-12-398296-4.00004-0

be integrated into optimization algorithms to determine optimum groundwater withdrawal or recharge strategies. Although this approach is an easy way of determining optimum pumping rates, it cannot be used for problems where pumping or injection rates and groundwater heads are not linearly correlated. On the other hand, in the embedding approach, a mathematical relationship between the individual pumping rates and the response of the aquifer (i.e., hydraulic heads) is not sought. Instead, the hydraulic head distribution for a given set of observation wells is determined by running the simulation model for the developed pumping scheme. The embedding approach provides more comprehensive results than the response matrix approach; however, its computational burden is usually greater, especially for large flow domains and/or transient flow simulations.

It should be noted that the solution of a groundwater management problem using simulation—optimization procedures requires the use of an independent optimization model. In this model, the solution can be obtained using deterministic and stochastic optimization algorithms. Deterministic algorithms are mostly gradient-based local search methods that require substantial information on the gradient of the objective function to find a solution. Among them, linear programming (LP), nonlinear programming (NLP), mixed-integer programming (MIP), and dynamic programming (DP) are the most important solution algorithms. In order to solve an optimization problem using these algorithms, partial derivatives of the objective function and constraint sets with respect to the decision variables must be calculated. Although these algorithms are computationally effective, finding a globally optimal solution using them is not an easy task unless the solution space of the problem is convex. On the other hand, many problems in practice have nonconvex solution spaces such that taking the partial derivatives may not be possible due to nondifferentiability, discontinuity, and other factors. For such problems, obtaining a globally optimal solution usually depends on the appropriateness of the initial solutions. Therefore, stochastic optimization algorithms are usually preferred over deterministic ones because of these potential pitfalls.

Stochastic optimization algorithms get their computational basis from processes observed in nature. For instance, in the case of genetic algorithms (GAs), the solution of an optimization problem is based on the mechanics of natural selection (Goldberg, 1989; Holland, 1975), whereas the particle swarm optimization (PSO) algorithm is adopted from social behaviors of bird or fish colonies (Kennedy and Eberhart, 1995). Most widely used stochastic optimization algorithms are GA, PSO, tabu search (TS Glover, 1977), simulated annealing (SA; Kirkpatrick et al., 1983), ant colony optimization (ACO Dorigo and Di Caro, 1999), and harmony search (HS; Geem et al., 2001). A vast number of published studies that provide mathematical structures and application areas of these algorithms exist in the literature. Although these algorithms are effective in exploring the entire solution space without requiring any special initial solution, they usually require long computational times to find precise, globally optimal solutions.

Recently, the number of applications of hybrid global—local optimization approaches to the solution of optimization problems increased. In these algorithms, the global search algorithm (stochastic optimization) starts the search process with multiple solution points and explores the entire solution space through heuristic optimization procedures. After this process, the local search algorithm gets the results of global search as the initial solution and precisely solves the problem through gradient-based optimization procedures. This solution sequence renders finding globally optimum solutions by virtue of the strong exploring capabilities and fine-tuning capabilities of global and local search methods, respectively. However, the implementation of these algorithms can require advanced programming skills since most of the local search methods requires the computation of partial derivatives, and subsequently, the generation of the Jacobian and/or Hessian matrices.

This chapter focuses on a recently proposed optimization algorithm, HS–Solver (Ayvaz et al., 2009), which is a viable tool to solve groundwater management problems. HS–Solver is a hybrid global–local optimization algorithm that combines the HS algorithm with the spreadsheet application called Solver (Frontline Systems, 2011). Solver is a built-in nonlinear optimization add-in that solves optimization problems based on information provided in the cells of the spreadsheet. This information includes user-defined constraints, lower and upper bounds of decision variables, and convergence criteria. The main advantage of using HS–Solver as a local optimizer is that it does not require advanced programming skills to perform the mathematical calculations mentioned earlier. The performance of the HS–Solver optimization algorithm is demonstrated here using two examples. In the first example, a groundwater-pumping maximization problem for a well-known hypothetical aquifer is solved. This problem was previously studied and solved by several researchers using a variety of solution methods. In the second example, the HS–Solver procedure is applied to a real-world groundwater flow model. The model was previously developed for the Tahtalı watershed, located near the city of Izmir in Turkey, which is a key component of the city's water supply system.

4.2   Development of the Hybrid HS–Solver Algorithm

4.2.1   The HS Optimization Algorithm

HS is a recently proposed stochastic optimization algorithm that is inspired from the improvisation process in music. In music, the purpose of improvisation is to seek a better harmony by making several trials. This process is analogous to the optimization process since the purpose of the optimization is to seek a better objective function value by making several iterations. In this analogy, the musicians in the orchestra are analogous to decision variables, whereas the notes in the musicians' memories are analogous to the possible values of these variables. When the musicians find a fantastic harmony through the notes in their memories, then, in mathematical terms, an optimal solution is obtained through the corresponding values of the decision variables. This is referred to as finding an *elite harmony*. In music, generation of a new harmony is usually performed by playing a note from memory, by playing a note that is close to another one in the memory, and by randomly playing a note from the possible note range. Adaptation of these musical rules to optimization problems is performed by selecting a decision variable from memory, by replacing the selected variable with another one that is close to the current value, and by randomly selecting a decision variable from the possible random number range. Figure 4.1 explains this analogy in detail.

As can be seen from Figure 4.1, we have a jazz trio in which each member has some notes in his memory. The main goal of the trio is to find a musically pleasing harmony by performing several improvisations. If we replace this trio with a decision variable set, the problem of finding a musically pleasing harmony can be converted to the problem of finding a globally optimum solution.

Figure 4.1 Analogy between the musical improvisation process and mathematical optimization.
Source: Adapted from Ayvaz (2010).

Depending on this analogy and the musical rules given earlier, solution of an optimization problem can be performed as follows (Ayvaz, 2010).

> Rule 1 (*Memory Consideration*): The first musician in Figure 4.1 has three notes, {Do,Si, La}, in his memory. Let us assume that these notes correspond to the numbers {2.7,3.1,1.2} in the optimization process. If this musician decides to select and play {La} from his memory, it is equivalent to selecting and using a value of {1.2} as the first decision variable.
>
> Rule 2 (*Pitch Adjusting*): The second musician in Figure 4.1 has also three notes, {Mi,Re, Do}, in his memory, and these notes correspond to the numbers {1.1,5.1,1.9} in the optimization process. Unlike the first one, this musician selects {Do} from his memory and plays its neighbor {Do#}. This note corresponds to {1.9}, which is a small random neighbor to {2.0}.
>
> Rule 3 (*Random Selection*): The third musician in Figure 4.1 has also three notes, {La,Fa, Sol}, in his memory. Although these notes were previously used, this musician decides to select and play a note randomly; in this case, {Mi}. As opposed to the possible data set stored in memory, {4.9} is randomly selected and used in this case, even if it does not exist among the possible values in memory. After generating a new harmony, {La,Do#, Mi}, through memory consideration, pitch adjusting, and random selection, the new decision variable set is generated as {1.2,2.0,4.9} in the optimization process.

Although HS is a recently proposed optimization algorithm, it has already been applied to solve a wide variety of problems, such as aquatic environment—related applications, structural design, operations research, information technology, transport-related problems, energy applications, and medical studies. The mathematical statement of HS algorithm can be given as follows.

Let $f(\vec{x})$ be an objective function to be minimized or maximized, N be the number of decision variables, x_k be a decision variable ($k = 1, 2,\ldots, N$), and $\vec{x}$ be a vector which contains x_k, such that $\vec{x} = [x_1, x_2, \ldots, x_N]^T$, where T is the transpose operator, M is the number of inequality constraints, $g_i(\vec{x})$ is an inequality constraint ($i = 1, 2,\ldots, M$), P is the number of equality constraints, $h_j(\vec{x})$ is an equality constraint ($j = 1, 2,\ldots, P$), and $x_{k,min}$ and $x_{k,max}$ are the lower and upper bounds of decision variables ($k = 1, 2,\ldots, N$).

Based on these definitions, the optimization problem can be formulated as follows:

$$\text{min or max } f(\vec{x}) \tag{4.1}$$

subject to

$$g_i(\vec{x}) \geq 0; \quad i = 1, 2, \ldots, M \tag{4.1a}$$

$$h_j(\vec{x}) = 0; \quad j = 1, 2, \ldots, P \tag{4.1b}$$

$$x_k \in [x_{k,min}, x_{k,max}]; \quad k = 1, 2, \ldots, N \tag{4.1c}$$

Just as the other heuristic algorithms do, HS includes some solution parameters, which are harmony memory size (HMS), harmony memory considering rate (HMCR), pitch adjusting rate (PAR), and the termination criterion. Harmony memory (HM) is a two-dimensional solution array that includes decision variables and calculated objective function values. HMCR and PAR parameters are the probabilities that are used to generate new solution vectors and maintain the diversity of the solution. After setting these parameters (recommended parameter ranges will be provided later), elements of the HM are randomly filled, such that $x_k^l = x_{k,\min} + r(0,1) \times (x_{k,\max} - x_{k,\min})$, $(k = 1, 2, \ldots, N; \, l = 1, 2, \ldots, \text{HMS})$ where r $(0,1)$ is a uniform random number between 0 and 1. After the filling process, the value of the objective function is calculated using Eq. (4.2):

$$\begin{bmatrix} x_1^1 & x_2^1 & \cdots & x_{N-1}^1 & x_N^1 & f(x^1) \\ x_1^2 & x_2^2 & \cdots & x_{N-1}^2 & x_N^2 & f(x^2) \\ \vdots & \vdots & \ddots & \vdots & \vdots & \vdots \\ x_1^{\text{HMS}-1} & x_2^{\text{HMS}-1} & \cdots & x_{N-1}^{\text{HMS}-1} & x_N^{\text{HMS}-1} & f(x^{\text{HMS}-1}) \\ x_1^{\text{HMS}} & x_2^{\text{HMS}} & \cdots & x_{N-1}^{\text{HMS}} & x_N^{\text{HMS}} & f(x^{\text{HMS}}) \end{bmatrix} \tag{4.2}$$

Note that the value of HMS depends on the problem type. Nevertheless, a value of $10 \leq \text{HMS} \leq 50$ is recommended, which suits the solution of many optimization problems. Normally, HS or other stochastic optimization methods cannot solve constrained optimization problems by themselves, and some auxiliary equations are needed to facilitate the optimization. Therefore, it is necessary to implement the so-called penalty functions. These functions are used to satisfy the inequality and equality constraints given in Eqs. (4.1a) and (4.1b). They also replace a constrained problem with a series of unconstrained problems whose solutions ideally converge to the solution of the constrained problem. There are many penalty approaches in literature, including static, dynamic, adaptive, and death penalties. For instance, a static penalty function can be used to calculate the objective function values subject to the given constraint sets as follows:

$$f'(\vec{x}) = f(\vec{x}) + \sum_{i=1}^{M} \alpha_i \times G_i + \sum_{j=1}^{P} \beta_j \times H_j \tag{4.3}$$

$$G_i = \min[0, g_i(\vec{x})]^2 \quad i = 1, 2, \ldots, M \tag{4.3a}$$

$$H_j = |h_j(\vec{x})|^2 \quad j = 1, 2, \ldots, P, \tag{4.3b}$$

where $f'(\vec{x})$ is the penalized objective function value; α_i and β_j are the penalty coefficients, which are mostly problem-dependent; and G_i and H_j are the constraint functions. After generating the initial HM and calculating the corresponding objective function values, the next task is to generate a new solution vector $x' = (x'_1, x'_2, x'_3, \ldots, x'_N)$, such that the elements of this vector are selected from either HM or any other possible random range. This process is controlled by the HMCR parameter. Note that this parameter is defined as the probability of selecting a decision variable from HM or otherwise. With this purpose, N uniform random numbers are generated (e.g., $r_i(0, 1)$, $i = 1, 2, \ldots, N$) and compared to HMCR. If the condition of r_i $(0,1) < \text{HMCR}$ is satisfied, it means that the value of the ith decision variable is randomly selected from HM such that $x'_i \in [x_i^1, x_i^2, \ldots, x_i^{\text{HMS}}]$. Otherwise, it is randomly selected from the possible random range such that $x'_i \in (x_{i,\min}, x_{i,\max})$. After this process, decision variables that are selected from HM are evaluated to determine whether pitch adjusting is necessary. This procedure is controlled by the PAR parameter, which is the probability of pitch adjusting. Therefore, N uniform random numbers are generated (e.g., $r_j(0, 1)$, $j = 1, 2, \ldots, N$) and compared to PAR. If r_j $(0,1) < \text{PAR}$, then the jth decision variable is subject to a slight change. Otherwise, nothing is done. These procedures can be stated as follows:

$$
\begin{aligned}
&\text{If } r_i(0, 1) < \text{HMCR} \\
&\quad x'_i \leftarrow x'_i \in [x_i^1, x_i^2, \ldots, x_i^{\text{HMS}}] \\
&\quad \text{If } r_j(0, 1) < \text{PAR} \\
&\quad\quad x'_j \leftarrow x'_j \pm r_j(0, 1) \times bw \\
&\quad \text{Else} \\
&\quad\quad x'_j \leftarrow x'_j \\
&\quad \text{End If} \\
&\text{Else} \\
&\quad x'_i \leftarrow x'_i \in (x_{i,\min}, x_{i,\max}) \\
&\text{End If}
\end{aligned}
\tag{4.4}
$$

where bw is a bandwidth that is used to perform slight changes. The values of HMCR and PAR largely affect the convergence behavior. For instance, if the value selected from HMCR is too low, only a few elite harmonies are selected and the algorithm may converge slowly; on the other hand, if the value selected is too high, the pitches in HM are mostly used and other random possibilities may not be explored (Yang, 2009). However, if the value selected from PAR is too low, a slow convergence is observed due to the exploration of a small subspace; a high value from PAR may cause the algorithm to work in a random search (Yang, 2009). According to experiences in practice, $0.70 \leq \text{HMCR} \leq 0.95$ and $0.20 \leq \text{PAR} \leq 0.50$ are plausible parameter ranges for solving optimization problems. After generating a new solution vector, the value of the objective function is calculated and compared to the worst one in HM. If the new objective function value is better than the worst one, the newly generated solution vector is replaced with the worst one in HM. Note that HS algorithm searches for an optimum solution by iterating from steps 3 to 5 until the given termination criterion is satisfied. The termination criterion is problem specific and is determined by the programmer. In some studies, terminating the solution after a given number of iterations is also possible, as is done in the example given in this chapter.

4.2.2   The Hybrid HS–Solver Optimization Algorithm

Stochastic optimization algorithms are effective in finding global or near-global optimum solutions. However, they can require long computational times to find a solution precisely. Therefore, the use of hybrid optimization algorithms is usually preferred due to their ability to find global optimum solutions. The main idea of hybrid algorithms is to combine the strong exploration of the global search approach and the fine-tuning capabilities of the local search approach. Although these algorithms provide better results than both global and local searches by themselves, they may require high-level programming skills; therefore, implementing them may not be an easy task for less experienced users.

HS–Solver, first proposed by Ayvaz et al. (2009), is a new global–local optimization algorithm, which does not require knowledge about programming gradient-based optimization algorithms. Solver is a gradient-based optimization add-in, and most commercial spreadsheet products (e.g., Microsoft Excel, which is used in the example in this chapter) include it. It solves optimization problems using the GRG2 algorithm (Lasdon et al., 1978) by using quasi-Newton and conjugate gradient methods in finding the search directions. Since Solver works with Excel, the development of a hybrid solution approach in Excel is necessary. Therefore, three separated Microsoft Visual Basic for Applications (VBA) codes have been developed using the Visual Basic Editor in Excel. The first VBA code is for the HS algorithm and aims to solve an optimization problem on the VBA platform. The second code calls the Solver add-in, which is created with the macro recording feature in Excel. A *macro* is a series of commands to accomplish a task automatically, and the source code of a macro can be easily modified in the Visual Basic Editor. The main advantage of using the macro recorder is that the recorded VBA code runs Solver automatically instead of having to call it manually. Finally, the developed final VBA code is used to link both running codes. In this linkage, if the newly improvised solution vector is better than the worst one stored in HM, this vector is subjected to a local search using Solver with a probability of P_c. The reason to use P_c is to prevent the application of local searching to all the solutions, since this can lead to slow convergence. Our trials, as well as the recommendations of other researchers (Fesanghary et al., 2008) indicate that $P_c = 0.10$ is sufficient to solve many optimization problems. After the selected convergence criterion of Solver is satisfied (default convergence criteria are used), the solution is updated as outlined in the previous section.

4.3   Formulation of the Management Problem

Groundwater management can be defined as selecting the best management strategy to maximize the economic, environmental, and hydraulic benefits (Willis and Yeh, 1987). In this maximization process, some managerial and technical constraints (e.g., well capacities, hydraulic heads, water demands, and drawdowns) must be taken into account by the optimization model. The response of the groundwater system is determined by running the groundwater flow model for a particular management scenario. For this purpose, the analytical or numerical solution of the governing groundwater flow equation is linked to the optimization model. Groundwater management problems are usually studied at the basin scale; therefore, two-dimensional areal groundwater models are generally used, where the thickness of the aquifer is negligible compared to the lateral dimension of the model domain. The governing equation for two-dimensional, isotropic, transient groundwater flow for an unconfined aquifer can be given as follows:

$$\frac{\partial}{\partial x}\left(Kh\frac{\partial h}{\partial x}\right) + \frac{\partial}{\partial y}\left(Kh\frac{\partial h}{\partial y}\right) + R - \sum_{i=1}^{N} Q_i \delta(x_i, y_i) = S_y \frac{\partial h}{\partial t} \qquad (4.5)$$

where x and y are the Cartesian coordinates, K is the hydraulic conductivity, h is the hydraulic head, R is the areal recharge rate, N is the number of pumping/injection wells, Q_i is the pumping/injection rate at well i, $\delta(x_i,y_i)$ is a Dirac delta function evaluated at point (x_i,y_i), and S_y is the specific yield. As indicated previously, the solution of Eq. (4.5) can be obtained with analytical and numerical solution methods. However, analytical solutions can be used only for simplified cases where the model domain has a simple geometry and aquifer properties are homogeneous. Therefore, numerical solution methods are usually preferred to solve this equation. Although numerical solutions do not provide the exact solutions that analytical solutions do, in a practical sense, they are usually a very efficient way to solve optimization problems. Note that for the case studies presented in this chapter, Eq. (4.5) is solved within the framework of MODFLOW (Harbaugh, 2000), which is a modular, FD-based groundwater flow model developed by the US Geological Survey (USGS). This model, which is an open-source code, is widely used by hydrogeologists to simulate the flow of groundwater through aquifer systems. In MODFLOW, the flow domain is subdivided into grid blocks where soil properties and sinks/sources are assumed to be uniform. Combination of these grid blocks in three dimensions generates a block-centered grid structure, as shown in Figure 4.2. As can be seen, the generated grid structure consists of some grid blocks with dimensions of $\Delta x_j \times \Delta y_i \times \Delta z_k$, where the subscripts i, j, and k correspond to the row, column, and layer numbers, respectively. The dashed lines on the grid structure represents the flow domain, where each block inside this domain corresponds to active FD cells. In order to simulate the groundwater flow process, the FD form of Eq. (4.5) is assigned to all the active cells (e.g., the black circles) and thereby, the partial differential equation governing groundwater flow is replaced with a set of much simpler algebraic equations. The resulting algebraic, linear equation set is solved by the built-in matrix solvers of MODFLOW. As indicated previously, groundwater flow problems usually are solved in two dimensions for cases where variations in geometry and aquifer properties in the vertical dimension are not significant. So, only one layer of Figure 4.2 is considered for all the solutions given in this chapter. After developing the two-dimensional MODFLOW

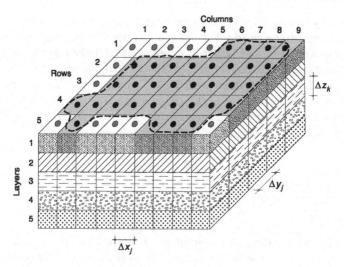

Figure 4.2 MODFLOWs three-dimensional grid structure.
Source: Adapted from Harbaugh (2000).

model, it is integrated into the HS−Solver-based optimization model to get the aquifer's response for the generated pumping rates. With this purpose, the optimization model calls MODFLOW in each iteration to solve for hydraulic heads. Then the MODFLOW solution results are passed to the optimization model to evaluate the objective function.

Groundwater management problems can be classified as either the hydraulic management or the groundwater quality management model. In the hydraulic management model, the main objective is to maximize the groundwater-pumping rate or minimize the total pumping cost. In either scenario, the decision variables are the pumping rates of available wells. The constraints for pumping rates can be set as a range consisting of lower and upper bounds. Similarly, hydraulic head values at certain wells can be constrained by defining a specified lower bound or a water demand can be specified that must be met. On the other hand, for the groundwater quality management model, the main objective is usually to minimize the remediation cost to satisfy any given water quality standards at the pumping wells. In this chapter, the solution of the pumping maximization management problem (hydraulic management) is illustrated.

Depending on the constraints, the management model can be stated in the following manner:

$$z = \max \left\{ \sum_{i=1}^{N} Q_i - P(h) \right\} \tag{4.6}$$

subject to

$$h_i \geq h_{i,\min} \quad i = 1, 2, 3, \ldots, N \tag{4.6a}$$

$$Q_{i,\min} \leq Q_i \leq Q_{i,\max} \quad i = 1, 2, 3, \ldots, N \tag{4.6b}$$

$$P(h) = \begin{cases} \lambda |h_{i,\min} - h_i| & \text{if } h_i < h_{i,\min} \\ 0 & \text{if } h_i \geq h_{i,\min} \end{cases} \quad i = 1, 2, 3, \ldots, N \tag{4.6c}$$

where $h_{i,\min}$ is the minimum permissible hydraulic head value at well i; $Q_{i,\min}$ and $Q_{i,\max}$ are the lower and upper bounds of the pumping rates at well i; $P(h)$ is the penalty function, which takes a zero value if no hydraulic head constraint is violated (otherwise, it varies linearly with the magnitude of constraint violation); and λ is a problem-dependent penalty coefficient. Note that a larger λ means that greater emphasis will be put on satisfying the constraints. Equation (4.6) states in simpler terms that the amount of pumped groundwater from the aquifer is maximized until the predefined hydraulic head limits at certain well locations are triggered. Figure 4.3 shows the flowchart of the proposed solution model.

4.4 Numerical Applications

In this section, the application of the hybrid HS—Solver optimization algorithm is demonstrated with two examples. The first example deals with the solution of a groundwater-pumping maximization problem. The model domain is a hypothetical aquifer model that was studied previously by other researchers for illustrating different solution approaches. The second example is a real-world case, which aims to solve the same optimization problem on the Tahtalı watershed (in Izmir, Turkey). For both case studies, HS solution parameters are set as follows: HMS = 10, HMCR = 0.95, and PAR = 0.50.

4.4.1 Example 1

This example deals with the solution of groundwater-pumping maximization problem on a hypothetical unconfined aquifer system. This problem was first studied by McKinney and Lin (1994) using LP and GA approaches. Several years later, the same problem was solved using GA and SA (Wang and Zheng, 1998), the GA-based SA penalty function approach (GASAPF) (Wu et al., 1999), the shuffled complex evolution (SCE) (Wu and Zhu, 2006), and HS (Ayvaz, 2009). Figure 4.4 shows the plan view and cross-section of the hypothetical aquifer system.

As can be seen from Figure 4.4, the aquifer system has no-flow boundary conditions in the north and south and a specified head condition in the east and west. Hydraulic conductivity distribution is assumed to be homogeneous with a value of 50 m/day. It is also assumed that a constant recharge of 0.001 m/day is applied

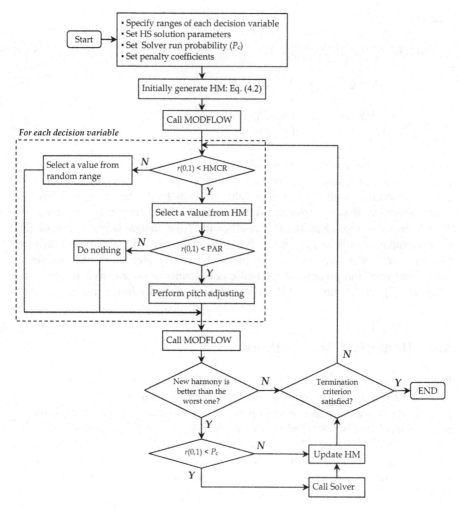

Figure 4.3 Flowchart of the proposed solution model.

throughout the flow domain. There are 10 pumping wells in the aquifer and the main objective is to maximize the pumping rates of these wells using the hybrid HS−Solver algorithm subject to certain constraints. The solution for hydraulic head is obtained using a MODFLOW model by discretizing the flow domain into rectangular FD-grid blocks. Maximization of the pumping rates of 10 wells is performed subject to $h_i^{min} = 0$, $Q_i^{min} = 0$, and $Q_i^{max} = 7000$ ($i = 1, 2, \ldots, 10$). Value of the weighting factor (λ) and the maximum number of iterations are both set to 10,000. For these settings, comparison of identification results with those obtained by other solution approaches is given in Table 4.1.

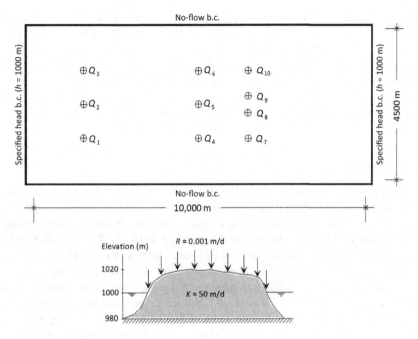

Figure 4.4 Hypothetical aquifer system used in the example.

Table 4.1 Comparison of the Identified Pumping Rates (m³/day)

Well No.	LP[1]	GA[1]	GA[2]	SA[2]	GASAPF[3]	SCE—UA[4]	HS[5]	HS—Solver
1	7,000	7,000	7,000	7,000	7,000	7,000	7,000	7,000
2	7,000	7,000	7,000	7,000	7,000	7,000	7,000	7,000
3	7,000	7,000	7,000	7,000	7,000	7,000	7,000	7,000
4	6,000	7,000	5,000	6,200	6,056	5,987	5,904	6,197
5	4,500	2,000	5,000	4,700	4,290	4,477	4,590	3,973
6	6,000	6,000	6,000	6,200	6,056	5,986	5,904	6,197
7	6,800	7,000	7,000	6,650	6,774	6,814	6,821	6,800
8	4,100	4,000	4,000	4,000	4,064	4,094	4,121	4,157
9	4,100	4,000	4,000	4,000	4,064	4,094	4,120	4,156
10	6,800	7,000	7,000	6,650	6,774	6,814	6,820	6,800
Total Pumping	**59,300**	**58,000**	**59,000**	**59,400**	**59,078**	**59,266**	**59,279**	**59,281**
Number of Simulations	N/A	640	27,800	17,200	N/A	N/A	8,843	3,260

[1]McKinney and Lin (1994)
[2]Wang and Zheng (1998)
[3]Wu et al. (1999)
[4]Wu and Zhu (2006)
[5]Ayvaz (2009)

It can be observed from Table 4.1 that although the pure HS-based solution model found an objective function value of 59,279 m^3/day, this value improved slightly to 59,281 m^3/day with the addition of Solver to the optimization model. This result agreed closely with LP (59,300 m^3/day), the global optimum solution for this problem. More important, comparison of the required number of simulations also implies that HS−Solver is more efficient because HS required 8843 iterations to find a solution. This number was reduced to 3260 (this number is the sum of HS and Solver iterations) with the inclusion of Solver.

4.4.2 Example 2

In this example, the applicability of the HS−Solver approach is demonstrated for a groundwater management problem of the semiurban Tahtalı watershed located in Izmir, Turkey. This watershed has a drainage area of 550 km^2 and is a subwatershed of the K. Menderes River watershed. As of 2010, the Tahtalı dam reservoir (38°08′N; 27°06′E) provides about 36% of Izmir's total water supply.

An areal, steady-state groundwater flow model was previously developed by Elçi et al. (2010) to calculate groundwater fluxes and the water budget for the surficial aquifer of the Tahtalı watershed. This one-layered model is set up for steady-state groundwater flow conditions and is based on MODFLOW. The spatial resolution of the FD grid is 150 × 150 m^2. The boundaries of the model were defined such that they encompass the entire area of interest and coincide with hydrological boundaries (e.g., sea, lake, or watershed boundaries). Figure 4.5 shows the general location map of the study area, boundaries of the groundwater flow model, and the locations of the pumping wells. The watershed is under environmental stress, in particular with respect to groundwater. There are many small communities, greenhouses, and farms in the study area, and each of them relies on groundwater to satisfy their water demand. Although many water supply wells were drilled in the surficial aquifer, only 17 wells with high pumping rates are considered since the solution with all the pumping wells may not be feasible (Figure 4.5). Therefore, the number of decision variables for this example is 17.

The solution of this problem by the HS−Solver algorithm is subject to $Q_i^{min} = 0$, $Q_i^{max} = 20,000$, and $h_i^{min} = 0.75\tilde{h}_i$ where $\tilde{h}_i$ represents the hydraulic head value with no pumping at the ith well, ($i = 1,2,\ldots,17$). The values of λ and the maximum number of iterations is the same as for Example 1. It should be noted that this problem is solved using both pure HS and HS−Solver algorithms in order to compare the identification results. For both solutions, same initial HM is used, which means that the optimization procedures begin with the same initial solutions. This situation can be seen in the convergence plot given in Figure 4.6, where both HS and HS−Solver start out with the exact same objective function value. Because the constraint set given in Eq. (4.6c) is not satisfied in the first iteration steps, the objective function inherently takes on some penalty values. Reaching a solution without penalty values requires 92 iterations in HS, while this number is reduced to 13 in HS−Solver. For these solutions, the comparison of the identified pumping rates is shown in Table 4.2.

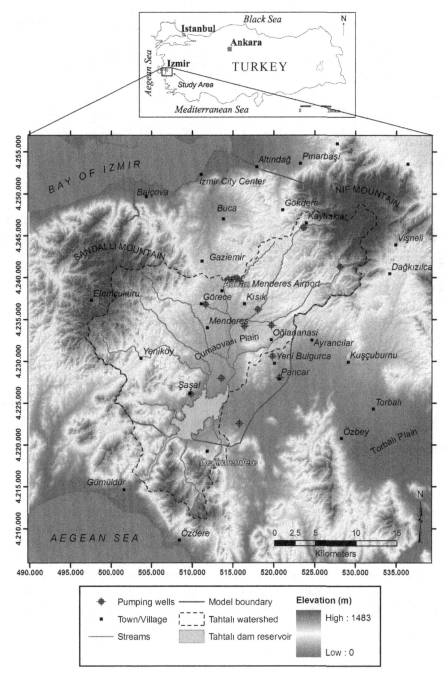

Figure 4.5 General location map of the study area and the boundaries of regional groundwater flow.
Source: Adapted from Elçi et al. (2010).

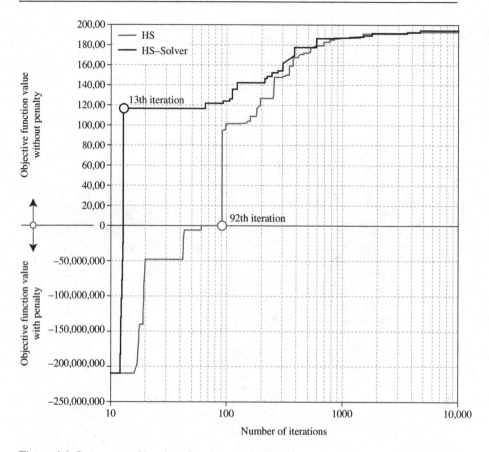

Figure 4.6 Convergence histories of pure HS and HS−Solver solutions.

The results given in Table 4.2 show that the orders of magnitude of the identified pumping rates are comparable and that there are some differences between the solutions of HS and HS−Solver, which can be associated with the used solution schemes. While HS resulted with an objective function value of 192,878 at the end of the 10,000th iteration, HS−Solver obtained a function value of 194,478. Note that HS−Solver found approximately the same objective function value with HS at the end of the 4761st iteration, which corresponds to a reduction of approximately 50% in the required number of iterations.

4.5   Conclusions

In this chapter, a linked simulation−optimization procedure to solve groundwater management problems is introduced and illustrated. In this procedure, groundwater

Table 4.2 Comparison of the Identified Pumping
Rates (m³/day)

Well No.	HS	HS—Solver
1	20,000	20,000
2	20,000	20,000
3	4,513	4,924
4	20,000	20,000
5	20,000	20,000
6	20,000	20,000
7	3,043	3,166
8	3,150	4,121
9	519	1,632
10	0	0
11	12,663	12,807
12	7,507	6,392
13	9,415	9,263
14	9,454	9,368
15	20,000	20,000
16	19,843	20,000
17	2,771	2,806
Total Pumping	**192,878**	**194,478**

flow processes are simulated using MODFLOW as the flow modeling code. This code is then coupled to an optimization model where the hybrid HS—Solver optimization algorithm is used as the solution method. The performance of this procedure is illustrated with two examples with hypothetical and real-world model domains. The following conclusions can be drawn from the results of this chapter:

- Most of the local search algorithms require programming of some mathematical operations such as partial derivatives and Jacobian or Hessian matrices. However, HS—Solver does not require vast knowledge of programming local search methods, which is an important advantage of this method.
- The optimization procedure outlined in this chapter can solve some optimization problems within reasonable computation times. The applications given in this chapter are appropriate examples for this type of optimization problem. Larger flow domains, finer finite-difference grids, or the consideration of transient flow conditions result in having a higher computational burden.
- Typically, problems do not occur in HS—Solver when taking the partial derivatives of the objective function with respect to the decision variables, as it is the case for any other gradient-based algorithm. However, it is possible to face difficulties in taking the partial derivatives when working with different problem setups. This problem can be observed, for instance, if the decision variables are discrete as opposed to continuous. In such cases, the inclusion of Solver into a HS-based optimization model cannot improve the objective function value.

Acknowledgments

The work given in this chapter is supported by the Turkish Academy of Sciences (TÜBA)—the Young Scientists Award Programme (GEBIP). The first author thanks TÜBA for their support of this study.

References

Ayvaz, M.T., 2009. Application of harmony search algorithm to the solution of groundwater management models. Adv. Water Resour. 32 (6), 916−924.

Ayvaz, M.T., 2010. Solution of groundwater management problems using harmony search algorithm. In: Geem, Z.W. (Ed.), Recent Advances in Harmony Search Algorithm. Springer, Berlin/Heidelberg.

Ayvaz, M.T., Kayhan, A.H., Ceylan, H., Gurarslan, G., 2009. Hybridizing harmony search algorithm with a spreadsheet solver for solving continuous engineering optimization problems. Eng. Optim. 41 (12), 1119−1144.

Dorigo, M., Di Caro, G., 1999. Ant colony optimisation: a new meta-heuristic. In: Proceedings of the Congress on Evolutionary Computation, Vol. 2, pp. 1470−1477.

Elçi, A., Karadaş, D., Fı, O., 2010. The combined use of MODFLOW and precipitation-runoff modeling to simulate groundwater flow in a diffuse-pollution prone watershed. Water Sci. Tech. 62 (1), 180−188.

Fesanghary, M., Mahdavi, M., Minary-Jolandan, M., Alizadeh, Y., 2008. Hybridizing harmony search algorithm with sequential quadratic programming for engineering optimization problems. Comput. Meth. Appl. Mech. Eng. 197, 3080−3091.

Frontline Systems, 2011. Frontline Systems's web site. <http://www.solver.com/probtype5.htm> (accessed 12.01.2011).

Geem, Z.W., Kim, J.H., Loganathan, G.V., 2001. A new heuristic optimization algorithm: harmony search. Simulation. 76 (2), 60−68.

Glover, F., 1977. Heuristic for integer programming using surrogate constraints. Decision Sci. 8 (1), 156−166.

Goldberg, D.E., 1989. Genetic Algorithms in Search, Optimization, and Machine Learning. Addison-Wesley, New York.

Harbaugh, A.W., 2000. MODFLOW-2000, U.S.G.S. modular ground-water model user guide to modularization concepts and the ground-water flow process. U.S.G.S. Report, Reston, VA.

Holland, J.H., 1975. Adaptation in Natural and Artificial Systems: An Introductory Analysis with Applications to Biology, Control, and Artificial Intelligence. University of Michigan Press, Michigan.

Kennedy, J., Eberhart, R., 1995. Particle swarm optimization. In: Proceedings of the IEEE International Conference on Neural Networks, Piscataway, NJ.

Kirkpatrick, S., Gelatt, C., Vecchi, M., 1983. Optimization by simulated annealing. Science. 220 (4598), 671−680.

Lasdon, L.S., Waren, A.D., Jain, A., Ratner, M., 1978. Design and testing of a generalized reduced gradient code for nonlinear programming. Trans. Math. Software. 4 (1), 34−49.

McKinney, D.C., Lin, M.D., 1994. Genetic algorithm solution of groundwater management models. Water Resour. Res. 30 (6), 1897—1906.

Wang, M., Zheng, C., 1998. Ground water management optimization using genetic algorithms and simulated annealing: formulation and comparison. J. Am. Water Resour. Assoc. 34 (3), 519—530.

Willis, R.L., Yeh, W.W.G., 1987. Groundwater Systems Planning and Management. Prentice-Hall, Englewood Cliffs, NJ.

Wu, J., Zhu, X., 2006. Using the shuffled complex evolution global optimization method to solve groundwater management models. Lect. Notes Comput. Sci. 3841, 986—995.

Wu, J., Zhu, X., Liu, J., 1999. Using genetic algorithm based simulated annealing penalty function to solve groundwater management model. Sci. China Ser. E: Technol. Sci. 42 (5), 521—529.

Yang, X-S., 2009. Harmony search as a metaheuristic algorithm. In: Geem, Z.W. (Ed.), Music-Inspired Harmony Search Algorithm: Theory and Applications. Springer, Berlin/ Heidelberg.

5 Water Distribution Networks Designing by the Multiobjective Genetic Algorithm and Game Theory

Ali Nikjoofar and Mahdi Zarghami

Faculty of Civil Engineering, University of Tabriz, Iran

5.1 Introduction

Traditionally, optimization of water distribution networks (WDNs) was done based on the simulation as well as experience, but in recent decades, scientists have used mathematical programming methods to optimize the networks. The optimal design of WDN is the real problem of optimization that includes finding the best way to transfer the water from the sources to the users and satisfy their requirements. Many researchers have reported algorithms for minimizing the network cost by applying a large variety of techniques, such as linear programming, nonlinear programming, and global optimization methods. However, a reliable and efficient method has not been found yet. Most results of the researches are scientific and have been tested on small WDNs, but some of the methods have been tested on large-scale real networks (Banos et al., 2010). In this research, optimization of WDNs by biobjective genetic algorithm (GA) (minimizing cost and maximizing pressure) is done. There is a trade-off between the benefit and the pressure; thus by increasing the pressure, the benefit will be decreased. Classical optimization methods usually produce one solution, while GA analysis produces all the possible solutions and finally will present the best solution.

For example, Goulter and Bouchart (1990) optimized a WDN with two objectives, including maximizing of the reliability and minimizing the cost. They concluded that the WDN with the objective function of reliability is a complex problem, as there is no exact definition for reliability in the network. Also Simpson et al. (1994) have compared using GAs with other methods in optimization of water pipelines and pointed out the advantages of this algorithm. Savic and Walters (1997) have used GA for minimizing WDN cost. Todini (2000) presented a heuristic method considering cost function and resilience index, a reliability measure, as the objectives. This method solves for minimum cost networks, heuristically, by fixing a value of resilience index between 0 and 1. Walski (2001) stressed the need for the development of new models, which addresses not only the minimization of network cost but also the maximization of net benefits. In recent studies, Behzadian (2008) presents optimal methods for sampling in designing the WDN with multiobjective decision-making approaches and also the GA. Matijasevic et al. (2010) have used the MATLAB software to optimize a WDN. Wu et al. (2010) have done a study on greenhouse gas emissions and their effect on the optimization of water distribution systems, in which reducing costs and the effect of the greenhouse gases were considered.

There is more than one output solution of GA; therefore, the optimum point selection will be the next step. As this optimization problem has two objective functions, game theory has been used in this study to select the optimum solution. Lippai and Heaney (2000) have used the game theory to solve the water conflict problem in a WDN. Madani (2010) studied the applicability of game theory to water resource management and conflict resolution through a series of noncooperative water resource games. The chapter illustrates the dynamic structure of water resource problems and the importance of considering the game's evolution path while studying such problems. Wei et al. (2010) applied game theory—based models to analyze and solve water conflicts concerning water allocation and its quality in the Middle Route of the South-to-North Water Transfer Project in China. The game simulation comprised two levels, including one main game with five players and four subgames where every game contains three subplayers. Salazar et al. (2007) applied the game theory to a multiobjective conflict problem for the Alto Rio Lerma Irrigation District, located in the state of Guanajuato in Mexico, where economic benefits from agricultural production should be balanced with associated negative environmental impacts. The short period of rainfall in this area, combined with high groundwater withdrawals from irrigation wells, has produced severe aquifer overdraft. In addition, current agricultural practices of applying high loads of fertilizers and pesticides have contaminated regions of the aquifer. The net economic benefit to this agricultural region in the short term lies with increasing crop yields, which require large pumping extractions for irrigation as well as high chemical loading. In the longer term, this can produce economic loss due to higher pumping costs, or even loss of the aquifer as a viable source of water. Negative environmental impacts are continued diminishment of groundwater quality and declining groundwater levels in the basin, which can damage surface water systems that support environmental habitats.

The aim of this study is to optimize the WDN by using the biobjective GA based on the earlier studies. In this study, unlike previous studies, the Pareto frontier is derived to reveal the best solution and the most preferred point is selected by using game theory. The section 2 describes the biobjective optimization model. The case study is introduced in the section 3, and in the section 4, the different optimum points are obtained with the GA. Finally, the preferred point is determined with game theory. In fact, the innovation of this study is combining two main methods by using GA and game theory in a WDN optimization problem.

5.2   The Objectives of WDN Optimization

In recent decades, in most regions of the world, especially in hot and dry climates, water shortage crisis have caused major water resource problems. These problems influence the use of demand management techniques. Water shortage is one of the main factors that limit the development of economic activities in Iran. Therefore, balancing water supply and demand is the essential principle. Regarding urbanization and new construction of residential and industrial towns, especially in big cities, the design of water supply network requires a comprehensive management of water distribution in urban areas due to shortage of drinking water.

This chapter describes the cost and pressure management of a WDN. A biobjective optimization problem with minimizing the cost and balancing the pressure in the network is the result of this optimization. In WDN projects, the main part of the cost is related to the pipes that are used in construction of the network. Therefore, minimizing the pipe diameter while considering the allowable pressure range will be the main concern. High pressure increases water loss, and low pressure causes water to return to the network so that the pressure optimization will be important. Cost and pressure are the main components in designing a WDN; if these components are the objective functions, then the length and diameter of the pipes will be the decision variables of the problem.

In Iran, most of the WDNs are designed by Loop or EPANET (hydraulic simulation software). Modern software programs like WaterGEMS[1] or WaterCAD may have been used as well, to a lesser extent. Traditionally, the pressure and cost optimization have been done by the designer without using any special software. In this case, the experience and skill of the engineer are key factors for optimization, and different engineers will get different results. The optimization tool in WaterGEMS solves this problem.

5.3   The Hydraulic of WDN

In WDN, the numerous pipes generally join and transfer the water from the reservoir to the place of consumption. The reservoir, tank, pumps, control valves, and mechanical and electrical tools are the components of WDN.

For simplicity, they are generally described as follows:

- *Pipes*: Transfer water from one node to another.
- *Nodes*: The junction of the pipes and the water consumption points. In general, supplying normal water pressure for these points is important.
- *Reservoirs and tanks*: Used for water storage in the network. The hydraulic gradient and the primary conditions are defined for them.
- *Pump*: When the gravity transfer is not possible, the pump supplies the required energy to increase the water head.
- *Valves*: Have many functions, controlling flow, and regulating water pressure in the pipes.

---

[1] www.haestad.com

With the aforementioned components, modeling of the network will be possible, but it should be mentioned that for network simulation, the mass and energy conservation laws must be used. The distribution network can be simulated by two assumptions: the steady state and the extended period. Steady-state analyses determine the operating behavior of the system at a specific point of time or under steady-state conditions (flow rates and hydraulic grades remain constant over time). This type of analysis is suitable for determining pressures and flow rates under minimum, average, peak, or short-term effects on the system due to fire flows. For this type of analysis, the network equations are solved with tanks being treated as fixed grade boundaries. The results of this type of analysis are instantaneous and may or may not be representative of the values of the system a few hours (or even a few minutes) later. When the variation of the system attributes over time is important, an extended period simulation is needed. This type of analysis succeeds in modeling tank filling and draining, regulating valves opening and closing, and changing pressures and flow rates throughout the system in response to varying demand conditions and automatic control strategies formulated by the modeler.

5.3.1 The Energy Equation

The energy equation in WDN problems is composed of three components in the length dimension:

1. Pressure head (L): (P/γ))—P: pressure (pa), γ: water-specific gravity (pa/m)
2. Elevation (L): (Z)
3. Velocity head (L): $(V^2/2g)$—V: velocity.

Then, the energy equation between two nodes is represented as

$$\frac{P_1}{\gamma} + Z_1 + \frac{V_1^2}{2g} + h_p - h_t = \frac{P_2}{\gamma} + Z_2 + \frac{V_2^2}{2g} + h_l \qquad (5.1)$$

where

h_p: the added head with pump,
h_t: the consumable head,
h_l: headloss between two nodes.

In this equation, the pressure head and velocity head are inversely related, which means that when water velocity increases, the pressure head decreases. Also, the friction between water and pipe increases, so that the water velocity and headloss have a direct relation (Figure 5.1).

5.3.2 The Principle of Mass Conservation

According to mass conservation law, in a specified time period (Δt), the total input water flow to any pipe in a WDN is equal to the total output water flow,

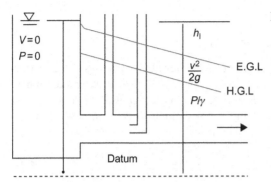

Figure 5.1 Water transfer pipeline.

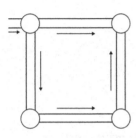

Figure 5.2 The paths between two nodes.

considering the total water losses along the way (Δvs). The definition of mass conservation law has been represented with the following equation:

$$\Sigma Q_{in}.\Delta t = \Sigma Q_{out}.\Delta t + \Delta vs \tag{5.2}$$

5.3.3 Energy Conservation Law

There will be different paths for transferring the water from a node to another; the water energy headloss will be similar between two nodes. This is the definition of the energy conservation law (Figure 5.2).

5.3.4 Water Headloss

Hazen−Williams, Manning, Chezy and Darcy−Weisbach are equations that are used to calculate water headloss. In this study, the Hazen−Williams headloss equation is used.

$$Q = K \cdot C \cdot A \cdot R^{0.63} \cdot S^{0.54} \tag{5.3}$$

where

Q: discharge in the section (m³/s, cfs)
C: Hazen−Williams coefficient
A: flow area (m², ft²)

R: hydraulic radius (m, ft)
S: friction slope (m/m, ft/ft)
K: constant (0.85 for SI units, 1.32 for US units).

The values for the roughness coefficient in the headloss equations are given in the literature for different pipe material. Also, if energy and hydraulic gradient lines descend in a section, minor energy losses occur. The magnitude of these losses depends primarily upon the shape of the fitting, which directly affects the flow lines in the pipe. The equation commonly used for minor losses is as follows:

$$h_m = k \, V^2/2g \tag{5.4}$$

where

h_m: minor losses (m)
V: velocity (m/s)
k: loss coefficient for the specific fitting
g: gravitational acceleration constant (m/s^2).

5.4   Basic Concepts: GA, Multiobjective Optimization, and Game Theory

During the past 10 years, more than 200 Ph.D. dissertations have been published on the studies of GA and applications in multidisciplinary areas. The GA was developed originally as a technique of function optimization derived from evolutionary ideas. In recent years, it has been found to be very useful for solving many problems in different fields. Combinatorial optimization problems are a major research aim in the GA community. GA has demonstrated great power, with very promising results for many difficult problems. Engineering design occupies a major body of research and applications of GA, e.g., topological structural design, network design, dynamic system design and integration, manufacturing cell design, pollution control, and reservoir operation. Since the early 1970s, when Holland (1962) first specified the GA, enormous effort has been devoted to theoretical investigation of GA to explain why and how it works. The foundational issues are coding and representation, variation, and recombination. The other items include the fitness land scopes and genetic operators, selection and convergence, parallelization, deception, genetic diversity, and parameter adaptation (Gen and Cheng, 2000). Today, GA is used as a searching and optimization tool in most commercial and engineering problems. The vast domain of applications and easy operation of this algorithm are some reasons for its success. Holland (1962) have been used this method for the first time, then the algorithm has been extended by him and his students.

Some of the basic ideas of genetics have been derived and then applied artificially in the GA. This algorithm is very powerful and needs only minimal information to solve problems. Darwin's natural theory method has been used for finding the optimization formula in this algorithm. Using GA as a forecasting technique on the basis of regression is also a good choice, and the nonparametric term could be used for this algorithm. As an evolutionary algorithm, it uses operators like choice, crossover, and mutation in solution collection in any generation.

5.4.1 Advantages of Using Evolutionary Algorithms

The classical optimization methods are separated into two categories: direct search methods and gradient-based methods. In the direct search method, optimization problems can be solved by using only objective function and constraints, while in the gradient-based method, the first and second derivative of objective function and constraints could be used. The first method takes too long to achieve the solution process; and the deficiency of the second method is in solving discontinuous functions. Also, in some classical methods, the convergence of the optimized solution depends on the initial selected solution and most of the algorithms prefer to find out the optimum solution. The efficiency of this algorithm regarding various optimization problems is different, and normally, for problems with discrete frontiers, misses their efficiency. The evolutionary algorithm does not use the differentiation, and it is categorized as the direct method and their application amplitude increases.

In this approach, unlike the classic optimization methods, which update one solution in each iteration (a point approach), lots of solutions are used, in any iteration. This point of view is called the *population approach*; the use of the population approach has a number of advantages as follows (Branke et al., 2008):

1. It provides an evolutionary optimization with a parallel processing control, achieving a computationally quick overall search.
2. It lets the evolutionary optimization to find multiple optimal solutions, thereby facilitating the solution of multimodal and multiobjective optimization problems.
3. It gives the evolutionary optimization with the ability to normalize decision variables (as well as objective and constraint functions) within an evolving population using the population-best minimum and maximum values.

However, the flip side of working with a population of solutions is the computational cost and memory associated with executing each iteration.

Also, the evolutionary optimization uses stochastic operators, while most classic methods use deterministic operators. Therefore, this algorithm has a global perspective in its search and good operation in multiobjective optimization problems. The operators in this method tend to solve problems using biased probability distributions to achieve desirable results, as opposed to using predetermined and fixed transition rules (Branke et al., 2008). Also, by definition, all the possible solutions will be accepted. But the evolutionary algorithm has a disadvantage. To find the answer with the evolutionary algorithm, an objective function must be used. Determination of the objective function is the most important part of the evolutionary algorithm. Having an unsuitable objective function results in mistakes, and the best solution is not found.

5.4.2 Biobjective Optimization

Optimization deals with the problem of identifying optimum solutions from a set of possible choices by satisfying certain criteria. If there is only one criterion to consider, it becomes a single-objective optimization problem, which has been studied in the past 50 years. If there is more than one objective function, it will become a multiobjective optimization problem. Multiobjective problems are faced in the design modeling and planning of many complex real-life systems in many areas like industrial production, urban transportation, capital budgeting, and reservoir management. Almost every important decision problem involves multiple and conflicting objectives that need to be considered while respecting various constraints, leading to overwhelming problem complexity. The GA has received considerable attention as a novel approach to multiobjective optimization problems, resulting in many research and applications known as genetic multiobjective optimizations (Gen and Cheng, 2000).

Optimization of a WDN has two subordinate objectives covering minimizing the cost and maximizing the pressure. As already mentioned, increasing the pressure requires pipes with large diameters; and in opposition, minimizing the cost requires pipes with small diameters. Therefore, in the optimization problem of selecting the preferred pipes, there are counteracting objectives. Actually, it is considered a biobjective problem and needs an algorithm with multiobjective solvability. The multiobjective GA is an optimization evolutionary algorithm, and it has the capability of solving complex, nonlinear problems. So in this study, this algorithm will be used as the optimization method. For starting the process, the objective function and constraints must be defined.

5.4.3 Biobjective GA

The multiobjective optimization problems have more than one objective function. Traditionally, due to lack of the appropriate method, they were solved as single-objective problems. All of the objective functions must be considered to find an optimized solution. Therefore, in biobjective optimization, the optimum point considers both of the objective functions. This solution may be just to supply the maximized pressure, or it may be the opposite. If the objective function determines all the solutions, then the results are drawn on the diagram. The resulting curve will be the Pareto frontier curve.

There is a striking difference between single-objective and multiobjective optimization. In multiobjective optimization, the objective functions and the usual decision variable spaces consist of multidimensional space. In single-objective optimization, the decision to accept or reject the solutions is based on the value of the objective function, and there is just one search space.

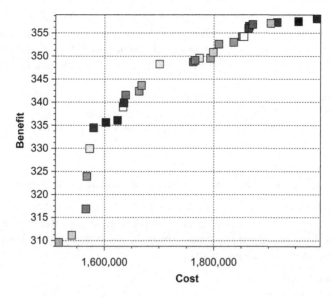

Figure 5.3 An example of a Pareto frontier.

Traditionally, for solving multiobjective problems, you convert the objective functions to one objective function. The designers used various methods to do this. For example, the weighing method creates a complex objective function. This problem is dispelled in multiobjective optimization. Some extra principles that have been used in multiobjective optimization have not been applied to single-objective optimization.

Using different methods in multiobjective problems creates a trade-off between different objectives. The final solution must be compatible with other objectives, and it is not correct to select an optimization point considering just one objective function. Multiobjective optimization follows two objectives. The first is finding the solution that determines the limits of the Pareto curve, and the second is finding the final optimized solutions that are on the Pareto frontier. Figure 5.3 represents an example of the Pareto optimization frontier.

5.4.4 Definition of Domination

In principle, multiobjective optimization problems are very different than single-objective optimization problems. In the single-objective case, one attempts to obtain the best solution, superior to all other alternatives. In the case of multiple objectives, there is not necessarily a solution that is better than all other objectives because of incommensurability and conflict among objectives. A solution may be best for one objective but worst for other objectives. Therefore, there usually exists a set of solutions for the multiobjective case, which cannot simply be compared to each other. In such solutions, called nondominated solutions or Pareto optimal solutions, no improvement in any objective function is possible without sacrificing at least one of the other objective functions. For a given nondominated point in the criterion space Z, its image point in the decision space S is called efficient or noninferior. A point in S is efficient if its image in Z is nondominated (Gen and Cheng, 2000).

Multiobjective algorithms have mostly been used to find domination. The domination between two solutions is defined as follows (Branke et al., 2008).

A solution p is said to dominate the other solution of q, if both of the following conditions are true:

1. In all objectives, the solution p is no worse than q.
2. The solution p is strictly better than solution q in at least one objective.

In domination relations, any solution could not be dominated by itself and the relation is not reflexive; it is asymmetric and antisymmetric. But it is a transitive relation, and if the p solution is dominant over the q solution and q is dominant over the r solution, so p is dominant over r:

($\blacktriangle$: *domination symbol*)
If $p \blacktriangle q$ *and* $q \blacktriangle r \rightarrow p \blacktriangle r$
and there is an important relation:
If $p \blacktriangle q \rightarrow q \blacktriangle p$ *or* $q \blacktriangledown p \rightarrow$ *then* q *will not necessarily dominate* p

5.4.5   Game Theory

Game theory is based on mathematics and is usually applied where selecting the preferred solution on the Pareto frontier is the problem. It studies the problems with different objective functions that are in a trade-off with each other. Outcomes predicted by game theory often differ from results suggested by optimization methods, which assumes that all parties are willing to achieve the best systemwide outcome. In a specific game, each individual player tries to maximize his or her benefit. This theory predicts the behavior of players and investigates the member's strategy to achieve better results.

Using game theory in multiobjective problems has several advantages, since it could consider different objective functions for finding the optimum point, but in classical optimization methods, they combine different objectives for defining only one objective function. Another advantage of game theory over traditional quantitative simulation and optimization methods is its ability to simulate different aspects of the conflict, incorporate various characteristics of the problem, and predict the possible resolutions in absence of quantitative payoff information (Madani, 2010).

5.5   Methodology

In this study, data of the WDN first are collected considering the principles of mass and energy balance and hydraulics rules, and are imported into the hydraulic simulation software. In this case, the optimization problem is solved with two objective functions: minimizing the cost and maximizing the optimum pressure. It means optimizing the pressure function considering the limits and standards. If the objective function is only to minimize the cost, then the smallest diameters will be chosen, and the water network will lose pressure. Therefore, a multiobjective problem is considered. In the first objective function, the stockholder is Water and Wastewater Company, and the construction cost must be paid by them; but in the second objective function, the stockholders are the consumers and they will pay the water tariffs, but they ask for appropriate pressure in the water network. The equations of the objective functions and constructs are as follows:

$$\text{Minimize } F_1 = \sum_{i=1}^{n}(C_i(D_i, L_i)) \tag{5.5}$$

$$\text{Maximize } F_2 = \sum_{k=1}^{N_D}\left[\sum_{i=1}^{N_j}\left(\frac{JQ_{i,k}}{JQ_{\text{total}k}}\right)\left(\frac{P_{ik}-P_{ik}^{\text{REF}}}{P_{ik}^{\text{REF}}}\right)^b\right] \tag{5.6}$$

subject to:

$$g_j(H, D) = 0, \quad j = 1, 2, \ldots, nn \tag{5.7}$$

$$H_j \geq H_j^{\text{min}}, \quad j = 1, 2, \ldots, nn \tag{5.8}$$

$$D_i \in \{A\}, \quad i = 1, 2, \ldots, np \tag{5.9}$$

where

$C_i (D_i, L_i)$: pipe cost with various diameters and lengths
N_D: number of design events
N_j: numbers of nodes
$JQ_{i,k}$: demand at junction i for demand alternative k
$JQ_{\text{total } k}$: total junction demand for demand alternative k
$P_{i,k}$: post-rehabilitation pressure at junction i for demand alternative k
$P_{i,k}^{\text{REF}}$: reference junction pressure is defined by the user to evaluate the pressure improvement. The reference pressure is taken as the minimum required for the junction.

Equations (5.5) and (5.6) are showing the cost and pressure objective functions. As shown in Eq. (5.6), this function is defined as the difference of real pressure and the minimum pressure is considered by the designer. The demand factor is used to normalize this difference. The b coefficient is the parameter indicating the importance of the difference. If $b = 1$, the importance of pressure increasing is not specified; however, it usually is selected as 0.5. In Eq. (5.7), the continuity equation (5.10) and energy conservation law (5.11) are applied to nodes. Equation (5.8), which refers to the water head, must be more than the specific value. Equation (5.9) refers to determining the pipe diameters, and it should use the series of the specific and commercial sizes.

$$\sum_{i=1}^{n} Q_{\text{in}} - \sum_{i=1}^{n} Q_{\text{out}} = 0 \tag{5.10}$$

where

Q_{in}: entrance discharge
Q_{out}: outlet discharge.

$$\sum_{i=1}^{n} \Delta H_i = 0 \tag{5.11}$$

where

n: number of loops
ΔH_i: energy conservation.

In Eq. (5.7), h_f refers to the headloss in pipes, and it is computed with the Hazen–Williams equation, which causes Eq. (5.7) to be nonlinear (Chu et al., 2008; Kapelan et al., 2005). E_p shows the amount of energy needed by pumps. This model is a complex and nonlinear optimization model that can be solved with an evolutionary algorithm, and in this model, the multiobjective GA has been used. Considering all of the constraints is the main limitation of the GA. To solve this problem, we will use WaterGEMS, hydraulic simulation software which considers the mass and energy conservation laws.

5.6   Case Study

The longitude and latitude of the case study (set in the city of Sahand, in Iran) are between E 46°30′, E 46°15′ and 37°53′, 37°59′. This area is situated southwest of Tabriz, and its elevation difference from its neighbors is more than 100 m. The average elevation of this area is about 1600 m above sea level (Figure 5.4).

The maximum water pressure in Iran's water networks has been recommended to be 5 atmospheres (atm). Since the area topography has limitations and it produces major costs, it is considered realistically to be between 5 and 7 atm. The minimum allowable water pressure for the first floor is 1.4 atm, and for each extra floor, 0.4 atm must be added. According to water network standards, the maximum allowable water velocity is 2 m/s, and during fires, 2.5 m/s is recommended. The minimum water velocity in pipes is assumed to be 0.3 m/s (Florescu et al., 2010).

In addition, regarding the allowable limits, shrinking the diameter in the network should not reduce the water head. The minimum allowable diameter in pipes with hydrants is 100 mm, and for those without hydrants, it is 60 mm. Also in this study, for the pipes of the WDN, several design groups by different diameters are determined. For example, the diameters in design group of the main pipeline are larger than 500 mm. The main pipelines transfer water from the reservoir to the entrance of the network. This process improves the operation of the GA for finding the optimum solution.

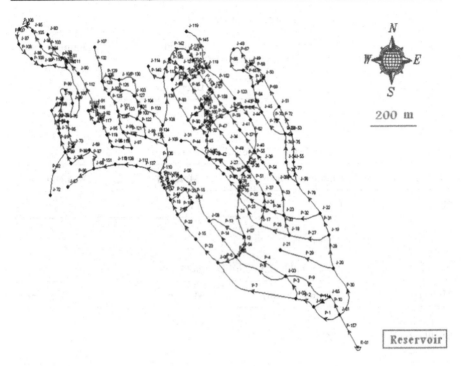

Figure 5.4 The water network in Sahand.

5.7 The Biobjective Optimization Problem

To define the problem, the constraints must be determined. The first constraint is the pressure limits of the nodes. For example, in Iran, it must not be less than 22 m (for the buildings with three floors) and more than 70 m. The second constraint is water velocity limits at pipes. In this study, the maximum velocity is proposed to be 3 m/s.

After determining the hydraulic constraints, the materials for the pipes of the network and the available diameters must be defined. In this study, the proposed material for the pipes with large diameters is ductile iron, and for the pipes with small diameters, it is polyethylene (PE). In this method, for any pipe, the design groups are defined with different sizes, diameters, and material of the pipes. In this study, for different pipes, three separate groups have been defined as in Tables 5.1−5.3.

Table 5.1 shows the design groups for the main pipeline that transfers water from the reservoir to the city. Table 5.2 refers to groups that have pipes with small diameters. Table 5.3 refers to the groups that cover the other pipes of WDN in this study. It must be noted that, because of the large thickness of PE pipes, the internal diameters of the pipes must be considered. After determining the designing groups

Table 5.1 First Design Group

Material	Diameter (mm)	Material	Diameter (mm)
Ductile iron	600	Ductile iron	450
Ductile iron	500	Ductile iron	400

Table 5.2 Second Design Group

Material	Diameter (mm)	Material	Diameter (mm)
PE	131	PE	90

Table 5.3 Third Design Group

Material	Diameter (mm)	Material	Diameter (mm)
Ductile iron	450	Ductile iron	350
Ductile iron	400	Ductile iron	300
Ductile iron	250	Ductile iron	200

and defining them for the network pipes, the GA parameters should be specified. In the results section, the GA parameters and defined ranges are mentioned.

Any solution provides the determined diameters for the pipes in the network. Considering the objectives, these solutions are different. The solutions that are positioned in the beginning and end of the Pareto curve are undesirable and would like to supply only one objective.

5.7.1 The Pareto Frontier

To obtain the Pareto frontier, objective function values must be calculated for all solutions. One of the axes is the cost function values; another is pressure function values. Then the calculated points will be located from a scatter diagram. The line that is fitted to the points will be the Pareto curve. In this study, the values of the objective function are calculated, and after normalization of numbers, the Pareto curve is drawn as shown in Figure 5.5.

To find the optimum solution using the usual game theory, the Pareto curve is better if it is convex. Therefore, both of the objective functions must be maximized, as in this study, and one of the objective functions is cost minimizing; thus, benefit maximizing is used to solve this problem.

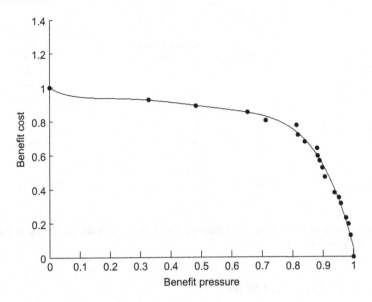

Figure 5.5 Pareto curve with normalized values.

5.7.2   Preparing the GA

In this study, GA was used to solve the biobjective problem. This algorithm tests all the available solutions and gives all the dominant solutions, while the classical optimization algorithms give only one solution. We also applied the classical GAMS software, and probably its result was not technically acceptable (nonoptimal). It must be mentioned that if GAs parameters are selected incorrectly, its operation will not be correct (Tabesh et al., 2011). The maximum era number and population size are the parameters with the most affection in the biobjective GA. By increasing these parameters, the number of solutions also will increase. Another parameter is the penalty factor. For the maximum era number, parameters starting from number 1 and number 2 also will be accepted for other alternatives. Considering the population size, the value will be increased. Also, the parameters of population size start from the number 150 and will increase to the next alternatives. These parameters and their values are shown in Table 5.4. Also, the stopping criterion for GAs is important, and it covers the maximum trials and nonimprovement generations. The first one determines the maximum number of trials before the GA stops. The value of that is the product of three parameters, including the numeric value of the maximum starting point, the nonimprovement numeric value, and the population size.

Whenever penalty factor is large, the GA concentrates to solutions that are in the defined limits. If this factor is less than 5000, then the results will be impractical solutions. The practical solution is the same as the applied solution. Next, convergence analysis must be done on the objective functions of GA. In this study, the

Table 5.4 The GAs Parameters

Parameter	Lower Bound	Upper Bound
Maximum era number	1	10
Population size	50	150
Era generation number	1	Larger than 1
Cut probability	1%	10%
Mutation probability	1%	10%
Splice probability	50%	90%
Random seed	0	1
Penalty factor	1000	5000

convergence analysis is done, and the results of this analysis show that the operation of GA is correct.

5.7.3 Convergence Test of GA

After the initial design, the biobjective optimization process will be done by the GA. But at first, the convergence test should be applied to the results of the biobjective GA. This test determines whether, with increasing the generation number, the resulted values of objective functions with multiobjective GA will be converged or not.

In this method, at any level, with an increased generation number parameter with different intervals, the function values are determined, and the convergence curve, related to the two functions, is drawn. The convergence curves are shown in Figures 5.6 and 5.7, and the results are shown in Table 5.5.

5.7.4 Curve Fitting and Selecting the Optimum Point

For selection of the optimum solution, drawing the fitting curve and finding the function are necessary. Usually, the calculated values of the objective function for various solutions are large numbers, so integration from this function will be difficult. When using normalized numbers, the problem must be solved. In this study, the values of the objective functions first are calculated and normalized. The result is plotted in Figure 5.8.

After importing the data to the hydraulic solver software and the initial design, the WDN is optimized by GA, and 20 of the optimization results are selected. After the normalization of the objective function values, the fitted Pareto curve (Eq. (5.12)) is fitted on the solutions (Figure 5.8). In this figure, the horizontal axis values are related to the pressure objective function (F_1) and vertical axis values are related to the cost objective function (F_2):

$$F_2 = -12.92F_1{}^5 + 26.31F_1{}^4 - 19.93F_1{}^3 + 6.633F_1{}^2 - 1.038F_1 + 1 \qquad (5.12)$$

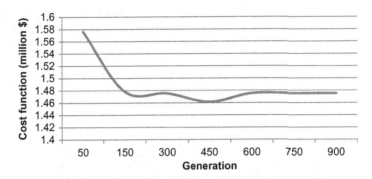

Figure 5.6 Convergence curve of cost function.

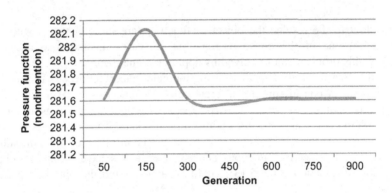

Figure 5.7 Convergence curve of pressure function.

Table 5.5 The Results of Biobjective Optimization and Game Theory

Weight of Pressure Function	Weight of Cost Function	x	$g(x)$	Answer	Cost Thousand (dollars)	Pressure Function	Solution Number
0.8	0.2	0.943	0.373	0.938	16,380	282.33	1
0.7	0.3	0.911	0.502	0.906	15,638	279.97	2
0.9	0.1	0.968	0.248	0.975	17,620	285.04	4
0.5	0.5	0.817	0.728	0.818	13,592	273.46	8
0.2	0.8	0.541	0.897	0.484	12,183	248.71	11
0.1	0.9	0.435	0.904	0.484	12,183	248.71	11
0.6	0.4	0.871	0.621	0.881	14,244	278.09	14
0.3	0.7	0.646	0.852	0.654	12,398	261.28	16
0.4	0.6	0.742	0.808	0.713	12,878	265.69	20

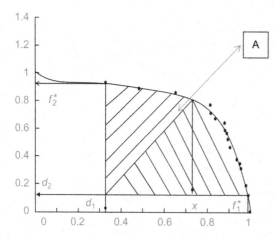

Figure 5.8 The fitted curve drawn on the solutions.

According to the results, the objective functions are in competition with each other, and if one of the objective functions increases, another one decreases. Therefore, this problem can be solved by applying game theory.

With the area monotonic method, using the game theory allows us to select the optimum point by considering the different objective functions together. In this section, the area monotonic method is used to select the optimum point as an innovation for solving the water network optimization problem. This method is based on the linear segment that divides the areas under the Pareto frontier into two subsets of equal area. If objective functions are asymmetric and have different weights, then nonlinear equation (Eq. (5.13)) could be used to find the optimum point as follows:

$$w_2\left[\int_{d_1}^{x} g(t)\mathrm{d}t - 1/2(x - d_1)(g(x) + d_2)\right]$$
$$= w_1\left[\int_{x}^{f_1^*} g(f)\mathrm{d}t - (f_1^* - x)d_2 + 1/2(x - d_1)(g(x) - d_2)\right] \tag{5.13}$$

where d_1, d_2 are the undesirable points and the f_1^*, f_2^* are the desirable points.

By determining the desirable and undesirable points in the objective functions, we can calculate the value of the optimum points for different weight percentages. The points transfer to the Pareto curve to determine the values of objective functions. The results of this process are represented in Table 5.5.

According to the results, when the weight coefficients of the objective function are not close together, the answers will be undesirable. For example, when the weight coefficient of the pressure objective function is 10%, the answer will be close to the minimum cost (or maximum benefit) and minimum pressure, but when the weight coefficient of benefits objective function is 10%, the answer will be close to maximum cost (minimum benefit) and maximum pressure (Nikjoofar et al., 2012).

By trial and error, we found that the best answers will result when the weight coefficients are either the same or close to each other. So the best selected answers

Table 5.6 The Numerical Values of Equation Variables of Eq. (5.14)

Variable	Numerical Value	Variable	Numerical Value
d_1	0.33	f_1^*	0.99
d_2	0.13	f_2^*	0.93

are 8, 20, and 14. Then, by using hydraulic simulation software, the selected diameters for pipes and the network will be solved again. According to the result, answer 14 is selected as the most preferred one. The final cost for this new city network is about 14,870,000,000 IR Rial, which is 13% less than the cost of the traditional network (Nikjoofar et al., 2012).

With the Kalai−Smorodinsky method, the Kalai−Smorodinsky solutions can be described as follows. Consider the linear segment between the disagreement point (d_1, d_2) and the "ideal" point (f_1^*, f_2^*). Then the solution is the unique intercept of this segment with the Pareto frontier. Hence, we have to compute the unique solution of Eq. (5.14) in interval (d_1, f_1^*). by solving Eq. (5.14):

$$d_2 + [(f_2^* - d_2)/(f_1^* - d_1)]\, (f_1 - d_1) - g(f_1) = 0 \qquad (5.14)$$

If both objectives are normalized, then $d_1 = d_2 = 0$ and f_1^* and $f_2^* = 1$. In the case of importance weights, the more important objective has to be improved more rapidly. This idea leads to the nonsymmetric Kalai−Smorodinsky solution, which computes the unique intercept between the Pareto frontier and the straight line where the two coordinate directions are the normalized objective functions (Salazar et al., 2007):

$$g(f_1) = (w_2/w_1)f_1 \qquad (5.15)$$

In this study, F_1 is the pressure objective function. The numerical values of the variables are presented in Table 5.6.

The resulting value for the F_1 function is 0.78, according to Table 5.5, and the selected optimum solutions are 14, 8, and 20. The area monotonic method shows that solutions 8 and 14 are closer to the optimum solution. In this study, w_1 and w_2 are weight coefficients and are assumed to be equal together. Determining the proper weight coefficient depends on the designer's viewpoint. If the weight coefficients are not equal, Eq. (5.15) could be used (Nikjoofar et al., 2012).

5.7.5 Discussion and Conclusion

Based on the convergence test, the proper operator for the GA is selected, and regarding the monotonic area and Kalai−Smorodinsky methods, solutions 8, 14, and 20 are determined to be the most preferred solutions. Then, the relevant diameters for these solutions must be used in hydraulic simulation software, and by this iteration, the final point will be determined.

Pressure is the most important function of the network; without supplying the requested pressure, other functions like cost minimizing could not be accepted. After simulating the three mentioned solutions, the pressure head must be controlled, and if the three solutions could supply the requested pressure of the network, the solution with the lower cost would be selected as the optimum solution. In addition to pressure, other factors, including the water rate and headloss in the pipes, must be controlled.

It must be mentioned that in both approaches of the game theory, the weight coefficients are the same. Therefore, solution 8 is selected as the optimum point. However, solution 14 has more cost, and solution 20 supplies lower pressure. Therefore, solution 8 is best.

In this research, a WDN as part of the city of Sahand is optimized. This model includes two objectives: minimizing the cost and maximizing the pressure. The simulation–optimization model of WaterGEMS, which benefits the GA, is used to design the network; and for selecting the final preferred point, game theory is used as an innovation. The results indicate a reduction of 13% in cost and a fair increase in the pressure. Then, for the new WDN design, using this evolutionary multiobjective model is recommended.

Acknowledgments

The assistance of Dr. M.T. Aalami and Mrs. Milani is kindly appreciated.

References

Banos, R., Gil, C., Reca, J., Montoya, F.G., 2010. A memetic algorithm applied to the design of water distribution networks. J. Appl. Soft Comput. 10 (1), 261–266.

Behzadian, K., 2008. Development of optimal sampling design methods for calibration of water distribution networks by using multi-criteria decision making, Ph.D. thesis, Amirkabir University of Technology.

Branke, G., Deb, K., Miettinen, K., Słowinski, R., 2008. Multiobjective Optimization. Springer-Verlag, Berlin Heidelberg.

Chu, Ch.W., Lin, M.D., Liu, G.F., Sung, Y.H., 2008. Application of immune algorithms on solving minimum-cost problem of water distribution network. Math. Comput. Model. 48, 1888–1900.

Florescu, C., Mirel, I., Carabet, A., Pode, V., 2010. Modelling flow processes in urban distribution networks. Rev. Chim. 61 (11), 1125–1129.

Gen, M., Cheng, R., 2000. Genetic Algorithms and Engineering Optimization. Ashikaga Institute of Technology, Japan.

Goulter, I.C., Bouchart, F., 1990. Reliability constrained pipe networks model. J. Hydraulics Eng. 116 (2), 211–229.

Holland, J.H., 1962. Concerning efficient adaptive systems. In: Yovits, M.C., Jacobi, G.T., Goldstein, G.B. (Eds.), Self-Organizing Systems. Spartan Press, New York, pp 215−230.

Kapelan, Z.S., Savic, D.A., Walters, G.A., 2005. Multiobjective design of water distribution systems under uncertainty. Water Resour. Res. 41, W11407.

Lippai, I., Heaney, J.P., 2000. Efficient and equitable impact fees for urban water systems. J. Water Resour. Planning Manage. 126 (2), 75−84.

Madani, K., 2010. Game theory and water resources. J. Hydrol. 381, 225−238.

Matijasevic, L., Dejanovic, I., Spoja, D., 2010. A water network optimization using MATLAB—A case study. Resour., Conserv. Recycl. 54 (12), 1362−1367.

Nikjoofar, A., Zarghami, M., Aalami, M., 2012. Bi-objective optimization of the water distribution networks (Case Study of Sahand City, Iran). J. Water Wastewater. (in press, in Persian).

Salazar, R., Szidarovszky, F., Coppola, E., Rojano, A., 2007. Application of game theory for a groundwater conflict in Mexico. J. Environ. Manage. 84 (2007), 560−571.

Savic, D., Walters, G., 1997. Genetic algorithms for least cost design of water distribution networks. J. Water Resour. Planning Manage. 123 (2), 67−77.

Simpson, A.R., Dandy, G.C., Murphy, L.J., 1994. Genetic algorithms compared to other techniques for pipe optimization. J Water Resour. Planning Manage. 120 (4), 423−443.

Tabesh, M., Azadi, B., Rouzbahani, A., 2011. Optimization of chlorine injection dosage in water distribution networks using a genetic algorithm. J. Water Wastewater. 22 (1), 2−11 (in Persian).

Todini, E., 2000. Looped water distribution networks design using a resilience index based heuristic approach. Urban Water. 2 (3), 115−122.

Walski, T.M., 2001. The wrong paradigm—Why water distribution optimization doesn't work. J. Water Resour. Planning Manage. 127 (4), 203−205.

Wei, Sh., Yang, H., Wei, Sh., Abbaspour, K., Mousavi, J., Gnauck, A., 2010. Game theory based models to analyze water conflicts in the middle route of the south-to-north water transfer project in China. J. Water Res. 44, 2499−2516.

Wu, W., Simpson, A., Maier, H., 2010. Accounting for greenhouse gas emissions in multiobjective genetic algorithm optimization of water distribution systems. J. Water Resour. Planning Manage. 136 (2), 146−155.

6 Ant Colony Optimization for Estimating Parameters of Flood Frequency Distributions

Siamak Talatahari[1], Vijay P. Singh[2] and Yousef Hassanzadeh[3]

[1]Marand Faculty of Engineering, University of Tabriz, Tabriz, Iran,
[2]Department of Biological and Agricultural Engineering and Department of Civil and Environmental Engineering, Texas A&M University, College Station, TX,
[3]Department of Civil Engineering, University of Tabriz, Tabriz, Iran

6.1 Introduction

Relating the magnitudes of extreme events to their frequencies of occurrence with the use of probability distributions constitutes the main objective of frequency analysis (Chow et al., 1988). Hydrologic frequency analysis entails selecting a probability distribution, estimating distribution parameters and fitting the distribution to hydrologic data, selecting the most appropriate distribution function based on goodness of fit tests, and determining desired quantiles from the selected frequency distribution. Therefore, the type of statistical distribution and parameter estimation technique will affect the accuracy of hydrologic frequency analysis. Although there are many probability distributions to choose among, selecting a suitable model is still one of the major problems in frequency analysis. In addition, there are several methods of parameter estimation, among which the most popular are the methods of maximum likelihood, moments, and probability-weighted moments (PWMs) (Hassanzadeh et al., 2011).

There are two common sources of errors associated with quantile estimation. The first source of errors results from the selection of a distribution from among different ones or the inability to determine the real unknown distribution. The second one is the error in parameter estimation when using sample data. There is a possibility of error in the sample data, and then the method of fitting should minimize these errors (Kite, 1977).

The optimization methods are generally divided into two groups: classical methods and metaheuristic approaches. Classical optimization methods are often based on mathematical programming and have computational drawbacks (Kaveh and Talatahari, 2009, 2011a). The metaheuristic search techniques often avoid these problems by using the ideas inspired from nature.

Metaheuristics in Water, Geotechnical and Transport Engineering. DOI: http://dx.doi.org/10.1016/B978-0-12-398296-4.00006-4

The genetic algorithm (GA) (Goldberg, 1989), the particle swarm optimizer (Eberhart and Kennedy, 1995), ant colony optimization (ACO) (Dorigo et al., 1996), the imperial competitive algorithm (Atashpaz-Gargari and Lucas, 2007; Kaveh and Talatahari, 2010a; Talatahari et al., 2012), charged system search (CSS) (Kaveh and Talatahari, 2010b,c,d, 2011b,c, 2012), Big Bang-Bug Crunch (BB-BC) (Erol and Eksin, 2006), the firefly algorithm (Gandomi et al., 2011), Cuckoo Search (Gandomi et al., 2012), and the bat algorithm (Yang and Gandomi, 2012) are some of the most familiar metaheuristic methods. These methods are applied to optimization problems because of their high potential for modeling engineering problems in environments that have not been solvable by classic techniques.

ACO is a discrete metaheuristic approach that has been applied in recent years to different engineering optimization problems. It is a multiagent randomized search technique in which a number of search space points are tested in each cycle. The random selection and the information obtained in each cycle are used to choose new points in subsequent cycles. Thus, in ACO, it is not necessary for a given function to be differentiable (Kaveh et al., 2008).

In this chapter, an improved algorithm based on ACO is developed for estimating the parameters of commonly used flood frequency distributions. Results are compared with some conventional methods using annual maximum discharge data of 14 rivers from East Azerbaijan, Iran.

6.2 A Review of Previous Work

There have been a number of investigations on parameter estimation using conventional approaches. Singh (1998) compared for a large number of probability distributions with different methods of parameter estimations, including the method of moments (MOM), PWMs, L-moments, maximum likelihood estimation (MML), and entropy method. He found that no method was uniformly superior to other methods. The superiority of a method depended on the goodness of fit, such as bias, root-mean-square error (RMSE), or robustness, as well as the sample size. Nevertheless, in most cases, the entropy and MML methods were better. The entropy-based method with the MOM, L-moments, and MML estimation were compared by Singh and Deng (2003) using four data sets of annual maximum rainfall and annual peak flow discharge. Results demonstrated that both the entropy and L-moments methods enabled the four-parameter Kappa (KAP) distribution to fit the data well. The generalized PWM, generalized moments, and MML estimation methods were applied by Mahdi and Ashkar (2004) to estimate parameters of the Weibull distribution and the MML produced better results than the other two. Öztekin (2005) estimated parameters of the three-parameter generalized Pareto (GPAR) distribution for observed annual maximum discharge data for 50 rivers, most of them in Turkey, by using the MOM, PWM, MML, principle of maximum entropy, and least squares. It is concluded that the MOM was superior to all other methods employed. Ashkar and Tatsambon (2007) proposed methods of MML,

MOM, PWM, and generalized PWMs for fitting the GPAR distribution. The 42 low-flow events for hydrometric station on the Fish River in New Brunswick, Canada, were used and the MML method was seen to provide the best fit for that particular data set. Hassanzadeh et al. (2008) investigated six distributions, including generalized extreme value (GEV), Pearson type 3 (PE3), generalized logistic (GLOG), GPAR, normal, and exponential, for 14 data series of annual maximum discharges. They showed that the MML method was the best parameter estimation method.

As an alternative to conventional methods, the GA has been known to be a useful tool in solving optimization engineering problems. For example, Karahan et al. (2007) developed and used a GA to predict the rainfall intensities for given return periods. For the known problem formulation, GA found the solution. Results showed that the proposed GA can be used to solve the rainfall intensity−duration−frequency relations with the lowest mean-squared error between measured and predicted intensities. They concluded that predicted intensities were in good agreement with measured values for given return periods. Optimizing the parameters of the instantaneous unit hydrograph was performed by Dong (2008) using GA and an approximate formula method, as well as the moment method. Results showed that GA was more effective than the other two methods. Optimization of looped water distribution systems by several metaheuristic techniques containing GA, simulated annealing, tabu search, and iterative local search was performed in Reca et al. (2008). The medium-sized benchmark networks, as well as a large irrigation water distribution network, were used as numerical examples. Results showed that GA was more efficient when dealing with a medium-sized network, but other methods outperformed it when dealing with a real complex one.

Recently, Rai et al. (2009) employed GA to derive the unit hydrograph (UH). Nine different distributions, such as beta, exponential, gamma, normal, lognormal, Weibull, logistic, GLOG, and PE3, were used for the determination of UH. Parameters of nine distribution functions were estimated using the real-coded GA optimization technique, and the distributions were tested on 13 watersheds of different characteristics. It was observed that except for the exponential distribution function, most distribution functions produced UHs that were in satisfactory agreement with the observed UHs. Also, Reddy and Adarsh (2010) used two metaheuristic search algorithms containing GAs and the PSO to obtain optimal solutions to the design of irrigation channels.

Similarly, ACO has been applied to different engineering problems. Water distribution system optimization (Maier et al., 2003), optimal design of open channels (Kaveh and Talatahari, 2010e; Nourani et al., 2009), optimization of soil hydraulic parameters (Abbaspour et al., 2001), identifying optimal sampling networks that minimize the number of monitoring locations in groundwater design optimization (Li and Chan, 2006), and determining the optimum design of skeletal structures containing optimum weights (Kaveh and Talatahari, 2010f) are some recent examples. The advantages of applying ACO to engineering problems are similar to those of GA. Both are multiagent randomized search techniques in which a number of search space points are selected and tested in each cycle. The random selection and the information obtained in each cycle are used to choose new points (design vectors) in subsequent cycles. ACO and GA, contrary to many conventional algorithms, are global optimizers.

ACO was used to derive operating policies for a multipurpose reservoir system by Kumar and Reddy (2006). To formulate the ACO for reservoir operation, they modified the problem by considering a finite time series of inflows, classifying the reservoir volume into several class intervals, and determining the reservoir release for each period with respect to a predefined optimality criterion. ACO was then applied to a case study of Hirakud reservoir, which is a multipurpose reservoir system located in India. Results of the two models indicated good performance of the ACO model in terms of higher annual power production, satisfying irrigation demands, and flood control restrictions. Applying ACO to estimate parameters of flood frequency distributions was performed by Hassanzadeh et al. (2011). They used the GA and ACO methodologies to estimate parameters of flood frequency distributions and compared them with conventional methods.

On the other hand, since the ACO algorithms generally target discrete optimization problems, there have been a few adaptations of ACO to continuous space function optimization problems until now. One of the first attempts to apply an ant-related algorithm to the continuous optimization problems was the continuous ACO (CACO; Bilchev and Parmee, 1995). Although the authors of CACO claim that they draw inspiration from the original ACO formulation, CACO employs the notion of nest, which does not exist in the ACO approach. Also, CACO does not perform an incremental construction of solutions, which is one of the main characteristics of ACO. Another ant-related approach to continuous optimization is the continuous interacting ant colony (CIAC; Dreo and Siarry, 2002). CIAC uses two types of communication between ants: indirect communication (spots of pheromone deposited in the search space) and direct communication. CIAC has many differences with the original concept of ACO. There is a direct communication between ants and no incremental construction of solutions. The other ant-based approach is ACO_R, introduced by Socha and Dorigo (2008). ACO_R tries to use all operators of the original ACO, but in each construction step, an ant chooses a value for variables using the Gaussian kernel probability density function (PDF) composed of a number of regular Gaussian functions that does not exist in the original ACO. In ACO_R, the pheromone information is stored as a solution archive, and pheromone update is accomplished by adding the set of newly generated solutions to the solution archive and removing the same number of worst solutions while in the original ACO, pheromone matrix contains the information of all possible states that an ant can select, and pheromone updating is done in a different manner. Kaveh and Talatahari (2010e) introduced an improved CACO to optimize the overall reliability and cost-effectiveness of composite channels. The models were developed to minimize the total cost, while satisfying the specified probability of channel capacity being greater than the design flow. Also, Madadgar and Afshar (2009) proposed an improved CACO to water resource problems. Results of a few well-known benchmark problems and a real-world water resource problem emphasized the robustness of the ACO in searching the continuous space more efficiently.

6.3   Standard ACO

6.3.1   General Aspects

In 1992, Dorigo developed a paradigm known as ACO, a cooperative search technique that mimics the foraging behavior of ant colonies (Dorigo, 1992; Dorigo et al., 1996). The ant algorithms mimic the techniques employed by real ants to rapidly establish the shortest route from food source to their nest and vice versa. Ants start searching the area surrounding their nest in a random manner. Ethologists observed that ants can construct the shortest path from their colony to the feed source and back using pheromone trails (Deneubourg and Goss, 1989; Goos et al., 1990), as shown in Figure 6.1A. When ants encounter an obstacle (Figure 6.1B), at first, there is an equal probability for all ants to move right or left, but after a while (Figure 6.1C), the number of ants choosing the shorter path increases because of the increase in the amount of pheromone on that path. With the increase in the number of ants and pheromone on the shorter path, all of the ants will choose and move along the shorter one, as shown in Figure 6.1D.

In fact, real ants use their pheromone trails as a medium for communication of information among them. When an isolated ant comes across some food source in its random sojourn, it deposits a quantity of pheromone on that location. Other randomly moving ants in the neighborhood can detect this marked pheromone trail. Further, they follow this trail with a very high degree of probability and simultaneously enhance the trail by depositing their own pheromone. More and more ants follow the pheromone-rich trail, and the probability of the trail being followed by other ants is further enhanced by the increased trail deposition. This is an autocatalytic (positive feedback) process, which favors the path along which more ants previously traversed. The ant algorithms are based on the indirect communication capabilities of ants. In the ACO algorithms, virtual ants are deputed to generate

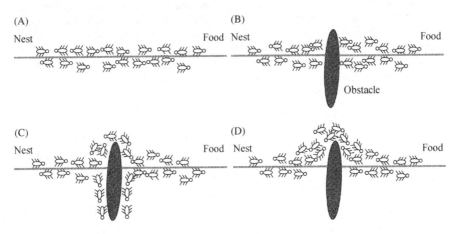

Figure 6.1 Ants finding the shortest path around an obstacle. (A) Ants from nest to food, (B) ants encounter an obstacle, (C) ants choose all paths, and (D) ants choose the shortest path.

rules by using heuristic information or visibility and the principle of indirect pheromone communication capabilities for the iterative improvement of rules.

ACO was initially used to solve the traveling salesman problem (TSP). The aim
of TSP is to find the shortest Hamiltonian graph, $G = (N,E)$, where N denotes the
set of nodes and E is the set of edges. The general procedure of the ACO algorithm
manages the scheduling of three steps (Dorigo and Caro, 1999):

Step 1. Initialization. The initialization of the ACO includes two parts: the first consists
mainly of the initialization of the pheromone trail. Second, a number of ants are arbi-
trarily placed on the nodes chosen randomly. Then, each of the distributed ants will perform a tour on the graph by constructing a path according to the node transition rule
described next.

Step 2. Solution construction. In the iteration, each ant constructs a complete solution to
the problem according to a probabilistic state transition rule. The state transition rule
depends mainly on the state of the pheromone and visibility of the ants. Visibility is an
additional ability used to make this method more efficient. For the path between i to j, it
is represented as η_{ij} and in TSP, it has a reverse relation with the distance between i to j.
The node transition rule is probabilistic. The ant decision table is obtained by combining
the visibility and pheromone trails as

$$a_{ij}(t) = \frac{[\tau_{ij}(t)]^{\alpha} \cdot [\eta_{ij}]^{\beta}}{\sum_{l \in N_i}[\tau_{il}(t)]^{\alpha} \cdot [\eta_{il}]^{\beta}} \quad \forall j \in N_i \qquad (6.1)$$

Here, the ant is in the city i, $\tau_{ij}(t)$ is the amount of pheromone in the i to j path, N_i is the
set of neighboring cities from city i, and parameters α and β represent constants that
determine the relative influence of pheromone and visibility, respectively. For the kth ant
on node i, the selection of the next node j to follow is according to the node transition
probability:

$$P_{ij}^k(t) = \frac{a_{ij}(t)}{\sum_{l \in N_i^k} a_{il}(t)} \quad \forall j \in N_i^k \qquad (6.2)$$

where $N_i^k \in N_i$ is the list of neighboring nodes from node i available to ant k at time t.

Step 3. Pheromone updating rule. When every ant has constructed a solution, the intensity of pheromone trails on each edge is updated by the pheromone updating rule (global
pheromone updating rule). The global pheromone updating rule is applied in two phases.
First, an evaporation phase where a fraction of the pheromone evaporates, and then a
reinforcement phase where the elitist ant, which has the best solution among others,
deposits an amount of pheromone:

$$\tau_{ij}(t + d) = (1 - \rho) \tau_{ij}(t) + \rho \, \Delta\tau_{ij}^{+} \qquad (6.3)$$

where ρ $(0 < \rho < 1)$ represents the persistence of pheromone trails $((1 - \rho)$ is the
evaporation rate), d is the number of variables or movements an ant must take to
complete a tour, and $\Delta\tau_{ij}^{+}$ is the amount of pheromone increase for the elitist ant
and equals:

$$\Delta\tau_{ij}^{+} = \frac{1}{L^{+}} \qquad (6.4)$$

where L^{+} is the length of the solution found by the elitist ant.

At the end of each movement, the local pheromone update reduces the level of the pheromone trail on paths selected by the ant colony during the preceding iteration. When an ant travels to node j from node i, the local update rule adjusts the intensity of pheromone on the path connecting these two nodes as follows:

$$\tau_{ij}(t+1) = \xi \cdot \tau_{ij}(t) \tag{6.5}$$

where ξ is an adjustable parameter between 0 and 1 representing the persistence of the pheromone.

This process is iterated until a stopping criterion is met.

6.3.2 Implementation for Solving Engineering Optimization Problems

In order to use the ACO method for design of engineering problems, the method explained in the previous section must be modified. Since ACO is a discrete optimization method, discrete values for each design variable (x_i) should be defined, and any amount of this discrete value for each variable is considered as a virtual path for the ants. In order to fulfill this goal, the permitted range and accuracy for the variables are determined. Unlike TSP, which had only one path between two nodes, in engineering optimization problems, the number of virtual paths between two nodes equals the number of allowable values. The target of the optimization of an engineering problem is to find the best path among all available virtual paths.

The length of each path equals a permissible amount of a variable:

$$x_{i,j} = x_{i,\min} + (j-1)x_i^* \quad \begin{cases} i = 1, 2, \ldots, d \\ j = 1, 2, \ldots, \mathrm{nm}_i \end{cases} \tag{6.6}$$

where x_i^* is the accuracy rate of the ith design variable, j is the number of virtual path from 1 to nm_i, and nm_i is the maximum number of virtual paths for the ith variable selected considering the required accuracy for the solved problem.

The amount of visibility for each path is

$$\eta_{ij} = \frac{1}{x_{i,j}} \quad \begin{cases} i = 1, 2, \ldots, d \\ j = 1, 2, \ldots, \mathrm{nm}_i \end{cases} \tag{6.7}$$

For each variable, a vector called the *pheromone vector* developed to record the amount of pheromone trails upon each path and is defined as $T_i(t) = [\tau_{ij}(t)]$. The initial amount of pheromone for all paths can be written as

$$\tau_{ij}(0) = \frac{1}{f_{\text{cost}}(\{x_{i,\min}\})} \tag{6.8}$$

where $f_{\text{cost}}(\{x_{i,\min}\})$ is obtained by setting the minimum values for the variables in the cost function.

For engineering problems, the ant decision table and probability function are the same as (Kaveh et al., 2008):

$$P_{ij}^k(t) = \frac{a_{ij}(t)}{\sum_{l\in N_i^k} a_{il}(t)} = \frac{\left([\tau_{ij}(t)]^{\alpha}\cdot[\eta_{ij}]^{\beta}/\sum_{l\in N_i}[\tau_{il}(t)]^{\alpha}\cdot[\eta_{il}]^{\beta}\right)}{\sum_{l\in N_i^k}\left([\tau_{ij}(t)]^{\alpha}\cdot[\eta_{ij}]^{\beta}/\sum_{ll\in N_i}^{n}[\tau_{ill}(t)]^{\alpha}\cdot[\eta_{ill}]^{\beta}\right)}$$

$$= \frac{\left([\tau_{ij}(t)]^{\alpha}\cdot[\eta_{ij}]^{\beta}/\sum_{l\in N_i}[\tau_{il}(t)]^{\alpha}\cdot[\eta_{il}]^{\beta}\right)}{\left(\sum_{l\in N_i^k}[\tau_{il}(t)]^{\alpha}\cdot[\eta_{il}]^{\beta}/\sum_{ll\in N_i}[\tau_{ill}(t)]^{\alpha}\cdot[\eta_{ill}]^{\beta}\right)}$$

$$= \frac{[\tau_{ij}(t)]^{\alpha}\cdot[\eta_{ij}]^{\beta}}{\sum_{l\in N_i^k}[\tau_{il}(t)]^{\alpha}\cdot[\eta_{il}]^{\beta}} = \frac{[\tau_{ij}(t)]^{\alpha}\cdot[\eta_{ij}]^{\beta}}{\sum_{l\in N_i}[\tau_{il}(t)]^{\alpha}\cdot[\eta_{il}]^{\beta}} = a_{ij}(t) \qquad (6.9)$$

Therefore, it is enough to calculate the ant decision table. As the denominator of the decision table is constant for each member in a cycle, it can be concluded that

$$a_{ij}(t) = [\tau_{ij}(t)]^{\alpha}\cdot[\eta_{ij}]^{\beta} \qquad (6.10)$$

The position of ants at the start of each cycle is expressed by the scatter vector. In this vector, subscript i is an integer and random digit between 1 and the number of variables that shows the first location of each ant. Using Eq. (6.10), the transition vector $[P_{ij}^k(t)]_{nm_i}$ determines the movements of ants at time t, and each ant selects a path for the first location (variable) and then moves to the next location. This movement is achieved by considering the number of variables. It means that when an ant is located in the ith variable, the next location is $i + 1$, and when it is in the last variable, the next location is the first variable. This process is continued until all ants select a value for each variable.

Equations (6.3) and (6.5) are used for the pheromone updating in design problems. Since the shortest Hamiltonian graph in TSP is analogous with the minimum cost function in the engineering optimization problems, in Eq. (6.4), the minimum amount of cost function in the kth iteration $[\min(f_{cost}^k)]$ may be used instead of the shortest length of the graph as

$$\Delta\tau_{ij}^+ = \frac{1}{\min(f_{cost}^k)} \qquad (6.11)$$

6.4 Improved ACO

6.4.1 Suboptimization Mechanism Added to ACO

Here, we add the suboptimization mechanism (SOM) to ACO. SOM is based on the principles of the finite element method (Kaveh and Talatahari, 2010e).

SOM divides the search space into subdomains and performs optimization into these patches, and then, based on the resulting solutions, the undesirable parts are deleted, and the remaining space is divided into smaller parts for more investigation in the next stage. This process continues until the remaining space gets smaller than the required accuracy.

SOM can be supposed to be the repetition of the following steps for definite times, nc (in stage k of the repetition) (Kaveh and Talatahari, 2010e):

Step 1. Calculate permissible bounds for each variable. If $x_i^{(k-1)}$ is the solution obtained from the previous stage $(k-1)$ for the ith variable, then:

$$
\begin{cases}
\text{If} & x_i^{(k-1)} < (1 - \alpha_1) \cdot x_{i,\min}^{(k-1)} + \alpha_1 \cdot x_{i,\max}^{(k-1)} \Rightarrow \begin{cases} x_{i,\min}^{(k)} = x_{i,\min}^{(k-1)} \\ x_{i,\max}^{(k)} = x_{i,\min}^{(k-1)} + 2 \cdot \alpha_1 \cdot (x_{i,\max}^{(k-1)} - x_{i,\min}^{(k-1)}) \end{cases} \\[1em]
\text{If} & x_i^{(k-1)} > \alpha_1 \cdot x_{i,\min}^{(k-1)} + (1 - \alpha_1) \cdot x_{i,\max}^{(k-1)} \Rightarrow \begin{cases} x_{i,\min}^{(k)} = x_{i,\max}^{(k-1)} - 2 \cdot \alpha_1 \cdot (x_{i,\max}^{(k-1)} - x_{i,\min}^{(k-1)}) \\ x_{i,\max}^{(k)} = x_{i,\max}^{(k-1)} \end{cases} \\[1em]
\text{Else} & \Rightarrow \begin{cases} x_{i,\min}^{(k)} = x_i^{(k-1)} - \alpha_1 \cdot (x_{i,\max}^{(k-1)} - x_{i,\min}^{(k-1)}) \\ x_{i,\max}^{(k)} = x_i^{(k-1)} + \alpha_1 \cdot (x_{i,\max}^{(k-1)} - x_{i,\min}^{(k-1)}) \end{cases}
\end{cases}
$$

(6.12)

where i = 1, 2,..., d; k = 2, ..., nc, α_1 is an adjustable factor that determines the amount of the remaining search space, nc is the maximum number of repetitious stages for SOM, and $x_{i,\min}^{(k)}$ and $x_{i,\max}^{(k)}$ are the minimum and the maximum allowable values for the ith variable at stage k, respectively. In stage 1, the amounts of $x_{i,\min}^{(1)}$ and $x_{i,\max}^{(1)}$ are set at

$$ x_{i,\min}^{(1)} = x_{i,\min}, \quad x_{i,\max}^{(1)} = x_{i,\max} \quad i = 1, 2, ..., d $$

(6.13)

Step 2. Determine the accuracy for the variables. In each stage, the number of permissible values for each variable is considered to be α_2, and therefore, the amount of the accuracy rate of each variable equals

$$ x_i^{*(k)} = \frac{(x_{i,\max}^{(k)} - x_{i,\min}^{(k)})}{(\alpha_2 - 1)} \quad i = 1, 2, ..., d $$

(6.14)

where $x_i^{*(k)}$ is the amount of increase in the ith variable, and α_2 is the number of subdomains considered instead of nm_i, and it has less value than nm_i in SOM.

Step 3. Create the series of allowable values for the variables. The set of allowable values for variable i can be defined by using Eqs. (6.12) and (6.14) as

$$ x_{i,\min}^{(k)}, \quad x_{i,\min}^{(k)} + x_i^{*(k)}, \quad ..., x_{i,\min}^{(k)} + (\alpha_2 - 1) x_i^{*(k)} = x_{i,\max}^{(k)} \quad i = 1, 2, ..., d $$

(6.15)

Step 4. Determine the optimum solution of the current stage. The last step is performing an optimization process using the ACO algorithm when Eq. (6.15) is considered as permissible values for the variables.

SOM ends when the amount of accuracy rate of the last stage (i.e., $x_i^{*(\text{nc})}$) is less than the amount of accuracy rate of the primary problem (i.e., x_i^*):

$$x_i^{*(\text{nc})} \leq x_i^* \quad i = 1, 2, \ldots, d \tag{6.16}$$

SOM improves the search process with updating the search space from one stage to the next stage. By applying this mechanism, the size of pheromone vectors and decision vectors decreases from nm_i to α_2. The search space reduces from $\Pi_{i=1}^d \text{nm}_i$ to $\alpha_2^d \times \text{nc}$. As an example, if $\text{nm}_i = 10^4$, $d = 4$, $\alpha_2 = 30$, and $\text{nc} = 16$, then $\Pi_{i=1}^d \text{nm}_i = 10^{16}$ while $\alpha_2^d \times \text{nc} = 1.3 \times 10^7$ (Kaveh and Talatahari, 2010e).

6.4.2 Parameter Setting

The standard ACO is parameterized by α, β, ρ, ξ, and the number of ants. Parameters α and β represent constants that control the relative contribution between the intensity of pheromone laid on edge (i,j) reflecting the previous experiences of ants about this edge, and the value of visibility determined by a greedy heuristic for the original problem. In this study, α is set to 1.0, but β is set to 0.4 (Kaveh et al., 2008).

Parameter ρ determines the pheromone evaporation in the global updating rule, and ξ represents the persistence of the pheromone trail in the local updating rule. These parameters have an influence on the exploratory behavior of ants. Herein, the values of ρ and ξ are set to $\rho = 0.2$ and $\xi = 0.1$ according to Nourani et al. (2009).

For engineering problems, the number of ants can be set to 20, because with smaller values, the success rates decrease and with greater values, the number of function evaluations and the running times increase (Kaveh et al., 2008).

In addition to the previous parameters, for improved ACO (IACO), the value of α_1 and α_2 should be determined. It can be shown that $0 < \alpha_1 < 0.5$ (Kaveh and Talatahari, 2010e). In the first stage, where there is less information about the search space, it is necessary that α_1 has a large value. Also, in the last stage where the aim of continuing the search process is to improve the previous solutions (as a local search process), the large value for α_1 may perform more appropriately than a small one. Meanwhile, whenever α_1 is close to 0.5, the number of function evaluations increases. Instead, if α_1 is selected as very small, probably the optimum solution is lost. Therefore, in this chapter, α_1 is set to 0.3.

The amount of α_2 highly influences the IACO performance. If α_2 is too small, the search process will end rapidly; on the contrary, if selected α_2 is too large, IACO will perform similarly to the original ACO algorithm and the effect of SOM will be eliminated, and a desirable solution cannot be obtained in less evaluations. In addition, α_2 can greatly affect the optimization time. A vast number of simulation investigations show that $\alpha_2 = 30$ is suitable (Kaveh and Talatahari, 2010e).

6.5   Other Well-Known Methods of Parameter Estimation

Results of the present algorithm are compared with some well-known parameter estimation methods; here, a brief review of these algorithms is presented (Hassanzadeh et al., 2011).

6.5.1   Probability-Weighted Moments

The PWMs for a random variable x and its cumulative distribution function (CDF) $F(x)$ can be defined as (Greenwood et al., 1979)

$$M_{p,r,s} = E\{x^p[F(x)]^r[1-F(x)]^s\} = \int_0^1 x^p[F(x)]^r[1-F(x)]^s dF \qquad (6.17)$$

where p, r, and s are real numbers. One of its special cases relates to $s = 0$, $p = 1$:

$$\beta_r = M_{1,r,0} = E\{x[F(x)]^r\} = \int_0^1 x[F(x)]^r dF, \quad r = 0, 1, 2, \ldots \qquad (6.18)$$

where β_r is linear in x and of sufficient generality for parameter estimation (Hosking, 1986).

Hosking (1990) introduced L-moments, which are linear functions of PWMs. For any distribution, the rth L-moment λ_r is calculated as follows:

$$\lambda_{r+1} = \sum_{j=0}^{r} P^*_{r,j}\beta_j, \quad r = 0, 1, 2, \ldots \qquad (6.19)$$

$$P^*_{r,j} = (-1)^{r-j}\binom{r}{j}\binom{r+j}{j} = \frac{(-1)^{r-j}(r+j)!}{j!(r-j)!} \qquad (6.20)$$

For example, the first four moments expressed as linear combinations of PWMs are (Hosking, 1990):

$$\lambda_1 = \beta_0 \qquad (6.21)$$

$$\lambda_2 = 2\beta_1 - \beta_0 \qquad (6.22)$$

$$\lambda_3 = 6\beta_2 - 6\beta_1 + \beta_0 \qquad (6.23)$$

$$\lambda_4 = 20\beta_3 - 30\beta_2 + 12\beta_1 - \beta_0 \qquad (6.24)$$

Also, ratios of L-moments are expressed as $\tau_r = \lambda_r/\lambda_2$ for $r = 3, 4, \ldots$.

In the above-mentioned relations, λ_1 is the mean, λ_2 is the L-standard deviation, λ_2/λ_1 is the L-coefficient of variation (L-C_v), τ_3 is the L-coefficient of skewness

(L-C_s), and τ_4 is the L-coefficient of kurtosis (L-C_k). The relationships for evaluating sample L-moments are defined as

$$b_r = \frac{1}{n}\sum_{i=1}^{n}\frac{(i-1)(i-2)\cdots(i-r)}{(n-1)(n-2)\cdots(n-r)}x_i, x_1 \leq x_2 \leq \cdots \leq x_n, \quad r=0,1,\ldots,n-1$$

(6.25)

$$l_{r+1} = \sum_{j=0}^{r} P_{r,j}^{*} b_j, \quad r=0,1,\ldots,n-1$$

(6.26)

Also, the ratio of L-moments is introduced by $t_r = l_r/l_2$ estimators considering $r = 3, 4,\ldots$.

6.5.2 Method of Moments

Estimates of parameters of a probability distribution function are obtained in the MOM by equating the moments of the sample to the moments of the probability distribution function. For a distribution with k parameters, $\alpha_1, \alpha_2, \ldots, \alpha_k$, which are to be estimated, the first k sample moments are set equal to the corresponding population moments that are given in terms of unknown parameters. These k equations are then solved simultaneously for the unknown parameters $\alpha_1, \alpha_2, \ldots, \alpha_k$.

For a distribution with PDF $f(x)$, the moment of r rank about the origin is

$$\mu_r' = \int_{-\infty}^{+\infty} x^r f(x)dx$$

(6.27)

and the corresponding central moments will be

$$\mu_r = \int_{-\infty}^{+\infty} (x-\mu_1')^r f(x)dx$$

(6.28)

6.5.3 Method of Maximum Likelihood

Estimation by the MML involves the choice of parameter estimates that produce a maximum probability of occurrence of observations. For a distribution with a PDF given by $f(x)$ and parameters $\alpha_1, \alpha_2, \ldots, \alpha_k$, the likelihood function is defined as the joint PDF of observations conditional on given values of parameters $\alpha_1, \alpha_2, \ldots, \alpha_k$ in the form

$$L(\alpha_1, \alpha_2, \ldots, \alpha_k) = \prod_{i=1}^{n} f(x_i; \alpha_1, \alpha_2, \ldots, \alpha_k)$$

(6.29)

The values of $\alpha_1, \alpha_2, \ldots, \alpha_k$ that maximize the likelihood function are computed by partial differentiation with respect to $\alpha_1, \alpha_2, \ldots, \alpha_k$ and the setting of these partial derivatives to zero, as in Eq. (6.30). The resulting set of equations is then solved simultaneously to obtain the values of $\alpha_1, \alpha_2, \ldots, \alpha_k$:

$$\frac{\partial L(\alpha_1, \alpha_2, \ldots, \alpha_k)}{\partial \alpha_i} = 0, \quad i = 1, 2, \ldots, k \tag{6.30}$$

In many cases, it is easier to maximize the natural logarithm of the likelihood function by using

$$\frac{\partial \ln L(\alpha_1, \alpha_2, \ldots, \alpha_k)}{\partial \alpha_i} = 0, \quad i = 1, 2, \ldots, k \tag{6.31}$$

6.6 Frequency Distributions

The utilized statistical distributions are presented in this section.

6.6.1 Generalized Extreme Value

The PDF of the GEV distribution can be expressed as

$$f(x) = \frac{1}{\alpha}\left[1 - k\left(\frac{x-u}{\alpha}\right)\right]^{1/k-1} \exp\left\{-\left[1 - k\left(\frac{x-u}{\alpha}\right)\right]^{1/k}\right\} \tag{6.32}$$

The range of variable x depends on the sign of parameter k. When k is negative, variable x can take on values in the range $u + \alpha/k < x < \infty$, which make it suitable for flood frequency analysis. However, when k is positive, variable x becomes upper bounded and takes on values in the range $-\infty < x < u + \alpha/k$, which may not be acceptable for analyzing floods unless there is sufficient evidence that such an upper bound does exist. When $k = 0$, the GEV distribution reduces to the type I extreme value distribution (EV1). The GEV CDF is of the form

$$F(x) = \exp\left\{-\left[1 - k\left(\frac{x-u}{\alpha}\right)\right]^{1/k}\right\} \tag{6.33}$$

6.6.2 Pearson Type 3

The PDF of the PE3 distribution is given as

$$f(x) = \frac{1}{\alpha \Gamma(\beta)}\left(\frac{x-\gamma}{\alpha}\right)^{\beta-1} \exp\left(-\left(\frac{x-\gamma}{\alpha}\right)\right) \tag{6.34}$$

The variable x can take on values in the range $\gamma < x < \infty$. Generally, α can be positive or negative, but for negative values of α, the distribution becomes upper bounded and therefore is not suitable for analyzing flood maxima. The PE3 CDF is of the form

$$F(x) = \frac{1}{\alpha\Gamma(\beta)} \int_\gamma^x \left(\frac{x-\gamma}{\alpha}\right)^{\beta-1} \exp\left(-\left(\frac{x-\gamma}{\alpha}\right)\right) \tag{6.35}$$

The Wilson–Hilferty approximation is quite accurate for $C_s \leq 1$ and may be sufficiently accurate for C_s values as high as 2:

$$K_T = \frac{2}{C_s}\left\{\left[\frac{C_s}{6}\left(u - \frac{C_s}{6}\right) + 1\right]^3 - 1\right\}, \quad C_s > 0 \tag{6.36}$$

where C_s is the skewness coefficient of the data, and u is the standardized normal variable:

$$u = \frac{\log(x - a) - \mu_y}{\sigma_y} \tag{6.37}$$

6.6.3   Lognormal Type 3

The PDF of the lognormal type 3 (LN3) distribution is given as

$$f(x) = \frac{1}{(x - a)\sigma_y\sqrt{2\pi}} \exp\left\{-\frac{1}{2\sigma_y^2}[\log(x-a)-\mu_y]^2\right\} \tag{6.38}$$

where μ_y and σ_y^2 are the location and scale parameters, respectively, which correspond to the mean and variance of the logarithm of the shifted variable $(x - a)$.

6.6.4   Generalized Logistic

The PDF of the generalized logistic (GLOG) distribution is given as

$$f(x) = \frac{1}{\alpha}\left[1 - k\left(\frac{x-\xi}{\alpha}\right)\right]^{1/k-1}\left\{1 + \left[1 - k\left(\frac{x-\xi}{\alpha}\right)\right]^{1/k}\right\}^{-2} \tag{6.39}$$

The variable x takes on values in the range $\xi + \alpha/k \leq x < \infty$ for $k \leq 0$, and $-\infty < x \leq \xi + \alpha/k$ for $k > 0$. The GLOG CDF is of the form

$$F(x) = \left\{1 + \left[1 - k\left(\frac{x-\xi}{\alpha}\right)\right]^{1/k}\right\}^{-1} \tag{6.40}$$

6.6.5 Generalized Pareto

The PDF of the GPAR distribution is given as

$$f(x) = \frac{1}{\alpha}\left[1 - k\left(\frac{x-\xi}{\alpha}\right)\right]^{1/k-1} \tag{6.41}$$

The CDF is written as

$$F(x) = 1 - \left[1 - k\left(\frac{x-\xi}{\alpha}\right)\right]^{1/k} \tag{6.42}$$

Variable x takes on values in the range: $\xi \le x < \infty$ for $k \le 0$ and $\xi \le x \le \xi + \alpha/k$ for $k > 0$. The special case of k being 0 yields the exponential distribution, whereas the special case of $k = 1$ yields the uniform distribution on $(\xi, \xi + \alpha)$.

6.6.6 Four-Parameter KAP Distribution

The KAP distribution has received only limited attention from the hydrologic community. Hosking and Wallis (1993) were probably the first to employ this distribution to generate artificial data for assessing the goodness of fit of different frequency distributions. Hosking (1994) described the properties of the KAP distribution, derived using the L-moments method of parameter estimation, and discussed an application for modeling maximum precipitation data (Singh and Deng, 2003). The PDF of the KAP distribution is given as

$$f(x) = \frac{1}{\alpha}\left[1 - k\left(\frac{x-\xi}{\alpha}\right)\right]^{1/k-1}\left\{1 - h\left[1 - k\left(\frac{x-\xi}{\alpha}\right)\right]^{1/k}\right\}^{1/h-1} \tag{6.43}$$

The distribution function is given as

$$F(x) = \left\{1 - h\left[1 - k\left(\frac{x-\xi}{\alpha}\right)\right]^{1/k}\right\}^{1/h} \tag{6.44}$$

The lower and upper bounds of random variable x will also depend on the parameter values as follows:

$$\begin{array}{ll} \xi + [\alpha(1 - 1/h^k)/k] \le x \le \xi + \alpha/k & \text{if } h > 0, \quad k > 0 \\ \xi + \alpha \log h \le x < \infty & \text{if } h > 0, \quad k = 0 \\ \xi + [\alpha(1 - 1/h^k)/k] \le x < \infty & \text{if } h > 0, \quad k < 0 \\ -\infty < x \le \xi + \alpha/k & \text{if } h \le 0, \quad k > 0 \\ -\infty < x < \infty & \text{if } h \le 0, \quad k = 0 \\ \xi + \alpha/k \le x < \infty & \text{if } h \le 0, \quad k < 0 \end{array} \tag{6.45}$$

Of the four parameters in Eq. (6.44), ξ is a location parameter, α is a scale parameter, and k and h are shape parameters. Apart from the restriction $\alpha > 0$, all parameter values yield valid distribution functions (Hosking, 1994). Equation (6.44) specializes into the following distribution functions, depending on the values of the shape parameters h and k, introduced in Table 6.1 (Parida, 1999).

6.6.7   Five-Parameter Wakeby Distribution

The quantile function of the Wakeby (WAK) distribution is given by

$$x(F) = \xi + \alpha[1 - (1-F)^{\beta}]/\beta - \gamma[1 - (1-F)^{-\delta}]/\delta \tag{6.46}$$

The WAK distribution is analytically defined only in the inverse form in Eq. (6.46). Therefore, explicit expressions cannot be obtained for either the PDF or the distribution function. Although moments of x can be obtained as functions of the parameters, the inverse relationship cannot be readily derived. Consequently, moment estimates of the parameters are not feasible. Similarly, maximum likelihood estimates of the parameters are not easily obtained. Only the PWM method is presently considered for this distribution.

The WAK distribution was proposed for flood frequency analysis by Houghton (1978), and was considered to be a superior distribution for this purpose. The PWM method was used to estimate the parameters of the WAK distribution by Landwehr et al. (1979).

The WAK distribution is potentially useful in flood frequency analysis for several reasons, as discussed by Greenwood et al. (1979). One of these is the large number of parameters in the WAK distribution, which permits better fitting of data than by distributions characterized by fewer parameters. Another reason is that it can accommodate a variety of flows ranging from low flows to floods.

In Eq. (6.46), ξ is a location parameter, α and γ are scale parameters, and β and δ are shape parameters. The range of x is such that $\xi \leq x < \infty$ if $\delta \geq 0$ and $\gamma > 0$; $\xi \leq x \leq \xi + \alpha/\beta - \gamma/\delta$ if $\delta < 0$ and $\gamma = 0$.

Table 6.1 Family of Distributions Generated by the KAP Distribution with Different Values of h and k

h	k	Distribution
1	$\neq 0$	GPAR distribution
0	$\neq 0$	GEV distribution
-1	$\neq 0$	GLOG distribution
1	0	Exponential distribution
0	0	Gumbel distribution
-1	0	Logistic distribution
1	1	Uniform distribution
0	1	Reverse exponential distribution

6.7   Simulation and Application

To evaluate the efficiency of the new algorithm, a region in Iran was considered (Hassanzadeh et al., 2011). The present study was carried out for the catchments of East Azerbaijan, which lie between latitude 36° to 39° North and longitude 45° to 48° East. The total geographical area spans over 45,491 km². Annual maximum discharge data from 14 stream-flow gauging sites lying in East Azerbaijan, northwest of Iran (Figure 6.2), and varying over 27−50 years in record length were obtained from the East Azerbaijan regional water corporation of the Iranian Ministry of Energy (Table 6.2).

Various criteria can be employed to evaluate the suitability of a probability distribution for describing a set of data. Statistical goodness of fit tests are used to determine whether selected distributions are consistent with the given set of observations (Stedinger et al., 1993). Three test criteria, such as coefficient of determination (CD), coefficient of efficiency (CE), and RMSE, are well-known statistical test criteria (Kite, 1977; Nash and Sutcliffe, 1970; Wang et al., 2009):

$$CD = \frac{\sum_{i=1}^{n}[(Q_{oi} - \overline{Q}_o)(Q_{ci} - \overline{Q}_c)]}{[\sum_{i=1}^{n}(Q_{oi} - \overline{Q}_o)^2 \sum_{i=1}^{n}(Q_{ci} - \overline{Q}_c)^2]^{1/2}}, \quad -1 \leq CD \leq 1 \qquad (6.47)$$

Figure 6.2 Plan of rivers and hydrometrical stations.

Table 6.2 Characteristics of Stations Selected for the Study

No.	Station	Sample Size (Year)	Geographical Characteristics		
			Height (m)	Latitude	Longitude
1	Bostan Abad	33	1725	37° 50′	46° 50′
2	Daryan	33	1616	38° 14′	45° 36′
3	Ghirmizi Ghol	40	1800	37° 43′	46° 06′
4	Gheshlaghe Amir	29	1520	37° 19′	46° 17′
5	Hervi	32	1920	37° 55′	46° 29′
6	Lighvan	50	2150	37° 50′	46° 26′
7	Maghanjigh	27	1650	37° 20′	46° 25′
8	Pole Sanikh	39	1352	38° 11′	46° 09′
9	Sahzab	29	1855	37° 59′	47° 39′
10	Saeed Abad	29	1850	37° 59′	46° 35′
11	Shishavan	38	1270	37° 28′	45° 53′
12	Shirin Kandi	38	1365	37° 01′	46° 16′
13	Tazekand	30	1610	37° 29′	46° 16′
14	Zinjanab	38	2100	37° 51′	46° 19′

$$CE = 1 - \frac{\sum_{i=1}^{n}(Q_{oi}-Q_{ci})^2}{\sum_{i=1}^{n}(Q_{oi}-\overline{Q}_o)^2}, \quad -\infty < CE \le 1 \tag{6.48}$$

$$RMSE = \left[\frac{1}{n}\sum_{i=1}^{n}(Q_{oi}-Q_{ci})^2\right]^{1/2} \tag{6.49}$$

where $\overline{Q}_o$ is the average of observed discharges, $\overline{Q}_c$ the average of computed discharges, Q_{oi} the ith observed discharge, and Q_{ci} the discharge computed from the selected distribution.

Here, for calculating statistical distribution parameters using the IACO algorithm, the objective function was calculated as

$$\text{Minimize} \quad \left(\frac{\sum_{i=1}^{n}(Q_{oi}-Q_{ci})^2}{\sum_{i=1}^{n}(Q_{oi}-\overline{Q}_o)^2}\right) \tag{6.50}$$

After the estimation of distribution parameters, quantiles $[x(F)]$ were estimated by using the equations (Rao and Hamed, 2000) in Table 6.3, where F represents the cumulative probability of nonexceedance, and $u,k,h,\alpha,\beta,\xi,\gamma,$ and δ are parameters of statistical distributions.

6.8   Results and Discussion

A ranking scheme was developed to evaluate the overall goodness of fit of each distribution by comparing the three categories of test criteria described earlier.

Table 6.3 Quantile Functions of Various Distributions Studied

No.	Distribution	Quantile Estimation
I	GEV	$X(F) = u + \alpha/k[1 - (-\ln F)^k]$
II	PE3	$X(F) = \alpha\beta + \gamma + K_T\sqrt{\alpha^2\beta^a}$
III	LN3	$X(F) = a + \exp(\mu + \sigma u)^b$
IV	GLOG	$X(F) = \xi + \alpha/k[1 - \{(1 - F)/F\}^k]$
V	GPAR	$X(F) = \xi + \alpha/k[1 - \{1/(1 - F)\}^{-k}]$
VI	KAP	$X(F) = \xi + \alpha/k[1 - \{(1 - F^h)/h\}^k]$
VII	WAK	$X(F) = m + a[1 - (1 - F)^b] - c[1 - (1 - F)^{-d}]$

[a]K_T: Frequency factor.
[b]u: Standardized normal variable.

Ranking was assigned to each distribution for every test category according to the relative magnitude of the test statistic. Table 6.4 presents the values of estimated parameters of different distributions obtained by IACO for all stations. Results of the CE for selection of the most appropriate distribution function for IACO, as well as conventional methods, are presented in Table 6.5.

A distribution with the highest CE was given the best rank. From Table 6.5, the results of goodness of fit tests for selection of the most appropriate distribution function showed that the proposed method was the best for parameter estimation. The goodness of fit assessment for all 14 stations reveals that the WAK distribution and then KAP distribution produced better fits.

Table 6.6 compares the values of CE obtained by IACO, standard ACO, and GA for the Shirin Kandi station. When the required time to find the optimum parameters was compared, the superiority of IACO was proved. IACO can find the results almost 2.2 times faster than the standard ACO. The reason is the use of SOM, which causes a decrease in the size of the pheromone vectors, decision vector, and search space; the number of function evaluations; and finally the required optimization time. SOM performs as a search-space-updating rule, and it can exchange discrete-continuous search domains with each other.

6.9 Conclusions

The main content of this chapter is an application of an improved ACO-based method for estimating parameters of statistical distributions, and an evaluation of the performance of this algorithm by comparison with the standard ACO and the classical methods, such as MML, MOM, and PWM. Various statistical distributions are used to find the most suitable distribution for annual maximum discharge.

IACO using SOM can exchange a continuous problem for a discrete one and continue the search process until reaching a solution with the required accuracy. IACO, contrary to previous CACO approaches, does not change the ACO-based principles; instead, it uses SOM to make handling continuous problems possible. Therefore, IACO has the capacity to deal with continuous as well as discrete problems.

Table 6.4 Values of Parameters of Different Distributions Estimated by IACO

Distribution	Parameter	Station						
		1	2	3	4	5	6	7
GEV	u	5.5276	1.2379	3.0020	8.1817	1.8887	2.0945	1.7810
	α	3.1155	0.7742	1.2631	3.4291	0.8296	0.8309	0.6874
	k	0.0854	−0.0366	0.2057	−0.0582	−0.0637	0.0380	0.1209
PE3	γ	−1.6465	−0.2039	−9.1688	2.62450	0.2768	0.3209	−0.4999
	α	1.3253	0.5571	0.1378	2.8186	0.5931	0.4456	0.2472
	β	6.4658	3.3002	92.0433	2.7121	3.5858	4.9543	10.527
LN3	α	−4.8884	−1.3462	−11.8079	−0.6312	−0.5596	−0.4155	−1.8772
	μ	2.4393	1.0380	2.7249	2.2979	1.0293	1.0278	1.3652
	σ	0.2912	0.3354	0.0887	0.4112	0.3687	0.3368	0.1995
GLOG	ξ	6.6227	1.4903	3.5153	9.5574	2.2426	2.3745	2.0636
	α	1.9472	0.5180	0.7201	2.3739	0.5940	0.5242	0.4341
	k	−0.1302	−0.2179	−0.0125	−0.1902	−0.1591	−0.1473	−0.0950
GPAR	ξ	2.1685	0.3442	1.3863	4.8109	1.0136	1.2046	1.0902
	α	6.5789	1.8885	3.8447	6.7179	1.7387	1.8487	1.4375
	k	0.3747	0.4048	0.8122	0.2292	0.2683	0.3909	0.4022
KAP	ξ	3.4496	0.3618	2.7263	7.7746	2.3944	1.5112	2.2554
	α	5.4397	1.7843	1.7106	3.9098	0.5182	1.4522	0.3305
	k	0.3329	0.3434	0.3725	0.0109	−0.2084	0.2790	−0.1711
	h	0.7181	1.0229	0.3050	0.1868	−1.7316	0.7475	−2.5928
WAK	ξ	1.0831	0.4688	0.6718	3.3277	0.6368	1.11691	0.6652
	α	1.3026	4.9806	1.2125	3.7230	1.1070	0.1301	1.1753
	β	21.448	0.48247	12.2331	4.4810	3.9714	15.713	3.5327
	γ	−16.022	0.5707	−4.9000	−152.949	16.2558	−4.6832	5.5794
	δ	−0.4147	−2.0202	−0.5080	−0.0266	0.0522	−0.3946	0.0864

Distribution	Parameter	Station						
		8	9	10	11	12	13	14
GEV	u	1.9118	2.1068	0.6322	6.7286	4.8687	11.9697	0.8526
	α	1.2598	0.9002	0.3207	3.4011	2.5513	5.6448	0.4331
	k	−0.1592	0.1932	0.0674	−0.3682	−0.1119	0.0858	0.0907
PE3	γ	0.7266	−0.7362	−0.7358	5.3474	0.9888	−2.5865	2.2506
	α	1.7801	0.3260	0.0932	13.658	2.4130	2.5138	−0.2051
	β	1.2152	9.8048	16.5640	0.3769	1.9653	6.9417	5.7609
LN3	α	−0.3778	−2.9630	−1.3150	1.5725	−0.3406	−11.1816	−1.7256
	μ	1.0174	1.6728	0.7360	1.8461	1.8068	3.2302	1.0071
	σ	0.5768	0.1836	0.1749	0.8081	0.5158	0.2462	0.1826
GLOG	ξ	2.3924	2.4416	0.7590	8.1908	5.9130	14.173	1.0238
	α	0.9365	0.5411	0.2073	3.0248	1.8229	3.6161	0.2732
	k	−0.2953	−0.0594	−0.0842	−0.3995	−0.2366	−0.1154	−0.0918
GPAR	ξ	0.8346	0.8578	0.2815	3.7317	2.5363	6.2379	0.4081
	α	2.0748	2.9484	0.7892	5.1864	4.3815	11.9856	0.8973
	k	0.0221	0.8231	0.5293	−0.2259	0.0830	0.4022	0.3846
KAP	ξ	2.2539	0.1684	0.7620	7.7715	5.0597	10.7596	0.7731
	α	0.9576	3.9176	0.2239	3.0673	2.2974	6.9258	0.5279
	k	−0.2743	0.9254	−0.0433	−0.3951	−0.1542	0.1622	0.1741
	h	−0.6149	1.2931	−0.7923	−0.6639	−0.1527	0.3022	0.2745
WAK	ξ	0.1211	0.8804	−0.2431	−0.6724	1.1280	4.6413	0.1362
	α	1.1908	2.8548	0.7488	5.4457	2.9236	10.4458	0.3715
	β	8.4682	0.8796	13.0670	18.3524	5.4645	1.7040	13.3292
	γ	9.9072	−0.9625	−2.3967	12.7496	20.8401	21.573	−2.2273
	δ	0.1453	−0.3088	−0.1742	0.3212	0.1301	0.14195	−0.3586

Table 6.5 Values of CE for Various Distributions at 14 Hydrometrical Stations in East Azerbaijan

No.	Station	Estimator	Statistical Distribution						
			GEV	PE3	LN3	GLO	GPA	KAP	WAK
1	Bostan Abad	MML	0.9798	0.9790	0.9762	0.8927	0.9841	–	–
		MOM	0.9838	0.9835	0.9823	0.9648	0.9909	–	–
		PWM	0.9815	0.9845	0.9823	0.9672	0.9905	0.9888	0.9920
		IACO	0.9840	0.9794	0.9830	0.9697	0.9888	0.9920	0.9926
2	Daryan	MML	0.9352	0.9556	0.9419	NaN	0.9793	–	–
		MOM	0.9663	0.9679	0.9653	0.9421	0.9840	–	–
		PWM	0.9590	0.9693	0.9632	0.9434	0.9818	0.9789	NaN
		IACO	0.9646	0.9650	0.9631	0.9433	0.9829	0.9840	0.9855
3	Ghirmizi Gol	MML	0.9831	0.9810	0.9812	0.9632	NaN	–	–
		MOM	0.9844	0.9816	0.9816	0.9680	0.9773	–	–
		PWM	0.9834	0.9810	0.9808	0.9667	0.9774	0.9824	0.9879
		IACO	0.9836	0.9815	0.9808	0.9667	0.9781	0.9872	0.9867
4	Gheshlaghe Amir	MML	0.9869	0.9877	0.9878	0.9755	0.9666	–	–
		MOM	0.9853	0.9857	0.9847	0.9729	0.8120	–	–
		PWM	0.9884	0.9881	0.9884	0.9846	0.9767	0.9874	0.9884
		IACO	0.9882	0.9882	0.9876	0.9852	0.9806	0.9887	0.9864
5	Hervi	MML	0.9827	0.9826	0.9833	0.9862	0.9422	–	–
		MOM	0.9832	0.9825	0.9828	0.9776	0.9658	–	–
		PWM	0.9842	0.9827	0.9839	0.9859	0.9565	0.9865	0.9867
		GA	0.9855	0.9835	0.9850	0.9869	0.9667	0.9873	0.9874
		IACO	0.9838	0.9834	0.9844	0.9865	0.9663	0.9876	0.9871
6	Lighvan	MML	0.9845	0.9865	0.9822	0.8832	0.9843	–	–
		MOM	0.9884	0.9889	0.9874	0.9700	0.9923	–	–
		PWM	0.9867	0.9903	0.9879	0.9723	0.9918	0.9925	0.9943
		IACO	0.9887	0.9901	0.9885	0.9742	0.9923	0.9942	0.9921
7	Maghanjigh	MML	0.9823	0.9836	0.9836	0.9824	0.9562	–	–
		MOM	0.9831	0.9833	0.9832	0.9776	0.9634	–	–
		PWM	0.9827	0.9834	0.9834	0.9814	0.9583	0.9825	0.9892
		IACO	0.9829	0.9841	0.9838	0.9812	0.9626	0.9871	0.9894
8	Pole Sanikh	MML	0.9834	0.9707	0.9790	NaN	0.9496	–	–
		MOM	0.9763	0.9780	0.9776	0.9616	0.9746	–	–
		PWM	0.9870	0.9773	0.9845	0.9879	0.9669	0.9880	0.9885
		IACO	0.9884	0.9797	0.9853	0.9886	0.9773	0.9912	0.9914
9	Sahzab	MML	0.9651	0.9092	0.9395	0.7879	0.9846	–	–
		MOM	0.9680	0.9644	0.9640	0.9412	0.9926	–	–
		PWM	0.9640	0.9639	0.9624	0.9400	0.9915	0.9866	0.9928
		IACO	0.9655	0.9644	0.9640	0.9418	0.9904	0.9935	0.9913
10	Saeed Abad	MML	0.9680	0.9687	0.9690	0.9735	0.8696	–	–
		MOM	0.9694	0.9698	0.9701	0.9730	0.9313	–	–
		PWM	0.9698	0.9699	0.9704	0.9749	0.9285	0.9741	0.9785
		IACO	0.9678	0.9694	0.9703	0.9733	0.9357	0.9749	0.9761

(Continued)

Table 6.5 (Continued)

No.	Station	Estimator	Statistical Distribution						
			GEV	PE3	LN3	GLO	GPA	KAP	WAK
11	Shishavan	MML	0.9351	0.8869	0.9126	NaN	0.8891	–	–
		MOM	0.9433	0.9580	0.9512	0.9194	0.9559	–	–
		PWM	0.9719	0.9539	0.9707	0.9695	0.9614	0.9682	0.9794
		IACO	0.9828	0.9687	0.9763	0.9820	0.9771	0.9831	0.9874
12	Shirin Kandi	MML	0.9691	0.9637	0.9675	0.9749	0.9217	–	–
		MOM	0.9685	0.9667	0.9678	0.9607	0.9558	–	–
		PWM	0.9714	0.9660	0.9697	0.9736	0.9436	0.9739	0.9762
		IACO	0.9735	0.9673	0.9716	0.9751	0.9575	0.9734	0.9766
13	Tazekand	MML	0.9830	0.9789	0.9836	0.9695	0.9740	–	–
		MOM	0.9833	0.9831	0.9826	0.9716	0.9750	–	–
		PWM	0.9840	0.9844	0.9841	0.9777	0.9715	0.9833	0.9858
		IACO	0.9840	0.9846	0.9838	0.9776	0.9754	0.9850	0.9864
14	Zinjanab	MML	0.9892	0.9896	0.9893	0.9765	0.9460	–	–
		MOM	0.9900	0.9893	0.9890	0.9800	0.9788	–	–
		PWM	0.9897	0.9897	0.9894	0.9819	0.9767	0.9866	0.9921
		IACO	0.9902	0.9893	0.9872	0.9828	0.9787	0.9902	0.9916

Table 6.6 Values of CE for Various Distributions at the Shirin Kandi Station Obtained by Metaheuristic Methods

Method	Statistical Distribution						
	GEV	PE3	LN3	GLOG	GPAR	KAP	WAK
GA (Hassanzadeh et al., 2011)	0.9735	0.9676	0.9713	0.9750	0.9577	0.9756	0.9773
ACO (Hassanzadeh et al., 2011)	0.9734	0.9672	0.9717	0.9754	0.9574	0.9760	0.9765
IACO	0.9735	0.9673	0.9716	0.9751	0.9575	0.9734	0.9766

The obtained results show that the IACO method can efficiently identify parameters of statistical distributions, which in turn can provide reasonable estimates of various flood quantiles at 14 sites located in East Azerbaijan, Iran. Comparing to the standard ACO, investigations prove the robustness of the proposed method in determining parameters of statistical flood frequency distributions.

References

Abbaspour, K.C., Schulin, R., van Genuchten, M.Th., 2001. Estimating unsaturated soil hydraulic parameters using ant colony optimization. Adv. Water Res. 24, 827–841.
Ashkar, F., Tatsambon, C.N., 2007. Revisiting some estimation methods for the generalized Pareto distribution. J. Hydrol. 346, 136–143.

Atashpaz-Gargari, E., Lucas, C., 2007. Imperialist competitive algorithm: an algorithm for optimization inspired by imperialistic competition. In: IEEE Congress on Evolutionary Computation. Singapore, pp. 4661−4667.

Bilchev, G., Parmee, I.C., 1995. The ant colony metaphor for searching continuous design spaces. In: Fogarty, T.C. (Ed.), Proceedings of the AISB Workshop on Evolutionary Computation, 993. Springer-Verlag, Berlin, pp. 25−39, LNCS.

Chow, V.T., Maidment, D.R., Mays, L.W., 1988. Applied Hydrology. McGraw-Hill, New York, NY.

Deneubourg, J.L., Goss, S., 1989. Collective patterns and decision-making. Ethnol. Ecol. Evol. 1, 295−311.

Dong, S.H., 2008. Genetic algorithm based parameter estimation of Nash model. Water Res. Manage. 22, 525−533.

Dorigo, M., 1992. Optimization, learning and natural algorithms. Ph.D. thesis. Dip. Elettronica e Informazione, Politecnico di Milano, Milano.

Dorigo, M., Di Caro, G., 1999. Ant colony optimization: a new meta-heuristic. In: Proceedings of the 1999 Conference on Evolutionary Computation, 2, pp. 1470−1477.

Dorigo, M., Maniezzo, V., Colorni, A., 1996. The ant system: optimization by a colony of cooperating agents. IEEE Trans. Syst. Man Cybern. Part B. 26 (1), 1−13.

Dreo, J., Siarry, P., 2002. A new ant colony algorithm using the hierarchical concept aimed at optimization of multiminima continuous functions. In: Dorigo, M., Di Caro, G., Sampels, M. (Eds.), Proceedings of the Third International Workshop on Ant Algorithms (ANTS'2002), 2463. Springer-Verlag, Berlin, pp. 216−221, LNCS.

Eberhart, R.C., Kennedy, J., 1995. A new optimizer using particle swarm theory. In: Proceedings of the Sixth International Symposium on Micro Machine and Human Science. Nagoya, Japan.

Erol, O.K., Eksin, I., 2006. New optimization method: Big Bang-Big Crunch. Adv. Eng. Software. 37, 106−111.

Gandomi, A.H., Yang, X.S., Alavi, A.H., 2011. Mixed variable structural optimization using firefly algorithm. Comput. Struct. 89 (23−24), 2325−2336.

Gandomi, A.H., Yang, X.S., Alavi, A.H., 2012. Cuckoo search algorithm: a metaheuristic approach to solve structural optimization problems. Eng. Comput. doi 10.1007/s00366-011-0241-y.

Goldberg, D.E., 1989. Genetic Algorithms in Search, Optimization and Machine Learning. Addison-Wesley, Boston, MA.

Goss, S., Beckers, R., Deneubourg, J.L., Aron, S., Pasteels, J.M., 1990. How trail laying and trail following can solve foraging problems for ant colonies. In: Hughes, R.N. (Ed.), Behavioural Mechanisms in Food Selection, NATO-ASI Series. vol. G 20, Berlin.

Greenwood, J.A., Landwehr, J.M., Matalas, N.C., Wallis, J.R., 1979. Probability weighted moments: definition and relation to parameters of several distributions expressible in inverse form. Water Res. Res. 15 (5), 1049−1054.

Hassanzadeh, Y., Abdi, A., Abdi, A., 2008. A comparison of the methods used for the estimation statistical distribution parameters. In: Proceedings of the Fifth International Engineering and Construction Conference (ASCE). Irvine, CA, USA, pp. 1009−1016.

Hassanzadeh, Y., Abdi, A., Talatahari, S., Singh, V.P., 2011. Meta-heuristic algorithms for hydrologic frequency analysis. Water Res. Manage. 25 (7), 1855−1879.

Hosking, J.R.M., 1986. The theory of probability weighted moments. IBM Math Research Report, RC 12210. Yorktown Heights, New York, NY.

Hosking, J.R.M., 1990. L-moments: analysis and estimation of distributions using linear combinations of order statistics. J. R. Stat. Soc. Ser. B. 52, 105−124.

Hosking, J.R.M., 1994. The four-parameter kappa distribution. IBM J. Res. Dev. 38 (3), 251−258.

Hosking, J.R.M., Wallis, J.R., 1993. Some statistics useful in regional frequency analysis. Water Res. Res. 29 (2), 271−281.

Houghton, J.C., 1978. Birth of a parent: the Wakeby distribution for modeling flood flows. Water Res. Res. 14 (6), 1105−1109.

Karahan, H., Ceylan, H., Ayvaz, M.T., 2007. Predicting rainfall intensity using a genetic algorithm approach. Hydrolog. Process. 21, 470−475.

Kaveh, A., Talatahari, S., 2009. Hybrid algorithm of harmony search, particle swarm and ant colony for structural design optimization. Stud. Comput. Intell. 239, 159−198.

Kaveh, A., Talatahari, S., 2010a. Optimum design of skeletal structures using imperialist competitive algorithm. Comput. Struct. 88 (21−22), 1220−1229.

Kaveh, A., Talatahari, S., 2010b. A novel heuristic optimization method: charged system search. Acta Mech. 213 (3−4), 267−289.

Kaveh, A., Talatahari, S., 2010c. Optimal design of skeletal structures via the charged system search algorithm. Struct. Multidiscip. Optim. 41 (6), 893−911.

Kaveh, A., Talatahari, S., 2010d. Charged system search for optimum grillage systems design using the LRFD-AISC code. J. Construct. Steel Res. 66 (6), 767−771.

Kaveh, A., Talatahari, S., 2010e. An improved ant colony optimization for constrained engineering design problems. Eng. Comput. 27 (1), 155−182.

Kaveh, A., Talatahari, S., 2010f. An improved ant colony optimization for design of planar steel frames. Eng. Struct. 32 (3), 864−873.

Kaveh, A., Talatahari, S., 2011a. A general model for meta-heuristic algorithms using the concept of fields of forces. Acta Mech. 221 (1−2), 119−132.

Kaveh, A., Talatahari, S., 2011b. Geometry and topology optimization of geodesic domes using charged system search. Struct. Multidiscip. Optim. 43 (2), 215−229.

Kaveh, A., Talatahari, S., 2011c. An enhanced charged system search for configuration optimization using the concept of fields of forces. Struct. Multidiscip. Optim. 43 (3), 339−351.

Kaveh, A., Talatahari, S., 2012. Charged system search for optimal design of frame structures. Appl. Soft Comput. 12 (1), 382−393.

Kaveh, A., Farhmand Azar, B., Talatahari, S., 2008. Ant colony optimization for design of space trusses. Int. J. Space Struct. 23 (3), 167−181.

Kite, G.V., 1977. Frequency and Risk Analysis in Hydrology. Water Resources Publication, Fort Collins, CO.

Kumar, D.N., Reddy, M.J., 2006. Ant colony optimization for multi-purpose reservoir operation. Water Res. Manage. 20, 879−898.

Landwehr, J.M., Matalas, N.C., Wallis, J.R., 1979. Estimation of parameters and quantiles of Wakeby distributions. Water Res. Res. 15 (6), 1361−1379.

Li, Y., Chan Hilton, A.B., 2006. Reducing spatial sampling in long-term groundwater monitoring using ant colony optimization. Int. J. Comput. Intell. Res. 1 (1), 19−28.

Madadgar, S., Afshar, A., 2009. An improved continuous ant algorithm for optimization of water resources problems. Water Res. Manage. 23, 2119−2139.

Mahdi, S., Ashkar, F., 2004. Exploring generalized probability weighted moments, generalized moments and maximum likelihood estimation methods in two-parameter Weibull model. J. Hydrol. 285, 62−75.

Maier, H.R., Simpson, A.R., Zecchin, A.C., Foong, W.K., Phang, K.Y., Seah, H.Y., et al., 2003. Ant colony optimization for design of water distribution systems. J. Water Res. Planning Manage. 129 (3), 200−209.

Nash, J.E., Sutcliffe, J.V., 1970. River flow forecasting through conceptual models I: a discussion of principles. J. Hydrol. 10 (3), 282−290.

Nourani, V., Talatahari, S., Monadjemi, P., Shahradfar, S., 2009. Application of ant colony optimization to investigate velocity profile effect on optimal design of open channels. J. Hydraul. Res. 47 (5), 656−665.

Öztekin, T., 2005. Comparison of parameter estimation methods for the three parameter generalized Pareto distribution. Turk. J. Agric. For. 29 (6), 419−428.

Parida, B.P., 1999. Modeling of Indian summer monsoon rainfall using a four-parameter kappa distribution. Int. J. Climatol. 19, 1389−1398.

Rai, R.K., Sarkar, S., Singh, V.P., 2009. Evaluation of the adequacy of statistical distribution functions for deriving unit hydrograph. Water Res. Manage. 23, 899−929.

Rao, A.R., Hamed, K.H., 2000. Flood Frequency Analysis. CRC Press, Boca Raton, FL.

Reca, J., Martínez, J., Gil, C., Baños, R., 2008. Application of several meta-heuristic techniques to the optimization of real looped water distribution networks. Water Res. Manage. 22, 1367−1379.

Reddy, M.J., Adarsh, S., 2010. Chance constrained optimal design of composite channels using meta-heuristic techniques. Water Res. Manage. 24, 2221−2235.

Singh, V.P., 1998. Entropy-Based Parameter Estimation in Hydrology. Kluwer Academic, Dordrecht, the Netherlands.

Singh, V.P., Deng, Z.Q., 2003. Entropy-based parameter estimation for kappa distribution. J. Hydrol. Eng. 8 (2), 81−92.

Socha, K., Dorigo, M., 2008. Ant colony optimization for continuous domains. Eur. J. Oper. Res. 185, 1155−1173.

Stedinger, J.R., Vogel, R.M., Foufoula-Georgiou, E., 1993. Frequency analysis of extreme events. Handbook of Hydrology. McGraw-Hill, New York, NY.

Talatahari, S., Farahmand Azar, B., Sheikholeslami, R., Gandomi, A.H., 2012. Imperialist competitive algorithm combined with chaos for global optimization. Commun. Nonlinear Sci. Numer. Simulat. 17 (3), 1312−1319.

Wang, W.C., Chau, K.W., Cheng, C.T., Qiu, L., 2009. A comparison of performance of several artificial intelligence methods for forecasting monthly discharge time series. J. Hydrol. 374, 294−306.

Yang, X.S., Gandomi, A.H., 2012. Bat algorithm: a novel approach for global engineering optimization. Eng. Comput. 29 (5), 464−483.

7 Optimal Reservoir Operation for Irrigation Planning Using the Swarm Intelligence Algorithm

A. Vasan

Department of Civil Engineering, Birla Institute of Technology and Science, Pilani—Hyderabad Campus, Andhra Pradesh, India

7.1 Introduction

The growing demand for water has required efficient utilization in the irrigation sector. In many countries, efforts to raise levels of agricultural production have led to a greater dependence not only on irrigation but also on other resources. This pressure has been most severe in developing countries, where many irrigation systems are primitive (Bouwer, 2002). Frederiksen (1996) opined that innovations are needed in both the technological and policy dimensions of water resource management to achieve the gains in productivity required to feed the world's increasing population. Also, it is required to formulate mathematical models and introduce new techniques to plan and manage efficient strategies (Ranjithan, 2005).

The escalating complexity of real-world applications similar to the one stated above has demanded that researchers find possible ways of solving such problems. This has motivated the researchers to take ideas from the nature and implant it in the engineering sciences. This way of thinking has led to the emergence of many biologically inspired algorithms that have proved to be efficient in handling the computationally complex problems with competence such as evolutionary algorithms and swarm intelligence (SI) algorithms (Kennedy and Eberhart, 2001). Recent studies have emphasized that evolutionary algorithms and SI algorithms are attractive solutions to many practical optimization problems because they are independent of the problem types (Deb, 2001). In the present study, a heuristic optimization algorithm (namely, the particle swarm optimization (PSO) algorithm) is applied to a case study of the Mahi Bajaj Sagar Project (MBSP) in Rajasthan, India, with the objective of optimizing annual net benefits that gives the optimum cropping pattern, storage, and release policy with consideration of conjunctive use of surface water and groundwater; study the applicability of the algorithm in irrigation planning and assess its capability in solving high-dimensional problems; and explore it as an alternative

Metaheuristics in Water, Geotechnical and Transport Engineering. DOI: http://dx.doi.org/10.1016/B978-0-12-398296-4.00007-6

methodology. The study is divided into a review of the research literature, a description of the SI algorithm, a description of the case study, mathematical modeling, results and discussion, and finally a conclusion.

7.2　Literature Review

For the purpose of planning and management of water resources, systems analysis techniques have been increasingly and extensively used for the last few decades. The important literature pertaining to the new techniques that are relevant to the present study are presented briefly in the following section, with relevance to optimal reservoir operation for irrigation planning aspects.

Chang and Chen (1998) applied real-coded and binary-coded algorithms for the case study of a flood control reservoir model. It is concluded that both variations of the genetic algorithm (GA) are more efficient and robust than the random search technique. It is also observed that real-coded GA performs better in terms of efficiency and precision than binary-coded GA. Wardlaw and Sharif (1999) evaluated several alternative formulations of a GA for the four-reservoir, deterministic, and finite-horizon problem. They concluded that real-value coding, tournament selection, uniform crossover, and modified uniform mutation are suitable for the planning problem. It is also concluded that real-value coding operates significantly faster than binary coding and produces better results. In addition, a nonlinear four-reservoir problem and a 10-reservoir problem are also considered with previously published results. They concluded that the GA approach is robust and is easily applied to complex systems. Similar studies are reported by Sharif and Wardlaw (2000) and Wardlaw and Bhaktikul (2004). Kuo et al. (2000) developed on-farm irrigation scheduling and the GA optimization model in irrigation project planning with the objective of optimizing economic profits. The model was applied to an irrigation project (namely, the Wilson canal system) located in Delta, Utah. Two other optimization techniques, namely simulated annealing (SA) and iterative improvement techniques, were also used, and the results were compared with those of GA. It was concluded that GA and SA consistently obtained near-optimal values, whereas the iterative improvement technique occasionally found the local optimum values.

Mardle and Pascoe (2000) emphasized the basic features, advantages, and disadvantages of the use of the evolutionary techniques, specifically GA. Ranjithan (2005) stressed on the role of evolutionary computation in environmental, water resources system analysis, and briefly discussed the various techniques; namely, SA, tabu search, GAs, evolutionary strategies, the PSO technique, and ant colony optimization (ACO). He concluded that new areas that shape the direction of a beneficial integration of evolutionary computation into environmental and water resources systems are essential. Raju and Nagesh Kumar (2004) applied binary-coded GA for irrigation planning to a case study of the Sri Ram Sagar Project in Andhra Pradesh, India. The GA technique is used to derive the cropping pattern, a reservoir operating policy that yields optimum annual net benefits. The results obtained by GA were compared to those of linear programming, and it was concluded that GA is an effective optimization tool for irrigation planning and can be used for any similar irrigation system. Vasan and Raju (2007) explored the applicability of differential evolution (DE) in irrigation planning and concluded that DE can be used as a successful alternative methodology. Similar studies are reported by Gupta et al. (2009), Vasan and Simonovic (2010), Raju et al. (2012), and Schardong et al. (2012).

Ramesh and Simonovic (2002) applied SA to a case study of a four-reservoir system that had previously been solved using a linear programming formulation to maximize the benefits. SA is also applied to a system of four hydropower-generating reservoirs in Manitoba, Canada, to derive optimal operating rules with an objective of minimizing the cost of power generation. Results obtained from these two applications suggest that SA can be used as an alternative approach for solving reservoir operation problems that are computationally intractable. Rao et al. (2003) developed a management model within a simulation (a sharp interface-flow model) optimization (SA algorithm) framework to determine the optimal groundwater extraction in a hypothetical deltaic region with specified Indian conditions. The objective is to determine optimal configuration of pump rates and their locations. They concluded that the model provided near-optimal solutions. Similar studies are reported by Cunha and Sousa (1999) and Rao et al. (2004). Application of SA for the case study in the irrigation planning context is limited except those reported by Kuo et al. (2000) and Vasan and Raju (2009).

Application of the PSO algorithm for irrigation planning is relatively new. Wegley et al. (2000) successfully applied this algorithm to optimize pump operations in water distribution systems. Coelho et al. (2005) showed that it was more efficient than the GA and sequential quadratic programming (SQP) to solve constrained optimization air temperature control problems. Shawn Matott et al. (2006) opined that PSO is a potential optimization algorithm for solving problems related to plume containment using pump and treat technology, compared to other algorithms such as GA and SA. Janga Reddy and Nagesh Kumar (2007) applied elitist-mutation particle swarm optimization (EMPSO) to optimize an operational model for short-term reservoir operation for irrigation of multiple crops. Economic benefits in the objective function were considered by the water allocation decisions for multiple crops per unit area. Afshar and Rajabpour (2009) used the PSO algorithm for the optimal design and operation of irrigation pumping systems. The results showed that this algorithm was a more competent tool than GA. The PSO algorithm has been credibly adopted for many optimization problems. It has performed superior as compared with other optimization techniques with fewer parameter adjustments and a significantly lower number of iterations for its approximation. It is observed from this literature review that very few studies have been reported of irrigation planning using the SI algorithm. The description of the various nontraditional optimization methods are given in the next section.

7.3   Method Description

Nontraditional optimization methods are gaining importance due to their advantage of handling nondifferentiable, nonlinear, multimodal functions having a complex search space with many local optimal solutions in a systematic and effective way. However, the difficulty of using nontraditional methods as seen by researchers is the number of input parameters and the determination of their precise values, thus making the solution process rather complex. In the present section, four heuristic optimization methods (namely, GA, DE, SA, and SI) are described in detail.

7.3.1   Genetic Algorithm

The GA is based on the mechanics of natural selection and natural genetics. They combine survival of the fittest with a structured but randomized information exchange to form a search algorithm (Goldberg, 1989). The GA works with an initial population of a string of variables known as *chromosomes*, which hold the parameters or genes and the population size. The chromosome can be represented using binary code or decimals and accordingly, it is termed as binary-coded GA or real-coded GA. There are three operators (namely, selection, crossover, and mutation) to generate a new population of points from the old population. In the selection operator, a set of chromosomes is selected as initial parents at the reproduction stage on the basis of their fitness, subject to the constraints posed by the problem. The fittest are given a greater chance of survival as well as a greater probability of reproducing more offspring. The process of mating is implemented through the crossover operator. Mutation, an arbitrary change of the genes, is implemented to preserve the genetic diversity in the population. Mutation probability of occurrence can be kept low because it can disrupt the good solution.

A stochastic selection process, biased toward the fitter individuals, is implemented to select the new population set for the next generation. In the present study, the tournament selection operator is used to select the good solutions. The newly created population is further evaluated and tested for termination, to decide the maximum number of generations. If the termination criterion is not met, the population is iteratively operated further by these three operators and evaluated. One cycle of these operations and its subsequent evaluation is known as a generation. This process is continued until the termination criterion of a preset maximum number of generations is met. The main feature of GA is its ability to operate on many solutions simultaneously, thereby exploring the search space of the objective function thoroughly. This resolves the problem of trapping in the local minimum.

A difficulty with population-based optimizers is that once the search has narrowed such that it is near the previous optimal solution, the diversity in the population may not be enough for the search to come out and proceed toward the new optimal solution. To overcome this problem, the self-adaptive behavior of real-coded genetic algorithms (RGAs) with a simulated binary crossover (SBX) operator and parameter-based mutation operator are explored in the present planning problem (Deb, 2001).

The SBX operator uses a probability distribution around two parents to create two child solutions (Deb and Agarwal, 1995). Unlike other real-parameter crossover operators, SBX uses a polynomial probability distribution that is similar in principle to the probability of creating child solutions in crossover operators used in binary-coded GA. The value of the distribution index for SBX controls the distance of the child solutions from the parents. Similarly, the distribution index for parameter-based mutation decides the effect of perturbance in the parent solutions. There are two aspects that give RGA with SBX their self-adaptive power: (1) child solutions closer to parent solutions are more likely to be created and (2) the span of child solutions is proportional to the span of parent solutions. Both the properties are essential for a crossover operator to exhibit self-adaptive behavior in GA. This is because with these properties, the diversity in child solutions is directly controlled by the diversity in parent solutions. Movement of the parent population in the search space is dictated by the fitness function through the selection operator. This selection operator (i.e., SBX crossover) allows the GA to search for a region near the parent population that exhibits self-adaptation. The population size, crossover probability, and mutation probability are the three important parameters that govern the successful working of the RGA.

7.3.2 Differential Evolution

The DE algorithm is a population-based search technique that uses population size NP as the population of D-dimensional parameter vectors for each generation. DE maintains two arrays, each of which holds a population of NP, D-dimensional, real-valued vectors. The primary array holds the current vector population, while the secondary array accumulates vectors that are selected for the next generation. In each generation, NP competitions are held to determine the composition of the next generation. Every pair of vectors (X_a, X_b) defines a vector differential as $(X_a - X_b)$. When X_a and X_b are chosen randomly, their weighted differential is used to perturb another randomly chosen vector X_c. This process can be mathematically expressed as

$$X'_c = X_c + F(X_a - X_b) \tag{7.1}$$

The weighting factor or scaling factor F is a user-supplied constant in the optimal range of $0.5-1$ (Price et al., 2005). In every generation, each primary array vector X_i is targeted for crossover with a vector like X'_c to produce a trial vector X_t. Thus, the trial vector is the child of two parents, a noisy random vector and the target vector against which it must compete. Nonuniform crossover is used with a crossover constant CR, in the optimal range of $0.5-1$ (Price et al., 2005), which represents the probability that the child vector inherits the parameter values from the noisy random vector. Then the cost of the trial vector is compared with that of the target vector, and the vector that has the lower cost of the two would survive for the next generation. This process is continued until the termination criterion of a preset maximum number of generations is met and the difference of function values between two consecutive generations reaches a small value. In all, three factors control evolution under DE: the population size NP, the weight applied to the random differential F, and the crossover constant CR.

Different strategies can be adopted in the DE algorithm depending on the type of problem to which it is applied. The strategies can vary based on the vector to be perturbed, the number of difference vectors considered for perturbation, and the type of crossover used. Price and Storn (1997) gave the working principle of DE with single-strategy DE/rand/1/bin. Later, they added nine more different strategies, namely, DE/best/1/bin, DE/best/2/bin, DE/rand/2/bin, DE/rand-to-best/1/bin, DE/rand/1/exp, DE/best/1/exp, DE/best/2/exp, DE/rand/2/exp, and DE/rand-to-best/1/exp (Price et al., 2005). Here, in the name DE/x/y/z, DE indicates differential evolution, x represents a string denoting the vector to be perturbed (rand: random vector; best: best vector), y is the number of difference vectors considered for perturbation of x, and z stands for the type of crossover being used (exp: exponential; bin: binomial). The working algorithm outlined above is for the strategy DE/rand/1/bin.

7.3.3 Simulated Annealing

The SA solution methodology resembles the cooling process of molten metals through annealing. At high temperatures, the atoms in the molten metal can move freely with respect to each other, but as the temperature is reduced, the movement of the atoms gets restricted. The atoms start to get arranged and finally form crystals having the minimum possible energy. However, the formation of the crystal mostly depends on the cooling rate (CoR). If the temperature is reduced at a very fast rate, the crystalline state may not be achieved at all; instead, the system may end up in a polycrystalline state, which may have a higher energy state than the crystalline state. Therefore, in order to achieve the absolute minimum energy state, the temperature needs to be reduced slowly. The process of slow cooling is known as the *annealing process*.

The SA procedure simulates this process of slow cooling of molten metals to achieve the minimum function value in a minimization problem. The cooling phenomenon is simulated by controlling a parameter, namely, temperature T, that is introduced with the concept of the Boltzmann probability distribution. According to the Boltzmann probability distribution, a system in thermal equilibrium at a temperature T has its energy distributed probabilistically according to $P(E) = e^{(-E/k_bT)}$, where E is the energy of the system and k_b is the Boltzmann constant. This expression indicates that a system at a high temperature has almost uniform probability of being at any energy state, but at a low temperature, it has a small probability of being at a high energy state. Therefore, by controlling the temperature T and assuming that the search process follows the Boltzmann probability distribution, the convergence of an algorithm can be controlled. Metropolis (Kirkpatrick et al., 1983) suggested a way to implement the Boltzmann probability distribution in simulated thermodynamic systems that can also be used in the function minimization context. For example, at any instant, the current point is $x(t')$ and the function value at that point is $E(t') = f(x(t'))$. Using the Metropolis algorithm, the probability of the next point being $x(t' + 1)$ depends on the difference in the function values at these two points, or $\Delta E = E(t' + 1) - E(t')$. Probability value $P(E(t' + 1))$ is calculated using the Boltzmann probability distribution:

$$P(E(t' + 1)) = \min(1, e^{(-\Delta E / k_b T)}) \qquad (7.2)$$

If $\Delta E \leq 0$, the above probability is 1 and the point $x(t' + 1)$ is always accepted. In the function minimization context, this is meaningful because if the function value at $x(t' + 1)$ is better than at $x(t')$, the point $x(t' + 1)$ must be accepted. When $\Delta E > 0$, which implies that the function value at $x(t' + 1)$ is worse than that at $x(t')$. According to the Metropolis algorithm, there is a finite probability of selecting the point $x(t' + 1)$, even though it is worse than the point $x(t')$. This probability depends on the relative magnitude of ΔE and T values. If the parameter T is large, this probability is more or less high for points with largely different function values. Thus, any point is almost acceptable for a large value of T. On the other hand, if the parameter T is small, the probability of accepting an arbitrary point is small. Thus, for small values of T, the points with only a small deviation in function value are accepted. In order to simulate the thermal equilibrium at every temperature, a number of iterations are performed at a particular temperature before reducing the temperature. The algorithm is terminated when a sufficiently small temperature is obtained and a small-enough change in function values is found. The initial temperature, cooling rate, and number of iterations performed at a particular temperature are the three important parameters that govern the successful working of the SA procedure.

7.3.4 Swarm Intelligence

The SI algorithm is a new area of research inspired by the social behavior of bird flocking and shares many similarities with evolutionary algorithms such as the GA and the DE algorithms. PSO is a population-based stochastic optimization algorithm in SI (Kennedy and Eberhart, 2001). This algorithm is becoming popular due to its simplicity of implementation and ability to converge to a reasonably good solution quickly (Shi and Eberhart, 1998). The system is initialized with a population of random solutions and searches for optima by updating generations. However, unlike GA, PSO has no evolution operators such as crossover and mutation. In PSO, the potential solutions, called *particles*, fly through the search space by following the current optimum particles. Each particle keeps track of its coordinates in the problem space that are associated with the best solution (fitness) it has achieved so far. The fitness value is also stored and is called *pbest*. Another value that is tracked by PSO is the best value obtained so far by any particle in the neighborhood of the particle. This location is called *lbest*. When a particle takes the entire population as its topological neighbors, the best value is a global best and is called *gbest*. The PSO concept consists of changing the velocity of (accelerating) each particle toward its *pbest* and *lbest* locations at each time step. Acceleration is weighted by a random term, with separate random numbers being generated for acceleration toward the *pbest* and *lbest* locations. The velocity (v) and the position (x) of the ith swarm are manipulated according to the following two equations:

$$v_{ij}^{k+1} = \chi[\omega v_{ij}^k + C_1 R_1(p_{ij}^k - x_{ij}^k) + C_2 R_2(p_{gj}^k - x_{ij}^k)] \tag{7.3}$$

$$x_{ij}^{k+1} = x_{ij}^k + v_{ij}^{k+1} \tag{7.4}$$

where i denotes the number of particles; j denotes number of decision variables; k denotes the iteration counter; χ is the constriction factor that controls and constricts the magnitude of the velocity; g denotes the *gbest* of a particle; p denotes the *pbest* of a particle; ω denotes the inertia weight, which is often used as a parameter to control exploration and exploitation in the search space; R_1 and R_2 are random variables uniformly distributed within [0,1]; and C_1, C_2 are acceleration coefficients, also called the *cognitive* and *social* parameters, respectively. C_1 and C_2 are popularly chosen to vary within [0,2] (Chatterjee and Siarry, 2006).

The search is terminated if either of the following criteria is satisfied: (1) the number of iterations reaches the maximum allowable number or (2) the accuracy between the best solution of two successive generations reached a prespecified number. The flowchart of PSO is presented in Figure 7.1.

In the past several years, PSO has been successfully applied to many research and application areas. It is demonstrated that PSO gets better results in a faster, cheaper way than other methods. Another reason that PSO is attractive is that there are few parameters to adjust. One version, with slight variations, works well in a wide variety of applications. PSO has been used for approaches that can be used across a wide range of applications, as well as for specific applications focused on a specific requirement.

7.4 Case Study

The MBSP is situated near the village of Borkhera, about 16 km northeast of Banswara in the southern part of Rajasthan state, bordering the states of Madhya Pradesh and Gujarat in India. Global coordinates of the site are 24°22′ N latitude and 73°19′ E longitude (Water Resources Planning for Mahi River Basin, 2001). The project includes a dam, a system of canals, and two hydroelectric power houses, PH1, located near Banswara with an installed capacity of 2 × 25 MW, and PH2, near Lilvani village, with an installed capacity of 2 × 45 MW. Gross and live storage capacities of the reservoir are 2180.39 Mm³ and 1829.27 Mm³, respectively. The culturable command area (CCA) of the project (Phase 1) is 80,000 ha. Out of these, 57,531 ha have been opened for irrigation thus far. The MBSP has three main canal systems, namely, Left Main Canal (LMC), Right Main Canal (RMC), and Bhungra Canal (BC), with canal capacities of 62.53, 30.00, and

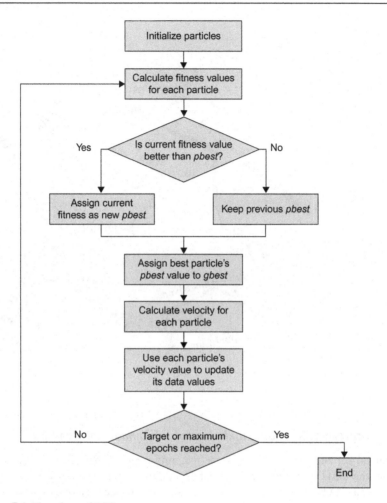

Figure 7.1 Flowchart of PSO.

3.19 cumecs, respectively. The principal crops grown in the command area in Kharif and Rabi seasons are paddy, cotton, wheat, gram, and pulses. An interstate agreement also exists to reserve 368.11 Mm^3 of Mahi water for upstream use in Madhya Pradesh and 1132.67 Mm^3 for downstream use in Gujarat (Mahi Bajaj Sagar Project Report, 1978; MBSP Report on Status June 2002 at a Glance, 2002). Figure 7.2 presents the index map of MBSP.

7.5   Mathematical Modeling

Mathematical modeling of the irrigation planning problem of MBSP command area is explained in the following sections.

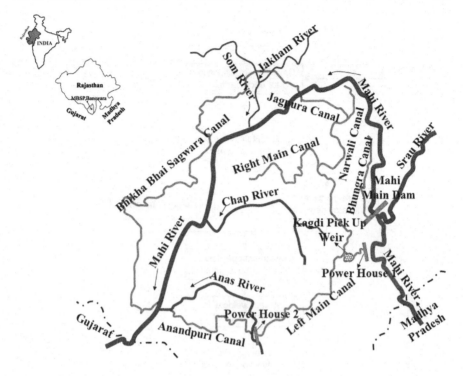

Figure 7.2 Index map of the MBSP.

7.5.1   Objective Function: Maximization of Annual Net Benefits

The annual net benefits (BEN) from the planning region under different crops after meeting the costs of seeds, fertilizer, labor, surface water, groundwater, and plant protection are to be maximized. Mathematically, this concept can be expressed as

$$\text{Max BEN} = \sum_{i=1}^{36} B_i \cdot A_i - P_{\text{GW}} \sum_{t=1}^{12} \text{GW}_t \tag{7.5}$$

where i = crop index [1 = maize (LMC), 2 = paddy (LMC), 3 = cotton (LMC), 4 = pulses (LMC), 5 = sugarcane (LMC), 6 = zaid Crop (LMC), 7 = wheat (LMC), 8 = barley (LMC), 9 = gram (LMC), 10 = barseen (LMC), 11 = mustard (LMC), 12 = fruits and vegetables (LMC), 13 = maize (RMC), 14 = paddy (RMC), 15 = cotton (RMC), 16 = pulses (RMC), 17 = sugarcane (RMC), 18 = zaid crop (RMC), 19 = wheat (LMC), 20 = barley (RMC), 21 = gram (RMC), 22 = barseen (RMC), 23 = mustard (RMC), 24 = fruits and vegetables (RMC), 25 = maize (BC), 26 = paddy (BC), 27 = cotton (BC), 28 = pulses (BC), 29 = sugarcane (BC), 30 = zaid crop (BC), 31 = wheat (BC), 32 = barley (BC), 33 = gram (BC), 34 = barseen (BC), 35 = mustard (BC), 36 = fruits and vegetables (BC)], LMC = Left Main Canal, RMC = Right Main Canal, BC = Bhungra Canal, t = time index (1 = January...12 = December). BEN = annual net benefits from the whole planning region (Rs);B_i = net return from ith crop excluding groundwater cost (Rs/ha); Ai = area of crop i grown in the command area (ha); P_{GW} = groundwater cost (Rs/Mm^3); GW$_t$ = monthly groundwater withdrawal (Mm^3).

The mathematical model is subjected to the following constraints:

Continuity Equation at the Main Reservoir

Reservoir operation includes water transfer, storage, inflow, and spillage activities. Water transfer activities consider the transport of water from the reservoir to the producing areas through canals to meet the water needs. A monthly continuity equation for the reservoir storage (Mm^3) can be expressed as

$$S_{t+1} = S_t + I_t - \text{IDB}_t - \text{PH1}_t - \text{EV}_t - \text{USMP}_t, \quad t = 1, \ldots, 12 \tag{7.6}$$

where S_{t+1} = reservoir storage volume at the end of month t or at the beginning of month $(t+1)$ (Mm^3); I_t = monthly inflows into the reservoir (Mm^3); IDB$_t$ = irrigation demand for BC for the month t (Mm^3); PH1$_t$ = water requirement for PH1 for the month t (Mm^3); EV$_t$ = monthly evaporation loss (Mm^3); USMP$_t$ = upstream requirement of water to Madhya Pradesh for the month t (Mm^3).

This constraint assumes that the monthly inflows into the reservoir are known with certainty. When uncertainty is incorporated in the inflow terms, Eq. (7.6) changes to

$$S_{t+1} - S_t + \text{IDB}_t + \text{PH1}_t + \text{EV}_t + \text{USMP}_t = I_t^{\alpha'}, \quad t = 1, \ldots, 12 \tag{7.7}$$

where $I_t^{\alpha'}$ is dependable inflow value at level α'.

Continuity Equation at Kagdi Pickup Weir

The Kagdi pickup weir acts as a balancing reservoir and water released from PH1 and the available groundwater potential should satisfy the demands of the LMC, the RMC, and PH2 for each month t. Conjunctive use of groundwater and surface water is also considered in the following equation:

$$\text{PH1}_t + \text{GW}_t \geq \text{IDL}_t + \text{IDR}_t + \text{PH2}_t, \quad t = 1, \ldots, 12 \tag{7.8}$$

where IDL$_t$ = irrigation demand for LMC for the month t (Mm^3); IDR$_t$ = irrigation demand for RMC for the month t (Mm^3); PH2$_t$ = water requirement for PH2 for the month t (Mm^3)

Command Area Limitations

The total area allocated for different crops in a particular season should be less than or equal to the CCA:

$$\sum_{i=1}^{6} A_i \leq p_{ki} \cdot \text{CCA}_L, \quad \text{Kharif season} \tag{7.9}$$

$$\sum_{i=7}^{12} A_i \leq p_{ri} \cdot \text{CCA}_L, \quad \text{Rabi season} \tag{7.10}$$

$$\sum_{i=13}^{18} A_i \leq p_{ki} \cdot \text{CCA}_R, \quad \text{Kharif season} \tag{7.11}$$

$$\sum_{i=19}^{24} A_i \leq p_{ri} \cdot \text{CCA}_R, \quad \text{Rabi season} \tag{7.12}$$

$$\sum_{i=25}^{30} A_i \leq p_{ki} \cdot \text{CCA}_B, \quad \text{Kharif season} \tag{7.13}$$

$$\sum_{i=31}^{36} A_i \leq p_{ri} \cdot \text{CCA}_B, \quad \text{Rabi season} \tag{7.14}$$

where $\text{CCA}_L = \text{CCA}$ under LMC (ha); $\text{CCA}_R = \text{CCA}$ under RMC (ha); $\text{CCA}_B = \text{CCA}$ under BC (ha); p_{ki} = percentage cropping intensity for each crop i in Kharif season (same percentage value for the three canal command areas); p_{ri} = percentage cropping intensity for each crop i in Rabi season (same percentage value for the three canal command areas).

Crop Diversion Requirements

Monthly crop diversion requirements CWR_{it} are calculated based on crop water requirements (the water required per hectare of crop activity i times the number of hectares of planted crop activity in that month t for each canal command area) and the overall efficiency. In the absence of any crop activity, CWR_{it} is taken as zero. Water releases from the reservoir must satisfy the irrigation demands of the command area:

$$\sum_{i=1}^{12} \text{CWR}_{it} A_i - \text{IDL}_t = 0, \quad t = 1, \ldots, 12 \tag{7.15}$$

$$\sum_{i=13}^{24} \text{CWR}_{it} A_i - \text{IDR}_t = 0, \quad t = 1, \ldots, 12 \tag{7.16}$$

$$\sum_{i=25}^{36} \text{CWR}_{it} A_i - \text{IDB}_t = 0, \quad t = 1, \ldots, 12 \tag{7.17}$$

where CWR_{it} = crop diversion requirements per hectare of crop i in month t (m).

Other constraints incorporated into the mathematical model are the water requirements for hydropower generation, upstream requirements, minimum and maximum areas of crops, groundwater withdrawals, canal capacity, and live storage restrictions.

7.6 Results and Discussion

A 75% dependable inflow level (Patra, 2002) is considered for the planning problem. Dependable inflow values into the reservoir for the months of June, July, August, September, and October are 14.73, 73.26, 669.13, 1066.69, and 1057.12 Mm^3, respectively. Inflows into the reservoir for other months are not significant and are disregarded. The present study is analyzed for CCA that is opened for irrigation and is targeted. Accordingly, the lower and upper limits are fixed based on cropping intensity values of 89%, i.e., 89% of 57,531 ha and 89% of 80,000 ha.

The developed mathematical model for irrigation planning is solved using the SI algorithm. The results obtained are compared with those by the DE algorithm (Vasan and Raju, 2007). The penalty function approach is used to convert the constrained problem into an unconstrained problem. Due to this, the solution falling outside the restricted solution region is given a high penalty, which forces the solution to adjust itself in such a way that after a few generations/iterations, it may fall into the restricted solution space. In the present study, a second-order penalty term is used (Deb, 2001). The total numbers of variables and constraints are 160 and 93, respectively. It has been inferred from the literature that the best set of parameters for reservoir operation problems for PSO are constriction coefficient $\chi = 0.9$, inertia weight $\omega = 1$, and acceleration coefficients $C_1 = 1$ and $C_2 = 0.5$ (Janga Reddy and Nagesh Kumar, 2009). The initial trials were tried with population sizes of 500, 750, 1,000, and 1,200. Each trial was tested with a different randomly chosen initial population. The number of iterations and accuracy between two successive iterations is set at 3,000 and 10^{-7}, respectively. The above chosen parameters are given as input to the developed model.

The results obtained by SI were compared with those of DE, and it was observed that SI produced the same results. Table 7.1 presents the cropping pattern. Table 7.2 presents the release policy for irrigation for LMC, RMC, and BC. Figure 7.3 presents the monthly storage policy values (including that of overflow), whereas Figure 7.4 presents the release policy for hydropower in PH1.

The following observations were made from the analysis of the results:

• It was observed from the LMC cropping pattern that paddy, cotton, pulses, and zaid crop reach the upper limit, whereas maize and sugarcane have a percentage deviation from the upper limit of 23.09 and 23.06, respectively. Similarly, wheat, gram, mustard, and fruits and vegetables reach the upper limit, whereas barley and barseen deviate 23.12% and 23.09%, respectively, from the upper limit. It is also observed that the ratio of total area to lower limit (expressed as a percentage) is 126.32.

Table 7.1 Optimal Cropping Pattern Obtained by SI and DE

Crops	Crop Area (ha)			Percentage Deviation from Upper Limit (%)
	Lower Limit	Upper Limit	Crop Area	
Maize (LMC)—K	9.36	12.17	9.36	23.09
Paddy (LMC)—K	12.48	16.23	16.23	0.00
Cotton (LMC)—K	28.07	36.51	36.51	0.00
Pulses (LMC)—K	15.60	20.29	20.29	0.00
Sugarcane (LMC)—K	6.24	8.11	6.24	23.06
Zaid Crop (LMC)—K	6.24	8.11	8.11	0.00
Area in Kharif (LMC)	77.99	101.42	96.74	4.61
Wheat (LMC)—R	93.57	121.71	121.71	0.00
Barley (LMC)—R	13.10	17.04	13.10	23.12
Gram (LMC)—R	78.60	102.24	102.24	0.00
Barseen (LMC)—R	5.93	7.71	5.93	23.09
Mustard (LMC)—R	6.86	8.93	8.93	0.00
Fruits & Veg. (LMC)—R	1.56	2.03	2.03	0.00
Area in Rabi (LMC)	199.62	259.66	253.94	2.20
Total area (Kharif + Rabi) (LMC)	277.61	361.08	350.68	2.88
Ratio of total area to lower limit (LMC) (%)			126.32	–
Maize (RMC)—K	7.54	10.78	7.54	30.06
Paddy (RMC)—K	10.06	14.38	14.38	0.00
Cotton (RMC)—K	22.63	32.35	32.35	0.00
Pulses (RMC)—K	12.57	17.97	17.97	0.00
Sugarcane (RMC)—K	5.03	7.19	5.03	30.04
Zaid Crop (RMC)—K	5.03	7.19	7.19	0.00
Area in Kharif (RMC)	62.86	89.86	84.46	6.01
Wheat (RMC)—R	75.42	107.82	107.82	0.00
Barley (RMC)—R	10.56	15.09	10.56	30.02
Gram (RMC)—R	63.36	90.57	90.57	0.00
Barseen (RMC)—R	4.78	6.83	4.78	30.01
Mustard (RMC)—R	5.53	7.91	7.91	0.00
Fruits & Veg. (RMC)—R	1.26	1.80	1.80	0.00
Area in Rabi (RMC)	160.91	230.02	223.44	2.86
Total area (Kharif + Rabi) (RMC)	223.77	319.88	307.90	3.75
Ratio of total area to lower limit (RMC) (%)			137.60	–
Maize (BC)—K	0.36	1.05	0.36	65.71
Paddy (BC)—K	0.48	1.40	0.48	65.71
Cotton (BC)—K	1.08	3.14	3.14	0.00
Pulses (BC)—K	0.60	1.75	0.60	65.71
Sugarcane (BC)—K	0.24	0.70	0.24	65.71
Zaid Crop (BC)—K	0.24	0.70	0.70	0.00
Area in Kharif (BC)	3.00	8.74	5.52	36.84
Wheat (BC)—R	3.60	10.47	3.60	65.62

(Continued)

Table 7.1 (Continued)

Crops	Crop Area (ha)			Percentage Deviation from Upper Limit (%)
	Lower Limit	Upper Limit	Crop Area	
Barley (BC)—R	0.50	1.47	0.50	65.99
Gram (BC)—R	3.02	8.79	8.10	7.85
Barseen (BC)—R	0.23	0.66	0.23	65.15
Mustard (BC)—R	0.26	0.77	0.77	0.00
Fruits & Veg. (BC)—R	0.06	0.17	0.06	64.71
Area in Rabi (BC)	7.67	22.33	13.26	40.62
Total area (Kharif + Rabi) (BC)	10.67	31.07	18.78	39.56
Ratio of total area to lower limit (BC) (%)			176.01	—
Total area (Kharif) (LMC + RMC + BC)	143.85	200.02	186.72	6.65
Total area (Rabi) (LMC + RMC + BC)	368.20	512.01	490.64	4.17
Total area (Kharif + Rabi) (LMC + RMC + BC)	512.05	712.03	677.36	4.87
Ratio of total area to lower limit (LMC + RMC + BC) (%)	132.28	—		

K, Kharif; R, Rabi; LMC, Left Main Canal; RMC, Right Main Canal; BC, Bhungra Canal.

Table 7.2 Release Policy for Irrigation for LMC, RMC, and BC (Mm^3)

Month	Surface Water			Groundwater
	LMC	RMC	BC	
January	56.14	49.28	2.82	3.99
February	59.11	51.92	3.01	2.34
March	16.37	14.31	0.67	0.00
April	0.90	0.73	0.03	0.00
May	2.40	1.94	0.09	0.00
June	12.58	10.84	0.57	0.00
July	22.67	19.69	1.05	24.74
August	28.16	24.52	1.46	0.00
September	26.74	23.31	1.51	0.00
October	18.75	16.38	1.25	0.00
November	14.73	12.95	0.86	0.00
December	45.11	39.63	2.30	0.00
Total	**303.66**	**265.50**	**15.62**	**31.07**

LMC, Left Main Canal; RMC, Right Main Canal; BC, Bhungra Canal.

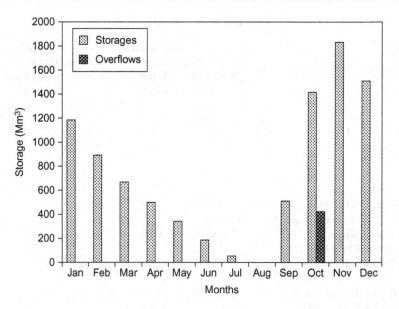

Figure 7.3 Monthly storages and overflows.

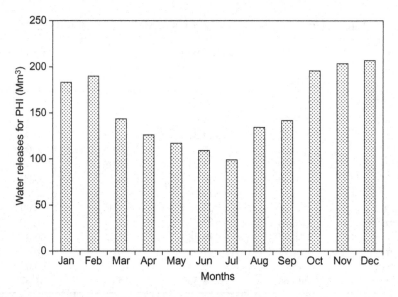

Figure 7.4 Release policy for hydropower generation in PH1.

- It was observed from the RMC cropping pattern that maize, sugarcane, barley, and barseen has an approximate deviation of 30% from the upper limit, whereas the other crops reach the upper limit. It was also observed that the ratio of total area to lower limit (expressed as a percentage) is 137.60.
- Similarly, in the case of the BC cropping pattern, only cotton, zaid crop, and mustard reach their upper limits. The ratio of total area to lower limit (expressed as a percentage) is 176.01.
- The total area irrigated is 67,736 ha, and the corresponding maximum annual net benefits from the project is Rs. 113.15 crores. The annual net benefits per ha is $\cong$ Rs. 16,705.
- It was observed from Figure 7.3 that empty storage is observed in the month of August and maximum live storage in the month of November.
- It was observed from Table 7.2 that releases for irrigation for LMC, RMC, and BC are high during the months of January, February, and December, whereas they are very low during the months of April and May.
- The annual releases (surface water) for LMC, RMC, and BC are 303.65, 265.50, and 15.63 Mm^3, respectively.
- It was observed that monthly irrigation releases to LMC, RMC, and BC are far less than the corresponding canal capacities.
- It was observed from Figure 7.4 that maximum and minimum requirements for PH1 are during December and July.
- It was observed that groundwater withdrawal is during the months of January, February, and July, which may help augment the surface water.

7.7 Conclusions

Based on the analysis of the results of the irrigation planning problem of MBSP, the following conclusions are drawn:

1. It is observed that the annual net benefits are Rs. 113.15 crores, resulting from the total irrigated area of 67,736 ha. The ratio of annual net benefits to irrigated area is Rs. 16,705 per hectare.
2. The maximum live storage is observed in the month of November, whereas significant overflows of around 424 Mm^3 are observed in the month of October. On the other hand, empty storage is observed in the month of August.
3. It is observed that the groundwater is used to its maximum potential of 31.07 Mm^3, of which 79% is in the month of July.
4. The annual releases into LMC, RMC, and BC are 303.65, 265.50, and 15.63 Mm^3, respectively. Monthly releases into the canal are always less than their maximum capacity.
5. It can be concluded that the SI algorithm can be used effectively as a potential alternative optimization algorithm to other heuristic algorithms for optimal use of a reservoir's available water resources.

References

Afshar, M.H., Rajabpour, R., 2009. Application of local and global particle swarm optimization algorithms to optimal design and operation of irrigation pumping systems. Irrigat. Drain. 58 (3), 321–331.

Bouwer, H., 2002. Integrated water management for the 21st century: problems and solutions. J. Irrigat. Drain. Eng. ASCE. 128 (4), 193−202.

Chang, F.J., Chen, L., 1998. Real-coded genetic algorithm for rule-based flood control reservoir management. Water Resour. Manag. 12 (3), 185−198.

Chatterjee, A., Siarry, P., 2006. Nonlinear inertia weight variation for dynamic adaptation in particle swarm optimization. Comput. Oper. Res. 33, 859−871.

Coelho, J.P., de Moura Oliveira, P.B., Boaventura Cunha, J., 2005. Greenhouse air temperature predictive control using the particle swarm optimisation algorithm. Comput. Electron. Agric. 49 (3), 330−344.

Cunha, M.D.C., Sousa, J., 1999. Water distribution network design optimization: simulated annealing approach. J. Water Resour. Plan. Manag. ASCE. 125 (4), 215−221.

Deb, K., 2001. Multi-Objective Optimization Using Evolutionary Algorithms. Wiley, Chichester.

Deb, K., Agrawal, R.B., 1995. Simulated binary crossover for continuous search space. Complex Syst. 9, 115−148.

Frederiksen, H.D., 1996. Water crisis in developing world: misconceptions about solutions. J. Water Res. Plan. Manag. ASCE. 122 (2), 79−87.

Goldberg, D.E., 1989. Genetic Algorithms in Search, Optimization and Machine Learning. Addison-Wesley, New York, NY.

Gupta, P., Vasan, A., Raju, K.S., 2009. Multiobjective differential evolution and differential evolution for irrigation planning. Proceedings of World Environmental and Water Resources Congress 2009, May 17−21, Kansas City, Missouri, ASCE, 1−8.

Janga Reddy, M., Nagesh Kumar, D., 2007. Optimal reservoir operation for irrigation of multiple crops using elitist-mutated particle swarm optimization. Hydrolog. Sci. J. 52 (4), 686−701.

Janga Reddy, M., Nagesh Kumar, D., 2009. Performance evaluation of elitist-mutated multiobjective particle swarm optimization for integrated water resources management. J. Hydroinform. 11 (1), 79−88 (IWA Publishing).

Kennedy, J., Eberhart, R.C., 2001. Swarm Intelligence. Morgan Kaufmann, San Mateo, CA.

Kirkpatrick, S., Gelatt Jr., C.J., Vecchi, M.P., 1983. Optimization by simulated annealing. Science. 220, 671−680.

Kuo, S.F., Merkley, G.P., Liu, C.W., 2000. Decision support for irrigation project planning using a genetic algorithm. Agric. Water Manag. 45, 243−266.

Mahi Bajaj Sagar Project Report, 1978. Government of Rajasthan, Banswara, Rajasthan, India.

Mardle, S., Pascoe, S., 2000. Use of evolutionary methods for bioeconomic optimization models: an application to fisheries. Agric. Syst. 66, 33−49.

MBSP Report on Status June 2002 at a Glance, 2002. Government of Rajasthan, Banswara, Rajasthan, India.

Patra, K.C., 2002. Hydrology and Water Resources Engineering. Narosa Publishing House, New Delhi.

Price, K., Storn, R., 1997. Differential evolution—a simple evolution strategy for fast optimization. Dr. Dobb's Journal. 22, 18−24 and 78.

Price, K., Storn, R., Lampinen, J.A., 2005. Differential Evolution: A Practical Approach to Global Optimization. Springer, New York, NY.

Raju, K.S., Nagesh Kumar, D., 2004. Irrigation planning using genetic algorithms. Water Resour. Manag. 18 (2), 163−176.

Raju, K.S., Vasan, A., Gupta, P., Ganesan, K., Mathur, H., 2012. Multiobjective differential application to irrigation planning. ISH J. Hydraulic Eng. 18 (1), 77−93 (Taylor & Francis).

Ramesh, S.V.T., Simonovic, S.P., 2002. Optimal operation of reservoir systems using simulated annealing. Water Resour. Manag. 16 (5), 401−428.

Ranjithan, S.R., 2005. Role of evolutionary computation in environmental and water resources systems analysis. J. Water Resour. Plan. Manag. ASCE. 131 (1), 1−2.

Rao, S.V.N., Bhallamudi, S.M., Thandaveswara, B.S., Mishra, G.C., 2004. Conjunctive use of surface and groundwater for coastal and deltaic systems. J. Water Resour. Plan. Manag. ASCE. 130 (3), 255−267.

Rao, S.V.N., Thandaveswara, B.S., Bhallamudi, S.M., Srinivasulu, V., 2003. Optimal groundwater management in deltaic regions using simulated annealing and neural networks. Water Resour. Manag. 17 (6), 409−428.

Schardong, A., Vasan, A., Simonovic, S.P., 2012. A multi-objective evolutionary approach to optimal reservoir optimization. J. Comput. Civ. Eng. ASCE.doi:http://dx.doi.org/10.1061/(ASCE)CP.1943-5487.0000213.

Sharif, M., Wardlaw, R., 2000. Multireservoir systems optimization using genetic algorithms: case study. J. Comput. Civ. Eng. ASCE. 14 (4), 255−263.

Shawn Matott, L., Rabideau, A., Craig, J.R., 2006. Pump and treat optimization using analytic element method flow models. Adv. Water Resour. 29 (5), 760−775.

Shi, Y., Eberhart, R.C., 1998. A modified particle swarm optimizer. Proceedings of the IEEE International Conference on Evolutionary Computation, IEEE Press, Picataway, NJ, 69−73.

Vasan, A., Raju, K.S., 2007. Application of differential evolution for irrigation planning: an Indian case study. Water Resour. Manag. 21 (2), 1393−1407.

Vasan, A., Raju, K.S., 2009. Comparative analysis of simulated annealing, simulated quenching and genetic algorithms for optimal reservoir operation. Appl. Soft Comput. 9 (1), 274−281.

Vasan, A., Simonovic, S.P., 2010. Optimization of water distribution networks using differential evolution. J. Water Resour. Plan. Manag. ASCE. 136 (2), 279−287.

Wardlaw, R., Bhaktikul, K., 2004. Comparison of genetic algorithm and linear programming approaches for lateral canal scheduling. J. Irrigat. Drain. Eng. ASCE. 130 (4), 311−317.

Wardlaw, R., Sharif, M., 1999. Evaluation of genetic algorithms for optimal reservoir system operation. J. Water Resour. Plan. Manag. ASCE. 125 (1), 25−33.

Water Resources Planning for Mahi River Basin, 2001. Government of Rajasthan, Investigation, Design and Research (Irrigation) Unit, Jaipur.

Wegley, C., Eusuff, M., Lansey, K., 2000. Determining pump operations using particle swarm optimization. ASCE Conference Proceedings, Proceedings of Joint Conference on Water Resource Engineering and Water Resources Planning and Management, Section 52, Chapter 3, ASCE, Minneapolis, USA.

Part Two

Geotechnical Engineering

8 Artificial Intelligence in Geotechnical Engineering: Applications, Modeling Aspects, and Future Directions

Mohamed A. Shahin

Department of Civil Engineering, Curtin University, Perth, WA, Australia

8.1 Introduction

Geotechnical engineering deals with materials (e.g., soil and rock) that, by their very nature, exhibit varied and uncertain behavior due to the imprecise physical processes associated with the formation of these materials. Modeling the behavior of such materials is complex and usually beyond the ability of most traditional forms of physically based engineering methods. Artificial intelligence (AI) is becoming more popular and particularly amenable to modeling the complex behavior of most geotechnical engineering materials because it has demonstrated superior predictive ability compared to traditional methods. Over the last decade, AI has been applied successfully to virtually every problem in geotechnical engineering. However, despite this success, AI techniques are still facing classical opposition due to some inherent reasons such as lack of transparency, knowledge extraction, and model uncertainty, which will be discussed in detail in this chapter.

Among the available AI techniques are artificial neural networks (ANNs), genetic programming (GP), evolutionary polynomial regression (EPR), support vector machines, M5 model trees, and K-nearest neighbors (Elshorbagy et al., 2010). In this chapter, the focus will be on three AI techniques, including ANNs, GP, and EPR. These three techniques are selected because they have been proved to be the most successful applied AI techniques in geotechnical engineering. Of these, ANN is by far the most commonly used one.

8.2 AI Applications in Geotechnical Engineering

In this section, the applications of the three selected AI techniques (i.e., ANNs, GP, and EPR) are briefly examined. Note that only post-2005 ANN applications

Metaheuristics in Water, Geotechnical and Transport Engineering. DOI: http://dx.doi.org/10.1016/B978-0-12-398296-4.00008-8

are acknowledged, for brevity; interested readers are referred to Shahin et al. (2001), where the pre-2001 applications are reviewed in some detail, and Shahin et al. (2009), where the post-2001 papers are briefly examined.

The behavior of foundations (deep and shallow) in soils is complex, uncertain, and not yet entirely understood. This fact has encouraged many researchers to apply the AI techniques to the prediction of behavior of foundations. For example, ANNs have been used extensively for modeling the axial and lateral load capacities of pile foundations in compression and uplift, including driven piles (Ahmad et al., 2007; Ardalan et al., 2009; Das and Basudhar, 2006; Pal and Deswal, 2008; Shahin, 2010), drilled shafts (Goh et al., 2005; Shahin, 2010), and ground anchor piles (Shahin and Jaksa, 2005, 2006). Predictions of the settlement and load–settlement response of piles have also been modeled by ANNs (Alkroosh and Nikraz, 2011b; Ismail and Jeng, 2011; Pooya Nejad et al., 2009). On the other hand, the prediction of the behavior of shallow foundations has been investigated by ANNs, including settlement estimation (Chen et al., 2006; Shahin et al., 2005a) and bearing capacity (Kuo et al., 2009; Padmini et al., 2008). The GP applications in foundations include the bearing capacity of piles (Alkroosh and Nikraz, 2011a; Gandomi and Alavi, 2012), uplift capacity of suction caissons (Gandomi et al., 2011), and settlement of shallow foundations (Rezania and Javadi, 2007). The single EPR application in foundations is the uplift capacity of suction caissons (Rezania and Javadi, 2008).

Classical constitutive modeling based on elasticity and plasticity theories has only a limited capability to simulate the behavior of geomaterials properly. This is attributed to reasons associated with the formulation complexity, idealization of material behavior, and excessive empirical parameters (Adeli, 2001). In this regard, AI techniques have been proposed as a reliable and practical alternative to modeling the constitutive monotonic and hysteretic behavior of geomaterials, including ANNs (Banimahd et al., 2005; Chen et al., 2010; Fu et al., 2007; Garaga and Latha, 2010; Johari et al., 2011; Najjar and Huang, 2007; Obrzud et al., 2009; Peng et al., 2008; Shahin and Indraratna, 2006), GP (Alkroosh and Nikraz, 2012; Cabalar et al., 2009; Shahnazari et al., 2010), and EPR (Javadi and Rezania, 2009).

Liquefaction during earthquakes is one of the very dangerous ground failure phenomena that can cause a large amount of damage to most civil engineering structures. Although the liquefaction mechanism is well known, the prediction of liquefaction potential is very complex (Baziar and Ghorbani, 2005). This fact has attracted many researchers to investigate the applicability of AI techniques, including ANNs, for predicting liquefaction (Alavi and Gandomi, 2011a; Baziar and Ghorbani, 2005; Hanna et al., 2007a,b; Javadi et al., 2006; Khozaghi and Choobbasti, 2007; Samui and Sitharam, 2011; Shuh-Gi and Ching-Yinn, 2009; Young-Su and Byung-Tak, 2006), GP (Alavi and Gandomi, 2011b, 2012; Baziar et al., 2011; Gandomi and Alavi, 2011, 2012; Javadi et al., 2006; Kayadelen, 2011), and EPR (Rezania et al., 2010, 2011).

Geotechnical properties of soils are controlled by factors such as mineralogy, fabric, and pore water, and the interactions of these factors are difficult to establish solely by traditional statistical methods due to their interdependence

(Yang and Rosenbaum, 2002). Based on the application of AI techniques, methodologies have been developed for estimating several soil properties including, for ANNs, preconsolidation pressure and soil compressibility (Celik and Tan, 2005; Jianping et al., 2011; Park and Lee, 2011), shear strength parameters and stress history (Baykasoglu et al., 2008; Byeon et al., 2006; Dincer, 2011; Gunaydin et al., 2010; Kaya, 2009; Kayadelen et al., 2009; Narendara et al., 2006; Tawadrous et al., 2009), soil swelling and swell pressure (Ashayeri and Yasrebi, 2009; Doostmohamadi et al., 2008; Erzin, 2007; Ikizler et al., 2009), lateral earth pressure (Das and Basudhar, 2005; Uncuoglu et al., 2008), soil permeability (Erzin et al., 2009; Park, 2011), and properties of soil dynamics (Baziar and Ghorbani, 2005; Garcia et al., 2006; Kamatchi et al., 2010; Kogut, 2007; Shafiee and Ghate, 2008; Singh and Singh, 2005; Tsompanakis et al., 2009). For GP, properties include hydraulic conductivity and shear strength (Johari et al., 2006; Kayadelen et al., 2009; Mollahasani et al., 2011; Narendara et al., 2006; Parasuraman et al., 2007), and for EPR, they include soil permeability (Ahangar-Asr et al., 2011).

Other applications of ANNs in geotechnical engineering include earth-retaining structures (Goh and Kulhawy, 2005; Kung et al., 2007; Yildiz et al., 2010), dams (Kim and Kim, 2008; Yu et al., 2007), blasting (Lu, 2005), mining (Singh and Singh, 2005), rock mechanics (Cevik et al., 2010; Garcia and Roma, 2009; Ma et al., 2006; Maji and Sitharam, 2008; Sarkar et al., 2010; Singh et al., 2005, 2007; Sitharam et al., 2008), site characterization (Caglar and Arman, 2007), tunneling and underground openings (Alimoradi et al., 2008; Boubou et al., 2010; Chen et al., 2009; Hajihassani et al., 2011; Neaupane and Adhikari, 2006; Santos et al., 2008; Tsekouras et al., 2010; Yoo and Kim, 2007), slope stability and landslides (Cho, 2009; Das et al., 2011a; Ferentinou and Sakellariou, 2007; Kanungo et al., 2006; Lee et al., 2008; Sakellariou and Ferentinou, 2005; Samui and Kumar, 2006; Wang and Sassa, 2006), deep excavation (Soroush et al., 2006), soil composition and classification (Bhattacharya and Solomatine, 2006; Kurup and Griffin, 2006), soil stabilization (Das et al., 2011b; Liao et al., 2011; Park and Kim, 2011; Tekin and Akbas, 2011), scouring of soils (Firat and Gungor, 2008; Zounemat-Kermani et al., 2009), and soil compaction and permeability (Abdel-Rahman, 2008; Sinha and Wang, 2008; Sivrikaya and Soycan, 2011; Sulewska, 2010). Other applications of GP include dams (Alavi and Gandomi, 2011b), slope stability (Adarsh and Jangareddy, 2010; Alavi and Gandomi, 2011b), tunneling (Alavi and Gandomi, 2011b; Gandomi and Alavi, 2012), soil classification (Alavi et al., 2010), and rock modeling (Feng et al., 2006). Other applications of EPR include slope stability (Ahangar-Asr et al., 2010) and compaction characteristics (Ahangar-Asr et al., 2011).

8.3 Overview of AI

AI is a computational method that attempts to mimic, in a very simplistic way, human cognition capability (e.g., emulating the operation of the human brain at the neural level) to solve engineering problems that have defied solution using conventional computational techniques (Flood, 2008). The essence of AI techniques in solving any engineering problem is to learn by examples of data inputs and outputs presented to them so that the subtle functional relationships among the data are captured, even if the underlying relationships are unknown or the physical meaning is difficult to explain. Thus, AI models are data-driven models (DDMs) that rely on the data alone to determine the structure and parameters that govern a phenomenon (or system) and do not make any assumptions about the physical behavior of the system. This is in contrast to most physically based models that use the first principles (e.g., physical laws) to derive the underlying relationships of the system and usually justifiably simplified with many assumptions, and require prior knowledge about the nature of the relationships among the data. This is one of the main benefits of AI techniques when compared to most physically based empirical and statistical methods.

The AI modeling philosophy is similar to a number of conventional statistical models, in the sense that both are attempting to capture the relationship between a historical set of model inputs and corresponding outputs. For example, imagine a set of x-values and corresponding y-values in two-dimensional space, where $y = f(x)$. The objective is to find the unknown function f, which relates the input variable x to the output variable y. In a linear regression statistical model, the function f can be obtained by changing the slope $\tan\phi$ and intercept β of the straight line in Figure 8.1A, so that the error between the actual outputs and the outputs of the straight line is minimized. The same principle is used in AI

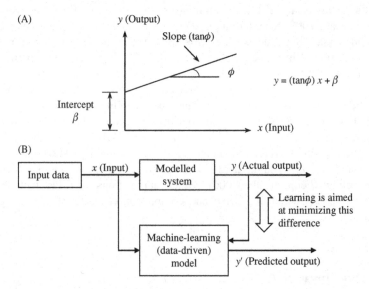

Figure 8.1 Linear regression versus AI modeling. (A) Linear regression modeling (Shahin et al., 2001); (B) AI data-driven modeling.
Source: Adapted from Solomatine and Ostfeld (2008).

models. AI can form the simple linear regression model by having one input and one output (Figure 8.1B). AI uses available data to map between the system inputs and the corresponding outputs using machine learning by repeatedly presenting examples of the model inputs and outputs (training) in order to find the function $y = f(x)$ that minimizes the error between the historical (actual) outputs and the outputs predicted by the AI model.

If the relationship between x and y is nonlinear, statistical regression analysis can be applied successfully only if prior knowledge of the nature of the nonlinearity exists. On the contrary, this prior knowledge of the nature of the non-linearity is not required for AI models. In the real world, it is likely that complex and highly nonlinear problems are encountered, and in such situations, traditional regression analyses are inadequate (Gardner and Dorling, 1998). In this section, a brief overview of three selected AI techniques (i.e., ANNs, GP, and EPR) is presented below.

8.3.1 Artificial Neural Networks

ANNs are a form of AI that attempt to mimic the function of the human brain and nervous system. Although the concept of ANNs was first introduced in 1943 (McCulloch and Pitts, 1943), research into applications of ANNs has blossomed since the introduction of the back-propagation training algorithm for feed-forward multilayer perceptrons in 1986 (Rumelhart et al., 1986). Many authors have described the structure and operation of ANNs (Fausett, 1994; Zurada, 1992). Typically, the architecture of an ANN consists of a series of processing elements (PEs), or nodes, that are usually arranged in layers: an input layer, an output layer, and one or more hidden layers, as shown in Figure 8.2.

The input from each PE in the previous layer x_i is multiplied by an adjustable connection weight w_{ji}. At each PE, the weighted input signals are summed and a threshold value θ_j is added. This combined input I_j is then passed through a nonlinear transfer function $f(\cdot)$ to produce the output of the PE y_j. The output of

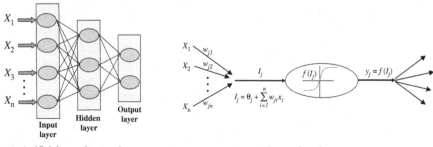

Artificial neural network   Processing element

Figure 8.2 Typical structure and operation of ANNs (Shahin et al., 2009).

one PE provides the input to the PEs in the next layer. This process is summarized in Eqs. (8.1) and (8.2) and illustrated in Figure 8.2:

$$I_j = \sum w_{ji}x_i + \theta_j \quad \text{summation} \tag{8.1}$$

$$y_j = f(I_j) \quad \text{transfer} \tag{8.2}$$

The propagation of information in an ANN starts at the input layer, where the input data are presented. The network adjusts its weights on the presentation of a training data set and uses a learning rule to find a set of weights that produces the input and output mapping that has the smallest possible error. This process is called *learning* or *training*. Once the training phase of the model has been successfully accomplished, the performance of the trained model needs to be validated using an independent validation set. The main steps involved in the development of an ANN, as suggested by Maier and Dandy (2000a), are illustrated in Figure 8.3, and several of these steps are discussed in some detail in the following section.

8.3.2 Genetic Programming

GP is an extension of genetic algorithms (GAs), which are evolutionary computing search (optimization) methods that are based on the principles of genetics and natural selection. In GA, some of the natural evolutionary mechanisms, such as reproduction, crossover, and mutation, are usually implemented to solve function identification problems. GA was first introduced by Holland (1975) and developed by Goldberg (1989), whereas GP was invented by Cramer (1985) and further developed by Koza (1992). The difference between GA and GP is that GA is generally used to evolve the best values for a given set of model parameters (i.e., parameter optimization), whereas GP generates a structured representation for a set of input variables and corresponding outputs (i.e., modeling or programming).

GP manipulates and optimizes a population of computer models (or programs) that have been proposed to solve a particular problem, so that the model that best fits the problem is obtained. A detailed description of GP can be found in many publications (e.g., Koza, 1992), and a brief overview is given herein. The modeling steps by GP start with the creation of an initial population of computer models (also called *individuals* or *chromosomes*) that are composed of two sets (i.e., a set of functions and a set of terminals) that are defined by the user to suit a certain problem. The functions and terminals are selected randomly and arranged in a tree-like structure to form a computer model that contains a root node, branches of functional nodes, and terminals, as shown by the typical example of GP tree representation in Figure 8.4. The functions can contain basic mathematical operators (e.g., $+$, $-$, $\times$, $/$), Boolean logic functions (e.g., AND, OR, and NOT), trigonometric functions (e.g., sin and cos), or any other user-defined functions. The terminals, on the other hand, may consist of numerical constants, logical constants, or variables.

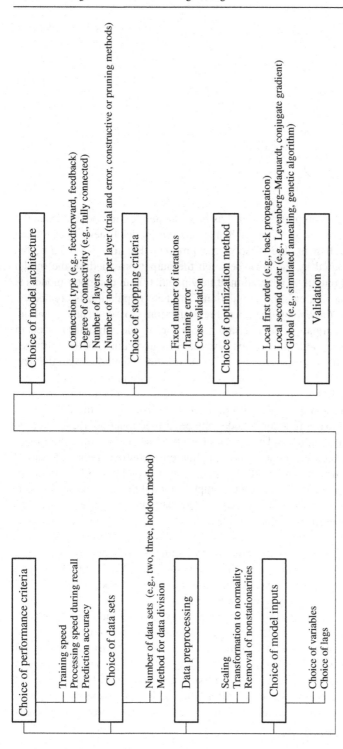

Figure 8.3 The main steps in ANN model development (Maier and Dandy, 2000a).

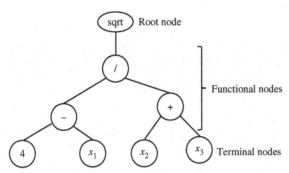

Figure 8.4 A typical example of GP tree representation for the function $[(4 - x_1)/(x_2 + x_3)]^2$.

Once a population of computer models has been created, each model is executed using available data for the problem at hand, and the model fitness is evaluated depending on how well it is able to solve the problem. For many problems, the model fitness is measured by the error between the output provided by the model and the desired actual output. A generation of new population of computer models is then created to replace the existing population. The new population is created by applying the following three main operations: reproduction, crossover, and mutation. These three operations are applied on certain proportions of the computer models in the existing population, and the models are selected according to their fitness. Reproduction is copying a computer model from an existing population into the new population without alteration. Crossover is genetically recombining (swapping) randomly chosen parts of two computer models. Mutation is replacing a randomly selected functional or terminal node with another node from the same function or terminal set, provided that a functional node replaces a functional node and a terminal node replaces a terminal node. The evolutionary process of evaluating the fitness of an existing population and producing new population is continued until a termination criterion is met, which can be either a particular acceptable error or a certain maximum number of generations. The best computer model that appears in any generation identifies the result of the GP process. There are currently three variants of GP available in the literature, including linear genetic programming, gene expression programming (GEP), and multi expression programming (Alavi and Gandomi, 2011b). More recently, multi-stage genetic programming (Gandomi and Alavi, 2011) and multi-gene genetic programming (Gandomi and Alavi, 2012) are also introduced. However, GEP is the most commonly used GP method in geotechnical engineering and is thus described in some detail next.

GEP was developed by Ferreira (2001) and utilizes the evolution of mathematical equations that are encoded linearly in chromosomes of fixed length and expressed nonlinearly in the form of expression trees (ETs) of different sizes and shapes. The chromosomes are composed of multiple genes, each of which is encoded in a smaller subprogram or subexpression tree (Sub-ET). Every gene has a constant length and consists of a head and a tail. The head can contain functions and terminals (variables and constants) required to code any expression, whereas the tail solely contains terminals.

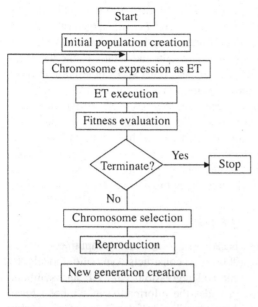

Figure 8.5 The algorithm of GEP (Teodorescu and Sherwood, 2008).

The genetic code represents a one-to-one relationship between the symbols of the chromosome and the function or terminal. The process of information decoding from chromosomes to ETs is called *translation*, which is based on sets of rules that determine the spatial organization of the functions and terminals in the ETs and the type of interaction (link) between the Sub-ETs (Ferreira, 2001). The main strength of GEP is that the creation of genetic diversity is extremely simplified as the genetic operators work at the chromosome level. Another strength is regarding the unique multigenetic nature of GEP, which allows the evolution of more powerful models/programs composed of several subprograms (Ferreira, 2001).

The major steps in the GEP procedure are schematically represented in Figure 8.5. The process begins with choosing sets of functions F and terminals T to create randomly an initial population of chromosomes of mathematical equations. One could choose, for example, the four basic arithmetic operators to form the set of functions, i.e., $F = \{ +, -, \times, / \}$, and the set of terminals will obviously consist of the independent variables of a particular problem; for example, for a problem that has two independent variables, x_1 and x_2 would be $T = \{x_1, x_2\}$. Choosing the chromosomal architecture, i.e., the number and length of genes and linking functions (e.g., addition, subtraction, multiplication, and division), is also part of this step. The chromosomes are then given in the form of ETs of different sizes and shapes, and the performance of each individual chromosome is evaluated by comparing the predicted and actual values of presented data. One could measure the fitness f_i of an individual chromosome i using the following expression:

$$f_i = \sum_{j=1}^{C_t}(M - |C_{(i,j)} - T_j|),$$
(8.3)

where M is the range of selection, $C_{(i,j)}$ is the value returned by the individual chromosome i for fitness case j (out of C_t fitness cases), and T_j is the target value for the fitness case j. There are, of course, other fitness functions available that can be appropriate for different problems. If the desired results (according to the measured errors) are satisfactory, the GEP process is stopped; otherwise, some chromosomes are selected and mutated to reproduce new chromosomes, and the process is repeated for a certain number of generations or until the desired fitness score is obtained.

Figure 8.6 shows a typical example of a chromosome with one gene, and its ET and corresponding mathematical equation. It can be seen that, while the head of a gene contains arithmetic and trigonometric functions (e.g., $+, -, \times, /, \sqrt{}$, sin, cos), the tail includes constants and independent variables (e.g., 1, a, b, c). The ET is codified reading the ET from left to right in the top line of the tree and from top to bottom.

8.3.3   Evolutionary Polynomial Regression

EPR is a hybrid regression technique based on evolutionary computing that was developed by Giustolisi and Savic (2006). It constructs symbolic models by integrating the soundest features of numerical regression, with GP and symbolic regression (Koza, 1992). This strategy provides the information in symbolic form, as usually defined in the mathematical literature. The following two steps roughly describe the underlying features of the EPR technique, which aimed to search for polynomial structures representing a system. In the first step, the selection of exponents for polynomial expressions is carried out, employing an evolutionary searching strategy by means of GAs (Goldberg, 1989). In the second step, numerical regression using the least squares method is conducted, aiming to compute the coefficients of the previously selected polynomial terms. The general form of expression in EPR can be presented as follows (Giustolisi and Savic, 2006):

$$y = \sum_{j=i}^{m} F(X, f(X), a_j) + a_0 \tag{8.4}$$

where y is the estimated vector of output of the process, m is the number of terms of the target expression, F is a function constructed by the process, X is the matrix

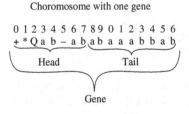

Choromosome with one gene

0 1 2 3 4 5 6 7 8 9 0 1 2 3 4 5 6
+ * Q a b − a b a b a a a b b a b

Head   Tail

Gene

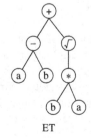

ET

Figure 8.6 Schematic representation of a chromosome with one gene and its ET and corresponding mathematical equation (Kayadelen, 2011).

$(a-b)+\sqrt{(a*b)}$

Corresponding mathematical equation

of input variables, f is a function defined by the user, and a_j is a constant. A typical example of EPR pseudo-polynomial expression that belongs to the class of Eq. (8.4) is as follows (Giustolisi and Savic, 2006):

$$\hat{Y} = a_0 + \sum_{j=i}^{m} a_j \cdot (X_1)^{\text{ES}(j,1)} \cdots (X_k)^{\text{ES}(j,k)} \cdot f[(X_1)^{\text{ES}(j,k+1)} \cdots (X_k)^{\text{ES}(j,2k)}] \qquad (8.5)$$

where $\hat{Y}$ is the vector of target values, m is the length of the expression, a_j is the value of the constants, X_i is the vector(s) of the k candidate inputs, ES is the matrix of exponents, and f is a function selected by the user.

EPR is suitable for modeling physical phenomena, based on two features (Savic et al., 2006): (1) the introduction of prior knowledge about the physical system/process, to be modeled at three different times, namely before, during, and after EPR modeling calibration; and (2) the production of symbolic formulas, enabling data mining to discover patterns that describe the desired parameters. In EPR feature (1), before the construction of the EPR model, the modeler selects the relevant inputs and arranges them in a suitable format according to their physical meaning. During the EPR model construction, model structures are determined by following user-defined settings such as general polynomial structure, user-defined function types (e.g., natural logarithms, exponentials, and tangential hyperbolics) and searching strategy parameters. The EPR starts from true polynomials and also allows for the development of nonpolynomial expressions containing user-defined functions (e.g., natural logarithms). After EPR model calibration, an optimum model can be selected from among the series of models returned. The optimum model is selected based on the modeler's judgment, in addition to statistical performance indicators such as the coefficient of determination. A typical flow diagram of the EPR procedure is shown in Figure 8.7, and a detailed description of the technique can be found in Giustolisi and Savic (2006).

8.3.4 Current Development and Future Directions in the Utilization of AI

Based on the author's experience, there are several factors in the use of AI techniques that need to be systematically investigated when developing AI models, so that model performance can be improved. These factors include the determination of adequate model inputs, data division, data preparation, model validation, model robustness, model transparency and knowledge extraction, model extrapolation, and model uncertainty. Some of these factors have received recent attention; others require further research. Each of these is discussed below.

Determination of Model Inputs

An important step in developing AI models is to select the model input variables that have the most significant impact on model performance. A good subset of

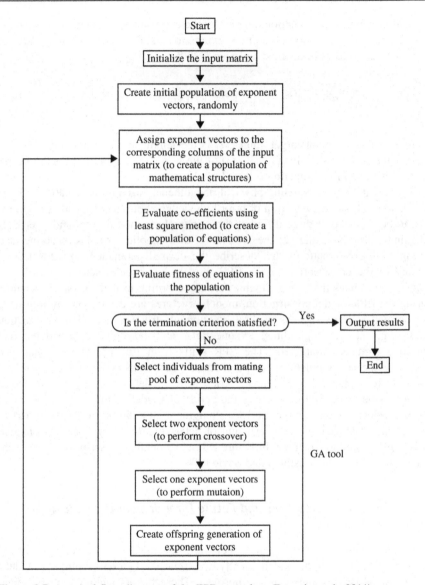

Figure 8.7 A typical flow diagram of the EPR procedure (Rezania et al., 2011).

input variables can substantially improve model performance. Presenting as large a number of input variables as possible to AI models usually increases the model size, resulting in a decrease in processing speed and model efficiency. A number of techniques have been suggested in the literature to assist with the selection of input variables. An approach that is usually utilized in the field of geotechnical engineering is that appropriate input variables can be selected in advance based on *a priori* knowledge. Another approach used by some researchers (Goh, 1994; Najjar et al., 1996; Ural and Saka, 1998) is to develop many models with different combinations of input variables and to select the model that has the best performance. A stepwise technique described by Maier and Dandy (2000b) can also be used in which separate models are trained, each using only one of the available variables as model inputs, and the model that performs the best is then retained, combining the variable that results in the best performance with each of the remaining variables. This process should be repeated for an increasing number of input variables, until the addition of additional variables results in no further improvement in model performance. Another useful approach is to employ a GA to search for the best sets of input variables (NeuralWare, 1997). For each possible set of input variables chosen by the GA, a model is trained and used to rank different subsets of possible inputs. A set of input variables derives its fitness from the model error obtained based on those variables. The adaptive spline modeling of observation data algorithm proposed by Kavli (1993) is also a useful technique that can be used for developing parsimonious models by automatically selecting a combination of model input variables that have the most significant impact on the outputs.

A potential shortcoming of these approaches is that they are model based. In other words, the determination as to whether a parameter input is significant or not is dependent on the error of a trained model, which is not only a function of the inputs but also model structure and calibration. This can potentially obscure the impact of different model inputs. In order to overcome this limitation, model-free approaches can be utilized, which use linear dependence measures, such as correlation, or nonlinear measures of dependence, such as mutual information, to obtain the significant model inputs prior to developing the AI models (Bowden et al., 2005, May et al., 2008).

Data Division

As described earlier, AI models are similar to conventional statistical models in the sense that model parameters are adjusted in the model calibration phase (training) so as to minimize the error between model outputs and the corresponding measured values for a particular data set (the training set). AI models perform best when they do not extrapolate beyond the range of the data used for calibration. Therefore, the purpose of AI models is to nonlinearly interpolate (generalize) in high-dimensional space between the data used for calibration. Unlike conventional statistical models, AI models generally have a large number of model parameters and can therefore overfit the training data, especially if the training data are noisy. In other words, if the number of degrees of freedom of the model is large compared with the number of data points used for calibration, the model might no longer fit the general trend, as desired, but might learn the idiosyncrasies of the particular data points used for calibration leading to memorization, rather than generalization. Consequently, a separate validation set is needed to ensure that the model can generalize within the range of the data used for calibration. It is a common practice to divide the available data into two subsets: a training set, to construct the model, and an independent validation set, to estimate the model performance in a deployed environment. Usually, two-thirds of the data are suggested for model training and one-third for validation (Hammerstrom, 1993). A modification of this data division method is cross-validation in ANNs (Stone, 1974), in which the data are divided into three sets: training, testing, and validation. The training set is used to adjust the model parameters, whereas the testing set is used to check the performance of the model at various stages of training and to determine when to stop training to avoid overfitting. The validation set is used to estimate the performance of the trained network in the deployed environment. In an attempt to find the optimal proportion of the data to use for training, testing, and validation in ANN models, Shahin et al. (2004) investigated the impact of the proportion of data used in various subsets on model performance for a case study of settlement prediction of shallow foundations and found that there is no clear relationship between the proportion of data for training, testing, and validation and model performance; however, they found that the best result was obtained when 20% of the data were used for validation and the remaining data were divided into two parts, 70% for training and 30% for testing.

In many situations, the available data are small enough to be solely devoted to model training, and collecting any more data for validation is difficult. In this situation, the *leave-k-out* method (Masters, 1993), which involves holding back a small fraction of the data for validation and using the rest of the data for training, can be used. After training, the performance of the trained network has to be estimated with the aid of the validation set. A different small subset of data is held back and the model is trained and tested again. This process is repeated many times with different subsets until an optimal model can be obtained from the use of all of the available data.

In the majority of AI applications in geotechnical engineering, the data were divided into their subsets on an arbitrary basis. However, some studies have found that the way the data are divided can have a significant impact on the results obtained (Tokar and Johnson, 1999). As AI models have difficulty extrapolating beyond the range of the data used for calibration, in order to develop the best AI models, given the available data, all of the patterns that are contained in the data need to be included in the calibration set. For example, if the available data contain extreme data points that were excluded from the calibration data set, the model cannot be expected to perform well because the validation data will test the model's extrapolation ability rather than its interpolation ability. If all of the patterns that are contained in the available data are contained in the calibration set, the toughest evaluation of the generalization ability of the model is if all the patterns (and not just a subset) are contained in the validation data. In addition, if cross-validation is used in ANN models, the results obtained using the testing set have to be representative of those obtained using the training set, as the testing set is used to decide when to stop training or, for example, which model architecture or learning rate is optimal. Consequently, the statistical properties (e.g., mean and standard deviation) of the various data subsets (e.g., training, testing, and validation) need to be similar to ensure that each subset represents the same statistical population

(Masters, 1993). If this is not the case, it may be difficult to judge the validity of AI models.

This fact has been recognized for some time (ASCE, 2000; Maier and Dandy, 2000b; Masters, 1993), and several studies have used ad hoc methods to ensure that the data used for calibration and validation have the same statistical properties (Braddock et al., 1998; Campolo et al., 1999; Ray and Klindworth, 2000; Tokar and Johnson, 1999). Masters (1993) strongly confirms the above strategy of data division as he says, "if our training set is not representative of the data on which the network will be tested, we will be wasting our time." However, it was not until a few years ago that systematic approaches for data division have been proposed in the literature. Bowden et al. (2002) used a GA to minimize the difference between the means and standard deviations of the data in the training, testing, and validation sets. While this approach ensures that the statistical properties of the various data subsets are similar, there is still a need to choose which proportion of the data to use for training, testing, and validation. Kocjancic and Zupan (2000) and Bowden et al. (2002) used a self-organizing map (SOM) to cluster high-dimensional input and output data in two-dimensional space and divided the available data so that values from each cluster were represented in the various data subsets. This ensures that data in the different subsets were representative of each other and had the additional advantage that there was no need to decide what percentage of the data to use for training, testing, and validation. The major shortcoming of this approach is that there are no guidelines for determining the optimum size and shape of the SOM (Cai et al., 1994; Giraudel and Lek, 2001). This has the potential to have a significant impact on the results obtained, as the underlying assumption of the approach is that the data points in one cluster provide the same information in high-dimensional space. However, if the SOM is too small, there may be significant intracluster variation. Conversely, if the map is too large, too many clusters may contain single data points, making it difficult to choose representative subsets. To overcome the problem of determining the optimum size of clusters associated with using SOMs, Shahin et al. (2004) introduced a data division approach that utilizes a fuzzy clustering technique so that data division can be carried out in a systematic manner.

Data Preparation

Data preparation is the process of presenting the data in a suitable form before they are presented to the AI techniques. Once the available data have been divided into their subsets (e.g., training and validation), it is important to preprocess the data to ensure that all variables receive equal attention during training. Preprocessing of the data also usually speeds up the learning process, and it can be in the form of data scaling or transformation (Masters, 1993). Scaling of the data is not necessary but almost always recommended (Masters, 1993). Transformation of the data into normal distribution or some known forms (e.g., linear, log, and exponential) may be helpful to improve the performance of AI models. The influence of data transformation was undertaken in a study carried out by Bowden et al. (2003) using different transformation methods, including linear, logarithmic, and seasonal trans-formations, histogram equalization, and a transformation to normality. In this study, it was found that the model using the linear transformation resulted in the smallest error, whereas more complex transformations did not improve model performance. Moreover, empirical trials carried out by Faraway and Chatfield (1998) showed that the model fits were the same, regardless of whether raw or transformed data were used. The author's own experience in geotechnical engi-neering is that data scaling is useful, but data transformation does not improve model performance.

Model Validation

Once the training phase of the model has been successfully accomplished, the performance of the trained model should be validated. The purpose of the model validation phase is to ensure that the model has the ability to generalize within the limits set by the training data in a robust fashion, rather than simply having memorized the input−output relationships that are contained in the train-ing data. The approach that is generally adopted in the literature to achieve this is to test the performance of trained AI models on an independent validation set that has not been used as part of the model building process. If such performance is adequate, the model is deemed to be able to generalize and is considered to be robust.

The choice of a suitable error function to investigate model validation is quite important, and the main measures that are often used in the literature to evaluate the performance of AI models include the coefficient of correlation, r; the root mean squared error, RMSE; and the mean absolute error, MAE. The formulas of these measures are as follows:

$$r = \frac{\sum_{i=1}^{N}(O_i - \overline{O})(P_i - \overline{P})}{\sqrt{\sum_{i=1}^{N}(O_i - \overline{O})^2 \sum_{i=1}^{N}(P_i - \overline{P})^2}} \tag{8.6}$$

$$\text{RMSE} = \sqrt{\frac{\sum_{i=1}^{N}(O_i - P_i)^2}{N}} \tag{8.7}$$

$$\text{MAE} = \frac{1}{N}\sum_{i=1}^{N}|O_i - P_i| \tag{8.8}$$

where N is the number of data points presented to the model; O_i and P_i are the observed and predicted outputs, respectively; and $\overline{O}$ and $\overline{P}$ are the mean of observed and predicted outputs, respectively.

The coefficient of correlation, r, is a measure that is used to determine the rela-tive correlation and the goodness-of-fit between the predicted and the observed data. Smith (1986) suggested the following guide for values of $|r|$ between 0.0 and 1.0:

- $|r| \geq 0.8$—Strong correlation exists between two sets of variables,
- $0.2 < |r| < 0.8$—Correlation exists between the two sets of variables, and
- $|r| \leq 0.2$—Weak correlation exists between the two sets of variables.

However, Das and Sivakugan (2010) argued that the use of r could be misleading because sometimes higher values of r may not necessarily indicate better model performance due to the tendency of the model to deviate toward higher or lower values, particularly when the data range is very wide and most of the data are distributed about their mean. It was suggested that the coefficient of efficiency, E, proposed by Nash and Sutcliffe (1970), can give an unbiased estimate and would be a better measure for model performance. E is calculated as follows:

$$E = 1 - \frac{\sum_{i=1}^{N} (O_i - P_i)^2}{\sum_{i=1}^{N} (O_i - \overline{O})^2} \tag{8.9}$$

According to Eq. (8.9), E may range from $-\infty$ to 1.0, where a value of 90% and above indicates very satisfactory performance and a value below 80% indicates unsatisfactory performance. However, Legates and McCabe (1999) raised the issue that E is oversensitive to extreme values (caused by squaring the difference terms), and introduced the modified coefficient of efficiency, E_1, which uses the absolute differences rather than their squares and can be computed as follows:

$$E_1 = 1 - \frac{\sum_{i=1}^{N} |(O_i - P_i)|}{\sum_{i=1}^{N} |(O_i - \overline{O})|} \tag{8.10}$$

The RMSE is the most popular error measure and has the advantage that large errors receive much greater attention than small errors (Hecht-Nielsen, 1990). However, as indicated by Cherkassky et al. (2006), there are situations when RMSE cannot guarantee that the model performance is optimal. Moreover, it was also argued by Das and Sivakugan (2010) that RMSE reflects only the short-term (overall) performance of the model information, showing the overall difference between the predicted and the measured values. Das and Sivakugan (2010) suggested that the use of the normalized mean biased error, NMBE, provides information with respect to overestimation or underestimation predictions and thus can give a better estimation in relation to the long-term model performance. In contrast with RMSE, MAE eliminates the emphasis given to large errors. Both RMSE and MAE are desirable when the evaluated output data are smooth or continuous (Twomey and Smith, 1997).

It is advised by Guven and Aytek (2009) that the combined use of RMSE, E, and E_1 provides a sufficient assessment of AI model performance and allows comparison of the accuracy of different AI modeling approaches. On the other hand, Elshorbagy et al. (2010) suggested that four different error statistics including

RMSE, mean absolute relative error (MARE), mean bias (MB), and coefficient of correlation (r), along with the visual comparison between the observed and the predicted output values, are sufficient to reveal any significant differences among the various modeling techniques with regard to their prediction accuracy. However, Elshorbagy et al. (2010) mentioned that sometimes conflicting results may arise due to the use of various measures and proposed a new error measure that combines the effects of the above-mentioned four error measures in one indicator. The new indicator is called the *ideal point error (IPE)*, and it is calculated as follows (Elshorbagy et al. 2010):

$$IPE_{ij} = \left\{ 0.25 \left[\left(\frac{RMSE_{ij} - 0.0}{\max\ RMSE_{ij}} \right)^2 + \left(\frac{MARE_{ij} - 0.0}{\max\ MARE_{ij}} \right)^2 \right. \right.$$

$$\left. \left. + \left| \frac{MB_{ij} - 0.0}{\max |MB_{ij}|} \right|^2 + \left(\frac{r_{ij} - 1.0}{1/\max\ r_{ij}} \right)^2 \right] \right\}^{1/2} \qquad (8.11)$$

where i and j denote model (i) and technique (j), respectively; and MARE and MB are calculated as follows:

$$MARE = \frac{1}{N} \sum_{i=1}^{N} \left| \frac{O_i - P_i}{O_i} \right| \qquad (8.12)$$

$$MB = \frac{1}{N} \sum_{i=1}^{N} (O_i - P_i) \qquad (8.13)$$

The IPE relies on identifying the ideal point in the four-dimensional error (space) that a model aims to reach. The ideal point should have the following coordinates: RMSE = 0.0, MARE = 0.0, MB = 0.0, and r = 1.0. The IPE measures how far a model performance is from the ideal point. All individual error measures are given equal relative weights and normalized using their maximum error, so the final IPE value ranges from 0.0 (for the best model performance) to 1.0 (for the worst model performance).

Model Robustness

Model robustness is the predictive ability of AI models to generalize over a range of data similar to that used for model training. With regard to ANNs, Kingston et al. (2005b) stated that if "ANNs are to become more widely accepted and reach their full potential..., they should not only provide a good fit to the calibration and validation data, but the predictions should also be plausible in terms of the relationship modeled and robust under a wide range of conditions," and that "while ANNs validated against error alone may produce accurate predictions for situations similar to those contained in the training data, they may not be robust under different conditions unless the relationship by which the data were generated has been adequately estimated." This agrees with the investigation into the robustness of ANNs carried out by Shahin et al. (2005b) for a case study of predicting the settlement of shallow foundations on granular soils. Shahin et al. (2005b) found that good performance of ANN models on the data used for model calibration and validation does not guarantee that the models will perform in a robust fashion over a range of data similar to those used in the model calibration phase. For this reason, Shahin et al. (2005b) proposed a method to test the robustness of the predictive ability of ANN models by carrying out a parametric study to investigate the response of ANN model outputs to changes in its inputs. The robustness of the model can then be determined by examining how well model predictions are in agreement with the known underlying physical processes of the problem in hand over a range of inputs. Shahin et al. (2005b) presented two different ANN models, which have the performance given in Table 8.1. Both the models were developed using the same software, model parameters, and architecture (i.e., five inputs: footing width, applied pressure, average Standard Penetration Test (SPT) blow count, footing geometry, and embedment ratio, and one hidden layer with two nodes and a single output: a foundation settlement), except that the models were optimized with different sets of random starting weights. It can be seen from Table 8.1 that both the models perform very well when assessed against traditional measures such as the coefficient of correlation, r, RMSE, and MAE. In the absence of any further information, one would normally adopt either of the two models and use it for predictive purposes within the range of the input data used to train the models.

Figure 8.8 shows the results of the parametric study performed to assess the generalization ability of both models. In order to carry out the parametric study, all input variables except one were fixed to the mean values used for training and a set of synthetic data (whose values lie between the minimum and the maximum values used for model training) were generated for the single input that was allowed to vary. The synthetic data were generated by increasing their values in increments equal to 5% of the total range between the minimum and the maximum values. These input values were then entered into both ANN models and the corresponding outputs were obtained. The robustness of the models was then determined by examining how well the predicted output (in this case, the footing settlement) agrees

Table 8.1 Performance of the ANN Models Developed by Shahin et al. (2005b)

Model No.	Data Set	r	RMSE (mm)	MAE (mm)
1	Training	0.92	10.8	7.4
	Testing	0.94	8.4	5.8
	Validation	0.88	12.9	9.8
2	Training	0.94	9.1	6.3
	Testing	0.94	9.1	6.8
	Validation	0.89	11.8	9.6

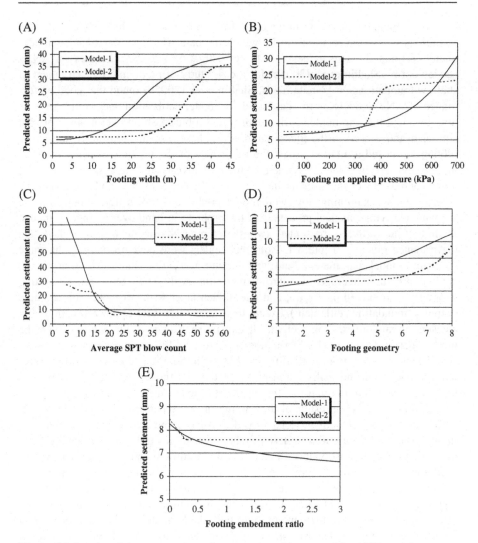

Figure 8.8 Results of the parametric study to test the robustness of the ANN models (Shahin et al., 2005b).

with the known underlying physical processes over the range of inputs examined. It can be seen that the results obtained for Model-1 agree with what one would expect based on the known physical behavior of the settlement of shallow foundations on granular soils. For example, in Figure 8.8A, B, and D, there is an increase in the predicted settlement, in a relatively consistent and smooth fashion, as the footing width, footing net applied pressure, and footing geometry, respectively, increase. On the other hand, in Figure 8.8C and E, the predicted settlement decreases, also in a consistent and smooth fashion, as the average SPT blow count and footing embedment ratio, respectively, increase. In contrast, it can be seen

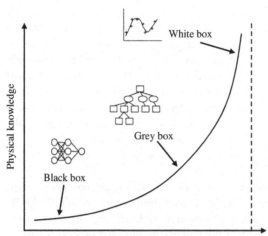

Figure 8.9 Graphical classification of modeling techniques.
Source: Adapted from Giustolisi et al. (2007).

from Figure 8.8 that the results obtained for Model-2 have an unexpected shape that is difficult to justify from a physical understanding of footing settlement. For example, there are abrupt changes in the predicted settlement in some instances and no change in predicted settlement for a range of inputs in others.

Shahin et al. (2005b) argued that since cross-validation (Stone, 1974) was adopted during the model development phase and an independent validation set was used to test the predictive ability of both models, the only plausible explanation for the different behaviors exhibited by both models was the connection weights included in each model. Shahin et al. (2005b) then advised that the connection weights should be examined as part of the interpretation of ANN model behavior, using, for example, the method suggested by Garson (1991). On the other hand, Kingston et al. (2005b) adopted the connection weight approach of Olden et al. (2004) for a case study in hydrological modeling in order to assess the relationship modeled by the ANNs. On the other hand, GP and EPR are claimed to provide better generalization ability than ANNs and therefore are worth further consideration in relation to achieving improved model robustness. However, it is also important to assess the relationship that has been modeled in the validation of AI models, rather than basing it on an error measure alone.

Model Transparency and Knowledge Extraction

Model transparency and knowledge extraction are the feasibility of interpreting AI models in a way that provides insights into how model inputs affect outputs. Figure 8.9 shows a representation of the classification of modeling techniques based on colors (Giustolisi et al., 2007) in which the higher the physical knowledge used during model development, the better the physical interpretation of the phenomenon that the model provides to the user. It can be seen that the color coding of mathematical modeling can be classified into white-, black-, and gray-box models, each of which can be explained as follows (Giustolisi et al., 2007). White-box models are systems that are based on first principles (e.g., physical laws) where model variables and parameters are known and have physical meaning by which the underlying physical relationships of the system can be explained. Black-box models are data-driven or regressive systems in which the functional form of relationships between model variables are unknown and need to be estimated. Black-box models rely on data to map the relationships between model inputs and corresponding outputs rather than to find a feasible structure of the model input−output relationships. Gray-box models are conceptual systems in which the mathematical structure of the model can be derived, allowing further information of the system behavior to be resolved.

According to the above classification of modeling techniques based on color, whereby meaning is related to three levels of prior information required, ANNs belong to the class of black-box models due to their lack of transparency and the fact that they neither consider nor explain the underlying physical processes explicitly. This is because the knowledge extracted by ANNs is stored in a set of weights that are difficult to interpret properly; and due to the large complexity of the network structure, ANNs fail to give a transparent function that relates the inputs to the corresponding outputs. Consequently, it is difficult to understand the nature of the input−output relationships that are derived. This issue has been addressed by many researchers with respect to hydrological engineering. For example, Jain et al. (2004) examined whether the physical processes in a watershed were inherent in a trained ANN rainfall-runoff model. This was carried out by assessing the strengths of the relationships between the distributed components of the ANN model, in terms of the responses from the hidden nodes, and the deterministic components of the hydrological process, computed from a conceptual rainfall-runoff model, along with the observed input variables, using correlation coefficients and scatter plots. They concluded that the trained ANN, in fact, captured different components of the physical process and a careful examination of the distributed information contained in the trained ANN can be informative about the nature of the physical processes captured by various components of the ANN model. Sudheer (2005) performed perturbation analysis to assess the influence of each individual input variable on the output variable and found it to be an effective means of identifying the underlying physical process inherent in the trained ANN. Olden et al. (2004), Sudheer and Jain (2004), and Kingston et al. (2006) also addressed this issue of model transparency and knowledge extraction.

In the context of geotechnical engineering, Shahin et al. (2002) and Shahin and Jaksa (2005) expressed the results of the trained ANNs in the form of relatively straightforward equations. This was possible due to the relatively small number of input and output variables and hidden nodes. Neuro-fuzzy applications are another means of knowledge extraction that facilitate model transparency via extraction of rules. Neuro-fuzzy networks use the fuzzy logic system to store knowledge acquired from a set of input variables $(x_1, x_2, \ldots, x_n)$ and the corresponding output variable (y) in a set of linguistic fuzzy rules that can be easily interpreted, such as IF $(x_1$ is high AND x_2 is low) THEN $(y$ is high), $c = 0.9$, where $(c = 0.9)$ is the rule confidence, which indicates the degree to which the above rule has contributed to the output. Examples of such applications in geotechnical engineering include Ni et al. (1996), Shahin et al. (2003), Gokceoglu et al. (2004), Provenzano et al. (2004), and Padmini et al. (2008). More recently, Cao and Qiao (2008) introduced the so-called neural network committee−based sensitivity analysis strategy to reveal the underlying relationships among the influential factors affecting a system through estimation of the relative contribution of each explicative (input) variable and dependent (output) variables. The strategy was applied to a case study of strata movement and provides employing a factor sensitivity analysis, instead of conventional single neural network analysis, to reveal the underlying mechanism of strata movement. This involves the following steps (Cao and Qiao, 2008): (i) the entire data set on strata movement is randomly split into two subsets, a training subset (4/5 of the samples) and a testing subset (1/5 of the samples); (ii) the model connection weights are adjusted using the training subset, and the model performance is tested using the testing subset; (iii) this process is repeatedly carried out many times so as to determine the best configuration of ANN, which captures the intrinsic mechanism of strata movement and transfers the observed data to implicit knowledge carried by the successfully trained neural network model.

Other researchers proposed the use of sensitivity analyses to explore the AI models by measuring the effects on the output of a given model when the inputs are varied through their range of values. This approach allows a ranking of the inputs based on the amount of output changes produced due to disturbances in a given input, enabling the model to be explained. The quantification of this process is determined by holding all input variables at a fixed baseline values (e.g., their average values), except one input attribute that is varied between its range ($x_a \in \{x_1, ..., x_n\}$), with ($j \in \{1, ..., L\}$) levels. The sensitivity response ($\hat{y}_{a,j}$) is determined for x_a to obtain the input relevance (R_a) using the sensitivity measure (S_a), as follows:

$$R_a = S_a / \sum_{i=1}^{n} S_i \times 100 \ (\%) \tag{8.14}$$

For continuous regression tasks, the sensitivity measures (S_a) can take one of the following three measures, including the range (r_a), gradient (g_a), and variance (v_a), as follows:

$$S_a = r_a = \max(\hat{y}_a) - \min(\hat{y}_a) \tag{8.15}$$

$$S_a = g_a = \sum_{i=2}^{L} |\hat{y}_{a,j} - \hat{y}_{a,j-1}/(L-1)| \tag{8.16}$$

$$S_a = v_a = \sum_{i=2}^{L} (\hat{y}_{a,j} - \bar{\hat{y}}_a)^2/(L-1) \tag{8.17}$$

For more input influence details, Cortez and Embrechts (2011) proposed the global sensitivity analysis algorithm in combination with several visualization techniques, such as the variable effect characteristics (VEC) curve. For a given input variable, the VEC curve plots the L level values on the x-axis versus the sensitivity analysis responses on the y-axis, enabling increased interpretability of the AI models. Another sensitivity method introduced by Francone (2001) for the GP-based models and applied in geotechnical engineering by Alavi et al. (2010) allows the determination of the contribution of input variables to predict target outputs in the form of frequency values of input variables. The frequency value evaluates the importance of an input variable by determining how many times the variable appears in the contribution of the fitness of the GP-evolved programs (Alavi et al., 2010). A frequency value of 1.0 indicates that the input variable appears in 100% of the best GP-evolved programs, indicating that the predictive model is more sensitive to this input variable.

The GP and EPR, on the other hand, can be classified as gray-box techniques (conceptualization of physical phenomena); despite the fact that they are based on observed data, they return a mathematical structure that is symbolic and usually uncomplicated. The nature of obtained GP/EPR models permits the global exploration of expressions, which provides insights into the relationship between the model inputs and the corresponding outputs; i.e., it allows the user to gain additional knowledge of how the system performs. An additional advantage of GP/EPR over ANNs is that the structure and network parameters of ANNs (e.g., the number of hidden layers and their number of nodes, transfer functions, and the learning rate) should be identified *a priori* and are usually obtained using ad hoc, trial-and-error approaches. However, the number and combination of terms, as well as the values of GP/EPR modeling parameters, are all evolved automatically during model calibration. At the same time, the prior physical knowledge based on engineering judgment or other human knowledge can be used hypothesize about the elements of the objective functions and their structure, hence enabling refinement of final models. It should be noted that while white-box models provide maximum transparency, their construction may be difficult to obtain due to many geotechnical engineering problems where the underlying mechanism is not entirely understood.

Model Extrapolation

Model extrapolation is the model's ability to appropriately predict outside the range of the data used for model calibration. It is generally accepted that DDMs perform best when they do not extrapolate beyond the range of the data used for model calibration, which is considered to be an important limitation of AI models because it restricts their usefulness and applicability. Extreme value prediction is of particular concern in several areas of civil engineering, such as hydrological engineering, when floods are forecast, as well as in geotechnical engineering, when liquefaction potential and the stability of slopes are assessed. Sudheer et al. (2003) highlighted this issue and proposed a methodology, based on the Wilson–Hilferty transformation, for enabling ANN models to predict extreme values with respect to peak river flows. Their methodology yielded superior predictions compared to those obtained from an ANN model using untransformed data. More recently, Ismail and Jeng (2011) suggested the use of nonasymptotic PEs such as high-order neural networks (HONs) in modeling the load−settlement behavior of piles. A HON uses polynomial functions to map inputs into outputs and can be trained through error back-propagation algorithm. It uses high-order neurons instead of summation neurons (e.g., sigmoid) as PEs and the advantage of this is that the input and output parameters do not have to be normalized within a certain range. This is because HON models are not asymptotic and do not have a limited dynamic range. To the author's knowledge, this type of neural networks has not been applied in geotechnical engineering and therefore is worth further consideration in relation to achieving improved model extrapolation.

Model Uncertainty

Finally, a further limitation of AI models is that the uncertainty in model predictions is seldom quantified. Failure to account for such uncertainty makes it impossible to assess the quality of AI model predictions, which may limit their efficacy. In addition, estimating uncertainty associated with predictions provided by DDMs is very important for decision making regardless of the generalization ability of the predictive model. This is because, from the point of view of a decision maker, the value of a prediction depends on the availability of additional information that helps to estimate the risks associated with decisions taken upon this prediction (Cherkassky et al., 2006).

In an effort to address the issue of model uncertainty, a few researchers have applied Bayesian techniques to ANN training (Buntine and Weigend, 1991; Kingston et al., 2005a, 2008; MacKay, 1992) in the context of hydrological engineering and Goh et al. (2005) did the same with respect to geotechnical engineering. In these studies, various Bayesian methods have been used to estimate the uncertainties in ANN parameters (weights) and Goh et al. (2005) observed that the integration of the Bayesian framework into the back-propagation algorithm enhanced neural network prediction capabilities and provided assessment of the confidence associated with network predictions. Research to date has demonstrated the value of Bayesian neural networks, although further work is needed in the area of geotechnical engineering. Shahin et al. (2005a) also incorporated uncertainty in the ANN process by developing a series of probabilistic design charts expressing the reliability of settlement predictions for shallow foundations on cohesionless soils. In the context of hydrological engineering, Shrestha and Solomatine (2006) introduced an approach to estimate model uncertainty using machine learning, and the method was tested in forecasting river flows. The idea is to build local models in which uncertainty is expressed in the form of the two quantiles (constituting the prediction interval) of the underlying distribution of prediction errors. Clustering and fuzzy logic are then used to model the propagation of integral uncertainty through the models.

8.4 Discussion and Conclusions

In the field of geotechnical engineering, it is possible to encounter some types of problems that are very complex and not well understood. In this regard, AI provides several advantages over more conventional computing techniques. For most traditional mathematical models, the lack of physical understanding is usually supplemented by either simplifying the problem or incorporating several assumptions into the models. Mathematical models also rely on assuming the structure of the model in advance, which may be less than optimal. Consequently, many mathematical models fail to simulate the complex behavior of most geotechnical engineering problems. In contrast, AI techniques are a data-driven approach in which the model can be trained on input−output data pairs to determine its structure and parameters. In this case, there is no need to either simplify the problem or incorporate any assumptions. Moreover, AI models can always be updated to obtain better results by presenting new training examples as new data become available. These factors combine to make AI techniques a powerful modeling tool in geotechnical engineering.

Despite the success of AI techniques in geotechnical engineering and other disciplines, they suffer from some shortcomings in relation to model transparency and knowledge extraction, ability of extrapolation, and model uncertainty, which need further attention in the future. For example, special attention should be paid to incorporating prior knowledge about the underlying physical process based on engineering judgment or human expertise into the learning formulation, checking of model robustness, and evaluation of model results. Furthermore, the standard RMSE error functions conventionally used in AI applications have to be updated and replaced with more representative error measures. Moreover, according to Flood (2008), ANNs in civil engineering (including geotechnical engineering) were used mostly as simple vector-mapping devices for function modeling of applications that require rarely more than a few tens of neurons without higher-order structuring. Together, improvements in these issues will greatly enhance the usefulness of ANN models and will provide the next generation of applied ANNs with the best way for advancing the field to the next level of sophistication and application.

The review of geotechnical engineering literature indicates that findings with regard to superiority of one AI technique over the other traditional methods are sometimes contradictory. Consequently, such findings should be treated as data specific and should not be generalized. The author suggests that for the time being, AI techniques might be treated as a complement to conventional computing techniques rather than as an alternative, or may be used as a quick check on solutions developed by more time-consuming and in-depth analyses.

References

Abdel-Rahman, A.H., 2008. Predicting compaction of cohesionless soils using ANN. Ground Improv. 161, 3−8.

Adarsh, S.A., Jangareddy, M., 2010. Slope stability modeling using genetic programming. Int. J. Earth Sci. Eng. 3, 1–8.

Adeli, H., 2001. Neural networks in civil engineering: 1989–2000. Comput. Aided Civ. Infrastruct. E. 16, 126–142.

Ahangar-Asr, A., Faramarzi, A., Javadi, A.A., 2010. A new approach for prediction of the stability of soil and rock slopes. Eng. Comput. Int. J. Comput. Aided Eng. Softw. 27, 878–893.

Ahangar-Asr, A., Faramarzi, A., Mottaghifard, N., Javadi, A.A., 2011. Modeling of permeability and compaction characteristics of soils using evolutionary polynomial regression. Comput. Geosci. 37, 1860–1869.

Ahmad, I., El Naggar, H., Kahn, A.N., 2007. Artificial neural network application to estimate kinematic soil pile interaction response parameters. Soil Dynam. Earthquake Eng. 27, 892–905.

Alavi, A.H., Gandomi, A.H., 2011a. Prediction of principal ground motion parameters using a hybrid method coupling artificial neural networks and simulated annealing. Comput. Struct. 89, 2176–2194.

Alavi, A.H., Gandomi, A.H., 2011b. A robust data mining approach for formulation of geotechnical engineering systems. Eng. Comput. Int. J. Comput. Aided Eng. Softw. 28, 242–274.

Alavi, A.H., Gandomi, A.H., 2011c. A robust data mining approach for formulation of geotechnical engineering systems. Eng. Comput. 28, 242–274.

Alavi, A.H., Gandomi, A.H., 2012. Energy-based numerical models for assessment of soil liquefaction. Geosci. Frontiers. 3, 541–555.

Alavi, A.H., Gandomi, A.H., Sahab, M.G., Gandomi, M., 2010. Multi expression programming: a new approach to formulation of soil classification. Eng. Comput. 26, 111–118.

Alimoradi, A., Moradzadeh, A., Naderi, R., Salehi, M.Z., Etemadi, A., 2008. Prediction of geological hazardous zones in front of a tunnel face using TSP-203 and artificial neural networks. Tunn. Undergr. Space Technol. 23, 711–717.

Alkroosh, I., Nikraz, H., 2011a. Correlation of pile axial capacity and CPT data using gene expression programming. Geotech. Geol. Eng. 29, 725–748.

Alkroosh, I., Nikraz, H., 2011b. Simulating pile load-settlement behavior from CPT data using intelligent computing. Cent. Eur. J. Eng. 1, 295–305.

Alkroosh, I., Nikraz, H., 2012. Predicting axial capacity of driven piles in cohesive soils using intelligent computing. Eng. Appl. Artif. Intell. 25, 618–627.

Ardalan, H., Eslami, A., Nariman-Zadeh, N., 2009. Piles shaft capacity from CPT and CPTU data by polynomial neural networks and genetic algorithms. Comput. Geotech. 36, 616–625.

ASCE, 2000. Artificial neural networks in hydrology. I: Preliminary concepts. J. Hydrol. Eng. 5, 115–123.

Ashayeri, I., Yasrebi, S., 2009. Free-swell and swelling pressure of unsaturated compacted clays; experiments and neural networks modelling. Geotech. Geol. Eng. 27, 137–153.

Banimahd, M., Yasrobi, S.S., Woodward, P.K., 2005. Artificial neural network for stress–strain behavior of sandy soils: knowledge based verification. Comput. Geotech. 32, 377–386.

Baykasoglu, A., Gullu, H., Canakci, H., Ozbakir, L., 2008. Prediction of compressive and tensile strength of limestone via genetic programming. Expert Syst. Appl. 35, 111–123.

Baziar, M.H., Ghorbani, A., 2005. Evaluation of lateral spreading using artificial neural networks. Soil Dynam. Earthquake Eng. 25, 1–9.

Baziar, M.H., Jafarian, Y., Shahnazari, H., Movahed, V., Tutunchian, M.A., 2011. Prediction of strain energy-based liquefaction resistance of sand−silt mixtures: an evolutionary approach. Comput. Geotech. 37, 1883−1893.

Bhattacharya, B., Solomatine, D., 2006. Machine learning in soil classification. Neural Networks. 19, 186−195.

Boubou, R., Emeriault, F., Kastner, R., 2010. Artificial neural network application for the prediction of ground surface movements induced by shield tunnelling. Can. Geotech. J. 47, 1214−1233.

Bowden, G.J., Maier, H.R., Dandy, G.C., 2002. Optimal division of data for neural network models in water resources applications. Water Resour. Res. 38, 2.1−2.11.

Bowden, G.J., Dandy, G.C., Maier, H.R., 2003. Data transformation for neural network models in water resources applications. J. Hydroinform. 5, 245−258.

Bowden, G.J., Dandy, G.C., Maier, H.R., 2005. Input determination for neural network models in water resources applications: Part 1—Background and methodology. J. Hydrol. 301, 75−92.

Braddock, R.D., Kremmer, M.L., Sanzogni, L., 1998. Feed-forward artificial neural network model for forecasting rainfall run-off. Environmetrics. 9, 419−432.

Buntine, W.L., Weigend, A.S., 1991. Bayesian back-propagation. Complex Systems. 5, 603−643.

Byeon, W.Y., Lee, S.R., Kim, Y.S., 2006. Application of flat DMT and ANN to Korean soft clay deposits for reliable estimation of undrained shear strength. Int. J. Offshore Polar Eng. 16, 73−80.

Cabalar, A.F., Cevik, A., Guzelbey, I.H., 2009. Constitutive modeling of Leighton Buzzard sands using genetic programming. Neural Comput. Appl. 19, 657−665.

Caglar, N., Arman, H., 2007. The applicability of neural networks in the determination of soil profiles. Bull. Eng. Geol. Environ. 66, 295−301.

Cai, S., Toral, H., Qiu, J., Archer, J.S., 1994. Neural network based objective flow regime identification in air−water two phase flow. Can. J. Chem. Eng. 72, 440−445.

Campolo, M., Soldati, A., Andreuss, I.P., 1999. Forecasting river flow rate during low-flow periods using neural networks. Water Resour. Res. 35, 3547−3552.

Cao, M., Qiao, P., 2008. Neural network committe-based sensitivity analysis strategy for geotechnical engineering problems. Neural Comput. Appl. 17, 509−519.

Celik, S., Tan, O., 2005. Determination of preconsolidation pressure with artificial neural network. Civ. Eng. Environ. Syst. 22, 217−231.

Cevik, A., Sezer, E.A., Cabalar, A.F., Gokceoglu, C., 2010. Modeling of the uniaxial compressive strength of some clay-bearing rocks using neural network. Appl. Soft. Comput. 11, 2587−2594.

Chen, Y., Azzam, R., Fernandez, T.M., Li, L., 2009. Studies on construction pre-control of a connection aisle between two neighbouring tunnels in Shanghai by means of 3D FEM, neural networks and fuzzy logic. Geotech. Geol. Eng. 27, 155−167.

Chen, Y., Azzam, R., Zhang, F., 2006. The displacement computation and construction pre-control of a foundation pit in Shanghai utilizing FEM and intelligent methods. Geotech. Geol. Eng. 24, 1781−1801.

Chen, Y., Hsieh, S., Liu, C., 2010. Simulation of stress−strain behavior of saturated sand in undrained triaxial tests based on genetic adaptive neural networks. Electron. J. Geotech. Eng. 15, 1814−1834.

Cherkassky, V., Krasnopolsky, V., Solomatine, D.P., Valdes, J., 2006. Computational intelligence in earth sciences and environmental applications: issues and challenges. Neural Networks. 19, 113−121.

Cho, S.E., 2009. Probabilistic stability analyses of slopes using the ANN-based response surface. Comput. Geotech. 36, 787−797.

Cortez, P., Embrechts, M.J., 2011. Opening black box data mining models using sensitivity analysis. In: Proceedings of the 2011 IEEE Symposium on Computational Intelligence and Data Mining, Paris, pp. 341−348.

Cramer, N.L., 1985. A representation for the adaptive generation of simple sequential programs. In: Proceedings of the International Conference on Genetic Algorithms and Their Applications, Carnegie-Mellon University, Pittsburgh, PA, pp. 183−187.

Das, S.K., Basudhar, P.K., 2005. Prediction of coefficient of lateral earth pressure using artificial neural networks. Electron. J. Geotech. Eng. 10.

Das, S.K., Basudhar, P.K., 2006. Undrained lateral load capacity of piles in clay using artificial neural network. Comput. Geotech. 33, 454−459.

Das, S.K., Sivakugan, N., 2010. Discussion of: intelligent computing for modeling axial capacity of pile foundations. Can. Geotech. J. 47, 928−930.

Das, S.K., Biswal, R.K., Sivakugan, N., Das, B., 2011a. Classification of slopes and prediction of factor of safety using differential neural networks. Environ. Earth Sci. 64, 201−210.

Das, S.K., Samui, P., Sabat, A.K., 2011b. Application of artificial intelligence to maximum dry density and unconfined compressive stregth of cement stabilized soil. Geotech. Geol. Eng. 29, 329−342.

Dincer, I., 2011. Models to predict the deformation modulus and the coefficient of subgrade reaction for earth filling structure. Adv. Eng. Softw. 42, 160−171.

Doostmohamadi, R., Moosavi, M., Araabi, B.N., 2008. Modelling time dependent swelling potential of mudrock using laboratory tests and artificial neural network. Min. Tech. 117, 32−41.

Elshorbagy, A., Corzo, G., Srinivasulu, S., Solomatine, D.P., 2010. Experimental investigation of the predictive capabilities of data driven modeling techniques in hydrology—part 1: concepts and methodology. Hydrol. Earth Syst. Sci. 14, 1931−1941.

Erzin, Y., 2007. Artificial neural networks approach for swell pressure versus soil suction behaviour. Can. Geotech. J. 44, 1215−1223.

Erzin, Y., Gumaste, S.D., Gupta, A.K., Singh, D.N., 2009. Artificial neural network (ANN) models for determining hydraulic conductivity of compacted fine-grained soils. Can. Geotech. J. 46, 955−968.

Faraway, J., Chatfield, C., 1998. Time series forecasting with neural networks: a comparative study using the airline data. Appl. Stat. 47, 231−250.

Fausett, L.V., 1994. Fundamentals Neural Networks: Architecture, Algorithms, and Applications. Prentice-Hall, Englewood Cliffs, NJ.

Feng, X.T., Chen, B., Yang, C., Zhou, H., Ding, X., 2006. Identification of visco-elastic models for rocks using genetic programming coupled with the modified particle swarm optimization algorithm. Int. J. Rock Mech. Min. Sci. 43, 789−801.

Ferentinou, M.D., Sakellariou, M.G., 2007. Computational intelligence tools for the prediction of slope performance. Comput. Geotech. 34, 362−384.

Ferreira, C., 2001. Gene expression programming: a new adaptive algorithm for solving problems. Complex Systems. 13, 87−129.

Firat, M., Gungor, M., 2008. Generalized regression neural networks and feed forward neural networks for prediction of scur depth around bridge piers. Adv. Eng. Softw. 40, 731−737.

Flood, I., 2008. Towards the next generation of artificial neural networks for civil engineering. Adv. Eng. Inform. 22, 4−14.

Francone, F.D., 2001. Discipulus Software Owner's Manual. Register Machine Learning Technologies Inc., Littleton, CO.

Fu, Q., Hashash, Y.M.A., Hung, S., Ghaboussi, J., 2007. Integration of laboratory testing and constitutive modelling of soils. Comput. Geotech. 34, 330–345.

Gandomi, A.H., Alavi, A.H., 2011. Multi-stage genetic programming: a new strategy to non-linear system modeling. Inf. Sci. 181, 5227–5239.

Gandomi, A.H., Alavi, A.H., 2012. A new multi-gene genetic programming approach to non-linear system modeling, Part II: Geotechnical and earthquake engineering problems. Neural Comput. Appl. 21, 189–201.

Gandomi, A.H., Alavi, A.H., Gun, J.Y., 2011. Formulation of uplift capacity of suction cassions using multi expression programming. KSCE J. Civ. Eng. 15, 363–373.

Garaga, A., Latha, G.M., 2010. Intelligent prediction of the stress–strain response of intact and jointed rocks. Comput. Geotech. 37, 629–637.

Garcia, S.R., Roma, M.P., 2009. Rockfill stregth evaluation using cascade correlation networks. Geotech. Geol. Eng. 27, 289–304.

Garcia, S.R., Romo, M.P., Figueroa-Nazuno, J., 2006. Soil dynamic properties determination: a neurofuzzy system approach. Contr. Intell. Syst. 34, 1–11.

Gardner, M.W., Dorling, S.R., 1998. Artificial neural networks (the multilayer perceptron)—a review of applications in the atmospheric sciences. Atmos. Environ. 32, 2627–2636.

Garson, G.D., 1991. Interpreting neural-network connection weights. AI Expert. 6, 47–51.

Giraudel, J.L., Lek, S., 2001. A comparison of self-organizing map algorithm and some conventional statistical methods for ecological community ordination. Ecol. Modell. 146, 329–339.

Giustolisi, O., Savic, D.A., 2006. A symbolic data-driven technique based on evolutionary polynomial regression. J. Hydroinform. 8, 207–222.

Giustolisi, O., Doglioni, A., Savic, D.A., Webb, B.W., 2007. A multi-model approach to analysis of environmental phenomena. Environ. Model. Softw. 22, 674–682.

Goh, A.T., Kulhawy, F.H., Chua, C.G., 2005. Bayesian neural network analysis of undrained side resistance of drilled shafts. J. Geotech. Geoenviron. Eng. 131, 84–93.

Goh, A.T.C., 1994. Seismic liquefaction potential assessed by neural network. J. Geotech. Geoenviron. Eng. 120, 1467–1480.

Goh, A.T.C., Kulhawy, F.H., 2005. Reliability assessment of serviceability performance of braced retaining walls using a neural network approach. Int. J. Numer. Anal. Methods Geomech. 29, 627–642.

Gokceoglu, C., Yesilnacar, E., Sonmez, H., Kayabasi, A., 2004. A neuro-fuzzy model for modulus of deformation of jointed rock masses. Comput. Geotech. 31, 375–383.

Goldberg, D.E., 1989. Genetic Algorithms in Search Optimization and Machine Learning. Addison-Wesley, Reading, MA.

Gunaydin, O., Gokoglu, A., Fener, M., 2010. Prediction of artificial soil's unconfined compression strength test using statistical analyses and artificial neural networks. Adv. Eng. Softw. 41, 1115–1123.

Guven, A., Aytek, A., 2009. New approach for stage–discharge relationship: gene-expression programming. J. Hydrol. Eng. 14, 812–820.

Hajihassani, M., Marto, A., Nazmi, E., Abad, S., Shahrbabaki, M., 2011. Prediction of surface settlements induced by NATM tunneling based on artificial neural networks. Electro. J. Geotech. Eng. 16, 1471–1480.

Hammerstrom, D., 1993. Working with neural networks. IEEE Spectr. 30, 46–53.

Hanna, A.M., Ural, D., Saygili, G., 2007a. Evaluation of liquefaction potential of soil deposits using artificial neural networks. Int. J. Comput. Aided Eng. Softw. 24, 5–16.

Hanna, A.M., Ural, D., Saygili, G., 2007b. Neural network model for liquefaction potential in soil deposits using Turkey and Taiwan earthquake data. Soil Dynam. Earthquake Eng. 27, 521–540.

Hecht-Nielsen, R., 1990. Neurocomputing. Addison-Wesley, Reading, MA.

Holland, J.H., 1975. Adaptation in Natural and Artificial Systems. Ann Arbor, Mich., University of Michigan Press.

Ikizler, S.B., Aytekin, M., Vekli, M., Kocabas, F., 2009. Prediction of swelling pressures of expansive soils using artificial neural networks. Adv. Eng. Softw. 41, 647–655.

Ismail, A.M., Jeng, D.S., 2011. Modelling load-settlement behaviour of piles using high-order neural network (HON-PILE model). Eng. Appl. Artif. Intell. 24, 813–821.

Jain, A., Sudheer, K.P., Srinivasulu, S., 2004. Identification of physical processes inherent in artificial neural network rainfall runoff models. Hydrolog. Process. 18, 571–581.

Javadi, A.A., Rezania, M., 2009. Intelligent finite element method: an evolutionary approach to constitutive modelling. Adv. Eng. Inform. 23, 442–451.

Javadi, A., Rezania, M., Mousavi, N.M., 2006. Evaluation of liquefaction induced lateral displacements using genetic programming. Comput. Geotech. 33, 222–233.

Jianping, J., Changhong, Y., Guangyun, G., 2011. Prediction of compressibility coefficient of soil based on RBF neural network. J. Jiangsu Univ. Natural Sci. Ed. 32, 232–235.

Johari, A., Habibagahi, G., Ghahramani, A., 2006. Prediction of soil–water characteristic curve using genetic programming. J. Geotech. Geoenviron. Eng. 132, 661–665.

Johari, A., Javadi, A., Habibagahi, G., 2011. Modelling the mechanical behaviour of unsaturated soils using a genetic algorithm-based neural network. Comput. Geotech. 38, 2–13.

Kamatchi, P., Rajasankar, J., Ramana, G.V., Nagpal, A.K., 2010. A neural network based methodology to predict site-specific spectral acce;eration values. Earthquake Eng. Vib. 9, 459–472.

Kanungo, D.P., Arora, M.K., Sarkar, S., Gupta, R.P., 2006. A comparative study of conventional ANN black box, fuzzy and combined neural and fuzzy weighting procedures for landslide susceptibility zonation in Darjeeling Himalayas. Eng. Geol. 85, 347–366.

Kavli, T., 1993. ASMOD—An algorithm for adaptive spline modelling of observation data. Int. J. Control. 58, 947–967.

Kaya, A., 2009. Residual and fully softened strength evaluation of soils using artificial neural networks. Geotech. Geol. Eng. 27, 281–288.

Kayadelen, C., 2011. Soil liquefaction modeling by genetic expression programming and neuro-fuzzy. Expert Syst. Appl. 38, 4080–4087.

Kayadelen, C., Gunaydin, O., Fener, M., Demir, A., Ozvan, A., 2009. Modeling of the angle of shearing resistance of soils using soft computing systems. Expert Syst. Appl. 36, 11814–11826.

Khozaghi, S.S.H., Choobbasti, A.J.A., 2007. Predicting of liquefaction potential in soils using artificial neural networks. Electron. J. Geotech. Eng.12.

Kim, Y., Kim, B., 2008. Prediction of relative crest settlement of concrete-faced rockfill dams analyzed using an artificial neural network model. Comput. Geotech. 35, 313–322.

Kingston, G.B., Lambert, M.F., Maier, H.R., 2005a. Bayesian parameter estimation applied to artificial neural networks used for hydrological modelling. Water Resour. Res. 41. doi:10.1029/2005WR004152.

Kingston, G.B., Maier, H.R., Lambert, M.F., 2005b. Calibration and validation of neural networks to ensure physically plausible hydrological modelling. J. Hydrol. 314, 158–176.

Kingston, G.B., Maier, H.R., Lambert, M.F., 2006. A probabilistic method to assist knowl-
    edge extraction from artificial neural networks used for hydrological prediction. Math.
    Comput. Model. 44, 499—512.
Kingston, G.B., Maier, H.R., Lambert, M.F., 2008. Bayesian model selection applied to
    artificial neural networks used for water resources modelling. Water Resour. Res. 44.
    doi:10.1029/2007WR006155.
Kocjancic, R., Zupan, J., 2000. Modelling of the river flow rate: the influence of the training
    set selection. Chemomet. Intell. Lab. Syst. 54, 21—34.
Kogut, J., 2007. Dynamic soil profile determination with the use of a neural network.
    Comput. Assist. Mech. Eng. Sci. 14, 209—217.
Koza, J.R., 1992. Genetic Programming: on the Programming of Computers by Natural
    Selection. MIT Press, Cambridge, MA.
Kung, G.T., Hsiao, E.C., Schuster, M., Juang, C., 2007. A neural network approach to esti-
    mating deflection of diaphram walls caused by excavation in clays. Comput. Geotech.
    34, 385—396.
Kuo, Y.L., Jaksa, M.B., Lyamin, A.V., Kaggwa, W.S., 2009. ANN-based model for predict-
    ing the bearing capacity of strip footing on multi-layered cohesive soil. Comput.
    Geotech. 36, 503—516.
Kurup, P.U., Griffin, E., 2006. Prediction of soil composition from CPT data using general
    regression neural network. J. Comput. Civil. Eng. 20, 281—289.
Lee, T.L., Lin, H.M., Jeng, D.S., Lu, Y.P., 2008. Back-propagation neural network for
    assessment of highway slope failure in Taiwan. Geotech. Eng. 39, 121—128.
Legates, D.R., Mccabe, G.J., 1999. Evaluating the use of goodness-of-fit measures in hydro-
    logic and hydroclimatic model validation. Water Resour. Res. 35, 233—241.
Liao, K., Fan, J., Huang, C., 2011. An artificial neural network for groutability prediction of
    permeation grouting with microfine cement grouts. Comput. Geotech. 38, 978—986.
Lu, Y., 2005. Underground blast induced ground shock and its modelling using artificial neu-
    ral network. Comput. Geotech. 32, 164—178.
Ma, S., Cao, L., Li, H., 2006. The improved neural network and its application for valuing
    rock mass mechanical parameter. J. Coal Sci. Eng. 12, 21—24.
Mackay, G., 1992. A practical Bayesian framework for backpropagation networks. Neural.
    Comput. 4, 448—472.
Maier, H.R., Dandy, G.C., 2000a. Applications of artificial neural networks to forecasting of
    surface water quality variables: issues, applications and challenges. In: Govindaraju, R.S.,
    Rao, A.R. (Eds.), Artificial Neural Networks in Hydrology. Kluwer, Dordrecht, The
    Netherlands.
Maier, H.R., Dandy, G.C., 2000b. Neural networks for the prediction and forecasting of
    water resources variables: a review of modelling issues and applications. Environ.
    Modell. Softw. 15, 101—124.
Maji, V.B., Sitharam, T.G., 2008. Prediction of elastic modulus of jointed rock mass using
    artificial neural networks. Geotech. Geol. Eng. 26, 443—452.
Masters, T., 1993. Practical Neural Network Recipes in C++. Academic Press, San Diego,
    CA.
May, R.J., Maier, H.R., Dandy, G.C., Fernando, T.M., 2008. Non-linear variable selection
    for artificial neural networks using partial mutual information. Environ. Model. Softw.
    23, 1312—1326
Mcculloch, W.S., Pitts, W., 1943. A logical calculus of ideas imminent in nervous activity.
    Bull. Math. Biophys. 5, 115—133.

Mollahasani, A., Alavi, A.H., Gandomi, A.H., 2011. Empirical modeling of plate load test moduli of soil via gene expression programming. Comput. Geotech. 38, 281–286.

Najjar, Y.M., Huang, C., 2007. Simulating the stress–strain behaviour of Georgia Kaolin via recurrent neuronet approach. Comput. Geotech. 34, 346–362.

Najjar, Y.M., Basheer, I.A., Naouss, W.A., 1996. On the identification of compaction characteristics by neuronets. Comput. Geotech. 18, 167–187.

Narendara, B.S., Sivapullaiah, P.V., Suresh, S., Omkar, S.N., 2006. Prediction of unconfined compressive strength of soft grounds using computational intelligence techniques: a comparative study. Comput. Geotech. 33, 196–208.

Nash, J.E., Sutcliffe, J.V., 1970. River flow forecasting through conceptual models. J. Hydrol. 10, 282–290.

Neaupane, K.M., Adhikari, N.R., 2006. Prediction of tunnelling-induced ground movement with the multi-layer perceptron. Tunn. Undergr. Space Technol. 21, 151–159.

NeuralWare, 1997. NeuralWorks Predict Release 2.1. NeuralWare Inc., Pittsburgh.

Ni, S.H., Lu, P.C., Juang, C.H., 1996. A fuzzy neural network approach to evaluation of slope failure potential. J. Microcomput. Civ. Eng. 11, 59–66.

Obrzud, R.F., Vulliet, L., Truty, A., 2009. A combined neural network/gradient-based approach for the identification of constitutive model parameters using self-boring pressuremeter tests. Int. J. Numer. Anal. Methods Geomech. 33, 817–849.

Olden, J.D., Joy, M.K., Death, R.G., 2004. An accurate comparison of methods for quantifying variable importance in artificial neural networks using simulated data. Ecol. Model. 178, 389–397.

Padmini, D., Ilamparuthi, K., Sudheer, K.P., 2008. Ultimate bearing capacity prediction of shallow foundations on cohesionless soils using neurofuzzy models. Comput. Geotech. 35, 33–46.

Pal, M., Deswal, S., 2008. Modeling pile capacity using support vector machines and generalized regression neural network. J. Geotech. Geoenviron. Eng. 134, 1021–1024.

Parasuraman, K., Elshorbagy, A., Bing, C.S., 2007. Estimating saturated conductivity using genetic programming. Soil Sci. Soc. Am. J. 71, 1676–1684.

Park, H., Lee, S.R., 2011. Evaluation of the compression index of soils using an artificial neural network. Comput. Geotech. 38, 472–481.

Park, H.I., 2011. Development of neural network model to estimate the permeability coefficient of soils. Mar. Georesour. Geotechnol. 29, 267–278.

Park, H.I., Kim, Y.T., 2011. Prediction of strength of reinforced lightweight soil using an artificial neural network. Eng. Comput. 28, 600–615.

Peng, X., Wang, Z., Luo, T., Yu, M., Luo, Y., 2008. An elasto-plastic constitutive model of moderate sandy clay based on BC-RBFNN. J. Cent. South Univ. Technol. 15, 47–50.

Pooya Nejad, F., Jaksa, M.B., Kakhi, M., Mccabe, B.A., 2009. Prediction of pile settlement using artificial neural networks based on standard penetration test data. Comput. Geotech.

Provenzano, P., Ferlisi, S., Musso, A., 2004. Interpretation of a model footing response through an adaptive neural fuzzy inference system. Comput. Geotech. 31, 251–266.

Ray, C., Klindworth, K.K., 2000. Neural networks for agrichemical vulnerability assessment of rural private wells. J. Hydrol. Eng. 5, 162–171.

Rezania, M., Faramarzi, A., Javadi, A., 2011. An evolutionary based approach for assessment of earthquake-induced soil liquefaction and lateral displacement. Eng. Appl. Artif Intell. 24, 142–153.

Rezania, M., Javadi, A., 2007. A new genetic programming model for predicting settlement of shallow foundations. Can. Geotech. J. 44, 1462–1472.

Rezania, M., Javadi, A., Giustolisi, O., 2010. Evaluation of liquefaction potential based on CPT results using evolutionary polynomial regression. Comput. Geotech. 37, 82−92.

Rezania, M., Javadi, A.A., 2008. An evolutionary-based data mining technique for assessment of civil engineering systems. Eng. Comput. Int. J. Comput. Aided Eng. Softw. 25, 500−517.

Rumelhart, D.E., Hinton, G.E., Williams, R.J., 1986. Learning internal representation by error propagation. In: Rumelhart, D.E., McClelland, J.L. (Eds.), Parallel Distributed Processing. MIT Press, Cambridge.

Sakellariou, M.G., Ferentinou, M.D., 2005. A study of slope stability prediction using neural networks. Geotech. Geol. Eng. 23, 419−445.

Samui, P., Kumar, B., 2006. Artificial neural network prediction of stability numbers for two-layered slopes with associated flow rule. Electron. J. Geotech. Eng. 11, <www.ejge.com>.

Samui, P., Sitharam, T.G., 2011. Determination of liquefaction susceptability of soil based on field test and artificial intelligence. Int. J. Earth Sci. Eng. 4, 216−222.

Santos, J., Ovidio, J., Celestino, T.B., 2008. Artificial neural networks analysis of Sao Paulo subway tunnel settlement data. Tunnell. Undegr. Space Technol. 23, 481−491.

Sarkar, K., Tiwary, A., Singh, T.N., 2010. Estimation of stregth parameters of rock using artificial neural networks. Bull. Eng. Geol. Environ. 69, 599−606.

Savic, D.A., Giutolisi, O., Berardi, L., Shepherd, W., Djordjevic, S., Saul, A., 2006. Modelling sewer failure by evolutionary computing. Proc. Inst. Eng. Water Manage. 159, 111−118.

Shafiee, A., Ghate, R., 2008. Shear modulus and damping ratio in aggregate-clay mixtures: an experimental study versus ANNs prediction. J. Appl. Sci. 8, 3068−3082.

Shahin, M.A., 2010. Intelligent computing for modelling axial capacity of pile foundations. Can. Geotech. J. 47, 230−243.

Shahin, M.A., Indraratna, B., 2006. Modelling the mechanical behaviour of railway ballast using artificial neural networks. Can. Geotech. J. 43, 1144−1152.

Shahin, M.A., Jaksa, M.B., 2005. Neural network prediction of pullout capacity of marquee ground anchors. Comput. Geotech. 32, 153−163.

Shahin, M.A., Jaksa, M.B., 2006. Pullout capacity of small ground anchors by direct cone penetration test methods and neural networks. Can. Geotech. J. 43, 626−637.

Shahin, M.A., Jaksa, M.B., Maier, H.R., 2001. Artificial neural network applications in geotechnical engineering. Aust. Geomech. 36, 49−62.

Shahin, M.A., Jaksa, M.B., Maier, H.R., 2002. Artificial neural network-based settlement prediction formula for shallow foundations on granular soils. Aust. Geomech. 37, 45−52.

Shahin, M.A., Maier, H.R., Jaksa, M.B., 2003. Settlement prediction of shallow foundations on granular soils using B-spline neurofuzzy models. Comput. Geotech. 30, 637−647.

Shahin, M.A., Maier, H.R., Jaksa, M.B., 2004. Data division for developing neural networks applied to geotechnical engineering. J. Comput. Civ. Eng. 18, 105−114.

Shahin, M.A., Jaksa, M.B., Maier, H.R., 2005a. Neural network based stochastic design charts for settlement prediction. Can. Geotech. J. 42, 110−120.

Shahin, M.A., Maier, H.R., Jaksa, M.B., 2005b. Investigation into the robustness of artificial neural network models for a case study in civil engineering. In: Proceedings of the International Congress on Modelling and Simulation, MODSIM 2005, Melbourne, Australia, pp. 79−83.

Shahin, M.A., Jaksa, M.B., Maier, H.R., 2009. Recent advances and future challenges for artificial neural systems in geotechnical engineering applications. J. Adv. Artif. Neural Syst. doi:10.1155/2009/308239.

Shahnazari, H., Dehnavi, Y., Alavi, A.H., 2010. Numerical modeling of stress−strain behavior of sand under cyclic loading. Eng. Geol. 116, 53−72.

Shrestha, D.L., Solomatine, D.P., 2006. Machine learning approacjes for estimation of prediction interval for the model output. Neural Networks. 19.

Shuh-Gi, C., Ching-Yinn, L., 2009. CPT-based simplified liquefaction assessment by using fuzzy-neural network. J. Mar. Sci. Technol. 17, 326−331.

Singh, T.N., Singh, V., 2005. An intelligent approach to prediction and control ground vibration in mines. Geotech. Geol. Eng.(23), 249−262.

Singh, T.N., Verma, A.K., Singh, V., Sahu, A., 2005. Slake durability of shaly rock and its predictions. Environ. Geol. 47, 246−253.

Singh, T.N., Verma, A.K., Sharma, P.K., 2007. A neuro-genetic approach for prediction of time dependent deformational characteristic of rock and its sensitivity analysis. Geotech. Geol. Eng. 25, 395−407.

Sinha, S.K., Wang, M.C., 2008. Artificial neural network prediction models for soil compaction and permeability. Geotech. Eng. J. 26, 47−64.

Sitharam, T.G., Samui, P., Anbazhagan, P., 2008. Spatial variability of rock depth in Bangalore using geostatistical, neural network and support vector machine models. Geotech. Geol. Eng. 26, 503−517.

Sivrikaya, O., Soycan, T.Y., 2011. Estimation of compaction parameters of fine-grained soils in terms of compaction energy using artificial neural networks. Int. J. Numer. Anal. Methods Geomech. 35, 1830−1841.

Smith, G.N., 1986. Probability and Statistics in Civil Engineering: An Introduction. Collins, London.

Solomatine, D.P., Ostfeld, A., 2008. Data-driven modelling: some past experience and new approaches. J. Hydroinform. 10, 3−22.

Soroush, A., Foroozan, R., Asadollahi, P., 2006. Simulation of 3D effect of excavation face advancement using an ANN trained by numerical models. Electron. J. Geotech. Eng. 11, <www.ejge.com>.

Stone, M., 1974. Cross-validatory choice and assessment of statistical predictions. J. Roy. Stat. Soc. B. 36, 111−147.

Sudheer, K.P., 2005. knowledge extraction from trained neural network river flow models. J Hydrol. Eng. 10, 264−269.

Sudheer, K.P., Jain, A., 2004. Explaining the internal behaviour of artificial neural network river flow models. Hydrol. Processes. 18, 833−844.

Sudheer, K.P., Nayak, P.C., Ramasastri, K.S., 2003. Improving peak flow estimates in artificial neural network river flow models. Hydrol. Processes. 17, 677−686.

Sulewska, M.J., 2010. Neural modelling of compactability characteristics of cohessionless soil. Comput. Assist. Mech. Eng. Sci. 17, 27−40.

Tawadrous, A.S., Degagne, D., Pierce, M., Masivars, D., 2009. Prediction of uniaxial compression PFC3D model micro-properties using artificial neural networks. Int. J. Numer. Anal. Methods Geomech. 33, 1953−1962.

Tekin, E., Akbas, S.O., 2011. Artificial neural networks approach for estimating the groutability of granular soils with cement-based grouts. Bull. Eng. Geol. Environ. 70, 153−161.

Teodorescu, L., Sherwood, D., 2008. High energy physics event selection with gene expression programming. Comupt. Phys. Commun. 178, 409−419.

Tokar, S.A., Johnson, P.A., 1999. Rainfall-runoff modeling using artificial neural networks. J. Hydrol. Eng. 4, 232–239.

Tsekouras, G.J., Koukoulis, J., Mastorakis, N.E., 2010. An optimized neural network for predicting settlements during tunneling excavation. WSEAS Trans. Syst. 9, 1153–1167.

Tsompanakis, Y., Lagaros, N.D., Psarropoulos, P.N., Georgopoulos, E.C., 2009. Simulating the seismic response of embankments via artificial neural networks. Adv. Eng. Softw. 40, 640–651.

Twomey, J.M., Smith, A.E., 1997. Validation and verification. In: Kartam, N., Flood, I., Garrett, J.H. (Eds.), Artificial Neural Networks for Civil Engineers: Fundamentals and Applications. ASCE, New York.

Uncuoglu, E., Laman, M., Saglamer, A., Kara, H.B., 2008. Prediction of lateral effective stresses in sand using artificial neural network. Soils Found. 48, 141–153.

Ural, D.N., Saka, H., 1998. Liquefaction assessment by neural networks. Electron. J. Geotech. Eng.3, <www.ejge.com>.

Wang, H.B., Sassa, K., 2006. Rainfall-induced landslide hazard assessment using artificial neural networks. Earth Surf. Processes Landforms. 31, 235–247.

Yang, Y., Rosenbaum, M.S., 2002. The artificial neural network as a tool for assessing geotechnical properties. Geotech. Eng. J. 20, 149–168.

Yildiz, E., Ozyazicioglu, M.H., Ozkan, M.Y., 2010. Lateral pressures on rigid retaining walls: a neural network approach. Gazi Univ. J. Sci. 23, 201–210.

Yoo, C., Kim, J., 2007. Tunneling performance prediction using an integrated GIS and neural network. Comput. Geotech. 34, 19–30.

Young-Su, K., Byung-Tak, K., 2006. Use of artificial neural networks in the prediction of liquefaction resistance of sands. J. Geotech. Geoenviron. Eng. 132, 1502–1504.

Yu, Y., Zhang, B., Yuan, H., 2007. An intelligent displacement back-analysis method for earth-rockfill dams. Comput. Geotech. 34, 423–434.

Zounemat-Kermani, M., Beheshti, A.A., Ataie-Ashtiani, B., Asabbagh-Yazdi, S.R., 2009. Estimation of current-induced scour depth around pile groups using neural network and adaptive neuro-fuzzy inference system. Appl. Soft Comput. 9, 746–755.

Zurada, J.M., 1992. Introduction to Artificial Neural Systems. West Publishing Company, St. Paul.

9 Hybrid Heuristic Optimization Methods in Geotechnical Engineering

Yung-Ming Cheng[1] and Zong Woo Geem[2]

[1]Department of Civil and Structural Engineering, Hong Kong Polytechnic University, Kowloon, Hong Kong, People's Republic of China, [2]Department of Energy and Information Technology, Gachon University, Seongnam, South Korea

9.1 Introduction

In many practical problems, the determination of the most economical and/or shortest path is required. Many engineering, scientific, and mathematical problems can also be cast into a form of optimum determination. For example, the minimum energy principle is one of the methods of solving engineering problems, and it also forms the fundamental principle of the finite element method. Due to the various needs for the determination of the optimum solution in various disciplines, different methods have been proposed and used. Resource allocation, packing, and scheduling, as well as many other similar problems, are traditionally analyzed using the linear and integer programming methods. Such methods usually require the objective function and constraints to be linear functions, but the optimum solution can usually be determined easily. The uses of gradient-type methods, which require the differentiability of the objective functions, have also found applications in many types of engineering problems. The differential of the objective function is also commonly formed by the finite difference method if the objective function cannot be expressed by simple mathematical expressions. These methods are limited, however, by the continuity requirement, and the global minimum may not be determined unless a good initial solution is used in the analysis. In geotechnical engineering as well as many other disciplines where multiple minima and discontinuity exist in the solution domain, the uses of the previous two groups of methods are seldom adopted. In general, the optimization process can be either unconstrained or constrained, and constrained analysis is actually more common in practice. The objective functions in many geotechnical or transportation problems are usually nonpolynomial hard

Metaheuristics in Water, Geotechnical and Transport Engineering. DOI: http://dx.doi.org/10.1016/B978-0-12-398296-4.00009-X

type problems (difficult to be proved, but is commonly believed) with the following features:

1. The objective functions are usually nonsmooth and nonconvex, and they may not be continuous over the whole solution domain (such as failure to converge in geotechnical problems and unacceptable traffic arrangement in transportation problems). It is also possible that the critical solution is not associated with the condition that the gradient of the objective function is zero.
2. Multiple local minima will exist in general, and many classical optimization methods can be trapped easily by the existence of a strong local minimum.
3. A good initial trial failure surface for a general global minimization problem is usually difficult to estimate for general complicated conditions. Furthermore, every heuristic global optimization method requires some kind of optimization parameters, and a good initial trial for these optimization parameters is difficult to establish for general conditions.

Due to the special difficulties of the objective functions and the needs of the engineers, many researchers have adopted different methods to search for the global minimum with various success and limitations, and a detailed discussion is given by Cheng et al. (2007a,b, 2008a,b, 2012). The modern heuristic global optimization methods that have evolved in recent years have attracted the attention of many geotechnical engineers recently. For slope stability problems, genetic algorithm (GA) has been used by McCombie and Wilkinson (2002), Zolfaghari et al. (2005), Cheng et al. (2007b), and Jianping et al. (2008). Bolton et al. (2003) have used the leap-frog optimization technique, while Cheng et al. (2008b) and Kahatadeniya et al. (2009) have applied the ant colony optimization (ACO) method in slope stability problems. Cheng et al. (2003, 2007a,b, 2008b) have also applied the simulated annealing (SA) method, particle swarm optimization (PSO), harmony search (HS), tabu search, and fish swarm methods for slope stability analysis. Cheng et al. (2007b) have also made a detailed comparison between six major types of heuristic global optimization methods for slope stability problems, and the sensitivity of these methods under different optimization parameters are investigated and compared. These efforts will not be reproduced here. Cheng et al. (2007b) have commented that no particular optimization method is superior in all cases, but some methods (ACO and tabu search) may be less effective for problems where the objective functions are highly discontinuous.

The authors have come across many practical slope stability problems from projects in different countries. One of the interesting projects worth mentioning is a complicated hydropower project in China where there are several strong local minima in the solution domain. Different methods have been used to search for the critical failure surface and the corresponding factor of safety of the foundation, and no method is found to be satisfactory by the engineers. To overcome these difficult cases, the authors propose a coupled optimization procedure based on the PSO and HS methods (Cheng et al., 2012). The authors will also introduce another coupling method based on the complex method and tabu search in this chapter. Actually, many heuristic optimization methods can be coupled to form a more stable solution algorithm. Since the performances between different coupling schemes are usually similar, the authors concentrate on two methods in this chapter, and part of the content of this chapter is based on Cheng et al. (2012).

9.2 Some Basic Heuristic Optimization Algorithms

9.2.1 Particle Swarm Optimization

Before the discussion of the coupling method, a brief review of PSO will be given. PSO is a heuristic global optimization algorithm developed by Kennedy and Eberhart (1995), which has been applied to many continuous and discrete optimization problems. PSO optimizes a problem by using a population of candidate solutions or particles, and the particles move stochastically around in the search space according to some simple mathematical formulas for the particle's position and velocity. The movement of each particle is influenced by its local best-known position, and the position and velocity are guided toward the best-known positions in the search space. The velocities and positions are updated if better positions are found by other particles, and this is expected to move the swarm toward the best solutions. Unlike the ACO method, where stigmergy is the main communication among the particles, the system communication between the particles that does not alter the environment is adopted in PSO because it is a population-based algorithm.

The PSO method is recognized as an effective method for global optimization and has received much attention in systems and control engineering, automatic recognition, radio systems, and other fields. Originally targeted toward simulating social behavior, the application of PSO has now been extended to many types of problems, with contributions from different researchers. Yin (2004) proposed a hybrid version of the PSO for the optimal polygonal approximation of digital curves; Salman et al. (2002) adopted PSO for the task assignment problem; and Ourique et al. (2002) used it for dynamic analysis in chemical processes. PSO does not require much computer memory, and the speed of computation is relatively fast.

In PSO, a group of particles (generally double the number of the control variables M), referred to as the *candidates* or *potential solutions* (given as X in Eq. (9.1)) enter the problem and search the search space to determine their optimum position. The optimum position is usually specified by the best solution of the objective function. Each "particle" is represented by a vector (X_i^k) in the multidimensional space to characterize its position, and another vector (V_i^k) to characterize its velocity at the current time step k. Conceptually, a group of birds determine the average direction and speed of flight during the search for food based on the amount of food normally found in certain regions of the search space. The results obtained at the current time step k will then be used to update the positions of the next time step. If a good source of food in a certain region of the space can be found, the group of particles will take this new piece of information into the consideration to formulate the "flight plan." Therefore, the best results obtained throughout the current time step are considered to generate the new set of positions for the whole group.

To evaluate the optimum of the objective function, the velocity V_i^k and the position X_i^k of each particle is adjusted in each time step according to the procedures as outlined below. The updated velocity V_i^{k+1} is a function of three major components:

1. the old velocity of the same particle (V_i^k),
2. the difference between the ith particle's best position found so far (called P_i) and the current position,
3. the difference between the best position of any particle within the context of the topological neighborhood of the ith particle found so far (called P_g—its objective function value is f_g) and the current position of the ith particle X_i^k.

For components 2 and 3, each component is stochastically weighted and added to component 1 to update the velocity of each particle with enough oscillations according to Eq. (9.1) to empower each particle to search for a better pattern within the problem space. Without sufficient oscillation, PSO can be trapped by the local minimum easily:

$$V_i^{k+1} = \omega V_i^k + c_1 r_1 (P_i - X_i^k) + c_2 r_2 (P_g - X_i^k)$$
$$X_i^{k+1} = X_i^k + V_i^{k+1} \quad i = 1, 2, \ldots, 2n \tag{9.1}$$

In Eq. (9.1), c_1 and c_2 are the stochastic weighting factors to components 2 and 3, respectively. These parameters are commonly given as 2, which will also be used in this study. Cheng et al. (2007b) has found that for normal problems, these optimization parameters are generally adequate. r_1 and r_2 are random numbers in the range [0,1], while ω is the inertia weight coefficient. A larger value for ω will enable the algorithm to explore the search space, while a smaller value of ω will lead the algorithm to exploit the refinement of the results. The flowchart for the PSO is shown in Figure 9.1.

Based on the experience on many slope stability problems, the authors have found that if the number of control variables is large, the number of trials required by the original PSO also will be large. To improve the efficiency of the solution for large-scale problems, a modified PSO (MPSO) is proposed by Cheng et al. (2007a). In MPSO, only several flights within the whole group of particles are allowed. In addition, particles with better objective function values are allowed to fly more within one iteration step than those with worse objective function values. The procedures for MPSO are:

1. Instead of one flight for each particle in the group, several flights are now performed. In addition, one particle can fly more than one time according to its objective function value. The better the objective function value of one particle, the more times it can fly. A parameter η ($0 < \eta < 1.0$) is used to implement this procedure of the flight. Suppose that N_a ($\leq M$) flights are allowed in MPSO within each iteration step. The current M particles are sorted in ascending order by the values of the objective function, and the probability of each particle flight (namely the flight probability) is determined according to Eq. (9.2):

$$\begin{cases} \mathrm{pr}_i = (1.0 - \eta)^{i-1} \times \eta \\ i = 1, 2, \ldots, M \end{cases} \tag{9.2}$$

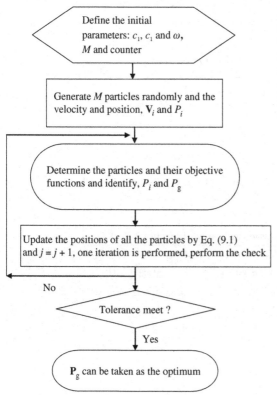

Figure 9.1 Flowchart for PSO (M is commonly taken to be twice the number of control variables and is related to the number of potential solutions; N_1 refers to the number of inner loop trials; and N_3 is a parameter to control the termination of the trials).

where pr_i means the flight probability of the ith particle. The accumulated probability is represented by the array $AP(i)$:

$$\begin{cases} AP(i) = \sum_{j=1}^{i} pr(j) \\ i = 1, 2, \ldots, M \end{cases} \tag{9.3}$$

A random number r_0, falling within the range from 0 to $AP(M)$, is generated in MPSO. If $AP(i-1) < r_0 \leq AP(i)$, then the ith particle will fly one time. It must be noted that after the ith particle flies to the new position, it will go back to the current position and the velocity will remain unchanged in this stage. The updating velocity will not be carried out until N_a flights have completed, which is a major difference with the original PSO, and this algorithm is more robust toward the presence of a local minimum.

2. After N_a flights are completed, the following procedure will be checked particle by particle. If a particle has the chance to fly more than one time, a new position and a new velocity are randomly chosen for the next iteration from the flights of that particle, and other randomly generated positions will be assigned to those having no chance to fly in the current iteration. This procedure is called the *updating rule* of the MPSO.

The procedures for MPSO for a given value of N_a are shown in Figure 9.2. f_{sf} and X_{sf} are used to restore the objective function value of the optimum solution and

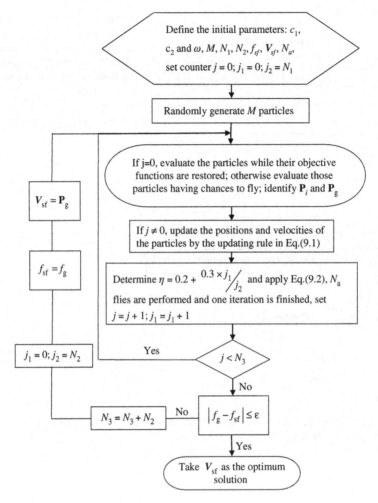

Figure 9.2 Flowchart for MPSO.

the optimum solution found so far. The initial value of the objective function (factor of safety) f_{sf} can be set to a large number, which is $1.0e + 10$ in this chapter. The parameters N_1, N_2, and N_3 are used to decide the termination criterion, and j_1 and j_2 are used to calculate the value of η. In the proposed coupled analysis, the MPSO by Cheng et al. (2007a) is used because a better performance is usually achieved by the MPSO compared to the original PSO.

9.2.2 HS Algorithm

Geem et al. (2001) and Lee and Geem (2005) have developed an HS metaheuristic algorithm (phenomenon-mimicking algorithm) that was inspired by the improvisation process of musicians in searching for a perfect state of harmony. Each musician (control variable) plays a note (objective function) to search for the harmony (global optimum). The harmony in music is analogous to the optimization solution vector, and the musician's improvisations are analogous to local and global search schemes in the optimization process. The HS algorithm is a population-based search method that uses a stochastic random search based on the harmony memory (HM) considering rate HR and the pitch adjusting rate PR. By its nature (discrete musical note), the HS algorithm does not require continuous control variables. An HM of size M is used to generate a new harmony, which is probably better than the optimum in the current HM. The HM consists of M harmonies (slip surfaces), and M harmonies are usually generated randomly. Consider $\mathbf{HM} = \{\mathbf{hm}_1, \mathbf{hm}_2, \ldots, \mathbf{hm}_M\}$ and

$$\mathbf{hm}_i = (v_{i1}, v_{i2}, \ldots, v_{im}) \tag{9.4}$$

where each element of $\mathbf{hm}_i$ corresponds to that in vector $\mathbf{V}$ described above. Consider the following function optimization problem, where $M = 6$, $m = 3$. Suppose that HR = 0.9 and PR = 0.1:

$$\begin{cases} \min \quad f(x_1, x_2, x_3) = (x_1 - 1)^2 + x_2{}^2 + (x_3 - 2.0)^2 \\ \text{s.t.} \quad 0 \le x_1 \le 2; \quad 1 \le x_2 \le 3; \quad 0 \le x_3 \le 2 \end{cases} \tag{9.5}$$

Six randomly generated harmonies comprise the HM shown in Table 9.1. The new harmony can be obtained by the HS algorithm with the following procedure. A random number in the range [0,1] is generated (e.g., 0.6 ($<$HR)), and one of the values from {1.0, 1.5, 0.5, 1.8, 0.9, 1.1} should be chosen as the value of x_1 in the new harmony. Take 1.0 as the value of x_1; then another random number of 0.95 ($>$HR) is obtained. A random value in the range [1,3] for x_2 is generated (say, 1.2), and similarly, 0.5 is chosen from the HM as the value of x_3; thus, a coarse new harmony $\mathbf{hm}'_n = (1.0, 1.2, 0.5)$ is generated. The improved new harmony is obtained by adjusting the coarse new harmony according to the parameter PR. Suppose that three random values in the range [0,1] (say, 0.7, 0.05, 0.8) are generated. Since the former value 0.7 is greater than PR, the value of x_1 in $\mathbf{hm}'_n$ remains unchanged. The second value 0.05 is lower than PR, so the value of 1.2 should be

Table 9.1 The Structure of the HM

HM	Control Variables			
	x_1	x_2	x_3	Objective Function
$\mathbf{hm}_1$	1.0	1.5	0.5	4.50
$\mathbf{hm}_2$	1.5	2.0	1.8	4.29
$\mathbf{hm}_3$	0.5	1.5	1.0	3.50
$\mathbf{hm}_4$	1.8	2.5	0.9	8.10
$\mathbf{hm}_5$	0.9	2.2	1.2	5.49
$\mathbf{hm}_6$	1.1	1.9	1.5	3.87

adjusted (say, to 1.10). These procedures continue until the final new harmony $hm_n = (1.0, 1.10, 0.5)$ is obtained. The objective function of the new harmony is determined as 3.46. The objective function value of 3.46 is better than that of the worst harmony, hm_4; thereby, hm_4 is excluded from the current HM, while hm_n is included in the HM. Up to this stage, one iteration step has finished. The algorithm will continue until the termination criterion is achieved.

The iterative steps of the HS algorithm in the optimization of Eq. (9.5) are as follows:

Step 1: Initialize the algorithm parameters: HR, PR, and M, randomly generate M harmonies (slip surfaces) and evaluate the harmonies.
Step 2: Generate a new harmony (shown in Figure 9.3) and evaluate it.
Step 3: Update the HM. If the new harmony is better than the worst harmony in the HM, the worst harmony is replaced with the new harmony. Take the ith value of the coarse harmony h'_n, v'_{ni} for reference. Its lower and upper bounds are named $v_{i,min}$ and $v_{i,max}$, respectively. A random number r_0 in the range [0,1] is generated. If $r_0 > 0.5$, then v'_{ni} is adjusted to v_{ni} using Eq. (9.6) to calculate the new value of v_{ni}:

$$v_{ni} = v'_{ni} + (v_{i,max} - v'_{ni}) \times \text{rand } r_0 > 0.5$$
$$v_{ni} = v'_{ni} - (v'_{ni} - v_{i,min}) \times \text{rand } r_0 \leq 0.5 \qquad (9.6)$$

where rand means a random number in the range [0,1].
Step 4: Repeat steps 2 and 3 until the termination criterion is achieved. The details of the HS method can be found in works by Geem et al. (2001).

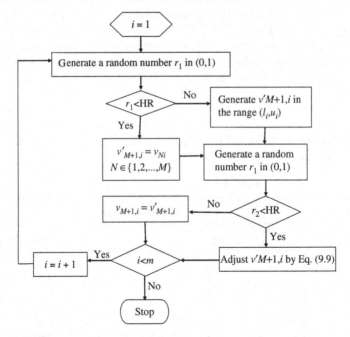

Figure 9.3 Generation of new harmony and the search procedure from Cheng et al. (2007a).

9.2.3   Tabu Search Algorithm

The tabu search algorithm, which searches for a solution using memory and expectation criteria, was introduced by Glover and Laguna (1996). The main idea of the algorithm is as follows: find the best local solution *ans* in a given solution and neighborhood, set the current best solution as $ans^* = ans$, and then search for the best local solution *ans'* in the current neighborhood. This best local solution, however, could be the same as the last one. In order to avoid such situations, a tabu form to record the recent operation has to be defined. If the current operation is in the tabu form, the search will stop; otherwise, *ans* will be replaced by *ans'*. The objective function value of *ans'* may be worse than *ans*, so the tabu search algorithm can accept a worse solution. For those useful operations that can improve the current best solution, they will be put at the top of the form to find a better solution rapidly.

In a tabu search, the neighborhood tabu is adopted instead of the solution tabu used in the discrete variable algorithm. The procedure of the spatial discretization is as follows: an *n*-dimensional hypercube can be formed from given upper and lower limits, as defined by $U = (u_1, u_2, \ldots, u_n)$, $L = (l_1, l_2, \ldots, l_n)$. In the variable spaces, divide the *i*th dimension into N_i pieces to form an $N_{adj} = N_1 \times N_2 \times \cdots N_n$ rectangle, which represents a neighborhood. Determine the length of the tabu from *Numtabu*; then the algorithm can tabu on *Numtabu* neighborhoods to the maximum.

If the newly found solution *ans'* is in a neighborhood that has not been tabu, then replace *ans* with *ans'*, and put this neighborhood into the tabu form. If *ans'* is better than the current best solution ans^*, then remove this neighborhood using the expectation criteria. Replace *ans* with *ans'*, or keep the current solution *ans* and search again until the termination condition is reached. The advantage of the tabu search algorithm is that it can avoid the duplication and push the algorithm to search a new space; however, it depends heavily on the initial solution. When a new solution is generated in the annealing process, the tabu search algorithm can avoid redundant search. Cheng et al. (2007b) have shown that tabu can be very efficient for some problems, but its performance in complicated problems is less reliable compared to other heuristic optimization methods.

9.2.4   Complex Method

The complex method, developed by Box (1966), is based on the simplex method. The main ideas of the complex method are: (1) choose some vertices in a bounded multidimensional space to form a polyhedron (complex); (2) compare the function value that corresponds to every vertex; (3) discard the vertex associated with inferior values and replace it with a new vertex which is not only within the constraints but also improves the solution; and (4) repeat these procedures until the minima is approached gradually. The complex method includes the production of initial complex, reflection, and contraction operators, and termination criteria similar to the simplex method. The flowchart for the complex method is shown in Figure 9.4.

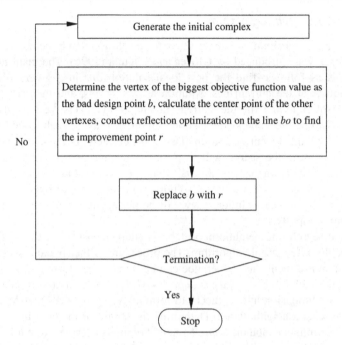

Figure 9.4 Flowchart for the original complex method described in this chapter.

The k initial vertices are $V_i = (x_1^i, \ldots x_n^i)$, $i = 1, 2, \ldots, k$. The center point is given by

$$V^o = \frac{1}{k-1} \sum_{\substack{i=1 \\ i \neq b}}^{k} V_i \qquad (9.7)$$

$$V^r = V^o + \alpha(V^o - V_b) \qquad (9.8)$$

Equation (9.8) is used to find the improve point r. α is the reflection coefficient, and the initial value α_{ini} is usually set to 1.3, and then it contracts to half repeatedly until the improvement point is found. If the improvement point is not found for $\alpha < \xi$ ($\xi < 10^{-5}$), then the reflection has failed. The termination criteria is the difference of the maximum and minimum in the objective functions of vertices being less than ε ($\varepsilon = 10^{-3}$). α is a positive number during the reflection and contraction, which means that the contraction operator searches the points on the dotted line only in the complex algorithm, as shown in Figure 9.5, but the points on the solid line are ignored. So the basic complex method is called the *partial scope complex method*. If a complete search need to be done on the line bo, the reflection coefficient has to be determined. Point b $V_b = (x_1^b, \ldots, x_n^b)$, center point $V_0 = (x_1^o, \ldots, x_n^o)$, and the reflection efficient α (α_{min} and α_{max}) can be determined as follows:

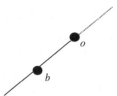

Figure 9.5 Illustration of the reflection and retraction line.

For $i = 1$ to n:

$$\alpha_{l_i} = (l_i - x_i^o)/(x_i^o - x_i^b) \quad \alpha_{u_i} = (u_i - x_i^o)/(x_i^o - x_i^b)$$
$$\alpha_{i,1} = \min(\alpha_{l_i}, \alpha_{u_i}) \quad \alpha_{i,2} = \max(\alpha_{l_i}, \alpha_{u_i})$$

next i:

$$\alpha_{\min} = \max\{\alpha_{i,1}\}_{i=1,2,...,n} \quad \alpha_{\max} = \min\{\alpha_{i,2}\}_{i=1,2,...,n} \tag{9.9}$$

First, set $\alpha = \alpha_{\max}$, and then reduce it to half repeatedly. Then the new solutions are on the dotted line. If the improvement point is not found when $\alpha < \xi$, then set $\alpha = \alpha_{\min}$ and reduces to half repeatedly. Then the new solutions are on the solid line, which is reverse reflection and contraction efficient. If the improvement point is not found when $abs(\alpha) < \xi$, then the forward reflection contraction efficient optimization has failed. This approach can expand the reflection and contraction forward and backward easily and form the full scope complex method.

The complex method that is adopted in the present chapter is slightly different from the basic complex method in the reflection contraction efficient and the bad design point b. The vertex that is most similar to the others is the bad design point b, a, and the authors find that such a minor change is usually more effective in the solution compared to the original scheme. The flowchart is shown in Figure 9.6.

9.2.5 PSO Coupled with HS

In the original PSO, the locations of the particles are updated by modifying the corresponding velocity vectors, and it is found that an incorrect value of ω may lead to the trap into the local minimum, which will be demonstrated in a later section. Generally speaking, a moderate value of 0.5 for ω is used for all the problems. Alternatively, a larger value of ω can be applied at the initial analysis to search the solution space, which is then reduced linearly to a small value to find better results near the existing best position. There is another way to simulate the PSO procedure in Eq. (9.1) as given by Wang and Liu (2008) which is based on: (1) the current positions of particles; (2) the best position found so far (P_i); (3) the best position of any particle within the context of the topological neighborhood of the ith particle found (P_g).

The HS method is another efficient and effective global optimization method when the number of control variables is less than 25 for many geotechnical problems, which is discussed by Cheng et al. (2008a). Cheng et al. (2008a) have also described a detailed procedure in the implementation of the modified HS (MHS) algorithm, which is adopted in the present coupling proposal. When the

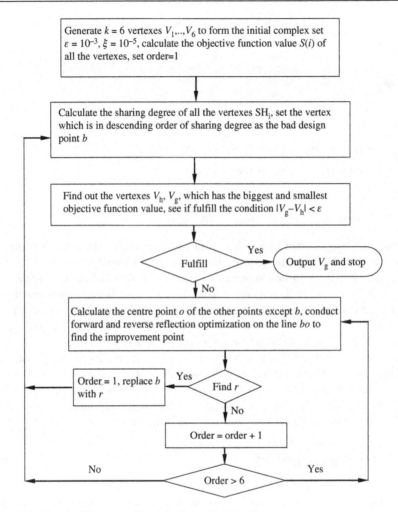

Figure 9.6 The modified complex method used in this chapter.

problem is large, it can be trapped by the local minimum easily, judging from various internal tests done by the authors, and Cheng et al. (2008a) have proposed an MHS search method to overcome the limitation of the original HS method. The utilization of the MHS is intuitively more exhaustive, generating several new harmonies rather than generating a new harmony during each iteration. Two parameters, HR and PR, for HS are required in the analysis, and the detailed procedure is shown in Figure 9.3.

If we take the above-mentioned positions (flights) from PSO as the harmonies in the HM in HS, a new position can be obtained by the HS procedure. In Figure 9.3, $z_{i,\text{min}}$ and $z_{i,\text{max}}$ are the minimum and maximum values of the ith element in vector X. z_{ij} is the jth element of X_i. Similar to the modified PSO, N_a ($\leq M$) flights within each iteration step are allowed with different approaches. It is possible to choose N_a particles randomly from the total generation rather than based on the fitness of the particles in the modified PSO. In this way, the choice of flight is controlled by the procedure in HM rather than the original procedure as outlined in Eqs. (9.2) and (9.3). This is a minor and simple trick to combine the two methods. Cheng et al. (2007b) have tried the GA, the SA method, PSO, HM, tabu search, and ACO, and have commented that no single method can outperform other methods in all cases. Each optimization method has its own merits and limitations, and the combination of two optimization methods may result in a better performance in difficult cases.

The flowchart of the coupled PSO and HS by Cheng et al. (2012), which is denoted as HMPSO, is shown in Figure 9.7. It should be noted that the flowchart in Figure 9.7 is a simple combination of those in Figures 9.2 and 9.3, and the authors do not attempt to propose a highly complicated procedure to combine these two methods (for simplicity). In Figure 9.2, the updating of the positions of all the

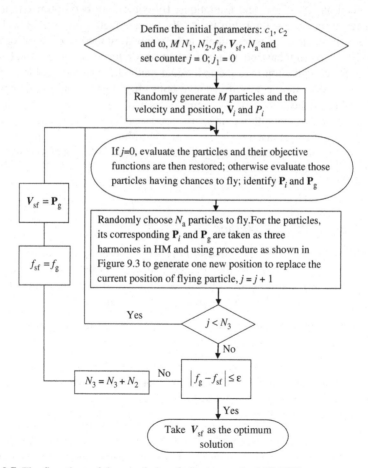

Figure 9.7 The flowchart of the coupled optimization method HMPSO.

particles is replaced by HS generation, as shown in Figure 9.3. Such a minor change can retain the simplicity of both optimization methods so that the proposed algorithm is easy to use and does not require a lot of computer memory. The authors have come across several very complicated cases in some projects, and the presently proposed coupled algorithms are more stable and robust for such problems. It is true that the present method will be less efficient than the simple method, and it is not recommended for such purposes because it is less efficient (though effective) for such cases. The proposed coupled algorithms are targeted toward those complicated problems (discontinuous objective functions with multiple strong local minima and sudden major changes in the material properties) in which the other algorithms may fail to perform satisfactorily.

Besides the coupling of PSO and HM, it is also possible to couple the tabu search, SA, the GA with the HS method, and the authors have also successfully implemented these coupling methods. As discussed by Cheng et al. (2007a,b), there is no single heuristic search method that can outperform other methods in all cases. Each method has its merits and limitations. By combining two optimization methods, the resulting search algorithm is usually more effective in dealing with more complicated problems and is less likely to be trapped by the presence of local minima. It should be emphasized that the adoption of the coupling method usually requires more computation than the individual method for simple problems. In this respect, there is no simple way to achieve both effectiveness and efficiency under all cases. With the advancement in computer technology and increasing complexities of the problems, the authors view that a more stable and effective algorithm is more important than a fast but less robust algorithm (keeping in mind that the increase in the computation is not significant for the coupling method, which will be illustrated in a later section).

9.2.6 Tabu SA Complex Method

To take advantage of tabu search and to improve its performance for more complicated problems, this method can be coupled with SA and the complex, denoted as TSAC, method in the following way. For example, to search the minimum safety factor of slip slope, the calculation steps can be put as follows:

1. Determine the number of design variables n (say, $n = 3$ for the circular slope stability problem) and the upper and lower limits of design variables $U = (u_1, u_2, \ldots, u_n)$, $L = (l_1, l_2, \ldots, l_n)$. Generate six vertices to form the initial complex and determine T_o (annealing temperature), N_x, N_y, N_z, and $Numtabu$, where N_x, N_y, N_z are the distances for equal partition of the three control variables. Meanwhile, put the neighborhoods of the six initial vertices into the tabu form, $order = 1$ and $T_c = T_o$.
2. Calculate the geometric center V_o of the other vertices except V_b and conduct optimization of the tabu annealing reflection contraction operator using these two points. If the optimization is successful, then set $order = 1$, $T_c = T_c \times 0.95$ and then turn back to find point b; if it failed, set $order = order + 1$ and check if $order$ is bigger than 6. If not, turn back to find point b. If $order$ is bigger than 6, then begin the calculation of the complex search.

In this method, there are five parameters to be defined. The tabu annealing complex method has five parameters T_o, N_x, N_y, N_z, and *Numtabu*. T_o and *Numtabu* are chosen to be 10 and 60 by the authors (relatively insensitive to these parameters for most problems), while the other three parameters will depend on the geometry requirement of the specific problem.

9.3 Demonstration of the Coupling Methods

To demonstrate the effectiveness of the present coupled optimization method against different versions of PSO methods, several standard test problems by Taillard (1993) and Lawrence (1984) are considered by Cheng et al. (2012), and the results are shown in Tables 9.2 and 9.3. In these standard problems, the coupled methods are always ranked as the best in the optimized results, compared to the GA and different versions of the PSO. These encouraging results have supported the applicability of the proposed coupled optimization method.

For the results as shown above, the parameters used for PSO are $c_1 = c_2 = 2.0$, $\omega = 0.5$, while M is taken as four times the number of control variables N. For the HS, the parameters are $M = 2N$, HR = 0.98, and PR = 0.1. For the TSAC, T_o and *Numtabu* are chosen to be 10 and 60, while the number of partition is set as 10, which will indirectly define the partition distance. The authors have also varied the parameters and found that the optimum results are practically insensitive to these parameters unless these parameters are set to an unreasonably large or small value. The efficiency of the analysis, however, is more affected by the

Table 9.2 Comparison of the Coupled Methods by Cheng et al. (2012) with GA and Different Versions of PSO by Taillard (1993)

Problem Size	GA	DPSO	DPSO1	DPSO2	DPSO3	HMPSO	TSAC
20 × 5	10.06	10.31	10.24	9.84	9.85	9.84	9.845
20 × 20	46.1	48.87	49.44	45.41	45.39	45.38	45.38
50 × 20	50.17	51.59	51.9	46.69	46.66	46.66	46.66

Discrete particle swarm optimization (DPSO), DPSO1, DPSO2, and DPSO3 are different variants of the PSO method.

Table 9.3 Comparison of the Coupled Methods by Cheng et al. (2012) with Different Versions of PSO by Lawrence (1984)

Problem	Problem Size	Best-Known Solution	Best from PSO	Worst from PSO	HMPSO	TSAC
LA21 (15,10)	15,10	1046	1046	1088	1046	1046
LA36 (15,15)	15,15	1268	1269	1297	1268	1268
LA26	20,10	1218	1218	1409	1218	1218

choice of these parameters. In modern computing, this issue appears to be less critical than obtaining a good optimum result.

9.4   Application of Coupling Methods in the Slope Stability Problem

The authors have worked on many types of geotechnical problems and found that most existing global optimization methods can work well for relatively simple problems. When the problem is complicated in geometry, with major differences in the soil parameters between different soils that have been experienced by the authors for several hydropower projects in China, the solution will be sensitive to the precise values of the control variables. Actually, the engineers have used different programs and solution methods for this problem, but the results from the existing programs and solution algorithms are still not satisfactory. The difficulty of this problem is that there are several strong local minima within the solution domain, and some commercial programs fail to escape from the local minima during the analysis. Furthermore, a good initial trial is difficult to be established for this problem. Due to this special case, the authors have developed several coupling methods, two of which are discussed in this chapter.

For the slope stability/foundation problem, it can be stated as a constrained global minimization problem as given by Eq. (9.10), where the factor of safety $f(\mathbf{X})$ is minimized subject to the coordinates of the slip surface being convex:

$$
\begin{aligned}
&\min \quad f(\mathbf{X}) \\
&\text{s. t.} \quad x_l \le x_1 \le x_u \quad x_L \le x_{n+1} \le x_U
\end{aligned}
\tag{9.10}
$$

In Eq. (9.10), the factor of safety function is evaluated by the Spencer method (1967). The requirement on the convexity is given by Cheng (2003), and the lower and upper bounds are dynamic in that the actual lower and upper bounds for variables $i + 1$ to N depends on the bounds for variables from 1 to i. That means, based on the trial x_1 and x_2, the upper and lower bounds for variable x_3 will be determined. There is no need to predefine the upper and lower bounds for variables from $i = 3$ to N, while the upper and lower bounds for the first two variables can be defined easily for the present problem (or use a conservatively large domain). On the other hand, classical optimization problems require the upper and lower bounds of all the control variables to be defined before the analysis. Cheng (2003) has demonstrated that this dynamic bound can give very high efficiency in the optimization analysis and is recommended for the slope stability problem. This dynamic bound requirement is also different from classical problems where the bounds are static.

For the first vertical retaining wall problem, shown in Figure 9.8, the ground surface behind the wall is inclined at an angle of 20° from the horizontal. The soil parameters are friction angle = 30° and cohesive strength = 0. For this problem,

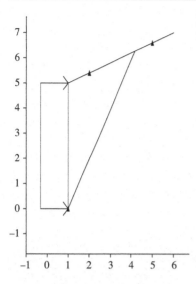

Figure 9.8 Generation of Coulomb earth pressure from optimization analysis.

the classical Coulomb solution gives an exact solution of the active pressure coefficient equal to 0.441, with a failure surface inclined at an angle of 52° from horizontal. Using HM, PSO, HMPSO, and TSAC, the same results are obtained as the classical analytical solutions. It is interesting, however, to note that the numbers of computations required to evaluate the critical solutions are 7260, 8234, 9546, and 9873 for HM, PSO, HMPSO, and TSAC, respectively, with 20 control variables. The results indicate that for simple problems, the use of the coupling method is effective but is not efficient for the analysis. This result is not surprising because the coupled methods explore more trials in the analysis.

For the second slope stability problem, shown in Figure 9.9A, there are three layers of soil, a water table, and a vertical surcharge on the slope. The soil parameters for this problem are given in Table 9.4. Soil layer 2 has a thickness of only 1 mm, which is very small; hence soil layer 2 appears to be missing from the figure. There is a sudden change of soil parameters within a narrow region, which is a very difficult problem for global optimization analysis. Such geotechnical conditions are possible in nature, however, and the Fei Tsui Road slope failure in Hong Kong is similar to that shown in Figure 9.9. The critical factor of safety for this problem is 0.495, which is obtained by both HMPSO and TSAC, while a factor of safety 0.552 is obtained by HM (PSO gives 0.543). Most of the failure surface lies within soil layer 2 with very low soil parameters, and this result is consistent with the simple sense of engineering. Furthermore, the critical failure surfaces given by HM or PSO are far from the critical solution shown in Figure 9.9A. Based on the simple sense of engineering, the critical solution must pass through soil layer 2, which is predicted by HMPSO and TSAC (which also happened in the Fei Tsui Road slope failure). On the other hand, the solution shown in Figure 9.9B is obviously not correct; the critical solution does not pass through soil layer 2, which is in conflict with the simple sense of engineering. On the other hand, HMPSO and TSAC require 13,627 and 14,264 computations, while only 7324 computations are

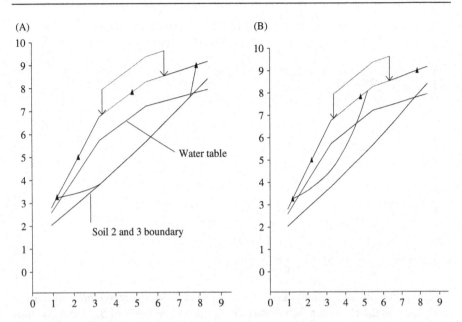

Figure 9.9 (A) Critical solution from HMPSO and TSAC; (B) critical solution from HM (20 control variables).

Table 9.4 Geotechnical Parameters for the Soils in Example 2

Layers	γ (kN/m^3)	c' (kPa)	ϕ' (degree)
1	19.0	3.0	30.0
2	19.0	0.0	20.0
3	19.0	2.0	32.0

γ = unit weight of soil, c' = effective cohesive strength, and ϕ' = effective angle of friction.

required for HM. Once again, coupled analysis requires more computation, but the results are much better than the single heuristic analysis. The results from these first two examples have clearly demonstrated that it is difficult to maintain both effectiveness and efficiency for general conditions. A more robust and stable algorithm will require more computations compared to a fast but less robust algorithm. The authors have uploaded this input file to http://www.cse.polyu.edu. hk/~ceymcheng/ so that readers can test the performance of single and coupled heuristic optimization analysis.

Before the discussion of a hydropower project in China, the authors would like to discuss a special problem covered by Cheng et al. (2012) and shown in Figure 9.10. This function is fluctuating rapidly about $x = 3.25$, and the relation between the dimensionless variables x and y are given in Table 9.5 and Figure 9.10.

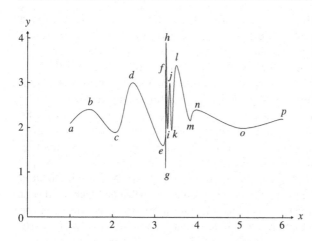

Figure 9.10 A simple one-dimensional function with the presence of several "strong" maxima and minima for the illustration of optimization by Cheng et al. (2012).

Table 9.5 Values for x and y in Figure 9.10 by Cheng et al. (2012)

Points	x	y
a	1.0	2.1
b	1.5	2.4
c	2.1	1.9
d	2.5	3.0
e	3.2	1.6
f	3.25	3.3
g	3.252	1.1
h	3.255	3.9
i	3.3	2.0
j	3.35	3.0
k	3.4	1.95
l	3.5	3.4
m	3.8	2.2
n	4.0	2.4
o	5.0	2.0
p	6.0	2.2

Actually, there are several fluctuating zones over a small region, and this condition can be taken as a simplification of the next example. This special function is given by Eq. (9.11), and the maximum and minimum of $f(y)$ are given by 25.45 and 0.461 over the solution domain of $x = 1-6$:

$$f(y) = \tan(y \times \pi/8) \tag{9.11}$$

To search for the critical one-dimensional solution, the authors have used the GA, HS, PSO, and ACO methods, which are independent of the initial trials in the solutions. The authors have also adopted a starting point of $x = 1.0$ in the SA method, and determined point *e* (with a value 0.727) from the above-mentioned heuristic optimization algorithms. These results are clearly far from the true critical solution, as point *e* can be obtained easily from all classical optimization methods, while point *g* is highly localized within a very narrow range and is shortly after point *e*. If the global minimum point *g* is to be determined, a very good initial trial is required for the SA, while specially tuned optimization parameters have to be used (which are obtained by trial and error). This situation occurs because the change from point *d* to *e* is less extreme as compared with the section from *f* to *g*. A trial within the region *d* to *e* is easily generated, while trials between points *f*, *g*, and *h* are difficult to generate because this region is too small. Because of this special geometrical requirement, the absolute minimum is missed in the optimization search unless special treatment (which may be problem dependent) is adopted. For the present simple one-dimensional problem, the absolute minimum can be obtained with ease by observation. For multidimensional problems, particularly when there are also regions of discontinuity, this situation will become highly complicated and will be missed in the optimization analysis. Although the situation shown in Figure 9.10 is a hypothetical problem, such conditions actually exist for the practical project that will be illustrated in the next example.

The next problem, shown in Figure 9.11, is one of the sections of a major hydropower project in China founded at a location with complicated ground conditions. In this site, there are several different layers of soft materials that are shaded in Figure 9.11, while the material parameters are shown in Table 9.6. For this project, several commercial programs have been used, giving different critical results.

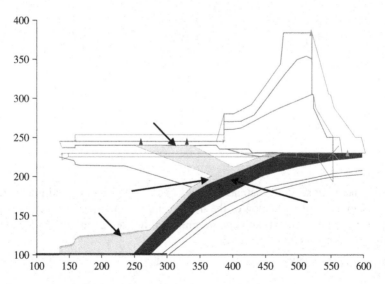

Figure 9.11 Soft soil in a shaded area for a dam project. The weak zones are shaded and marked with arrows for illustration.
Source: From Cheng et al. (2012).

Table 9.6 Geotechnical Parameters for the Problem in Figure 9.11 by Cheng et al. (2012)

Layer	γ (kN/m^3)	c' (kPa)	ϕ' (degree)
1	16.00	2000	56.31
2	24.00	2000	56.31
3	24.00	2000	56.31
4	26.00	1000	50.20
5	26.00	1400	54.50
6	26.00	1000	44.70
7	26.00	100.0	19.30
8	26.00	100.0	19.30
9	26.00	1000	44.70
10	26.00	1400	54.50
11	26.00	100.0	19.30
12	26.00	100.0	19.30
13	26.00	100.0	19.30
14	23.00	130.0	22.30
15	26.00	1400	54.50
16	26.00	100.0	19.30
17	26.00	1400	54.50

Mindful of the importance of this project, the authors have carried out a detailed study using the coupling method. For most of the sections of this project, classical heuristic optimization methods can still work properly. There are, however, some special sections for which no satisfactory results can be obtained, and a wide range of minimized results is obtained, as shown in Table 9.7. Conceptually, the nature of this problem is similar to that as shown in Figure 9.10. As shown in Table 9.6, there are several layers of soil that are thin, and the soil parameters are low. This geological condition corresponds to the presence of several strong local minima, similar to those in Figure 9.10. The ability of the algorithm to escape from strong local minima is a difficult task worth considering.

To carry out their analysis, Cheng et al. (2012) defined the left exit end of the failure surface to lie within the domain of $x = 260-330$ m, while the right exit end is defined to lie within the domain of $x = 520-575$ m. The bounds for the left and right exit ends were actually more than enough, as a good initial trial cannot be established easily. In Figure 9.12, the failure surfaces based on the MHS and MPSO are close to the original MHS and MPSO methods, and, for the purposes of clarity, they are not shown. It is interesting to note that the failure surfaces from different optimization methods were virtually the same at the right side. This is possibly due to the constraints from the local soil profiles and the geometry of this project. To the right exit end, there were major differences between the failure surfaces from different methods of optimization, which are shown in Figure 9.12. The first difference was that the starting point of the critical failure surface from HMPSO was $x = 278.0$, while it ranged from 320.25 to 320.38 for all the other methods. The second difference was that the exit angle of the failure surface for

Table 9.7 Minimum Factors of Safety for Example 3 Based on Spencer Method (41 Control Variables) by Cheng et al. (2012)

Method of Global Optimization	PSO	MPSO	AFSA	MHM	HMPSO	TSAC
Minimum factor of safety (FOS)	2.18	2.15	1.83	1.98	1.65	1.652
Number of trials	121,124	59,288	394,527	132,098	130,156	142,362
Minimum FOS at evaluation number	99,824	35,460	219,284	98,426	112,342	98,462

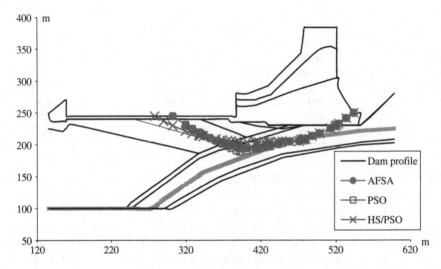

Figure 9.12 Critical failure surfaces by different global optimization methods based on the Spencer method by Cheng et al. (2012). Note that the critical failure surfaces by MHS and MPSO are not shown for clarity. The gray colored line represents the critical weak zone in soil.

HMPSO method was smaller than all the other methods. Finally, all the optimization methods except for the HMPSO were more attracted by soil 13 in the analysis, so the critical failure surfaces were deeper than those by the HMPSO. Table 9.7 shows that most of the global optimization methods are not satisfactory except for the artificial fish swarm algorithm (AFSA), which gives a factor of safety less than 2.0 (this is still not good enough, but HM and PSO perform more poorly) but requires 394,527 trials in the analysis. Actually, the authors (Cheng et al., 2008a) found that HS can be very inefficient and sometimes is not effective when the number of control variables is large, and this situation is also reflected in the current example. From these results, it can be seen that all the single optimization methods are attracted by the presence of strong local minima during the search, except for the coupled HMPSO analysis, which is less affected by the attraction of the local minima. Based on HMPSO, the minimum factor of safety is

1.65, with 130,156 evaluations, and the result is the best of all the five different global optimizations, as shown in Table 9.7. The authors have also applied TSAC to this problem, and the results are virtually the same as that achieved by HMPSO (but, for clarity, this is not shown in Figure 9.12). It is true that the coupled optimization method is less efficient for simple problems, which is demonstrated in Example 1 as shown in Figure 9.8, but the method is also more stable for problems where there are several strong local minima. For the present large-scale construction work, a good result is much more important than the time of computation required to find it, and the proposed coupled method has provided good results without excessive computations.

9.5 Discussion and Conclusions

Many geotechnical engineering problems are governed by critical solutions, and various optimization methods have been used in various computer programs to find such solutions. In the past, the classical simplex or gradient method has been used for to solve simple problems with regular geometry. These methods, however, are easily trapped by a local minimum and fail to work for discontinuous problems, which is the fundamental nature of many geotechnical problems. Many researchers are now turning to modern global optimization methods, which are not limited by the presence of a local minimum during the optimization process. Actually, some commercial programs have also adopted heuristic optimization methods in the search for critical solutions. In general, these programs can perform well for normal problems, which is reported from the experience of some engineers.

Cheng et al. (2007b) have clearly demonstrated that no single heuristic global optimization problem can outperform other methods in all cases, and every method has its own merits and drawbacks. While commercial programs usually adopt the basic heuristic optimization algorithms (usually with modifications), for complicated problems, particularly for those discontinuous problem or problems with several strong local minima, the authors have found that existing methods may sometimes fail to give the best solution. Due to the needs of some critical construction projects, coupling methods have been developed that can usually give a better performance with less dependence on the choice of parameters. The coupling methods presented in this chapter are new approaches to global optimization. The advantages of one optimization method can balance the disadvantages of another. The final outcome will be a more robust and stable solution algorithm, which is applicable over a wider range of problems. The new method has been used for several major construction projects in China where existing methods have been found to be ineffective. The problem, as shown in Figure 9.12, has created major difficulties in optimization analysis. The authors tried many exiting methods in vain before the development of the present coupled algorithms. The coupled methods are relatively simple to implement, and an arbitrary initial trial failure surface far from the critical solution can be used for the analysis. The authors have found that the present proposal works well in all the practical cases that they have encountered, and the method is particularly useful for problems governed by several strong local minima, similar to those shown in Figure 9.10. For more difficult problems, it has been demonstrated in the present study that the coupled method is stable in operation and can work well despite having poor initial trials. The coupled algorithm is actually suitable for general problems with multiple local minima within the solution domain, as the optimization method is not attached to the objective function in the present proposal.

This chapter has reviewed the basic formulation of coupled optimization analysis, and several examples (both theoretical and practical problems) are used to illustrate the capability of the coupled analysis. As demonstrated by the numerical examples, the proposed method is less efficient (though still effective) for simple problems and is not recommended for such cases, but it is more stable in the analysis because it is less attracted by the local minima for difficult problems. The authors believe that effectiveness and efficiency cannot be maintained simultaneously for such difficult problems. There is no simple way to ensure both effectiveness and efficiency in all cases. More importantly, there is also no simple way to assign suitable optimization parameters suitable for all general cases. The use of the coupling method, which employs the advantages of one optimization method to counterbalance the disadvantages of another, has been demonstrated to be more stable and robust, and this approach is recommended for generally difficult problems where the initial trials or parameters are difficult to establish.

Acknowledgment

The present project is funded from Research Grants Council through the project B-Q12G, PolyU 513808.

References

Bolton, H.P.J., Heymann, G., Groenwold, A., 2003. Global search for critical failure surface in slope stability analysis. Eng. Optim. 35, 51−65.

Box, M.J., 1966. A new method of constrained optimization and a comparison with other methods. Comput. J. 8 (1), 42−52.

Cheng, Y.M., 2003. Locations of critical failure surface and some further studies on slope stability analysis. Comput. Geotech. 30, 255−267.

Cheng, Y.M., Li, L., Chi, S.C., Wei, W.B., 2007a. Particle swarm optimization algorithm for location of critical non-circular failure surface in two dimensional slope stability analysis. Comput. Geotech. 34 (2), 92−103.

Cheng, Y.M., Li, L., Chi, S.C., 2007b. Performance studies on six heuristic global optimization methods in the location of critical failure surface. Comput. Geotech. 34, 462−484.

Cheng, Y.M., Li, L., Lansivaara, T., Chi, S.C., 2008a. Minimization of factor of safety using different slip surface generation methods and an improved harmony search minimization algorithm. Eng. Optim. 40, 95−115.

Cheng, Y.M., Li, L., Chi, S.C., Wei, W.B., 2008b. Determination of the critical slip surface using artificial fish swarms algorithm. J. Geotech. Geoenviron. Eng. ASCE. 134 (2), 244–251.

Cheng, Y.M., Li, L., Sun, Y.J., Au, S.K., 2012. A coupled particle swarm and harmony search optimization algorithm for difficult geotechnical problems. Struct. Multidiscip. Optim. 45, 489–501.

Geem, Z.W., Kim, J.H., Loganathan, G.V., 2001. A new heuristic optimization algorithm: harmony search. Simulation. 76 (2), 60–68.

Glover, F., Laguna, M., 1996. Tabu Search. Kluwer Academic, Dordrecht, The Netherlands.

Jianping, S., Li, J., Liu, Q., 2008. Search for critical slip surface in slope stability analysis by spline-based GA method. J. Geotech. Geoenviron. Eng. ASCE. 134 (2), 252–256.

Kahatadeniya, K.S., Nanakorn, P., Neaupane, K.M., 2009. Determination of the critical failure surface for slope stability analysis using ant colony optimization. Eng. Geol. 108, 133–141.

Kennedy, J., Eberhart, R., 1995. Particle swarm optimization. In: Proceeding of the IEEE International Conference on Neural Networks. Perth, Australia, pp. 1942–1948.

Lawrence, S., 1984. Resource Constrained Project Scheduling: An Experimental Investigation of Heuristic Scheduling Techniques. School of Industrial Administration, Carnegie Mellon University, Pittsburgh, PA.

Lee, K.S., Geem, Z.W., 2005. A new meta-heuristic algorithm for continuous engineering optimization: harmony search theory and practice. Comp. Meth. Appl. Mech. Eng. 194 (36–38), 3902–3933.

McCombie, P., Wilkinson, P., 2002. The use of the simple genetic algorithm in finding the critical factor of safety in slope stability analysis. Comput. Geotech. 29 (8), 699–714.

Ourique, C.O., Biscaia, E.C., Pinto, J.C., 2002. The use of particle swarm optimization for dynamic analysis in chemical processes. Comput. Chem. Eng. 26, 1783–1793.

Salman, A., Ahmad, I., Madani, S.A., 2002. Particle swarm optimization for task assignment problem. Microprocess. Microsyst. 26, 363–371.

Spencer, E., 1967. A method of analysis of the stability of embankments assuming parallel inter-slice forces. Geotechnique. 17, 11–26.

Taillard, E.D., 1993. Benchmarks for basic scheduling problems. Eur. J. Oper. Res. 64, 278–285.

Wang, L., Liu, B., 2008. Particle Swarm Optimization and Scheduling Algorithms. Tsinghua University Press, Beijing.

Yin, P.Y., 2004. A discrete particle swarm algorithm for optimal polygonal approximation of digital curves. J. Vis. Comm. Image Represent. 15, 241–260.

Zolfaghari, A.R., Heath, A.C., McCombie, P.F., 2005. Simple genetic algorithm search for critical non-circular failure surface in slope stability analysis. Comput. Geotech. 32, 139–152.

10 Artificial Neural Networks in Geotechnical Engineering: Modeling and Application Issues

Sarat Kumar Das

Department of Civil Engineering, National Institute of Technology, Rourkela, Odisha, India

10.1 Introduction

The evolution of computational geotechnical engineering analyses closely follows the development in computational methods. At the early stage of geotechnical engineering, analytical methods and the simple limit equilibrium method, coupled with engineering expertise, were used to develop physical models of geotechnical engineering problems. Over the years, finite element methods, finite difference methods, and discrete element methods are used for difficult and complex problems. Unlike other engineering materials, the success of the above-mentioned methods in applications in geotechnical engineering is hindered due to difficulty in obtaining an accurate constitutive model and spatial variability of soil, particularly for complex issues like liquefaction and pile capacity problems. Hence, based on case histories/field tests, statistically derived empirical methods and semiempirical methods based on analytical methods are more popular in such cases. The success of these empirical and semiempirical methods depends to a great extent on the chosen statistical/theoretical model for the system to be analyzed matching the input–output data, as well as on the statistical methods used to find out the model parameters (Das and Basudhar, 2006a). Very often, it is difficult to develop theoretical/statistical models due to the complex nature of the problem and uncertainty in soil parameters. These are situations where the data-driven approach has been found to be more appropriate than the model-oriented approach. To take care of such problems, artificial neural networks (ANNs) based on artificial intelligence (AI) have been developed in the computational methods. Within a short period, it found wide applicability, cutting across various disciplines. This has led to a growth in research activities into the art of applying such methods to solve real-life problems, highlighting the latent capabilities and drawbacks of such methods.

The application of ANNs in geotechnical engineering started in the early 1990s by Goh (1994) and Ghaboussi and Sidarta (1998). With this pioneering work,

Metaheuristics in Water, Geotechnical and Transport Engineering. DOI: http://dx.doi.org/10.1016/B978-0-12-398296-4.00010-6

Goh (1994) described the capability of ANNs to predict the highly complex lique-faction potential of soil and described the intrinsic constitutive relationships of sand using ANN. ANNs are receiving increased attention in geotechnical engineering as a powerful and flexible statistical modeling technique for solving some complex problems.

Shahin et al. (2001) and Das (2005) presented applications of ANNs to different geotechnical engineering problems. Recently, Shahin et al. (2009) presented the current status and future development in ANNs, which are becoming more reliable than statistical methods due to their special attributes of identifying complex systems when input and output are known from either laboratory or field experiments. However, there is no comprehensive literature on the critical evaluation of applying the modeling aspects of ANN in geotechnical engineering. The efficiency of all numerical methods is generally problem dependent, and no technique can be the universal tool for solving all types of problems. There are certain issues that need to be addressed in order to understand the ANN method and its successful application properly.

With this in mind, this chapter highlights the basic formulation, modeling, and application issues of ANNs in general. These issues have been described and explained with suitable examples of geotechnical engineering problems and explanations. An overview of the application of ANNs in geotechnical engineering is also discussed and presented.

10.2 Basic Formulation

10.2.1 Biological Model of a Neuron

ANN is a problem-solving algorithm modeled on the structure of the human brain. Neural network technology mimics the brain's own problem-solving process. The neuron (cell) is the fundamental unit of the biological nervous system. It is a simple processing unit, which receives and processes the signal (input) from other neurons through its input path, called a *dendrite*. The activity of a neuron is an all-or-nothing process. If the combined signal is strong enough, it generates the output signal to its output path (called an *axon*), which splits up and connects to other neurons' input paths through a junction referred to as a *synapse* (Figure 10.1). The amount of signal transferred depends on the synaptic strength of the junction, which is chemical in nature. This synaptic strength is modified during the learning process of the brain; therefore, it can be considered as a memory unit of each interconnection.

10.2.2 Mathematical Modeling of Neurons

The neurons are described as processing elements or nodes in the mathematical model of the ANN. A network with an input vector of elements x_l ($l = 1,...,N_i$) is

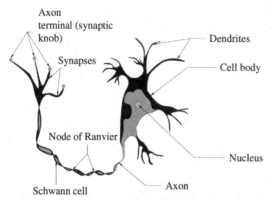

Figure 10.1 Simplified configuration of an organic neuron.

Axon terminal (synaptic knob)

Synapses

Node of Ranvier

Schwann cell

Dendrites

Cell body

Nucleus

Axon

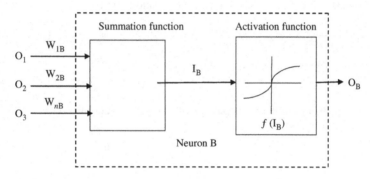

Figure 10.2 An artificial (mathematical) model of a neuron.

transmitted through a connection that is multiplied by weight w_{jl} to give the hidden unit z_j ($j = 1,\ldots,N_h$):

$$z_j = \sum_{l=1}^{N_i} w_{jl}x_l + b_{j0} \tag{10.1}$$

where N_h is the number of hidden units, and N_i is the number of input units. The hidden units consist of the weighted input and a bias (b_{j0}). A bias is simply a weight with constant input of 1, which serves as a constant added to the weight. This is similar to multilinear regression analysis in statistics. Figure 10.2 shows the basic operation of a single neuron. To incorporate nonlinearity in the input–output relationship, these inputs are passed through a layer of transfer function/activation function f, which produces

$$r_j = f\left[\sum_{l=1}^{N_i} w_{jl}x_l + b_{j0}\right] \tag{10.2}$$

Figure 10.3 shows some common activation functions used in ANNs. The most commonly used activation functions are the sigmoid, logistic sigmoid (Eq. (10.3)),

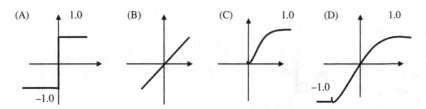

Figure 10.3 Different transfer functions: (A) stepped, (B) linear, (C) logistic sigmoid, and (D) hyperbolic tangent sigmoid.

and hyperbolic tangent sigmoid (Eq. (10.4)) functions. The basic properties of the sigmoid function are continuous, differentiable, and bounded:

$$f(z) = \frac{1}{1 + e^{-z}} \tag{10.3}$$

$$f(z) = \frac{e^z - e^{-z}}{e^z + e^{-z}} \tag{10.4}$$

The outputs from hidden units pass another layer of neurons:

$$v_k = \sum_{j=1}^{N_h} w_{kj} r_j + b_{k0} = \sum_{j=1}^{N_h} w_{kj} f \left[\sum_{l=1}^{N_i} w_{jl} x_l + b_{j0} \right] + b_{k0} \tag{10.5}$$

and are fed into another activation function F to produce output y ($k = 1,\ldots,N_o$):

$$y_k = F(v_k) = F \left[\sum_{j=1}^{N_h} w_{kj} f \left[\sum_{l=1}^{N_i} w_{jl} x_l + b_{j0} \right] + b_{k0} \right] \tag{10.6}$$

It continues in this way, depending upon the number of hidden layers and finally the output layer. The most common activation functions used in geotechnical engineering are either the logistic sigmoid function or the hyperbolic tangent sigmoid function. As these functions are bounded, the extrapolation is not recommended. However, if extrapolation is desired, the linear activation function may be used for the output layer (Maier and Dandy, 2000). This multilayer arrangement (hidden layer and output layer) with the nonlinear transfer function is termed as the *universal approximator*. But it gives rise to a highly nonlinear function with a number of unknown parameters in terms of weights and biases. Figure 10.4 shows the typical architecture of a three-layer ANN: input layer, hidden layer, and output layer. With four input-layer neurons, three hidden-layer neurons, and two output-layer neurons, it is called a 4-3-2 ANN architecture. It should be mentioned here that the human nervous system has approximately 3×10^{10} neurons, whereas the neurons in ANNs may number in a few hundred, and in geotechnical engineering, the number of neurons are even less than a hundred.

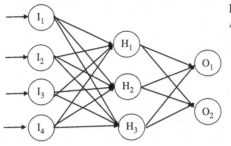

Figure 10.4 The typical architecture of an ANN.

10.2.3 ANN and Statistical Methods

Studies dealing with various engineering applications indicate that although an ANN mimics the human brain, ANN models are not significantly different from statistical models (Warner and Mishra, 1996). Table 10.1 shows ANN terminology and corresponding statistics. The statistician's primary objective is to develop universal methodology under strict statistical rules and guidelines. The rules governing sophisticated statistical models have been generally considered to be too restrictive, which makes it too difficult to use them for real-life applications. ANNs were developed by engineers and computer scientists in a process inspired by AI. The learning and training phase in ANNs are no different than the parameter estimation phase in conventional statistical models. The engineers and computer scientists have used this terminology to distinguish rule-based approaches, such as expert systems, from those that "learn" from empirical examples. In contrast, neural network practitioners are primarily concerned about prediction accuracy and finding methods that work. In general, the problems dealt by ANNs are more complex, and as such, the dimensionality of the models tends to be much higher. However, the interaction between statistical communities and the neural network experts is limited.

10.3   Modeling and Application Issues in General

10.3.1   The Basic ANN Architecture

As per architectural differences, ANNs can be classified as back-propagation neural networks (BPNNs), categorical learning (unsupervised) networks (self-organizing maps (SOMs)), and probabilistic neural networks (PNNs) (Hagan et al., 2002). BPNNs are better suited for prediction problems, while categorical learning ANNs are generally used for classification problems. Back propagation uses gradient descent laws, categorical, uses Kohonen learning laws and probabilistic neural network (PNN) uses both Kohonen and probabilistic learning laws. Determination of appropriate network architecture (geometry) is one of the most important and difficult tasks in the model building process. Figure 10.4 shows the typical architecture of a BPNN. The recurrent neural network is another form of BPNN. In BPNN, nodes in one layer are only connected to nodes in the next layer. However, in

Table 10.1 The Terminology Used in ANN and Corresponding Meanings in Statistical Methods (Warren, 2003)

Neural Network Jargon	Statistical Definition
Neuron, neurode	A linear or nonlinear computing element accepts one or more inputs, computes a function thereof, and may direct the result to one or more computing elements
Neural networks	A class of flexible nonlinear regression and discriminant models, data reduction models
Architecture	A model
Training, learning	Model parameter estimation
Classification	Discriminant analysis
Supervised learning	Regression
Unsupervised learning, self-organization	Cluster analysis
Training set	Construction set
Test set, validation set	Holdout sample

recurrent networks, the nodes in one layer can be connected to nodes in the next layer, the previous layer, or the same layer. Figure 10.5 shows a recurrent neural network. In geotechnical engineering, BPNN is the most common architecture. Recurrent networks are used to model stress–strain characteristics (Ghaboussi and Sidarta, 1998). The generalized regression neural network (GRNN) and radial basis neural network (RBNN) are part of the PNN (Hagan et al., 2002). GRNNs perform regression, where the target variable is continuous, whereas probabilistic networks perform classification where the target variable is categorical. Figure 10.6 shows a typical radial basis function neural network. The application of GRNN in geotechnical engineering is very much limited. Abu-Keifa (1998) used GRNN to figure the capacity of driven pile in cohesionless soil; Juang et al. (2001) and Kurup and Griffin (2006) used GRNN for site characterization based on cone penetration test (CPT) data; and Juang et al. (2003) used RBNN for site characterization using CPT data. A typical architecture of a SOM neural network is presented in Figure 10.7. Das and Basudhar (2009) used an unsupervised learning network (a SOM) for the clustering of CPT data for soil stratification. The soil stratification based on SOM is found to similar to that obtained using fuzzy-C clustering, but different from commonly used hierarchical clustering (Hegazy and Mayne, 2002). Ferentinou and Sakellariou (2007) used SOM for the slope stability analysis problem. In this chapter, only modeling and application issues related to BPNN are presented.

Once the type of network is selected, it is necessary to determine the optimum network geometry. The network geometry determines the number of connection weights and how they are arranged. This is generally done by fixing the number of hidden layers and choosing the number of nodes in each of these layers.

Small networks usually have better generalization ability than large networks, and this aspect is discussed later in this chapter. Small networks require fewer

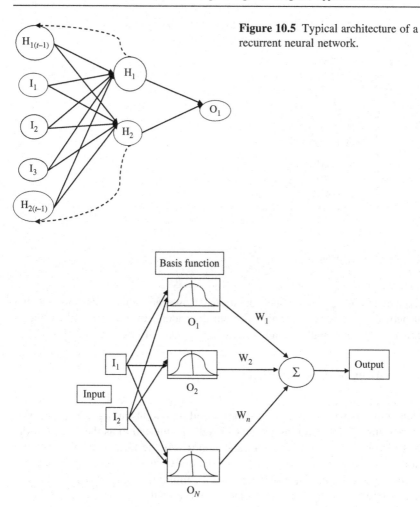

Figure 10.5 Typical architecture of a recurrent neural network.

Figure 10.6 Typical architecture of a radial basis function neural network.

storage space, have higher processing speed, can be implemented more easily, and make rule extraction simpler. However, the error surface of smaller networks is more complicated and contains more local minima (Maier and Dandy, 2000). This aspect has not been discussed properly in the application of ANNs in geotechnical engineering, where the number of data points is limited, and small architecture networks are used.

It has been found that ANNs with one hidden layer can approximate any function, assuming that sufficient degrees of freedom are provided. However, in practice, many functions are difficult to approximate with one hidden layer and require a prohibitive number of hidden layers. The use of more than one hidden layer provides greater flexibility and enables the approximation of complex functions with

Feature map

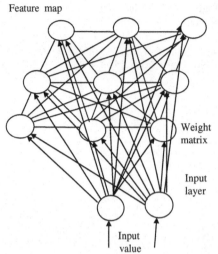

Figure 10.7 Typical architecture of a SOM neural network.

Weight matrix

Input layer

Input value

fewer connection weights in many situations. However, it must be stressed that optimal network geometry is highly problem dependent. In geotechnical engineering applications, ANNs with a single hidden layer generally have been adopted.

10.3.2 Learning Process—Training

The weights are adjustable parameters of the network and are determined from a set of data with known inputs and the corresponding output through the process of learning or training. The learning process in ANN is referred to as the ability of the network to learn from its environment and improve its performance. The learning techniques may be divided into two main categories: supervised learning and unsupervised learning. In the case of supervised learning, the weights are adjusted to match the output from the network to the known output (target). However, in the case of unsupervised learning, the output is unknown and the weights are adjusted based on another criterion, known as the Kohonen learning rule (Hagan et al., 2002).

So, in the case of supervised learning, the objective is to minimize the sum of the squares of the residuals between the measured and predicted output. The variables are the weights

$$E(W, U) = \sum_{l=1}^{N_s} \sum_{k=1}^{N_o} (\widehat{y}_{lk}(x_l) - y_{lk})^2 \tag{10.7}$$

where N_s is the number of samples; N_o is the number of outputs; W and U are the weights of the hidden and output layer, respectively; and $\widehat{y}(x)$ is the predicted output from inputs x. The most commonly used algorithm for this process is known as the *back-propagation algorithm*. Figure 10.8 shows the typical architecture of

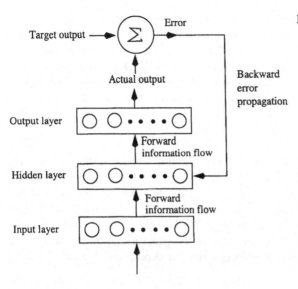

Figure 10.8 A typical BPNN.

a back-propagation algorithm. In back-propagation algorithm–based learning, the weights of connection are randomly chosen. Based on the initial weight values, the algorithm tries to minimize the square root of the above mean square error (MSE) (Eq. (10.7)).

In each subsequent training step, the initial set of weight vectors are adjusted toward the direction of maximum decrease of E, which is scaled by a learning rate lambda (λ). Mathematically, a weight is updated to its new value as follows:

$$w_{\text{new}} = w_{\text{old}} - \lambda \nabla E, \quad \text{where } \nabla E = \left(\frac{\partial E}{\partial w_1}, \frac{\partial E}{\partial w_2}, \ldots, \frac{\partial E}{\partial w_n} \right) \tag{10.8}$$

One useful property of sigmoid function is that

$$\frac{df(x)}{dx} = f(x)(1 - f(x)) \tag{10.9}$$

This means that the derivative (gradient) of the sigmoid function can be calculated by applying a simple multiplication and subtraction operator itself. This property simplifies the computation of new weights from initial random values. Most supervised learning applications use back propagation. However, when the number of layers, number of variables, and data point increase, the learning time tends to slow during neural network training. The learning time increases with the size of the problem. Again, as it is a gradient-based algorithm, it may reach a local minimum in weight space. These problems in back propagation have been taken care of by increasing the step size (learning rate) to increase the speed of the algorithm, and using a momentum factor to avoid the local minima.

The other method to increase the learning efficiency of the network is using various second-order optimization techniques, mostly a modified Newton's method like the Fletcher–Reeve, Davidon–Fletcher–Powell, Broydon–Fletcher–Goldfarb–Shanno, and Levenberg–Marquardt (LM) algorithm (Hagan et al., 2002). Other techniques, such as the least squares optimization method and the constrained optimization technique, have been used to improve the basic back-propagation problem.

10.3.3 Testing of the Network

At the end of the training phase, the associated trained weights of the neurons are stored in the ANN's memory. In the next phase, testing, the trained network is fed with a new set of data. The ANN predictions (using the trained weight) are compared to the target output values to assess the ability of the network to produce (generalize) correct responses to the testing patterns. This is similar to the validation stage of the statistical models. Once the training and testing phases are found to be successful, the corresponding ANN can be used in practical application.

10.3.4 Selection of Model Inputs

The statistical approaches are model driven, where the data points are used to find the model parameters only. In contrast, the ANN is a data-driven approach, i.e., the input and output data decided the type of model and the model parameters suitable for that particular problem. Data-driven approaches have the ability to determine which model inputs are critical. Thus, in ANNs, little attention is given to the selection of proper input variables. However, presenting a large number of inputs to ANNs usually increases the network size and the amount of data required to estimate the connection weights efficiently, thereby causing a decrease in the processing speed. Hence, there is a need in using analytical techniques to determine suitable inputs for the ANN models (Bowden et al., 2004; Guyon and Elisseeff, 2003; Olden et al., 2004).

The choice of input variable is based on *a priori* knowledge of causal variables in conjunction with the inspection of plots of potential inputs and outputs. If the relationship to be modeled is less understood, analytical techniques such as cross-correlation analysis and principal component analysis can be used. A stepwise approach can also be used in which separate networks are trained for each variable. The best-performing network is then retained, and the effect of adding each of the remaining inputs in turn is assessed. This is continued until the addition of an extra variable does not result in a significant improvement in model performance. However, this approach is computationally intensive and has the disadvantages of being unable to capture the importance of certain combinations of variables that might be insignificant on their own (Guyon and Elisseeff, 2003). In geotechnical engineering, in general, the important inputs to the ANNs are determined by trial and error. The combination of inputs with minimum error in the testing phase and better correlation between predicted and observed outputs are considered to be important inputs.

10.3.5 Division of Data and Preprocessing

To study the generalization of applying neural network models, it is a common practice to divide the data into two subsets: a training set and an independent testing set. However, depending upon the number of data points, a set of data may be used as a validation set to avoid overfitting (Shahin et al., 2002). It is also essential that the training, testing, and validation sets are representative of the same population (data set). ANNs are generally not used to extrapolate, i.e., they are not used to find the correlations for data values outside the range of values for which they were trained. In geotechnical engineering, generally data are randomly divided into different subsets. However, Shahin et al. (2002) have divided the data points in such a manner that the statistical parameters like mean, standard deviation, maximum, and minimum values of the input parameters are consistent for the three subsets (training, testing, and validation). Shi (2002) grouped the total data set into number of clusters based on fuzzy clustering and used ANNs separately in different clusters. Shahin et al. (2004) found that division of data based on SOM clustering and fuzzy clustering have advantages over the random division of data points.

Once the data have been divided into training, testing, and/or validation sets, it is important to preprocess the data to a suitable form before applying ANNs. The preprocessing helps in avoiding the dimensional dissimilarities of different input parameters. Figure 10.9 shows the bar chart of data for hydraulic conductivity of clay liner in (A) absolute and (B) logarithmic values. It can be noted that the logarithmic value shows a symmetric distribution of data points, and the skewness decreases from 2.69 to 0.203. The skewness is a measure of the degree of asymmetry of a distribution.

The variables have to be scaled in such a way as to be commensurate with the limits of the activation function used in the output layer. As described earlier, the commonly used activation functions are log sigmoid (logsig) (Figure 10.3C) and tangent hyperbolic (tanh) (Figure 10.3D) functions. It can be seen that logsig is between [0,1], and tanh is between $[-1,1]$. Hence, it is recommended to either normalize the data in $[-1,1]$ for the tanh activation function and in the range [0,1] for using the logistic sigmoid function. For example, if the outputs of the logistic functions are between 0 and 1, the data are generally scaled in the range 0.1−0.9 or 0.2−0.8. If the hyperbolic tangent sigmoid function is used, then the data need to be scaled in the range $[-1,1]$. In geotechnical engineering, data processing is mostly between 0.1−0.9 and 0, with 1 as the logistic sigmoid type of transfer functions used. However, Kurup and Dudani (2002) normalized the data −1 to 1.0 in commensurate with the hyperbolic tangent sigmoid function. Habibagahi (1998) used a different normalization procedure, from 0.0 to 2.4 for one variable and from 0 to 0.8 for others. Shi (2000) has used a nonlinear type normalization based on suitable statistical distribution of the data. The scaling is not strictly required, if the transfer function in the output layer is unbounded (linear). However, scaling to a uniform range is recommended for the efficient application of ANNs. The prediction of value beyond the training data range may be obtained by increasing the maximum value in the data set by a factor in excess of 1 (1.5 or 2.0) for scaling purposes (Maier and Dandy, 2000).

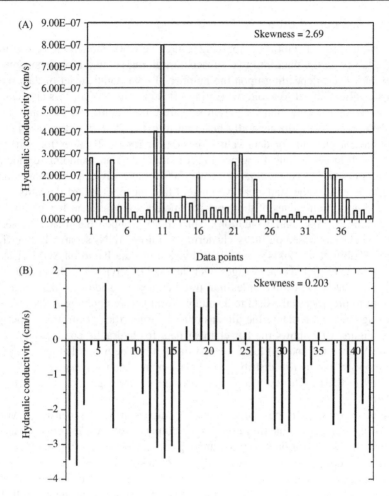

Figure 10.9 Variation of skewness: (A) actual value and (B) logarithmic value of hydraulic conductivity of clay liners.

10.3.6   Transfer/Activation Function

The transfer functions that are most commonly used are sigmoid type functions, such as logistic and hyperbolic tangent functions. However, another transfer function can be used so long as its derivative exists. It has been observed that the nonsigmoid type (polynomial, rational function, and Fourier series) transfer function performs better when the data were noiseless and contained a highly nonlinear relationship. However, when the data are noisy and contained a mildly nonlinear relationship, the performance of the polynomial transfer function is inferior and performances of others are comparable (Maier and Dandy, 2000). Maier and Dandy (1998) observed from empirical results that the hyperbolic tangent transfer function should be used.

Generally, the same transfer function is used in all layers. However, Rahman et al. (2001) have used hyperbolic tangent transfer function for connection between input layer and hidden layer and logistic sigmoid function between hidden layer and output layer. It should be noted that the type of transfer function used affects the size of the step taken in weight space. The sigmoid type function in hidden layers and linear transfer functions in the output layer can be an advantage when it is necessary to extrapolate beyond the range of training data (Maier and Dandy, 1998). But this has not been explored in geotechnical engineering, particularly for the problem where, standard penetration (N) values have been extrapolated for site characterization (Itani and Najjar, 2000).

10.3.7 Training—Optimization

The general learning or training process in ANN is a nonlinear optimization of an error function, and the terminology used to describe it has been borrowed from electrical science. This is equivalent to the parameter estimation phase in conventional statistical models, in which the parameters (weights and biases in ANN) are obtained by minimizing the error function. The error associated with weights and sigmoid function is a highly nonlinear optimization with many local minima (Shahin et al., 2002). Figure 10.10 shows a typical description of local minima, where point B is the local minima and whereas A is the true minima.

The aim is to find a global solution of a highly nonlinear optimization. As discussed earlier, the error surface of a smaller network, which appears very frequently in geotechnical engineering, is more complicated and contains more local minima. The error function, E, that is most commonly used is the MSE function. Local and global optimization methods are used to find the weight vectors. The local optimization methods are generally gradient-based algorithms of first-order and second-order methods. First-order methods are based on gradient descent, whereas second-order models are based partly on Newton's method. In both cases, iterative techniques are used to minimize the error function. The steepest descent algorithm, which is known as gradient descent algorithm, is mostly used in geotechnical engineering. The LM algorithm is the other optimization used in the implementation of ANNs in geotechnical engineering, because the training process is very fast compared to the gradient descent algorithm (Das and Basudhar, 2006b; Juang and Elton, 1997). The LM algorithm may be considered to be a hybrid

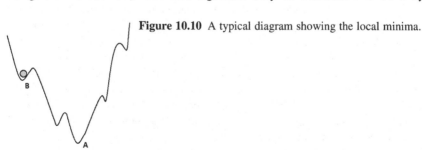

Figure 10.10 A typical diagram showing the local minima.

between the classical Newton and steepest descent algorithms. When far from a local minimum, the algorithm's behavior is similar to that of the steepest descent methods; however, in the vicinity of a local minimum, it has a convergence rate of the second order.

As the characteristics of the traditional nonlinear programming-based optimization method are initially point dependent, the results obtained using the back-propagation algorithm are sensitive to initial conditions (i.e., the weight vector) (Shahin et al., 2002). Generally, the weights are initialized randomly with adequate upper and lower bounds $(XL \cdot XU)$. If the range is too small, it may hinder the training. On the other hand, if the range is too wide, it may slow down training and result in the cessation of training at suboptimal levels.

These limits have not been discussed in the implementation of ANNs in geotechnical engineering. However, Das and Basudhar (2008) and Das et al. (2011c) have used the Bayesian regularization neural network (BRNN), where the error function is such that the function with higher weights are penalized to adapt to lower weight values. It is also reported that the magnitude of weights leads to overfitting, not the number of weights (Bartlett, 1998). The overfitting issue will be discussed in the following sections.

The increase in step size of the weight vector is controlled by the learning rate. Generally, the learning rate is kept fixed during training and optimal learning rates are found out by trial and error. The iteration of BPNN has been further augmented with the introduction of the momentum term, which reduces the oscillation in the error value and increases the speed of the convergence. The suitable combination of learning rate and momentum factor gives the optimum weight vector during training. The use of global optimization algorithms like genetic algorithm (GA), simulated annealing, and differential evolution (DE), though being widely used in other field of engineering (Ilonen et al., 2003; Jain and Srinivasulu, 2004; Morshed and Kaluarachchi, 1998), its use in geotechnical engineering is limited. Goh (2002) used GA to find the optimum spread of the probabilistic network for liquefaction analysis, and Goh et al. (2005) used GA for training the BPNN. Recently, Das et al. (2010, 2011a) used DE for the BPNN, while predicting swelling pressure of expansive soil and factor of safety of slope, respectively.

DE Neural Network

In the recent past, heuristic global optimization called DE, introduced by Storn and Price (1995), has been used successfully in aerodynamic shape optimization and mechanical design. The training of the feed-forward BPNN using DE optimization is known as the DE neural network (DENN) (Ilonen et al., 2003). DE optimization is a population-based heuristic global optimization method. Unlike other evolutionary optimization, in DE, the vectors in current populations are randomly sampled and combined to create vectors for the next generation. The real-valued crossover factor and mutation factor govern the convergence of the search process. The detail of DENN is available in Ilomen et al. (2003).

The prediction of factor of safety using the ANN model trained with DENN is found to be more efficient compared to traditional learning algorithms, the Bayesian regularization method (BRNN) and the LM-trained neural network (LMNN) (Das et al., 2011a). The database consisting of case studies of 23 dry and 23 wet slopes with 29 failed and 17 stable slopes, available in Sah et al. (1994), is considered. The input data consist of parameters like height of slope H (m), unit weight γ (kN/m^3), cohesion C (kPa), internal friction angle ϕ (°), slope angle β (°), and pore pressure parameters r_u. The output database consists of qualitative information (stable or failed) and quantitative information (factor of safety as per the limit equilibrium method). The ANN models developed with the output of the stable slope as 1 and that of the failed slope as 0. Out of 46 data points, 32 were used for training and 14 data points were used for testing.

As the efficiency of the model should be judged in terms of its performance to the new data set, the results pertaining to testing data are only presented in Table 10.2. It can be seen that DENN could exactly classify the stable and failed slope, but BRNN could misclassify it for one case. However, more prediction models need to be developed for comparison of the above algorithms.

10.3.8 Generalization

The aim of training is to minimize the error function to get the optimized weight vectors. However, when dealing with noisy data, reducing it beyond a certain point might lead to overtraining. The overtraining is referred to as the large error in the network when new data is presented to the trained network. The overfitting generally occurs when the data points in training set are scanty, but the error in the

Table 10.2 Performance of ANN Models for the Classification Problem Using the Testing Data Set

γ (kN/m^3)	C (kPa)	ϕ (°)	β (°)	H (m)	r_u	Field Condition	BRNN	LMNN	DENN
22.40	10.00	35.00	45.00	10.00	0.40	0	0	0	0
20.00	20.00	36.00	45.00	50.00	0.25	0	0	0	0
20.00	20.00	36.00	45.00	50.00	0.50	0	0	0	0
20.00	0.00	36.00	45.00	50.00	0.25	0	0	0	0
20.00	0.00	36.00	45.00	50.00	0.50	0	0	0	0
22.00	0.00	40.00	33.00	8.00	0.35	1	1	1	1
20.00	0.00	24.50	20.00	8.00	0.35	1	0	1	1
18.00	5.00	30.00	20.00	8.00	0.30	1	1	1	1
16.50	11.49	0.00	30.00	3.66	0.00	0	0	0	0
26.00	150.05	45.00	50.00	200.0	0.00	1	1	1	1
22.00	20.00	36.00	45.00	50.00	0.00	0	0	0	0
19.63	11.97	20.00	22.00	12.19	0.41	0	0	0	0
18.84	0.00	20.00	20.00	7.62	0.45	0	0	0	0
24.00	0.00	40.00	33.00	8.00	0.30	1	1	1	1

network is a very small value. Some of the rules are based on the concept that the number of weights should not exceed the number of training samples (Roger and Dowla, 1994), while others are based on the rule that the ratio of number of training samples to the number of connection weights should be 2 to 1 (Masters, 1993) or 10 to 1 (Maier and Dandy, 2000). Amari et al. (1997) suggest that overfitting does not occur if the number of training samples is at least 30 times the number of free parameters. The number of weights (w) in a network is defined as

$$w = (I + 1) \times H + (H + 1) \times O$$

where H is the number of neurons in the hidden layer, I is the number of inputs, and O is the number of outputs. Table 10.3 shows the number of weights and number of training data for some example problems in geotechnical engineering. It can be seen that except in a few cases (Das et al., 2011c; Goh, 1994; Hanna et al., 2004; Rahman et al., 2001; Shahin and Indraratna, 2006), the number of training samples is more than the number of weights. In these cases, overfitting has not been analyzed.

The network needs to be equally efficient for new data during testing and validation, which is called *generalization*. Generalization is the most important aspect for successful implementation of ANN. There are different methods for generalization, like early stopping or cross-validation (Basheer, 2001; Shahin et al., 2002). In the case of early stopping criteria, the error in the validation/testing set is monitored during the training process. The validation error normally decreases during the initial phase of training, as does the training set error. However, when the network begins to overfit the data, the error on the validation set will typically begin to rise. When the validation error increases for a specified number of iterations, the training is stopped, and the weights and biases at the minimum of the validation error are returned.

Das (2005) discussed the early stopping criteria with an example, shown in Figure 10.11, while developing the ANN model to predict the hydraulic conductivity of clay liners. It can be seen that as the number of epochs (iterations) increases, there is decrease in errors during training; but for the testing set data, initially there is a decrease in error up to certain iterations, and thereafter, the error continues to increase or remains constant. The correlation between the predicted and observed values of soil permeability for training and testing data with 100 iterations is shown in Figure 10.12. It can be seen that the correlation coefficient (R) value for training data is 1.0, whereas for testing data, it is 0.756. This shows poor generalization of the model for data outside the training set. Figure 10.13 shows the agreement between the predicted and observed permeability value when the network training is stopped after 10 iterations. Although there was a decrease in the value of R (0.962) for the training set, the results of the testing phase ($R = 0.914$) suggest that the ANN model was capable of generalization. This is known as early stopping criteria, i.e., the training is stopped when testing phase error increases, although errors during the training phase may go on decreasing.

In cross-validation, an independent test set is used to assess the performance of the model at various stages of learning. The available data need to be divided into three subsets: a training set, a testing set, and a validation set, which is very data intensive.

Table 10.3 Sample of Problems Showing the Number of Weights and Number of Training Data

S. No.	Problem Description	Network Architecture	Number of Weights	Number of Training Samples
1	Soil liquefaction potential (Goh, 1994)	8-8-1	81	59
2	Friction capacity of driven pile (Goh, 1995a,b)	4-3-1	19	45
3	Deflection of braced excavation (Goh et al., 1995)	7-3-1	28	196
4	Liquefaction potential (Goh, 1996b)	5-5-1	36	74
5	Compaction characteristic (Najjar et al., 1996b)	(i) 4-5-1 (ii) 11-1-1	(i) 31 (ii) 14	33
6	Swelling pressure (Najjar et al., 1996a)	3-2-1	11	310
7	Stress−strain of sand and volcanic soil (Zhu et al., 1998)	8-20-2	222	
8	Liquefaction-induced horizontal displacement (Wang and Rahman, 1999)	8-9-1	91	367
9	Limit state function for liquefaction (Juang et al., 2000)	5-4-1	29	163
10	Compaction curve (Basheer, 2001)	5-3-1	22	420
11	Uplift capacity of suction caisson (Rahman et al., 2001)	5-10-1	71	50
12	Soil liquefaction potential (Juang et al., 2003)	4-3-1	19	151
13	Stress−strain behavior of unsaturated soil (Habibagahi and Bamdad, 2003)	9-5-3	68	5731
14	Unsaturated shear strength (Lee et al., 2003)	5-2-1	15	20
15	Prediction of pile group efficiency (Hanna et al., 2004)	23-17-1	426	130
16	Coefficient of earth pressure at rest (K_0) (Das and Basudhar, 2005)	4-3-1	20	25
17	Settlement prediction of shallow foundation (Shahin et al., 2005)	5-2-1	15	106
18	Prediction of preconsolidation pressure (Celik and Tan, 2005)	6-4-1	33	53
19	Deviator stress and excess pore pressure (Banimahd et al., 2005)	(i) 9-10-1 (ii) 9-15-1	(i) 111 (ii) 166	107
20	Prediction of deviator stress and volumetric strain evaluation (Shahin and Indraratna, 2006)	10-10-2	132	24
21	Lateral load-carrying capacity of pile (Q_p) (Das and Basudhar, 2006b)	4-2-1	13	29
22	Liquefaction resistance of sand (Kim and Kim, 2006)	9-11-1	122	260
23	Hydraulic conductivity of clay liner (laboratory data) (k_l) (Das and Basudhar, 2007)	8-2-1	21	35

(Continued)

Table 10.3 (Continued)

S. No.	Problem Description	Network Architecture	Number of Weights	Number of Training Samples
24	Soil suction and swell pressure (Erzin, 2007)	(i) 2-9-9-1 (ii) 4-9-9-1	(i) 73 (ii) 109	(i) 87 (ii) 69
25	Prediction of maximum deflection of diaphragm walls (Kung et al., 2007)	5-7-1	50	2324
26	Kinematic soil pile interaction response parameters (Ahmad et al., 2007)	(i) 3-4-1 (ii) 3-5-1 (iii) 4-5-1 (iv) 2-2-1	(i) 21 (ii) 26 (iii) 31 (iv) 9	–
27	Residual friction angle of clay (ϕ_r) (Das and Basudhar, 2008)	2-4-1	17	39
28	Relative crest settlement of concrete-faced rock-fill dams (Kim and Kim, 2008)	3-4-1	21	21
29	Prediction of maximum dry density (MDD) and specific gravity (G) of fly ash (Das and Sabat, 2008)	(i) 4-3-1 (ii) 3-3-1	(i) 19 (ii) 16	(i) 25 (ii) 80
31	Prediction of swelling pressure (Das et al., 2010)	5-3-1	22	167
32	Prediction of factor of safety of slopes (Das et al., 2011a)	6-4-1	33	32
33	Hydraulic conductivity of clay liners (Das et al., 2011c)	9-4-1	45	32
34	Prediction of MDD and unconfined compressive strength (UCS) of cement stabilized soil (Das et al., 2011b)	7-4-1	37	37

Figure 10.14 shows a typical run of BPNN with a validation data set.

Poor validation can also be due to network architecture, a lack of inadequate data preprocessing, and normalization of training/validation data. However, a cross-validation method is not suitable if data points are scanty, and in geotechnical engineering, it is very difficult to get sufficient reliable data points. In such cases, another method to achieve good generalization is the BRNN method (Demuth and Beale, 2000).

BRNN Method

In BPNN, overfitting is due to unbounded values of weights (parameters) during minimization of the error function, MSE. The other method, called *regularization*, in which the performance function is changed by adding a term that consists of the MSE of weights and biases, as shown here:

$$\text{MSEREG} = \lambda \text{MSE} + (1 - \lambda)\text{MSW} \tag{10.10}$$

Figure 10.11 Performances of the proposed model during training and testing.

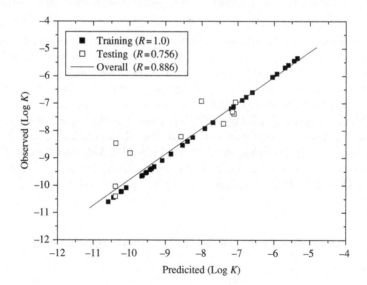

Figure 10.12 The observed soil permeability versus predicted soil permeability for 100 iterations.

where MSE is the mean square error of the network, λ is the performance ratio, and

$$MSW = \frac{1}{n} \sum_{j=1}^{n} w_j^2$$

(10.11)

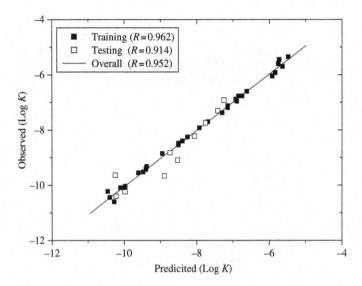

Figure 10.13 The observed soil permeability versus predicted soil permeability for 10 iterations.

This performance function will cause the network to have smaller weights and biases, thereby making networks less likely to be overfit. The optimal regularization parameter λ is determined through the Bayesian framework (Demuth and Beale, 2000) because the low value of λ will not adequately fit the training data and the high value of λ may result in overfitting. The number of network parameters (weights and biases) is being effectively used by the network and it can be found out by the above algorithm. The above combination works best when the inputs and targets are scaled in the range $[-1,1]$ (Demuth and Beale, 2000). Figure 10.15 shows the trials of BPNN in the prediction of the residual friction angle of clay, where Figure 10.15A shows 200 epochs and Figure 10.15B 400 epochs. But in both cases, it shows equal performances. The BRNN model is more stable in terms of variation with trials.

10.3.9 Choice of Performance Criteria for Comparison of ANNs

The acceptability and efficiency of ANNs are compared in terms of coefficient of correlation (Goh, 1994, 1995a,b, 1996b; Goh et al., 1995; Habibabaghi, 1998). But the coefficient of correlation is a biased estimate, so later the performance criteria like coefficient of determination/efficiency (R^2), root mean square error (RMSE), and mean absolute error (MAE) are being used. The RMSE is biased to large errors, whereas MAE measures the variation of error term by term. Dawson and Wilby (2001) noticed that performance statistics based on the squared error provide a measure of model performance, but they do not identify specific regions where a model is deficient. Therefore, it will be desirable to have certain other statistical measures that are unbiased and have different forms in order to test the effectiveness of the developed models in terms of their predictability criteria. Measures of prediction accuracy

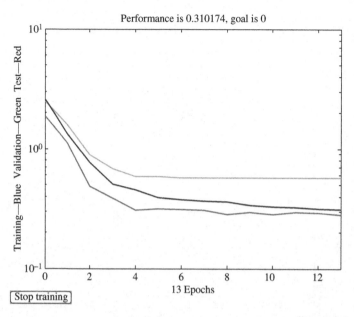

Figure 10.14 A typical run of BPNN with a validation data set in Matlab (MathWorks Inc., 2005).

that take parsimony into account include the Akaike's Information Criterion (AIC) and Schwarz's Bayesian Information Criterion (BIC) (Wilby et al., 2003). AIC and BIC scoring penalize the model with superfluous parameters. However, these criteria have not been considered in the application of ANNs in geotechnical engineering.

Although R has been widely used in geotechnical engineering problems, it is a biased estimate. Sometimes, higher values of R may not necessarily indicate better performance of the model because of the tendency of the model to be biased toward higher or lower values, particularly when the data range is very wide and most of the data are distributed about their mean. Lately, the unbiased estimate coefficient of efficiency (Nash and Sutcliffe, 1970) is being used as a better way to compare the ANN models (Das and Basudhar, 2006b, 2008), which is an unbiased estimate and defined as

$$E = \frac{E_1 - E_2}{E_1} \tag{10.12}$$

$$E_1 = \sum_{t=1}^{N}(Q_{um} - \overline{Q_{um}})^2$$

$$E_2 = \sum_{t=1}^{N}(Q_{uann} - Q_{um})^2$$

where Q_{uann} is the predicted value as per the ANN model, Q_{um} is the measured value, and $\overline{Q_{um}}$ is the average of measured values.

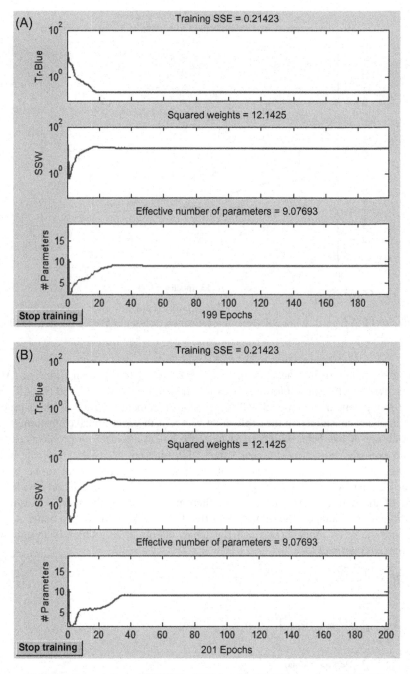

Figure 10.15 Implementation of BRNN in Matlab with (A) 200 epochs and (B) 400 epochs for the prediction of residual friction angle of clay.

The E is used to provide an assessment of overall model performance, but it is also sensitive to differences in the observed and predicted means and variances and insensitive to the size of the data set. For the hypothetical data set shown in Figure 10.16, the R values are comparable with 0.946 and 0.944 for data set A and B, respectively, but it can be seen that the A data set overpredicts. However, the E values are -0.546 and 0.890, respectively, for data set A and B, respectively, showing the advantage of using E.

10.3.10   Extraction of Knowledge

ANN is considered to be a black-box system, unable to explain the input—output relationship and the interpretation of its weights. However, several attempts have been made to explain the weights to express in terms of a model equation (Das and Basudhar, 2006b, 2008; Goh et al., 2005). The trained weights of the ANN model also have been used to perform sensitivity analysis in order to find the relationship between inputs and outputs. These aspects are discussed next.

Model Development Based on Trained Neural Networks

After the ANN is trained, a model equation can be established, with the weights as the model parameters. The mathematical equation relating input variables (X) and the output (Y) can be written as

$$Y = f_{\text{sig}}\left\{ b_0 + \sum_{k=1}^{h} \left[w_k \times f_{\text{sig}}\left(b_{hk} + \sum_{i=1}^{m} w_{ik} X_i \right) \right] \right\} \tag{10.13}$$

where

b_0 = the bias at the output layer,
w_k = the connection weight between the kth neuron of the hidden layer and the single output neuron,
b_{hk} = bias at the kth neuron of hidden layer,

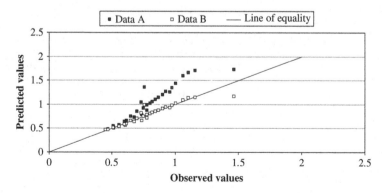

Figure 10.16 Example showing the importance of coefficient of efficiency (E) over correlation of coefficient (R).

h = the number of neurons in the hidden layer,
w_{ik} = the connection weight between ith input variable and kth neuron of hidden layer,
X_i = normalized input variable i, and
f_{sig} = the sigmoid transfer function.

The following example is presented for the lateral load capacity of piles under the undrained condition following Das and Basudhar (2006b). The weights obtained as per the trained model are shown in Table 10.4.

$$Q_{pn} = \frac{e^{C_1} - e^{-C_1}}{e^{C_1} + e^{-C1}} \tag{10.14}$$

The Q_p value as obtained from Eq. (5.10) is in the range $[-1,1]$, and this needs to be denormalized as

$$Q_p = 0.5(Q_{pn} + 1)(Q_{m,max} - Q_{m,min}) + Q_{m,min} \tag{10.15}$$

where $Q_{m,max}$ and $Q_{m,min}$ are the maximum and minimum values of Q_m, respectively, in the data set, and

$$C_1 = -0.8645 + B_1 + B_2 + B_3 + B_4 \tag{10.16}$$

$$B_1 = 2.2413 \times \frac{e^{A_1} - e^{-A_1}}{e^{A_1} + e^{-A_1}} \tag{10.17}$$

$$B_2 = 2.0593 \times \frac{e^{A_2} - e^{-A_2}}{e^{A_2} + e^{-A_2}} \tag{10.18}$$

$$A_1 = 2.9916 + 1.4121D + 0.2503L + 0.6842e + 3.7774C_u \tag{10.19}$$

$$A_2 = -1.6005 - 0.7871D + 1.9661L - 2.5228e - 1.8843C_u \tag{10.20}$$

Table 10.4 Connection Weights and Biases for Pile Capacity (Q_p)

Neuron	Weights (w_{ik})					Biases	
	Diameter (D)	Length (L)	Eccentricity (e)	Cohesion (C_u)	Q_p	b_{hk}	b_0
Hidden neuron 1 (k = 1)	1.4121	0.2503	0.6842	3.7774	2.2413	2.9916	−0.8645
Hidden neuron 2 (k = 2)	−0.7871	1.9661	−2.5228	−1.8843	2.0593	−1.6005	−

10.3.11   Sensitivity Analysis

The sensitivity analysis is an important aspect of model development to know input—output dependencies. As discussed earlier, the generalization of the ANN model depends upon the ratio of the number of training data to the number of ANN parameters, and the ANN parameters depend upon the number of input variables. So with limited data, there is a need to identify proper input variables for ANNs to increase the generalization of the model. However, as ANN is a data-driven approach rather than a statistical approach, the important inputs are selected based on the performances of the ANN models or by sensitivity analysis using Garson's algorithm (Garson, 1991) and the connection weight approach (Olden et al., 2004).

Garson's Algorithm

Garson (1991) proposed a method of partitioning the neural network weights to determine the relative importance of each input variable in the network which has been modified and used by Goh (1994), Shahin et al. (2002), and Das (2005), and other researchers. The input-hidden and hidden-output weights are partitioned, and the absolute values of the weights are taken to select the important input variables. The details of the algorithm with an example have been described in Goh (1994) and Das (2005).

Connection Weight Approach

Garson's algorithm (Garson, 1991) uses the absolute values of the connection weights when calculating variable contribution as described above. So it does not provide information on the effect of input variables in terms of directly or inversely related to the output. Olden et al. (2004) presented a connection weight approach in which the actual values of input-hidden and hidden-output weights are taken. This method sums the products across all the hidden neurons. The details of this approach, with examples, have been described in Das (2005).

The above methods have been described using examples as discussed above for the lateral load capacity of pile in clay under undrained conditions (Das and Basudhar, 2006b). Following the methodology described in Goh et al. (2005) and Das (2005), and using weights as per Table 10.4, the sensitivity analysis is presented in Table 10.5. Here, C_u is found to be the most important input parameter, followed by e, D, and L, as per Garson's method. It can also be seen that as per the method described in Olden et al. (2004), L and C_u are the most important input parameters, followed by e and D. So, it can be concluded that the interpretation of the weights to find the important input parameters based on Garson's algorithm and the connection weight approach matches the physical meaning for the lateral load-carrying capacity of piles. However, using the connection weight approach, it can be seen that the pile capacity increases with the increase in C_u with a positive S_j value (4.59) and decreases with eccentricity (e) with a negative S_j value (−3.66). Similarly, sensitivity analysis as per Garson's algorithm and the connection weight approach is presented for other problems in Table 10.6. It can be seen that

Table 10.5 The Relative Importance of Different Inputs as per Garson's Algorithm and the Connection Weight Approach

Parameters	Garson's Algorithm (%)		Connection Weight Approach	
	Relative Importance (%)	Ranking of Inputs as per Relative Importance	S_i Values as per Connection Weight Approach	Ranking of Inputs as per Relative Importance
D	17.02	3	1.54	4
L	15.78	4	4.61	1
e	23.20	2	− 3.66	3
C_u	44.0	1	4.59	2

in most cases, the most important input in both approaches match, but the connection weight approach has the added advantages of knowing the positive or negative effect of input on output.

Neural Interpretation Diagram

Ożesmi and Ożesmi (1999) proposed a neural interpretation diagram (NID) for visual interpretation of the connection weight among the neurons. In the NID, the lines joining the input-hidden and hidden-output neurons represent the magnitude of weights and their directions. Positive weights are represented by black lines and negative weights by gray lines, and the thickness of the lines is proportional to their magnitude. The relationship between the input and output is determined in two steps. The positive effect of the input variables is depicted by positive input-hidden and positive hidden-output weights, or negative input-hidden and negative hidden-output weights. The positive input-hidden and negative hidden-output and negative input-hidden and positive hidden-output weight indicates the negative effect of the input variables.

So, unlike absolute multiplication of weights, in this case, multiplication of actual weights of input-hidden and hidden-output indicates the effect of that input variable on the output. The input having positive effect on the output is represented with a gray circle and the input having negative effect is represented with a white circle. The connection weight approach as proposed by Olden et al. (2004) is based on this concept. Figure 10.17 explains the NID showing positive and negative weights and inputs that are directly or indirectly proportional to the output for the lateral load capacity of pile, as discussed above using weights and biases in Table 10.4.

10.3.12 Application of ANN in Geotechnical Engineering

ANNs have been successfully applied to difficult geotechnical engineering problems. Most of these applications include liquefaction analysis, pile foundations,

Table 10.6 Sensitivity Analysis of Inputs as per Garson's Algorithm and the Connection Weight Approach for Other Problems

Problem	Parameters	Garson's Algorithm		Connection Weight Approach	
		Relative Importance (%)	Ranking of Inputs as per Relative Importance	S_j Values	Ranking of Inputs as per Relative Importance
Undrained side resistance of drilled shafts (Goh et al., 2005)	σ'_{vm} (kPa)	19.44	2	−0.25	2
	s_u (CIUC) (kPa)	80.56	1	95.57	1
Liquefaction assessment (Baziar and Jafarian, 2007)	b'_{mean} (kPa)	9.843	5	−0.546	4
	D_r (%)	23.583	3	4.107	2
	FC (%)	27.477	1	−5.782	1
	C_u	12.481	4	−2.931	3
	D_{50} (mm)	26.614	2	−0.178	5
Settlement of shallow foundation (Shahin et al., 2002)	B	22.429	3	−7.123	3
	q	28.692	2	3.676	4
	N	30.862	1	−12.485	1
	L/B	3.143	5	1.476	5
	D_f/B	14.872	4	−7.412	2
Coefficient of lateral earth pressure at rest (Das and Basudhar, 2005)	I_D	25.300	2	1.638	2
	PI	22.050	4	0.097	4
	K_D	29.600	1	3.736	1
	s_u/σ_{v0}	23.020	3	0.272	3
Residual friction angle of clay (Das and Basudhar, 2008)	CF	33.6	2	−8.55	2
	ΔPI	66.33	1	−9.65	1

slope stability, and constitutive relations, particularly where finding analytical solutions is difficult. Other applications include settlement of foundation, soil properties, site characterization, parameter estimation, and prediction of movement of slopes. Table 10.7 presents a comprehensive list of application of ANNs to different geotechnical engineering problems. Because ANN is a data-driven approach, *in situ* data and reliable laboratory data have been used for model development. As it is a difficult and costly business to obtain reliable data in geotechnical engineering, validation data have been used in only a few cases (Alavi and Gandomi, 2011; Alavi et al., 2009, 2010; Basheer, 2001; Hanna et al., 2004, 2007; Kaya, 2009; Shahin and Jaksa, 2005; Shahin et al., 2002). The random data partition also was found to be more popular than other types of data partitioning. BPNNs are the most widely used ANNs, followed by GRNN and SOM.

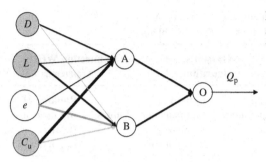

Figure 10.17 The NID showing axons representing the connection weights and effects of inputs on Q_p.

10.4 Future Challenges

ANN is still considered to be a black-box system with poor generalization, though various attempts have been made about refinement and explanations. Recently, a support vector machine (SVM), based on statistical learning theory and structural risk minimization is being used as an alternative prediction model (Das et al., 2010, 2011c). The SVM uses structural constrained minimization that penalizes the error margin during training (Vapnik, 1998). The error function being a convex function better generalization used to be observed in SVM (Das et al. 2010, 2011c) compared to ANN. Another technique, characterized as a gray box model (Giustolisi et al., 2007), is genetic programming (GP) (Koza, 1992), which mimics the biological evolution of living organisms and uses the principle of GA. Various attempts have been made recently to use GP to solve some geotechnical engineering problems (Gandomi and Alavi, 2011, 2012). GP helps in achieving a greatly simplified model formula compared to the ANN model, but a trade-off is made between the complexity of the formula and the accuracy of the model. Another class of model, which may be termed a white box model, is the multivariate adaptive regression spline (MARS), which was developed based on statistical model developed by Friedman (1991). MARS can adjust any functional form, and therefore it is suitable for exploratory data analysis. Samui et al. (2011) observed that the MARS model for uplift capacity of suction caisson has better statistical performance than the ANN and finite element method models. Hence, more research is required in ANN regarding the generalization, control on the model parameters, extrapolation, and depiction of the simplified model equation.

10.5 Conclusions

The basic formulation, modeling, and applications of ANN in geotechnical engineering have been discussed in this chapter. In geotechnical engineering, ANN has mostly been used as a prediction model, with the BPNN as a widely used algorithm, followed by the probabilistic algorithm with very limited use of categorical learning algorithms (SOM). Mostly, ANN has been applied to difficult geotechnical

Table 10.7 Summary of Applications of ANN in Geotechnical Engineering

S. No.	Reference	Network	Problem	Data Type	Data Partition	Number of Training Data	Number of Testing Data
1	Goh (1994)	BPNN	Soil liquefaction potential	Field	Random	59	26
2	Goh (1995a)	BPNN	Skin friction of driven pile in clay	Field	Random	45	20
3	Goh (1995b)	BPNN	(i) CPT cone resistance	Field/Lab	Random	(i) 93	(i) 74
			(ii) Hydraulic conductivity of clay liners			(ii) 31	(ii) 16
4	Goh et al. (1995)	BPNN	Deflection of braced excavation	Field	Random	196	57
5	Chan et al. (1995)	BPNN	Pile bearing capacity	Field	Random	34	34
6	Ellis et al. (1995)	BPNN, recurrent	Stress–strain relationship	Lab	Random	18	7
7	Goh (1996a)	BPNN	Liquefaction potential	Field	Random	74	35
8	Goh (1996b)	BPNN	Ultimate pile capacity in cohesionless soil	Field	Random	59	35
9	Lee and Lee (1996)	BPNN	Ultimate pile capacity in cohesionless soil	Lab/Field	Even for training odd for testing	14–21	14–7
10	Najjar et al. (1996b)	BPNN	Compaction characteristic (MDD, optimum moisture content (OMC))	Lab	Random	33	6
11	Najjar et al. (1996a)	BPNN	Swelling pressure	Lab	Random	310	103
12	Teh et al. (1997)	BPNN	Static pile capacity	Field	Random	27	4–10
13	Abu-Keifa (1998)	GRNN	Ultimate pile capacity	Field	Random	38	21
14	Ghaboussi and Sidarta (1998)	NANN	Soil stress–strain relationship	Lab	Random	15 sets	15 sets
15	Zhu et al. (1998)	Recurrent	Stress–strain of sand and volcanic soil	Lab	Random	160–240	80–120
16	Habibagahi (1998)	RBF	Reservoir-induced earthquake	Field	Random	25	5

(Continued)

Table 10.7 (Continued)

S. No.	Reference	Network	Problem	Data Type	Data Partition	Number of Training Data	Number of Testing Data
17	Wang and Rahman (1999)	BPNN	Liquefaction-induced horizontal displacement	Field	Random	367	99
18	Penumadu and Zhao (1999)	Recurrent	Stress–strain of sand and gravel	Lab	Random	(i) 81 (sand) (ii) 80 (Gravel)	(i) 20 (sand) (ii) 20 (gravel)
19	Itani and Najjar (2000)	BPNN	Spatial soil properties	Field	Random	4 sets	2 sets
20	Juang et al. (2000)	BPNN	Limit state function for liquefaction	Field	Random	163	62
21	Shi (2000)	BPNN	Settlement of tunnel	Field	Random	312	28
22	Romo et al. (2001)	Recurrent	Pore pressure–stress–strain of sand	Lab	Random	14 sets	4 sets
23	Juang et al. (2001)	GRNN (RBF), BPNN	Spatial soil characteristics	Field	Random	23 sets	9 sets
24	Basheer (2001)	BPNN	Compaction curve	Lab	Random	420	77 + 77[a]
25	Chiru-Danzer et al. (2001)	BPNN	Liquefaction-induced horizontal displacement	Field	Random	137	68
26	Rahman et al. (2001)	BPNN	Predict the uplift capacity	Field	Random	50	10
27	Rahman and Wang (2002)	BPNN	Liquefaction potential	Field	Random	176	28
28	Mayoraz and Vulliet (2002)	BPNN, recurrent	Slope movement prediction	Field	Random	–	–

#	Reference	Model	Application	Data	Sampling	Training	Testing
29	Kurup and Dudani (2002)	BPNN	Prediction of overconsolidation ratio (OCR)	Field	Random	(i) 122 (ii) 123	(i) 45 (ii) 72
30	Shahin et al. (2002)	BPNN	Settlement of shallow foundation	Field	Random	106	45 + 38[a]
31	Goh (2002)	PBNN	Liquefaction analysis: (i) CPT data (ii) Shear wave velocity	Field	Random	(i) 73 (ii) 125	(i) 36 (ii) 61
32	Juang et al. (2003)	BPNN	Soil liquefaction potential	Field	Random	151	75
33	Lee et al. (2003)	BPNN	Ultimate increment of apparent cohesion	Lab	Random	20	7
34	Habibagahi and Bamdad (2003)	Recurrent	Stress–strain behavior of unsaturated soil	Lab	Random	5,731	910
35	Shahin et al. (2004)	BPNN	Settlement of shallow foundation	Field	SOM, fuzzy clustering	106	45 + 38[a]
36	Hanna et al. (2004)	BPNN	Pile group efficiency	Field, Lab	Random	130	23 + 23[a]
37	Goh et al. (2005)	BPNN	Undrained side resistance of drilled shaft	Field	Random	85	42
38	Das and Basudhar (2005)	BPNN	Coefficient of lateral earth pressure at rest	Field	Random	25	11
39	Celik and Tan (2005)	BPNN	Preconsolidation pressure	Lab	Random	53	23
40	Shahin and Jaksa (2005)	BPNN	Prediction of ultimate pullout capacity of marquee ground anchors	Field	Random	67	29 + 23[a]
41	Shahin et al. (2005)	BPNN	Settlement of shallow foundation	Field	Random	106	45 + 38[a]
42	Singh and Singh (2005)	BPNN	Dominant frequency of blast vibration in mines	Field	Random	200	15

(Continued)

Table 10.7 (Continued)

S. No.	Reference	Network	Problem	Data Type	Data Partition	Number of Training Data	Number of Testing Data
43	Banimahd et al. (2005)	BPNN	Deviator stress, excess pore pressure	Lab	Random	107	12 + 10[a]
44	Kuzniar et al. (2005)	BPNN	Normalized acceleration response spectrum	Experimental data	Random	(i) 22,968 (ii) 20,196	(i) 5,940 (ii) 5,148
45	Dihoru et al. (2005)	BPNN	Displacement error in three directions	Lab	Random	36	–
46	Shahin and Indraratna (2006)	BPNN	Deviator stress and volumetric strain evaluation	Lab	Random	24	5
47	Shahin and Jaksa (2006)	BPNN	Pullout capacity of small ground anchors	Field	Random	119	–
48	Chen et al. (2006)	BPNN	Displacement of a foundation pit	Field	Random	24	–
49	Das and Basudhar (2006b)	BPNN	Undrained lateral load capacity of piles	Field	Random	29	9
50	Kim and Kim (2006)	BPNN	Cyclic resistance ratio (CRR) of sands	Lab	Random	260	86
51	Narendara et al. (2006)	BPNN, RBFN	UCS of soft grounds using cement stabilization	Lab	Random	154	32
52	Pradeep et al. (2006)	GRNN	Soil composition (coarse/fine grained)	Field	Random	100	42
53	Das and Basudhar (2007)	BPNN	Prediction of hydraulic conductivity of clay liners	Field	Random	32	10

No.	Reference	ANN type	Application			Training	Testing
54	Erzin (2007)	BPNN	(i) Soil suction	Lab	Random	(i) 87	(i) 5
			(ii) Swell pressure			(ii) 69	(ii) 5
55	Hanna and Saygili (2007)	GRNN	Liquefaction potential	Field	Random	413	112 + 95[a]
56	Kung et al. (2007)	BPNN	Maximum deflection of diaphragm walls	Numerical	Random	2,324	1,162
57	Najjar and Huang (2007)	Recurrent	Total stress and pore water pressure	Lab	Random	4,453	167
58	Ahmad et al. (2007)	BPNN	Kinematic soil pile interaction response parameters	Analytical	Random	50%	50%
59	Yoo and Kim (2007)	BPNN	Tunneling performance of crown settlement, maximum ground surface settlement	Numerical	Random	52	23 20*
60	Ferentinou and Sakellariou (2007)	BPNN, SOM	Factor of safety of slope	Field	Random	31	15
61	Baziar and Jafarian (2007)	BPNN	Logarithm of strain energy density required to trigger liquefaction	Lab	Random	199	85
62	Abdel-Rahman (2007)	BPNN	MDD, OMC of cohesionless soil	Lab	Random	150	25
63	Kim and Kim (2008)	BPNN	Relative crest settlement of concrete-faced rock-fill dams	Lab	Random	21	09
64	Das and Sabat (2008)	BPNN	MDD and G of fly ash	Lab	Random	(i) 25	(i) 15
						(ii) 80	(ii) 33
65	Das and Basudhar (2008)	BPNN	Residual friction angle of clay	Lab	Random	39	15
66	Padmini et al. (2008)	ANFIS, BPNN	Ultimate bearing capacity of shallow foundations	Lab	Random	78	19

(Continued)

Table 10.7 (Continued)

S. No.	Reference	Network	Problem	Data Type	Data Partition	Number of Training Data	Number of Testing Data
67	Sinha and Wang (2008)	BPNN	MDD, OMC of soil	Lab	Random	45	12
68	Maji and Sitharam (2009)	BPNN, RBF	Elastic modulus ratio of jointed rocks	Lab	Random	726	170
69	Kaya (2009)	BPNN	Secant residual friction angle	Lab	Random	51	26 + 25[a]
70	Alavi et al. (2009)	RBF	OMC and MDD	Lab	Random	100	46 + 46[a]
71	Alavi et al. (2010)	BPNN	OMC and MDD	Lab	Random	100	46 + 46[a]
72	Park and Cho (2010)	BPNN	Total resistance, shaft resistance, and tip resistance of driven piles	Field	Random	148	17
73	Alavi and Gandomi (2011)	BPNN	Ground motion parameter	Field	Random	1,971	563 + 281[a]
74	Mollahasani et al. (2011)	BPNN	Soil cohesion intercept	Lab	Random	69	12
75	Das et al. (2010)	BPNN	Swelling pressure of soil	Lab	Random	167	63
76	Das et al. (2011a)	BPNN	Factor of safety of slopes	Lab	Random	32	14
77	Das et al. (2011c)	BPNN	Hydraulic conductivity of clay liners	Field	Random	32	9
78	Das et al. (2011b)	BPNN	MDD and UCS of cement-stabilized soil	Lab	Random	37	14

[a]Validation data set.

engineering problems like liquefaction analysis, pile foundations, slope stability, and constitutive relations, particularly where finding an analytical solution is difficult. Although ANN is being used as an alternate statistical method, its prediction capability is described in terms of the correlation coefficient between predicted and observed values and RMSE. However, in this chapter, the need for other statistical performance criteria has been emphasized. One of the most important aspects of applying ANN is generalization, i.e., the performance of the model with a new data set. This topic was discussed extensively in this chapter, as well as the prediction model equations based on weights and biases, NID, and connection weight approaches to find the relationship between input and output variables. This chapter explained that ANN should not be considered a black-box system, and professionals can use the model equations developed by ANN with ease. Other prediction algorithms included SVM, GP, and MARS.

References

Abdel-Rahman, A.H., 2007. Predicting compaction of cohesionless soils using ANN. Proc. Inst. Civ. Eng., Ground Improve. 161, 3−8.

Abu-Keifa, M.A., 1998. General regression neural networks for driven piles in cohesionless soils. J. Geotech. Geoenviron. Eng. ASCE. 124 (12), 1177−1185.

Ahmad, I., El Naggar, H., Kahn, A.N., 2007. Artificial neural network application to estimate kinematic soil pile interaction response parameters. Soil Dynam. Earthquake Eng. 27 (9), 892−905.

Alavi, A.H., Gandomi, A.H., 2011. Prediction of principal ground-motion parameters using a hybrid method coupling artificial neural networks and simulated annealing. Comput. Struct. 89 (23−24), 2176−2194.

Alavi, A.H., Gandomi, A.H., Gandomi, M., Sadat Hosseini, S.S., 2009. Prediction of maximum dry density and optimum moisture content of stabilized soil using RBF neural networks. IES J. A Civ. Struct. Eng. 2 (2), 98−106.

Alavi, A.H., Gandomi, A.H., Mollahasani, A., Heshmati, A.A.R., Rashed, A., 2010. Modeling of maximum dry density and optimum moisture content of stabilized soil using artificial neural networks. J. Plant Nutr. Soil Sci. 173 (3), 368−379.

Amari, S.I., Murata, N., Müller, K.R., Finke, M., Yang, H.H., 1997. Asymptotic statistical theory of overtraining and cross-validation. IEEE Trans. Neural Networks. 8 (5), 985−996.

Banimahd, M., Yasrobi, S.S., Woodward, P.K., 2005. Artificial neural network for stress−strain behavior of sandy soils: knowledge based verification. Comput. Geotech. 32, 377−386.

Bartlett, P.L., 1998. The sample complexity of pattern classification with neural networks: the size of the weights is more important than the size of the network. IEEE Trans. Inform. Theor. 44 (2), 525−536.

Basheer, I.A., 2001. Empirical modeling of the compaction curve of cohesive soil. Can. Geotech. J. 38 (1), 29−45.

Baziar, M.H., Jafarian, V.Y., 2007. Assessment of liquefaction triggering using strain energy concept and ANN model: capacity energy. Soil Dynam. Earthquake Eng. 27, 1056−1072.

Bowden, G.J., Dandy, G.C., Maier, H.R., 2004. Input determination for neural network models in water resources application. Part 1—background and methodology. J. Hydrol. 301 (1−4), 75−92.

Celik, S., Tan, O., 2005. Determination of pre-consolidation pressure with artificial neural network. Civ. Eng. Environ. Syst. 22 (4), 217–231.

Chan, W.T., Chwo, Y.K., Liu, L.F., 1995. Neural network: an alternate to pile driving formulas. Comput. Geotech. 17 (2), 135–156.

Chen, Y., Azzam, R., Zhang, F., 2006. The displacement computation and construction pre-control of a foundation pit in Shanghai utilizing FEM and intelligent methods. Geotech. Geol. Eng. 24 (6), 1781–1801.

Chiru-Danzer, M., Juang, C.H., Christopher, R.A., Suber, J., 2001. Estimation of liquefied induced horizontal displacement using artificial neural network. Can. Geotech. J. 38 (1), 200–207.

Das, S.K., 2005. Applications of genetic algorithm and artificial neural network to some geotechnical engineering problems. Ph.D. Thesis. Indian Institute of Technology Kanpur, Kanpur, India.

Das, S.K., Basudhar, P.K., 2005. Prediction of coefficient of lateral earth pressure using artificial neural networks. Electron. J. Geotech. Eng. 10, 1.

Das, S.K., Basudhar, P.K., 2006a. Comparison study of parameter estimation techniques for rock failure criterion models. Can. Geotech. J. 43 (7), 764–771.

Das, S.K., Basudhar, P.K., 2006b. Undrained lateral load capacity of piles in clay using artificial neural network. Comput. Geotech. 33 (8), 454–459.

Das, S.K., Basudhar, P.K., 2007. Prediction of hydraulic conductivity of clay liners using artificial neural network. Lowland Technol. Int. 9 (1), 50–58.

Das, S.K., Basudhar, P.K., 2008. Prediction of residual friction angle of clays using artificial neural network. Eng. Geol. 100 (3–4), 142–145.

Das, S.K., Basudhar, P.K., 2009. Utilization of self-organizing map and fuzzy clustering for site characterization using piezocene data. Comput. Geotech. 36 (1–2), 241–248.

Das, S.K., Sabat, A.K., 2008. Using neural networks for prediction of some properties of fly ash. Electron. J. Geotech. Eng. 13, 1–13.

Das, S.K., Samui, P., Sabat, A.K., Sitharam, T.G., 2010. Prediction of swelling pressure of soil using artificial intelligence techniques. Environ. Earth Sci. 61, 393–403.

Das, S.K., Biswal, R.K., Sivakugan, N., Das, B., 2011a. Classification of slopes and prediction of factor of safety using differential evolution neural networks. Environ. Earth Sci. 64, 201–210.

Das, S.K., Samui, P., Sabat, A.K., 2011b. Application of artificial intelligence to maximum dry density and unconfined compressive strength of cement stabilized soil. Geotech. Geol. J. 29 (3), 329–342.

Das, S.K., Samui, P., Sabat, A.K., 2011c. Prediction of field hydraulic conductivity of clay liners using artificial neural network and support vector machine. Int. J. Geomech. 3, 1–22.

Dawson, C.W., Wilby, R.L., 2001. Hydrological modeling using artificial neural networks. Progr. Phys. Geogr. 25, 80–108.

Demuth, H., Beale, M., 2000. Neural Network Toolbox. MathWorks Inc., Natick, MA.

Dihoru, L., Muir Wood, D., Sadek, T., Lings, M., 2005. A neural network for error prediction in a true triaxial apparatus with flexible boundaries. Comput. Geotech. 32, 59–71.

Ellis, G.W., Yao, C., Zhao, R., Penumadu, D., 1995. Stress–strain modeling of sands using artificial neural networks. J. Geotech. Eng. ASCE. 121 (5), 429–435.

Erzin, Y., 2007. Artificial neural networks approach for swell pressure versus soil suction behavior. Can. Geotech. J. 44 (10), 1215–1223.

Ferentinou, M.D., Sakellariou, M.G., 2007. Computational intelligence tools for the prediction of slope performance. Comput. Geotech. 34 (5), 362–384.

Friedman, J., 1991. Multivariate adaptive regression splines. Ann. Stat. 19, 1–141.

Gandomi, A.H., Alavi, A.H., 2011. Multi-stage genetic programming: a new strategy to non-linear system modeling. Infor. Sci. 181 (23), 5227–5239.

Gandomi, A.H., Alavi, A.H., 2012. A new multi-gene genetic programming approach to non-linear system modeling. Part II: geotechnical and earthquake engineering problems. Neural Comput. Appl. 21 (1), 189–201.

Garson, G.D., 1991. Interpreting neural-network connection weights. Artif. Intell. Expert. 6 (7), 47–51.

Ghaboussi, J., Sidarta, D.E., 1998. New nested adaptive neural networks (NANN) for constitutive modeling. Comput. Geotech. 22 (1), 29–52.

Giustolisi, O., Doglioni, A., Savic, D.A., Webb, B.W., 2007. A multi-model approach to analysis of environmental phenomena. Environ. Model. Software. 22 (5), 674–682.

Goh, A.T.C., 1994. Seismic liquefaction potential assessed by neural network. J. Geotech. Eng. ASCE. 120 (9), 1467–1480.

Goh, A.T.C., 1995a. Empirical design in geotechnics using neural network. Geotechnique. 45 (4), 709–714.

Goh, A.T.C., 1995b. Modeling soil correlations using neural network. J. Comput. Civ. Eng. 9 (4), 275–278.

Goh, A.T.C., 1996a. Neural-network modeling of CPT seismic liquefaction data. J. Geotech. Eng. ASCE. 122 (1), 70–73.

Goh, A.T.C., 1996b. Pile driving records reanalyzed using neural networks. J. Geotech. Eng. ASCE. 122 (6), 492–495.

Goh, A.T.C., 2002. Probabilistic neural network for evaluating seismic liquefaction potential. Can. Geotech. J. 39, 219–232.

Goh, A.T.C., Wong, K.S., Broms, B.B., 1995. Estimation of lateral wall movements in braced excavations using neural networks. Can. Geotech. J. 32 (6), 1059–1064.

Goh, A.T.C., Kulhawy, F.H., Chua, C.G., 2005. Bayesian neural network analysis of undrained side resistance of drilled shafts. J. Geotech. Geoenviron. Eng. ASCE. 131 (1), 84–93.

Guyon, I., Elisseeff, A., 2003. An introduction to variable and feature selection. J. Mach. Learn. Res. 3, 1157–1182.

Habibagahi, G., 1998. Reservoir induced earthquakes analyzed via radial basis function networks. Soil Dynam. Earthquake Eng. 17 (1), 53–56.

Habibagahi, G., Bamdad, A., 2003. A neural network framework for mechanical behavior of unsaturated soils. Can. Geotech. J. 40 (3), 684–693.

Hagan, M.T., Demuth, H.B., Beale, M., 2002. Neural Network Design. Thomson Learning, Singapore.

Hanna, A.M., Morcous, G., Helmy, M., 2004. Efficiency of pile groups installed in cohesionless soil using artificial neural networks. Can. Geotech. J. 41 (6), 1241–1249.

Hanna, A.M., Ural, D., Saygili, G., 2007. Neural network model for liquefaction potential in soil deposits using Turkey and Taiwan earthquake data. Soil Dynam. Earthquake Eng. 27 (6), 521–540.

Hegazy, Y.A., Mayne, P.W., 2002. Objective site characterization using clustering of piezocone data. J. Geotech. Geoenviron. Eng. 128 (12), 986–996.

Ilonen, J., Kamarainen, J.K., Lampinen, J., 2003. Differential evolution training algorithm for feed-forward neural network. Neural Process. Lett. 17, 93–105.

Itani, O.M., Najjar, Y.M., 2000. Three-dimensional modeling of spatial soil properties via artificial neural network. Transport. Res. Rec. 1709, 50–59.

Jain, A., Srinivasulu, S., 2004. Development of effective and efficient rainfall–runoff models using integration of deterministic, real-coded genetic algorithms, and artificial neural network techniques. Water Resour. Res. 40, 4.

Juang, C.H., Elton, D.J., 1997. Prediction of collapse potential of soil with neural networks. Transport. Res. Rec. 1582, 22–28.

Juang, C.H., Chen, C.J., Tang, W.H., 2000. CPT based liquefaction analysis Part 1: determination of limit state function. Geotechnique. 50 (5), 583–592.

Juang, C.H., Jiang, T., Christopher, R.A., 2001. Three-dimensional site characterization: neural network approach. Geotechnique. 51 (9), 799–809.

Juang, C.H., Yuan, H., Lee, D.H., Lin, P.S., 2003. Simplified cone penetration test-based method for evaluating liquefaction resistance of soils. J. Geotech. Geoenviron. Eng. ASCE. 129 (1), 66–80.

Kaya, A., 2009. Residual and fully softened strength evaluation of soils using artificial neural networks. Geotech. Geol. Eng. 27, 281–288.

Kim, Y., Kim, B., 2006. Use of artificial neural networks in the prediction of liquefaction resistance of sands. J. Geotech. Geoenviron. Eng. ASCE. 132 (11), 1502–1504.

Kim, Y., Kim, B., 2008. Prediction of relative crest settlement of concrete-faced rockfill dams analyzed using an artificial neural network model. Comput. Geotech. 35 (3), 313–322.

Koza, J.R., 1992. Genetic Programming: On the Programming of Computers by Natural Selection. MIT Press, Cambridge, MA.

Kung, G.T., Hsiao, E.C., Schuster, M., Juang, C.H., 2007. A neural network approach to estimating deflection of diaphragm walls caused by excavation in clays. Comput. Geotech. 34 (5), 385–396.

Kurup, P.U., Dudani, N.K., 2002. Neural networks for profiling stress history of clays from PCPT data. J. Geotech. Geoenviron. Eng. ASCE. 128 (7), 569–579.

Kurup, P.U., Griffin, E.P., 2006. Prediction of soil composition from CPT data using general regression neural network. J. Comput. Civ. Eng. 20, 281–289.

Kuzniar, K., Maciag, E., Waszczyszy, Z., 2005. Computation of response spectra from mining tremors using neural networks. Soil Dynam. Earthquake Eng. 25, 331–339.

Lee, I.M., Lee, J.H., 1996. Prediction of pile bearing capacity using artificial neural networks. Comput. Geotech. 18 (3), 189–200.

Lee, S.J., Lee, S.R., Kim, Y.S., 2003. An approach to estimate unsaturated shear strength using artificial neural network and hyperbolic formulation. Comput. Geotech. 30 (6), 489–503.

Maier, H.R., Dandy, G.C., 1998. The effect of internal parameters and geometry on performances of back propagation neural networks: an empirical study. Environ. Model. Software. 13, 193–209.

Maier, H.R., Dandy, G.C., 2000. Neural networks for the prediction and forecasting of water resource variables: a review of modelling issues and applications. Environ. Model. Software. 15, 101–123.

Maji, V.B., Sitharam, T.G., 2009. Prediction of elastic modulus of jointed rock mass using artificial neural networks. Geotech. Geol. Eng. 26, 443–452.

Masters, T., 1993. Practical Neural Network Recipes in C++. Academic Press, San Diego, CA.

MathWorks Inc., 2005. Matlab User's Manual. Version 7.1. MathWorks Inc., Natick, MA.

Mayoraz, F., Vulliet, L., 2002. Neural networks for slope movement prediction. Int. J. Geomech. 2 (2), 153–173.

Mollahasani, A., Alavi, A.H., Gandomi, A.H., Rashed, A., 2011. Nonlinear neural-based modelling of soil cohesion intercept. KSCE J. Civ. Eng. 15 (5), 831–840.

Morshed, J., Kaluarachchi, J.J., 1998. Parameter estimation using artificial neural network and genetic algorithm for free-product migration and recovery. Water Resour. Res. 34 (5), 1101–1113.

Najjar, Y.M., Huang, C., 2007. Simulating the stress–strain behavior of Georgia kaolin via recurrent neuronet approach. Comput. Geotech. 34 (5), 346–362.

Najjar, M.Y., Basheer, I.A., McReynold, R., 1996a. Neural modeling of Kansas soil swelling. Transport. Res. Rec. 1526, 14–19.

Najjar, M.Y., Basheer, I.A., Naouss, W.A., 1996b. On the identification of compaction characteristics by neuronets. Comput. Geotech. 18 (3), 167–187.

Narendara, B.S., Sivapullaiah, P.V., Suresh, S., Omkar, S.N., 2006. Prediction of unconfined compressive strength of soft grounds using computational intelligence techniques: a comparative study. Comput. Geotech. 33 (3), 196–208.

Nash, J.E., Sutcliffe, J.V., 1970. River flow forecasting through conceptual models part I—A discussion of principles. J. Hydrol. 10 (3), 282–290.

Olden, J.D., Joy, M.K., Death, R.G., 2004. An accurate comparison of methods for quantifying variable importance in artificial neural networks using simulated data. Ecol. Model. 178 (3), 389–397.

Ozesmi, S.L., Ozesmi, U., 1999. An artificial neural network approach to spatial modeling with inter specific interactions. Ecol. Model. 116, 15–31.

Padmini, D., Ilamparuthi, K., Sudheer, K.P., 2008. Ultimate bearing capacity prediction of shallow foundations on cohesionless soils using neurofuzzy models. Comput. Geotech. 35 (1), 33–46.

Park, H.I., Cho, C.W., 2010. Neural network model for predicting the resistance of driven piles. Mar. Georesour. Geotechnol. 28, 324–344.

Penumadu, D., Zhao, R., 1999. Triaxial compression behavior of sand and gravel using artificial neural networks (ANN). Comput. Geotech. 24 (3), 207–230.

Pradeep, U., Kurup, P.E., Erin, P., Griffin, S.M., 2006. Prediction of soil composition from CPT data using general regression neural network. J. Comput. Civ. Eng. ASCE. 20 (4), 281–289.

Rahman, M.S., Wang, J., 2002. Fuzzy neural models for liquefaction prediction. Soil Dynam. Earthquake Eng. 22, 685–694.

Rahman, M.S., Wang, J., Deng, J.P., Carter, J.P., 2001. A neural network model for the uplift capacity of suction caisson. Comput. Geotech. 28, 269–287.

Roger, L.L., Dowla, F.U., 1994. Optimization of ground water remediation using artificial neural networks with parallel solute transport modeling. Water Resour. Res. 30 (2), 457–481.

Romo, M.P., García, S.R., Mendoza, M.J., Urtuzuástegui, V.T., 2001. Recurrent and constructive-algorithm networks for sand behavior modeling. Int. J. Geomech. 4, 371–387.

Sah, N.K., Sheorey, P.R., Upadhyana, L.N., 1994. Maximum likelihood estimation of slope stability. Int. J. Rock Mech. Min. Sci. Geomech. Abstr. 31, 47–53.

Samui, P., Das, S.K., Kim, D., 2011. Uplift capacity of suction caisson in clay using multivariate adaptive regression spline. Ocean Eng. 38 (17–18), 2123–2127.

Shahin, M.A., Indraratna, B., 2006. Modeling the mechanical behavior of railway ballast using artificial neural networks. Can. Geotech. J. 43 (1), 1144–1152.

Shahin, M.A., Jaksa, M.B., 2005. Neural network prediction of pullout capacity of marquee ground anchors. Comput. Geotech. 32 (3), 153–163.

Shahin, M.A., Jaksa, M.B., 2006. Pullout capacity of small ground anchors by direct cone penetration test methods and neural methods. Can. Geotech. J. 43 (6), 626–637.

Shahin, M.A., Jaksa, M.B., Maier, H.R., 2001. Artificial neural network applications in geotechnical engineering. Aust. Geomech. 36 (1), 49–62.

Shahin, M.A., Maier, H.R., Jaksa, M.B., 2002. Predicting settlement of shallow foundations using neural network. J. Geotech. Geoenviron. Eng. ASCE. 128 (9), 785–793.

Shahin, M.A., Maier, H.R., Jaksa, M.B., 2004. Data division for developing neural networks applied to geotechnical engineering. J. Comput. Civ. Eng. ASCE. 18 (2), 105–114.

Shahin, M.A., Jaksa, M.B., Maier, H.R., 2005. Neural network based stochastic design charts for settlement prediction. Can. Geotech. J. 42, 110–120.

Shahin, M.A., Jaksa, M.B., Maier, H.R., 2009. Recent advances and future challenges for artificial neural systems in geotechnical engineering applications. Adv. Artif. Neural Syst. 2009, 1–9. doi: 10.1155/2009/308239.

Shi, J.J., 2000. Reducing prediction error by transforming input data for neural networks. J. Comput. Civ. Eng. 14 (2), 109–116.

Shi, J.J., 2002. Clustering technique for evaluating and validating neural network performance. J. Comput. Civ. Eng. 16 (2), 152–155.

Singh, T.N., Singh, V., 2005. An intelligent approach to prediction and control ground vibration in mines. Geotech. Geol. Eng. 23, 249–262.

Sinha, S.K., Wang, M.C., 2008. Artificial neural network prediction models for soil compaction and permeability. Geotech. Eng. J. 26 (1), 47–64.

Storn R., Price K., 1995. Differential Evolution—A simple and efficient adaptive scheme for global optimization over continuous spaces. Technical Report TR-95-012, International Computer Science Institute, Berkeley, CA, USA.

Teh, C.I., Wong, K.S., Goh, A.T.C., Jaritngam, S., 1997. Prediction of pile capacity using neural networks. J. Comput. Civ. Eng. 11 (2), 129–138.

Vapnik, V.N., 1998. Statistical Learning Theory. Wiley, New York, NY.

Wang, J., Rahman, M.S., 1999. A neural network model for liquefaction-induced horizontal ground displacement. Soil Dynam. Earthquake Eng. 18 (8), 555–568.

Warner, B., Mishra, M., 1996. Understanding neural networks as statistical tools. Am. Stat. 50 (4), 284–293.

Warren, S.S., 2003. Neural Network and Statistical Jargon. <http://www.crmportals.com/Neural_Networks_Basics_presentation.pdf> Accessed 15.4.2012.

Wilby, R.L., Abrahart, R.J., Dawson, C.W., 2003. Detection of conceptual model rainfall–rainoff process inside an artificial neural network. Hydrolog. Sci. 48 (2), 163–181.

Yoo, C., Kim, J., 2007. Tunneling performance prediction using an integrated GIS and neural network. Comput. Geotech. 34 (1), 19–30.

Zhu, J.H., Zaman, M.M., Anderson, S.A., 1998. Modelling of soil behavior with a recurrent neural network. Can. Geotech. J. 35 (5), 858–872.

11 Geotechnical Applications of Bayesian Neural Networks

Anthony T.C. Goh[1] and Chai Guan Chua[2]

[1]School of Civil and Environmental Engineering, Nanyang Technological University, Singapore, [2]Jalan Putra Permai 8G, Taman Equine, Seri Kembangan, Selangor, Malaysia

11.1 Introduction

In geotechnical engineering design, empirical relationships are often employed to estimate design parameters and engineering properties, as well as predict the behavior of geotechnical structures. Generally, the behavior of the system, such as the performance of a braced retaining wall system, is characterized by a number of interacting factors in which the relationship between these factors is not precisely known. In addition, the data associated with these parameters are usually incomplete or erroneous (i.e., noisy). The extraction of knowledge from the data to develop these empirical relationships is a formidable task requiring sophisticated modeling techniques as well as human intuition and experience. This chapter demonstrates the use of Bayesian neural network learning to alleviate this problem. The neural network is a product of artificial intelligence research. Neural networks have been successfully used in pattern recognition and the modeling of nonlinear relationships involving a multitude of variables, in place of conventional techniques such as regression analysis. One of the strengths of neural networks is its capability to "learn" from example patterns and find meaningful solutions without the need to specify the relationship between variables. Therefore, they are useful for finding solutions for which there is a lack of understanding of the problem or the behavior of the problem. For example, in geotechnical engineering, this methodology has been successfully applied to seismic liquefaction (Goh, 1994; Juang et al., 1999), constitutive modeling (Banimahd et al., 2005; Ghaboussi et al., 1991), ground property estimation (Alavi and Gandomi, 2011; Alavi et al., 2009), dam analysis (Kim and Kim, 2008; Yu et al., 2007), and tunneling (Benardos and Kaliampakos, 2004). First, an overview of the conventional neural network and Bayesian neural network methodologies are presented. This is followed by some practical examples in geotechnical engineering to demonstrate the potential of this approach for capturing nonlinear interactions between variables in complex engineering systems.

Metaheuristics in Water, Geotechnical and Transport Engineering. DOI: http://dx.doi.org/10.1016/B978-0-12-398296-4.00011-8

11.2 Neural Networks

Neural networks are biologically inspired computer models that essentially mimic the operations of the human brain. By far, the most commonly used neural network model is the back-propagation algorithm. Detailed description of this algorithm can be found in the literature (Caudill and Butler, 1991; Rumelhart et al., 1986). A neural network has a parallel-distributed architecture with a number of interconnected nodes commonly referred to as *neurons*, as shown in Figure 11.1. The neurons interact with each other via weighted connections. Each neuron is connected to all the neurons in the next layer. Data is presented to the neural network through the input neurons, and an output neuron transmits the response of the network to the input. The processing of the inputs through the intermediate (hidden) neurons enables the network to represent and compute complicated associations between patterns.

In the back-propagation algorithm, the neural network is presented with a series of examples of associated input and target output values. Neural network "learning" involves presenting a pattern to the input layer, passing the signal through the hidden layer where the input data is transformed via a nonlinear transfer function, and determining the output. The actual output from the output neuron is then compared with the target value and any difference corresponds to an error. The main objective in "training" the neural network is to modify the connection weights to reduce the errors between the actual output values and the target output values through the minimization of the defined error function (e.g., sum-squared error) using the gradient descent approach. Validation of the performance of the neural network, to assess the generalization capability of the trained neural network model to produce the correct input−output mapping even when the input is different from the examples used to train the network, is carried out by "testing" with a separate set of data that was never used in training the neural network. Generalization is influenced

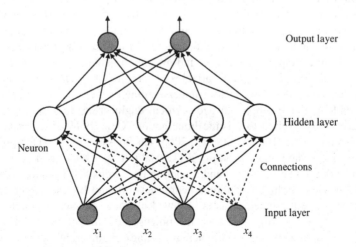

Figure 11.1 Neural network architecture.

by factors such as the size of the training data, how representative the data is of the problem to be considered, and the physical complexity of the problem. The neural network architecture is also important. Too few hidden neurons may mean that the network is unable to model the nonlinear problem correctly. An excessive number of neurons may cause a phenomenon called *overfitting*, in which the network learns insignificant aspects of the training set (i.e., the intrinsic noise in the data). Determining the optimal number of hidden neurons is commonly carried out by a trial-and-error approach through repeatedly increasing the number of hidden neurons until no further improvement in the network performance is obtained.

11.3   Bayesian Neural Network

To simplify the network architecture selection for the conventional back-propagation algorithm and to produce a network that generalizes well, Mackay (1991) and Neal (1992) proposed the use of Bayesian back-propagation neural networks. The method is based on the Bayesian statistical approach (Box and Tiao, 1973) and originated in the field of maximum entropy (Gull, 1988). The Bayesian back-propagation algorithm involves constraining the size of the network parameters through a regularizer that penalizes the more complicated weight functions in favor of simpler functions by adding a penalty term to the sum-squared error.

Instead of searching for a unique optimal value for the unknown weights (as is done in conventional back propagation), the Bayesian network models the weights with a probability density function, i.e., the uncertainty in the weight space is assigned a probability distribution representing the degree of belief in the different values of the weight vector. Mackay (1991) has shown that maximizing the posterior distribution corresponds to minimizing the regularized error function.

11.4   Evolutionary Bayesian Back-Propagation Neural Network

The two main components of the evolutionary Bayesian back-propagation (EBBP) developed by Chua and Goh (2003) are evolutionary training and Bayesian inference procedures. The following sections give an overview of these two procedures.

11.4.1   Evolutionary Training

In the evolutionary training phase, the genetic algorithms (GAs) and gradient descent methods are used to locate the most probable parameters. The GAs enhance the searching capacity in locating the global minima for the neural network model. First, a series of initial weight vectors $\mathbf{w}_1$, $\mathbf{w}_2$, $\mathbf{w}_3$,..., $\mathbf{w}_i$ is assigned to the population using the Nguyen−Widrow (1990) method, which distributes the active region of each neuron in the layer evenly across the layer's input space. Updating of the weights **w** is carried out using the stochastic gradient descent approach:

$$w_{new} = w_{old} - \rho\nabla E_p(w) + \psi\Delta w_{old} \qquad (11.1)$$

where ∇E_p denotes the gradient of the error function of every presentation of an input pattern and ρ and ψ are the learning rate and the momentum terms that control the step of change in weight for every updating. The modified Metropolis algorithm (Metropolis et al., 1953) is used to decide whether to accept a new search point based on the error function E_p performance:

$$w_{new} = \begin{cases} w_{old} - \rho\nabla E_p(w) + \alpha\Delta w_{old} & \text{If}(E_{pnew} - E_{pold}) < 0 \\ w_{old} - 0.9(\rho\nabla E_p(w) + \alpha\Delta w_{old}) & \text{otherwise} \end{cases} \qquad (11.2)$$

The testing error is then determined from an independent set of testing data.

The fitness function used to assess how well an individual $\mathbf{w}_i$ of the population performs is defined by

$$Fitness(w_i) = \frac{1}{0.5 * (MSE(w_i) + TSE(w_i)) + 0.01} \qquad (11.3)$$

where $MSE(\mathbf{w}_i)$ and $TSE(\mathbf{w}_i)$ are the mean-squared error for the training data and testing data, respectively. The small constant value of 0.01 is to prevent the fitness value from growing to infinity.

The GA operations consisted of four procedures: selection, crossover, mutation, and elitism. A roulette-wheel approach (Holland, 1975), based on the proportional fitness of individual weights $\mathbf{w}_i$ with respect to the fitness probability distribution of the population, is used as the selection operator. The crossover operation performs a multidirectional search and exchanges good subsolutions without deteriorating the learning process of neural networks. It replaces the poor individual with good offspring and ensures a fitter population at the end of the evolution. The mutation operation introduces some extra diversity to the population by introducing weights that were not present in the initial population. The elitism operation (De Jong, 1975) ensures that the fittest $\mathbf{w}_i$ survives in the evolutionary process. Before the crossover and mutation processes, the fitness of the population is ranked according to Eq. (11.3). As the population size is kept constant, in order to keep the good seeds of population and their offspring, the remaining (elite) members of the population are selected from the $\mathbf{w}_i$ that rank at the top in terms of fitness. The first part of learning is stopped when 5% of the population achieves the defined most probable state, the maximum iteration number has been reached, or the performance of the fittest individual deteriorates after 10 consecutive iterations.

11.4.2 Bayesian Regularization

In the next phase of EBBP, only the stronger population (20% of the top rank) determined from the first phase is selected to perform Bayesian regularization.

The Bayesian framework essentially provides better generalization and a statistical approach to deal with data uncertainty in comparison to the conventional back propagation. This is carried out using the Levenberg–Marquardt (LM; Levenberg, 1994; Marquardt, 1963) second-order gradient descent algorithm to minimize the regularized objective function:

$$M(w) = \beta E_D + \alpha_1 E_{w_1} + \alpha_2 E_{w_2} + \cdots + \alpha_c E_{w_c} \tag{11.4}$$

where β and α_1 to α_c are hyperparameters to regularize learning. E_D is defined by

$$E_D = \sum_{l=1}^{N} \frac{1}{2}(y_l - t_l)^2 \tag{11.5}$$

where y_l and t_l denote the lth network output and the lth target value, respectively, and N denotes the product of the number of output units and the number of input patterns.

$$E_{w_c} = \sum_{i=1}^{n_c} \frac{1}{2} w_i^2 \tag{11.6}$$

where n_c denotes the number of weights for the group of $\mathbf{w}_c$. The α_c hyperparameters correspond to the weight decay parameters for the weights of each input unit to the hidden layer, the biases to the hidden units, and the weights of the output layer that consist of its biases and all the weights in that particular layer. The updating rule for the LM algorithm is

$$w_{\text{new}} = w_{\text{old}} - [\beta \mathbf{J}^T \mathbf{J} + (\mu \mathbf{I} + \overline{\alpha})]^{-1} \mathbf{J}^T \overline{\varepsilon} \tag{11.7}$$

where $\mathbf{J}$ is the Jacobian matrix, $\overline{\varepsilon}$ is the error vector that defines the differences between output values and target values, $\mathbf{I}$ is the unit matrix, μ is the step control parameter, and $\overline{\alpha}$ is the diagonal matrix consisting of regularizers $\alpha_1, \alpha_2, \alpha_3, \dots,$ α_c. The $N \times n$ $\mathbf{J}$ matrix and the $\overline{\varepsilon}$ vector are as follows:

$$\mathbf{J} = \begin{bmatrix} \dfrac{\partial \varepsilon_1}{\partial w_1} & \dfrac{\partial \varepsilon_1}{\partial w_2} & \cdots & \dfrac{\partial \varepsilon_1}{\partial w_n} \\[2ex] \dfrac{\partial \varepsilon_2}{\partial w_1} & \dfrac{\partial \varepsilon_2}{\partial w_2} & \cdots & \dfrac{\partial \varepsilon_2}{\partial w_n} \\[2ex] \vdots & & & \\[1ex] \dfrac{\partial \varepsilon_N}{\partial w_1} & \dfrac{\partial \varepsilon_N}{\partial w_2} & \cdots & \dfrac{\partial \varepsilon_N}{\partial w_n} \end{bmatrix} \tag{11.8}$$

and

$$\bar{\varepsilon} = \begin{bmatrix} y_1 - t_1 \\ y_2 - t_2 \\ \vdots \\ y_N - t_N \end{bmatrix} \tag{11.9}$$

where y_N and t_N denote the network Nth output value and Nth target value.

The step control parameter μ is initially set to 0.01. The change in the regularized objective function $M(\mathbf{w})$ is monitored after each iteration. If the $M(\mathbf{w})$ decreases after an iteration using Eq. (11.7), the new weight vector is accepted, the value of μ is decreased by a factor of 10, and the process repeats. However, if the $M(\mathbf{w})$ increases, then μ is increased by a factor of 10, the old weight vector is restored, and a new weight update is computed. When the scalar μ is small, this is the exact Newton's method, using the approximate Hessian matrix. When μ is large, this becomes the gradient descent with a small step size.

During the minimization process, the $M(\mathbf{w})$ is monitored and compared to the defined most probable state. The most probable state is defined as a state that gives the maximum posterior probability. The Bayesian regularization process is triggered once the most probable state is reached. In this study, the most probable state is initially set to a value that is close to or less than the minimum value of the output unit or a tolerable value based on the problem domain to be learned. It is then adjusted to the average squared error obtained in the first phase of training.

Once the most probable (MP) state is reached, the log of the evidence (E) of the learning process is evaluated as follows:

$$\ln(E) = -\alpha_1 E_{W_1}^{MP} - \alpha_2 E_{W_2}^{MP} \dots - \alpha_c E_{W_c}^{MP} - \beta E_D^{MP} - \frac{1}{2}\ln|A|$$

$$+ \frac{W_1}{2}\ln\alpha_1 + \frac{W_2}{2}\ln\alpha_2 + \dots + \frac{W_c}{2}\ln\alpha_c + \frac{N}{2}\ln\beta - \frac{N}{2}\ln(2\pi) \tag{11.10}$$

where $W_1, W_2, \dots, W_c$ comprise the weights for each input unit, bias of the hidden layer, and weights and biases of the output layer.

$$\ln|A| = \sum_i \ln(\lambda_{i1} + \alpha_1) + \sum_i \ln(\lambda_{i2} + \alpha_2) + \dots + \sum_i \ln(\lambda_{ic} + \alpha_c) \tag{11.11}$$

where $\mathbf{A}$ is equal to $[\beta \mathbf{J}^T \mathbf{J} + \bar{\alpha}]$ and λ_{ic} is the eigenvalue corresponding to the particular w_i of regularizer α_c.

The log evidence is monitored during the learning process as well. The regularization parameters are updated as shown here each time the evidence is increased:

$$\gamma_c = W_c - \alpha_c(\text{Trace } \mathbf{A}^{-1})_{\text{sub}}$$

$$\gamma_c = W_c - \alpha_c \sum \frac{1}{\lambda_{ic} + \alpha_c} \tag{11.12}$$

and

$$\gamma = \sum_c \gamma_c \tag{11.13}$$

where γ is the number of well-determined parameters that measure the effective number of weights whose values are controlled by the data rather than the prior and $(\text{Trace } \mathbf{A}^{-1})_{sub}$ is the subdomain of $\text{Trace } \mathbf{A}^{-1}$:

$$\alpha_c = \frac{\gamma_c}{2E_{W_c}^{MP}} \tag{11.14}$$

and

$$\beta = \frac{N - \gamma}{2E_D^{MP}} \tag{11.15}$$

Because of the high computational effort involved in the second part of the EBBP, only 20% of the top-ranked population are selected to undergo this Bayesian process.

In practice, due to the intrinsic noise of the data and uneven distribution of the data density, it is unlikely that perfect learning (MSE = TSE = 0) will be achieved. Therefore, the following stopping criteria are used:

1. Neither the MSE or TSE nor evidence has shown improvement in five consecutive iterations. This is an indication that learning has started to deteriorate or is trapped in local minima with a long valley.
2. The total norm of the gradient of the error function E_D is smaller than 1×10^{-10}. This indicates that the minimization of the error function E_D has descended to the local minima, and that overall, the error is generally small.
3. The μ is greater than 1×10^{10}. The large μ will give a very small step in gradient descent. This implies that the learning is probably trapped in singular local minima or a saddle point.
4. The determinant of $\mathbf{A}$ is a negative value or zero. This indicates that poor conditioning exists in the matrix $\mathbf{A}$ and the inverse matrix is unattainable.

By assuming that the trained neural network has arrived at the most probable state and a Gaussian function for the posterior distribution of the connection weights, the error bar (standard deviation, σ_t) of every prediction y made by the model is given as

$$\sigma_t = \sqrt{\frac{1}{\beta} + g^T A^{-1} g} \tag{11.16}$$

where $g \equiv \nabla_w y|_{w_{MP}}$.

11.5 Examples

This section demonstrates the efficiency of the EBBP to locate the minima and handle the uncertainties in data for complicated nonlinear problems through reviewing some of the practical geotechnical applications previously studied by the authors.

11.5.1 Example 1—Pile Skin Friction for Driven Piles

A common method to estimate the skin friction capacity of driven piles in cohesive soils is the alpha (α) method developed by Tomlinson (1957). From field load test data mainly from driven piles, the skin friction was related to the undrained shear strength s_u by an empirical coefficient denoted α. Subsequent studies by Randolph and Murphy (1985) and Semple and Ridgen (1986) showed that α is also influenced by factors such as the mean effective overburden stress σ'_{vm}, the overconsolidation ratio, OCR, the effective stress friction angle ϕ', the pile width D, and the pile length L.

In this example, 65 data records of driven pile load tests taken from the literature were used to assess the skin friction f_s using the EBBP. The data records are summarized in Goh (1995). The four input parameters were the pile length, the pile diameter, the average effective vertical stress, and the average undrained shear strength. The training data consisted of 45 randomly selected patterns, and the remaining 20 patterns were used for the testing phase. The EBBP architecture consisted of four input neurons, three hidden neurons, and an output neuron.

A plot of the EBBP-predicted values versus the measured values is shown in Figure 11.2, with the corresponding coefficient of correlation R (between the

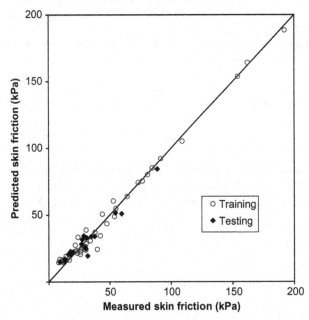

Figure 11.2 Predicted and measured skin friction values for driven piles in clay.

predicted and the measured values) of 0.993 (training data) and 0.968 (testing data). Using the conventional Semple and Rigden method, the corresponding coefficient of correlation R was 0.976 (training data) and 0.885 (testing data). The results indicate that the predictions using the EBBP were an improvement over those by the conventional method.

11.5.2 Example 2—Pile Skin Friction for Drilled Shafts

In the second example, the EBBP was used to model the skin friction f_s for drilled shafts. The training and testing data were based on the field load test data compiled by Chen and Kulhawy (1994), in which the measured undrained shear strength s_u values were converted to a consistent test type (triaxial CIUC) denoted as $s_u(CIUC)$ values.

Of the 127 field load test data patterns, 85 patterns were randomly selected as the training data, and the remaining 42 patterns were used for the testing data (Goh et al., 2005). The shaft diameters ranged from 0.18 to 1.8 m, while the shaft lengths were in the range of 1.62 to 77 m. The σ'_{vm} were in the range from 11 to 343 kPa, $s_u(CIUC)$ was in the range from 21 to 483 kPa, and α_{CIUC} was in the range from 0.24 to 1.03. The majority of the soils were overconsolidated. Most of the patterns had $s_u(CIUC)/\sigma'_{vm}$ ratios in the range from 0.49 to 6.9. The EBBP architecture used in the analyses consisted of two input neurons representing σ'_{vm} and $s_u(CIUC)$, four hidden neurons, and an output neuron representing α_{CIUC}.

The following regression equation proposed by Chen and Kulhawy (1994) was used for comparison:

$$\alpha_{CIUC} = 0.31 + 0.17/[s_u(CIUC)/p_a] \tag{11.17}$$

in which p_a is the atmospheric pressure.

A plot of the EBBP predicted values versus the measured α_{CIUC} values for both the training and the testing data is shown in Figure 11.3. A similar plot of predictions based on Eq. (11.17) is shown in Figure 11.4. The "measured α_{CIUC} value" is defined in the most general sense of having been inferred from the back-calculated field load test data. The EBBP predictions show less scatter compared to the predictions using Eq. (11.17). The corresponding coefficient of correlation R (between the predicted and the measured values) was 0.867 (training data) and 0.891 (testing data) using the EBBP method. Using Eq. (11.17), the corresponding coefficient of correlation R was 0.651 (training data) and 0.814 (testing data). The results indicate that the EBBP predictions were an improvement over those based on Eq. (11.17), particularly for the training data set.

For the EBBP, every neural network prediction is associated with an error bar. These error bars are the standard deviations for the predictions based on the data distribution and inherent noise. For clarity, they have been omitted in Figures 11.2–11.4. The calculation of the standard deviation of α_{CIUC} for the 42 testing data patterns ranged from 0.0896 to 0.1051.

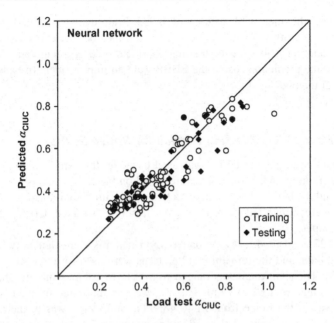

Figure 11.3 EBBP predicted and measured α_{CIUC} values.

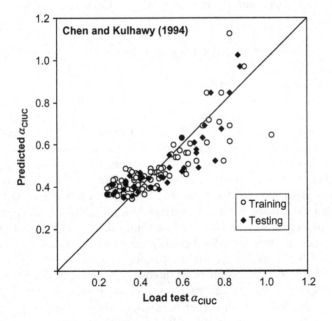

Figure 11.4 Predicted and measured α_{CIUC} values using Eq. (11.17).

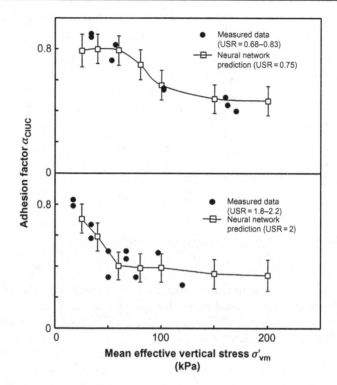

Figure 11.5 Neural network predicted α_{CIUC} for drilled shafts (USR of 0.75 and 2).

The variation of α_{CIUC} was further examined through a parametric study for σ'_{vm} in the range of 25–200 kPa, and undrained shear strength ratio (USR) $s_{\text{u}}/\sigma'_{\text{vm}}$ in the range of 0.75–3. The EBBP predictions for USR = 0.75 and 2.0, with the error bars, are plotted in Figure 11.5, along with the measured results that were close to the corresponding USR ratios. The plots show that the EBBP gives logical and consistent trends with the measured results. The general trend was for the α_{CIUC} value to decrease with increasing σ'_{vm} and with increasing USR.

11.5.3 Example 3—Retaining Wall Deflection

For fairly deep excavations, braced retaining wall support systems are commonly used to provide lateral support for the soil around the excavation. One of the major concerns in carrying out an excavation using a braced retaining wall system in a congested urban environment is that the excavation-induced ground movements will damage adjacent buildings and utilities. Some of the critical factors that influence the magnitude of ground movements are the width and depth of the excavation, the soil type and properties, the bracing system, and the stiffness of the wall. Various methods (Clough and O'Rourke, 1990; Hashash and Whittle, 1996; Long, 2001; Peck, 1969) have been proposed for estimating lateral wall

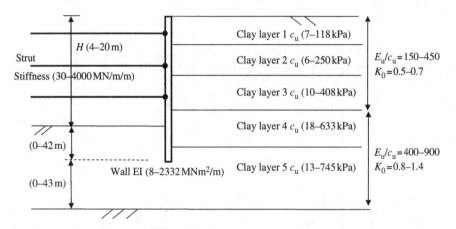

Figure 11.6 Cross section of braced excavation geometry.

deflections and ground deformations around deep excavations. Neural networks have also been successfully used for predicting the wall deflection of braced excavation systems (Goh et al., 1995; Jan et al., 2002). In both these studies, seven input variables were considered.

This study expands on the approach by Goh et al. (1995). The EBBP model consisted of 35 different input parameters, which take into account the wall length and stiffness (*EI*), width (*B*) and depth of excavation (*H*), support stiffness and location, *in situ* stress state, nonhomogeneous soil conditions, and variation of soil properties with depth (Chua and Goh, 2005). Figure 11.6 shows a schematic representation of the excavation geometry and the main parameters considered in the analyses. Five different soil layers were considered, as depicted in Figure 11.6. The inputs for each clay layer comprised the undrained shear strength c_u, the undrained elastic modulus E_u, the soil unit weight, the coefficient of earth pressure at-rest K_0 and the thickness of the layer. The output variable was the normalized maximum lateral wall deflection $\delta_{h,max}/H$. The training and testing data were obtained from finite element method (FEM) analysis using the program EXCAV97 (Wong and Goh, 1997). A large database of 6925 patterns was used. The data obtained from the analyses were randomly separated into 3844 training patterns and 3081 testing patterns.

The range of soil profiles considered encompassed excavations in soft clay, stiff clay, and mixed soil comprising both soft and stiff clay. The range of values of the variables is also shown in Figure 11.6 in parentheses. The excavation width ranged from 11 to 95 m. The retaining structures consisted of either flexible or stiff walls. Situations of walls either floating in the clay, resting on the stiff stratum, or penetrating into the stiff stratum were considered. Excavations with two or three strut levels to depths of up to 20 m were considered. The final optimal architecture of the EBBP model consisted of 35 input neurons, 14 hidden neurons, and 1 output neuron (35:14:1).

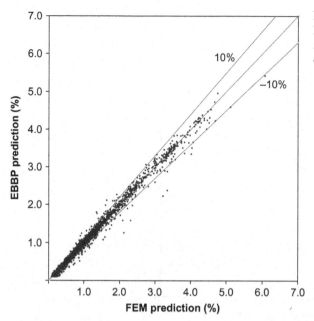

Figure 11.7 Comparison of EBBP- and FEM-normalized maximum wall deflection (training data).

The EBBP-predicted wall deflections versus the actual (FEM) wall deflections are shown in Figures 11.7 and 11.8. The corresponding coefficient of correlation R was 0.993 (training data) and 0.968 (testing data). The plot of the results in Figures 11.7 and 11.8 and the coefficient of correlation statistics show a good fit between the predicted and the actual output values. Seven instrumented case histories of braced excavations were also used to further validate the generalization capability of the trained EBBP model. The case histories were extracted from the literature and involved excavations in soft clay and stiff clay. The results of the EBBP predictions and measured results are also shown in Table 11.1. The range of predictions ($\pm$ one standard deviation) for the EBBP is shown in the far-right column of Table 11.1. Figure 11.9 shows the plot of the predicted wall deflection versus the average measured wall deflections. Generally, the predictions were in good agreement with the measured data.

11.6 Conclusions

The EBBP is a hybrid neural network that incorporates the GA search methodology with Bayesian neural network learning. The algorithm overcomes the overfitting problem by penalizing complicated weight models through a regularization term. With the Bayesian regularization, the uncertainty of data can be indicated as an error bar. The error bar gives the standard deviation of every prediction based on the data distribution and intrinsic noise. Three examples, all with high coefficient of correlations, were presented to demonstrate the capability of the EBBP to locate

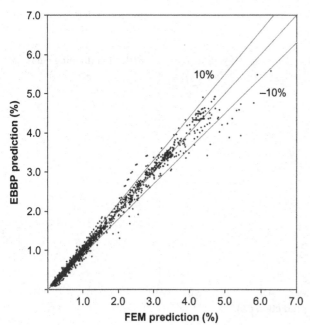

Figure 11.8 Comparison of EBBP- and FEM-normalized maximum wall deflection (testing data).

Table 11.1 Summary of Measured Deflections and EBBP Predictions for Case Histories

Case	B (m)	H (m)	Measured $\delta_{h,max}$ (mm)	Range of EBBP $\delta_{h,max}$ (mm)	Reference
(a) Vaterland	11.0	8.1	105−145	104−126	Mana (1978)
(b) New Palace	50.0	10	18.5	22−38	Burland and Hancock
		13	20.5	27−47	(1977)
(c) Lion Yard	45.0	8.2	11.5	8−21	Lings et al. (1991)
		10.2	14.5	10−26	
(d) Telecom	27.0	5.75	58−82	60−71	Lee et al. (1986)
		7.4	100−130	90−104	
(e) MOE	70.0	3.7	135	157−171	Tan et al. (1985)
		5.1	180	205−222	
		7.0	310	282−302	
(f) CH1167w	42.0	9.9	28	42−59	Poh (1996)
(g) Rochor	95.0	4.2	46−85	84−92	Lee and Ng (1994)
		6.3	125−150	137−140	

the minima and handle the uncertainties in data for nonlinear multivariate problems. EBBP can be applied to domains where there is incomplete understanding of the problem to be solved, but where training data are readily available. They are particularly useful for approximating complicated nonlinear problems. Unlike other

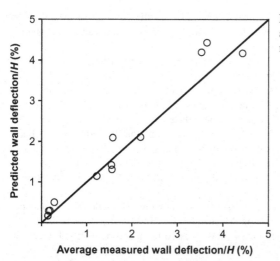

Figure 11.9 Comparison of EBBP and measured normalized maximum wall deflection.

nonlinear statistical modeling techniques, the relationships between the variables do not have to be specified in advance. Therefore, one can avoid making assumptions that may not be correct or relevant. It is a useful adjunct to other mathematical methods to model complex problems. A trained model is particular useful for parametric studies to demonstrate explicitly the captured relationships between each input parameter, as well as confidence level of each prediction. This helps to identify the range of low-density data domains that require improvement, as well as the influence of each input parameter.

References

Alavi, A.H., Gandomi, A.H., 2011. Prediction of principal ground-motion parameters using a hybrid method coupling artificial neural networks and simulated annealing. Comput. Struct. 89 (23—24), 2176—2194.

Alavi, A.H., Gandomi, A.H., Gandomi, M., Sadat Hosseini, S.S., 2009. Prediction of maximum dry density and optimum moisture content of stabilized soil using RBF neural networks. IES J. A Civ. Struct. Eng. 2 (2), 98—106.

Banimahd, M., Yasrobi, S.S., Woodward, P.K., 2005. Artificial neural network for stress—strain behavior of sandy soils: knowledge based verification. Comput. Geotech. 32 (5), 377—386.

Benardos, A.G., Kaliampakos, D.C., 2004. Modeling TBM performance with artificial neural networks. Tunnelling Underground Space Technol. 19 (6), 597—605.

Box, G.E.P., Tiao, G.C., 1973. Bayesian Inference in Statistical Analysis. Addison-Wesley, Reading, MA.

Burland, J.B., Hancock, R.J.R., 1977. Underground car park at the House of Commons, London: geotechnical aspects. Struct. Eng. 55 (2), 87—100.

Caudill, M., Butler, C., 1991. Naturally Intelligent Systems. MIT Press, Cambridge, MA.

Chen, Y.J., Kulhawy, F.H., 1994. Case history evaluation of the behavior of drilled shafts under axial and lateral loading. Report TR-104601. Electric Power Research Institute, Palo Alto, CA.

Chua, C.G., Goh, A.T.C., 2003. A hybrid Bayesian back-propagation approach to multivariate modelling. Int. J. Numer. Anal. Methods Geomech. 27 (8), 651–667.

Chua, C.G., Goh, A.T.C., 2005. Estimating wall deflections in deep excavations using Bayesian neural networks. Tunnell. Undergr. Space Technol. 20 (4), 400–409.

Clough, G.W., O'Rourke, T.D., 1990. Construction induced movements of in situ walls. In: Lambe P.C., Hansen L.A. (Eds.), Proceedings, Design and Performance of Earth Retaining Structures, ASCE Specialty Conference, Ithaca, New York, pp. 439–470.

De Jong, K., 1975. An Analysis of the Behavior of a Class of Genetic Adaptive Systems. Ph.D. Thesis. Department of Computer and Communications Sciences, University of Michigan, Ann Arbor, MI.

Ghaboussi, J., Garett J.H., Jr., Wu, X., 1991. Knowledge-based modeling of material behaviour with neural networks. J. Eng. Mech. 117 (1), 132–153.

Goh, A.T.C., 1994. Seisimic liquefaction potential assessed by neural networks. J. Geotech. Eng. 120 (9), 1467–1480.

Goh, A.T.C., 1995. Empirical design in geotechnics using neural networks. Geotechnique. 45 (4), 709–714.

Goh, A.T.C., Wong, K.S., Broms, B.B., 1995. Estimation of lateral wall movements in braced excavations using neural networks. Can. Geotech. J. 32 (6), 1059–1064.

Goh, A.T.C., Kulhawy, F.H., Chua, C.G., 2005. Bayesian neural network analysis of undrained side resistance of drilled shafts. J. Geotech. Geoenviron. Eng. 131 (1), 84–93.

Gull, S.F., 1988. Bayesian inductive inference and maximum entropy. In: Ericson, G.J., Smith, C.R. (Eds.), Maximum Entropy and Bayesian Methods in Science and Engineering, 1. Kluwer, Norwell, pp. 53–74.

Hashash, Y.M.A., Whittle, A.J., 1996. Ground movement prediction for deep excavation in soft clay. J. Geotech. Eng. 122 (6), 474–486.

Holland, J.H., 1975. Adaptation in Natural and Artificial Systems. University of Michigan Press, Ann Arbor, MI.

Jan, J.C., Hung, S.L., Chi, S.Y., Chen, J.C., 2002. Neural network forecast model in deep excavation. J. Comput. Civ. Eng. 16 (1), 59–65.

Juang, C.H., Chen, C.J., Tien, Y.M., 1999. Appraising CPT-based liquefaction resistance evaluation methods-artificial neural network approach. Can. Geotech. J. 36 (3), 443–454.

Kim, Y.S., Kim, B.T., 2008. Prediction of relative crest settlement of concrete-faced rockfill dams analyzed using an artificial neural network model. Comput. Geotech. 35 (3), 313–322.

Lee, C.W., Ng, C.K., 1994. Numerical Analysis of Braced Excavations. Final Year Project Report. Nanyang Technological University, Singapore.

Lee, S.L., Yong, K.Y., Karunaratne, G.P., Chua, L.H., 1986. Field instrumentation for a strutted deep excavation in soft clay. In: Goh A., Low B.K. (Eds.), Proceedings of Fourth International Geotechnical Seminar on Field Instrumentation and In Situ Measurements, Nanyang Technological Institute, Singapore, pp. 183–186.

Levenberg, K., 1994. A method for the solution of certain non-linear problems in least squares. Q. J. Appl. Math. II. 2, 164–168.

Lings, M.L., Nash, D.F.T., Ng, C.W.W., Boyce, M.D., 1991. Observed behaviour of a deep excavation in Gault clay, a preliminary appraisal. In: the organisers: Associazione

Geotecnica Italiana. Proceedings of 10th European Conference on Soil Mechanics and Foundation Engineering, vol. 2, pp. 467–470.

Long, M., 2001. Database for retaining wall and ground movements due to deep excavations. J. Geotech. Geoenviron. Eng. 127 (3), 203–224.

Mackay, D.J.C., 1991. Bayesian Methods for Adaptive Models. Ph.D. Thesis. California Institute of Technology, CA.

Mana, I.A., 1978. Finite Element Analyses of Deep Excavation Behaviour in Soft Clay. Ph.D. Thesis. Stanford University, Stanford, CA.

Marquardt, D.W., 1963. An algorithm for least-squares estimation of non-linear parameters. J. Soc. Ind. Appl. Math. 11 (2), 431–441.

Metropolis, N., Rosenbluth, A.W., Rosenbluth, M.N., Teller, A.H., Teller, E., 1953. Equation of state calculations by fast computing machines. J. Chem. Phys. 21 (6), 1087–1092.

Neal, R.M., 1992. Bayesian training of back-propagation networks by the hybrid Monte Carlo method. Technical report CRG-TG-92-1. Department of Computer Science, University of Toronto, Canada.

Nguyen D., Widrow, B., 1990. Improving the learning speed of 2-layer neural networks by choosing initial values of the adaptive weights. In: Organizing Committee. Proceedings of International Joint Conference on Neural Networks, vol. 3, pp. 21–26.

Peck, R.B., 1969. Deep excavations and tunneling in soft ground. In: Organizing Committee. Proceedings of 7th International Conference on Soil Mechanics and Foundation Engineering, Mexico City, State of the Art Volume, pp. 225–290.

Poh, T.Y., 1996. Deep excavations in stiff soils in Singapore. M.Eng. Thesis, Nanyang Technological University, Singapore.

Randolph, M.F., Murphy, B.S., 1985. Shaft capacity of driven piles in clay. In: Organizing Committee. Proceedings of 17th Offshore Technology Conference, Houston, pp. 371–378.

Rumelhart, D.E., Hinton, G.E., Williams, R.J., 1986. Learning internal representation by error propagation. In: Rumelhart, D.E., McClelland, J.L. (Eds.), Parallel Distributed Processing, 1. MIT Press, Cambridge, MA, pp. 318–362.

Semple, R.M., Ridgen, W.J., 1986. Shaft capacity of driven pipe piles in clay. Ground Eng. 19 (1), 11–17.

Tan, S.B., Tan, S.L., Chin, Y.K., 1985. A braced sheetpile excavation in soft Singapore marine clay. In: Organizing Committee. Proceedings of 11th International Conference on Soil Mechanics and Foundation Engineering, San Francisco, CA, vol. 3, pp. 1671–1674.

Tomlinson, M.J., 1957. Adhesion of piles driven in clay soils. In: Organizing Committee. Prcoeedings of 4th International Conference on Soil Mechanics and Foundation Engineering, London, vol. 2, pp. 66–71.

Wong, K.S., Goh, A.T.C., 1997. Excav97: a computer program for analysis of stresses and movements in excavations. Geotechnical Research Report NTU/GT/97-1. Nanyang Technological University, Singapore.

Yu, Y., Zhang, B., Yuan, H., 2007. An intelligent displacement back-analysis method for earth-rockfill dams. Comput. Geotech. 34 (6), 423–434.

12 Linear and Tree-Based Genetic Programming for Solving Geotechnical Engineering Problems

Amir Hossein Alavi[1], Amir Hossein Gandomi[2], Ali Mollahasani[3] and Jafar Bolouri Bazaz[4]

[1]School of Civil Engineering, Iran University of Science and Technology, Tehran, Iran, [2]Department of Civil Engineering, The University of Akron, Akron, OH, USA, [3]Department of Civil, Environmental, and Material Engineering, University of Bologna, Bologna, Italy, [4]Civil Engineering Department, Ferdowsi University of Mashhad, Mashhad, Iran

12.1 Introduction

Various methods can be employed for the behavioral modeling of geotechnical engineering systems. Owing to the large variety of methods available in this field, no one method can be considered as a universally applicable solution. The modeling of geotechnical engineering problems is a difficult task because of the need to estimate both the structure and the parameters of such systems. Different criteria can be characterized for model classification while dealing with a system-modeling task (Gandomi and Alavi, 2011; Torres et al., 2009). A model can be classified as phenomenological or behavioral (Metenidis et al., 2004). A phenomenological model is derived by considering the physical relationships governing a system. As a result, the structure of the model is selected according to prior knowledge about the system. It is not always possible to design phenomenological models for geotechnical engineering systems because of their complexity. To deal with this issue, the behavioral models are commonly employed. These models approximate the relationships between the inputs and the outputs on the basis of a measured set of data, without the need for prior knowledge about the mechanisms that produced the experimental data. Behavioral models can provide very good results with minimal effort (Gandomi and Alavi, 2011). Traditional statistical regression techniques are usually used

Metaheuristics in Water, Geotechnical and Transport Engineering. DOI: http://dx.doi.org/10.1016/B978-0-12-398296-4.00012-X

for behavioral modeling purposes. However, regression analysis can have large uncertainties. Further, it has major drawbacks in terms of idealization of complex processes, approximation, and averaging widely varying prototype conditions. Regression analysis often assumes linear (or in some cases nonlinear) relationships between the output and the predictor variables; these assumptions do not always hold (Gandomi and Alavi, 2011).

Several alternative computer-aided pattern recognition and data classification approaches have been developed for behavioral modeling. An example in this field is pattern recognition systems, which learn adaptively from experience and extract various discriminators. Artificial neural networks (ANNs) are the most widely used pattern recognition procedure. ANNs have been used for a wide range of geotechnical engineering problems (Alavi and Gandomi, 2011a, b; Alavi et al., 2009, 2010a; Cabalar and Cevik, 2009a, b; Goh, 1994; Juang et al., 2001; Kim and Kim, 2008; Shahin et al., 2001, 2008, 2009). Despite the acceptable performance of ANNs in most cases, they do not usually provide a definite function to calculate the outcome. In addition, ANNs require the structure of the network to be identified *a priori*. The ANN approach is mostly suited to being used as part of a computer program.

Empirical modeling of geotechnical engineering problems by genetic programming (GP; Banzhaf et al., 1998; Koza, 1992) can be regarded as an alternative approach to conventional methods (e.g., the finite element method). GP is based on the data alone to determine the structure and parameters of the model. It is a specialization of genetic algorithms (GAs), where the solutions are computer programs rather than fixed-length binary strings. The computer programs created by standard GP are represented as tree structures (Gandomi et al., 2011a; Koza, 1992). This classical approach is also referred to as tree-based genetic programming (TGP). Linear genetic programming (LGP; Brameier and Banzhaf, 2007) is a particular subset of TGP. LGP evolves programs of an imperative language or machine language instead of the standard TGP expressions of a functional programming language (Brameier and Banzhaf, 2001, 2007). LGP has shown to be an efficient alternative to traditional TGP (Oltean and Grossan, 2003).

This chapter illustrates the feasibility of using TGP and LGP paradigms to simulate the complex behavior of geotechnical engineering systems. The formulation capabilities of TGP and LGP are demonstrated by applying them to the formulation of effective angle of shearing resistance of soils. Further, a comparative study is conducted using the results obtained via TGP, LGP, and other existing methods. The chapter is organized as follows: Section 12.2 presents a brief review of the literature on the applications of TGP and LGP. Section 12.3 provides descriptions of the TGP and LGP methodologies. Section 12.4 outlines a numerical example and reviews the results. Section 12.5 presents a general discussion of the capabilities of TGP and LGP. Finally, Section 12.6 gives concluding remarks.

12.2   Previous Studies on Applications of TGP and LGP in Geotechnical Engineering

GA is a powerful stochastic optimization method based on the principles of genetics and natural selection. For nearly two decades, GA has been shown to be suitably robust for a wide variety of complex geotechnical problems (Levasseur et al., 2007, 2009; McCombie and Wilkinson, 2002; Pal et al., 1996; Simpson and Priest, 1993). In contrast with GA and ANN, application of TGP and LGP in the field of civil engineering is totally new and original. For the last 10 years, TGP and LGP have been pronounced as new methods for simulating the behavior of geotechnical engineering problems. The first application of TGP in the fields of geotechnical engineering is carried out by Yang et al. (2004) to analyze the stability of slopes. Afterward, TGP and its variants have been successfully applied to other geotechnical engineering problems, such as identification of nonlinear dynamics of landslides (Yang and Feng, 2005), formulation of the unconfined compressive strength of soft ground (Narendra et al., 2006), evaluation of liquefaction-induced lateral displacements (Javadi et al., 2006), prediction of the soil–water characteristic curve (Johari et al., 2006), predicting the settlement of shallow foundations (Rezania and Javadi, 2007), modeling of the angle of shearing resistance of soils (Kayadelen et al., 2009), constitutive modeling of Leighton Buzzard sands (Cabalar et al., 2009), modeling the damping ratio and shear modulus of sand–mica mixtures (Cevik and Cabalar, 2009), deriving attenuation relationships (Cabalar and Cevik, 2009a, b), prediction of uplift capacity of suction caissons (Alavi et al., 2010b), formulation of soil classification (Alavi et al., 2010c), prediction of the v_{max}/a_{max} ratio of strong ground motions (Jafarian et al., 2010), modeling of stress–strain behavior of sand under cyclic loading (Shahnazari et al., 2010), nonlinear system modeling of geotechnical engineering problems (Gandomi and Alavi, 2012), prediction of time-domain parameters of ground motions (Gandomi et al., 2011b), and formulation of soil deformation moduli obtained from plate load testing (Mousavi et al., 2011a).

LGP is a robust variant of the GP method. This linear variant of TGP makes a clear distinction between the genotype and the phenotype of an individual (Oltean and Grossan, 2003). More specifically, LGP operates on programs that are represented as linear sequences of instructions of an imperative programming language. Unlike TGP and other soft computing tools like ANNs, the LGP applications are even restricted to fewer geotechnical areas. LGP is first applied to geotechnical engineering problems by Alavi et al. (2008) to predict the performance characteristics of the stabilized soil. Thereafter, this powerful technique is used by researchers to solve problems in geotechnical engineering. Some important studies in this area include prediction of circular pile scour (Guven et al., 2009), formulation of geotechnical engineering systems (Alavi and Gandomi, 2011a, b), assessment of soil liquefaction (Alavi and Gandomi, 2012), simulation of soil shear strength parameters (Mousavi et al., 2011b), modeling of uplift capacity of suction caissons (Alavi et al., 2011), and modeling of soil deformation modulus using pressure meter test results (Rashed et al., 2012). However, recent studies have shown that the GP-based techniques possess some obvious superiority over ANN in dealing with geotechnical engineering problems (Alavi et al., 2010c; Rezania and Javadi, 2007).

12.3 Tree-Based Genetic Programming

GP uses the principle of Darwinian natural selection to create computer programs. A breakthrough in GP is made by conducting experiments of Koza (1992) on symbolic regression. This classical GP technique is also referred to as TGP (Koza, 1992). GP was introduced by Koza (1992) as an extension of GAs. Most of the genetic operators used in GA can also be implemented in GP with minor changes. The main difference between GP and GA is the representation of the solution. GA creates a string of numbers that represent the solution. The GP solutions are computer programs represented as tree structures and expressed in a functional programming language like List Processing (LISP) (Koza, 1992). In other words, the individuals (programs) evolved by GP are parse trees that can vary in length throughout the run, rather than fixed-length binary strings. Essentially, this is the beginning of computer programs that program themselves (Koza, 1992). Since GP often evolves computer programs, the solutions can be executed without post-processing, while coded binary strings typically evolved by GA require post-processing. Traditional optimization techniques like GA are generally used in parameter optimization to evolve the best values for a given set of model parameters. GP, on the other hand, gives the basic structure of the approximation model with the values of its parameters. GP optimizes a population of computer programs according to a fitness landscape determined by a program's ability to perform a given computational task. The fitness of each program in the population is evaluated using a fitness function. Thus, the fitness function is the objective function that GP aims to optimize (Gandomi et al., 2011c; Torres et al., 2009).

In tree-based GP (TGP) a random population of individuals (trees) is created to achieve high diversity. A population member in TGP is a hierarchically structured tree comprising functions and terminals. The functions and terminals are selected from a set of functions and a set of terminals. The functions and terminals are chosen at random and put together to form a computer model in a treelike structure with a root point with branches extending from each function and ending in a terminal (Gandomi et al., 2011a). An example of a simple tree representation of a TGP model is illustrated in Figure 12.1.

Creating an initial population is a blind random search for solutions in the large space of possible solutions. Once a population of models has been created at random, the TGP algorithm evaluates the individuals, selects individuals for reproduction, and generates new individuals by mutation, crossover, and direct reproduction. Finally, TGP creates the new generation in all iterations (Gandomi et al., 2011a). During the crossover procedure, a point on a branch of each program is selected at random and the set of terminals and/or functions from each program

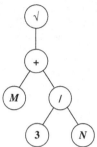

Figure 12.1 The tree representation of a GP model ($\sqrt{(M + 3/N)}$).

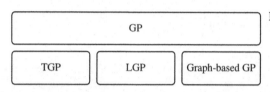

Figure 12.2 Different types of GP.

are then swapped to create two new programs. The evolutionary process continues by evaluating the fitness of the new population and starting a new round of reproduction and crossover. During the mutation process, the TGP algorithm occasionally selects a function or terminal from a model at random and mutates it (Gandomi et al., 2011a).

As shown in Figure 12.2, linear and graph-based GPs are types of GP other than traditional TGP (Alavi et al., 2011). The emphasis of the present study is placed on linear-based GP techniques. LGP is a robust linear-based GP method. There are some reasons for using LGP. Basic computer architectures are fundamentally the same now as they were 20 years ago, when GP began. Almost all architectures represent computer programs in a linear fashion. Moreover, computers do not naturally run tree-shaped programs. Hence, slow interpreters have to be used as part of GP (TGP). Conversely, by evolving the binary bit patterns actually obeyed by the computer, the use of an expensive interpreter (or compiler) is avoided, and GP can run several orders of magnitude faster (Alavi and Gandomi, 2011a, b). The enhanced speed of the linear variants of GP (e.g., LGP) permits conducting many runs in realistic time frames, which leads to deriving consistent, high-precision models with little customization (Francone and Deschaine, 2004).

12.3.1 Linear Genetic Programming

LGP is a subset of GP with a linear representation of individuals. The main characteristic of LGP in comparison with traditional TGP is that expressions of a functional programming language like LISP are substituted by programs of an imperative language like C/C++ (Brameier and Banzhaf, 2001, 2007). Figure 12.3 presents a comparison of the program structures in LGP and TGP. As shown in Figure 12.3A, an LGP can be seen as a data flow graph generated by

(A)

```
ƒ[0] = 0;
L0: ƒ[0] += v[1];
L1: ƒ[0] /= 3;
L2: ƒ[0] += v[4];
return ƒ[0];
```

(B)

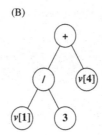

Figure 12.3 Comparison of the GP program structures: (A) LGP, (B) TGP (Gandomi et al., 2008).

$$y = ƒ[0] = (v[1]/3) + v[4]$$

void LGP (double r[5])   **Figure 12.4** An excerpt of an LGP.
```
{...
r[0] = r[5] + 70;
r[5] = r[0] − 50;
if (r[1] > 0)
if (r[5] > 2)
r[4] = r[2] × r[1];
r[2] = r[5] + r[4];
r[0] = sin(r[2]);
}
```

multiple usages of register content. That is, on the functional level, the evolved imperative structure denotes a special directed graph. As can be observed from Figure 12.3B, in TGP, the data flow is more rigidly determined by the tree structure of the program (Brameier and Banzhaf, 2001, 2007).

In the LGP system described here, an individual program is interpreted as a variable-length sequence of simple C instructions. The instruction set or function set of LGP consists of arithmetic operations, conditional branches, and function calls. The terminal set of the system is composed of variables and constants. The instructions are restricted to operations that accept a minimum number of constants or memory variables, called *registers* (*r*), and assign the result to a destination register (e.g., $r_0 = r_1 + 1$). A part of an LGP in C code is represented in Figure 12.4, in which register r[0] holds the final program output.

LGP allows structurally ineffective codes to coexist with effective codes in programs (Brameier and Banzhaf, 2001). An instruction of an LGP is called "effective" at its position if it affects the program output. The ineffective codes in genetic programs represent instructions without any influence on the program's behavior. These codes act as a protection, reducing the effect of variation on the effective code. Because of the program structure in LGP, the ineffective codes can be detected and eliminated much easier than in TGP and other comparable interpreting systems (Francone and Deschaine, 2004). Thus, the linear genetic code is interpreted more efficiently. Another feature of the LGP system is that the ineffective codes can be removed before an LGP is executed during fitness calculation. This is done by copying all effective instructions to a temporary program buffer, and it results in an enormous acceleration in the LGP execution speed.

Automatic Induction of Machine code by Genetic Programming (AIMGP) is a particular form of LGP. In AIMGP, evolved programs are stored as linear strings of native binary machine code and are directly executed by the processor during fitness calculation. The absence of an interpreter and complex memory handling results in a significant speedup in the AIMGP execution compared to TGP (Brameier and Banzhaf, 2001). This machine-code-based LGP approach searches for the computer program and the constants at the same time. Here are the steps the machine-code-based LGP system follows for a single run (Brameier and Banzhaf, 2007):

1. *Initializing a population of randomly generated programs and calculating their fitness values*: To evaluate the fitness of the evolved programs, mean absolute error is commonly used.
2. *Running a tournament*: In this step, four programs are selected from the population randomly. They are compared and based on their fitness, and two programs are designated the winners and two the losers.
3. *Transforming the winner programs*: After that, two winner programs are copied and transformed probabilistically into two new programs via crossover and mutation operators.
4. *Replacing the loser programs in the tournament with the transformed winner programs*: The winners of the tournament remain without changing.
5. *Repeating steps 2 through 4 until termination or convergence conditions are satisfied.*

Crossover occurs between instruction blocks. During this process, a segment of random position and arbitrary length is selected in each of the two parents and exchanged. If one of the two children would exceed the maximum length, crossover is aborted and restarted with exchanging equal-sized segments (Brameier and Banzhaf, 2001). The mutation operation occurs on a single instruction. Comprehensive descriptions of the basic parameters used to direct a search for an LGP can be found in Brameier and Banzhaf (2007).

12.4   Application to Geotechnical Engineering Problems

12.4.1   Modeling of the Effective Angle of Shearing Resistance

One of the most important engineering properties of soil is its ability to resist sliding along internal surfaces within a mass. The stability of structures built on soil depends upon the shearing resistance offered by the soil along the probable surfaces of slippage. The shear strength of geotechnical materials is generally represented by the Mohr–Coulomb theory. According to this theory, the soil shear strength varies linearly with the applied stress through two shear strength components known as the *cohesion intercept* and *angle of shearing resistance*. The tangent to the Mohr–Coulomb failure envelopes is represented by its slope and intercept. The slope expressed in degrees is the angle of shearing resistance, and the intercept is cohesion (Mousavi et al., 2012; Murthy, 2008). The cohesion intercept and angle of shearing resistance are treated as constants over the range of normal stresses. The values of these empirical parameters for any soil depend upon several factors, such as the soil's textural properties, its past history, its initial state, its permeability characteristics, and the conditions of drainage allowed to take place during the test (Mousavi et al., 2012; Murthy, 2008). If the cohesion intercept and angle of shearing resistance are determined using the total stresses, they are named the total or undrained cohesion intercept (c) and the angle of shearing resistance (ϕ). The effective stress is the difference between the total stress and the excess pore water pressure. If the pore water pressures are measured during the test, the effective circles can be plotted and the effective strength parameters (c' and ϕ') are obtained.

Accurate determination of ϕ' is a major concern in the design of different geotechnical structures such as foundations, slopes, underground chambers, and open excavations. This key parameter can be determined either in the field or in the laboratory. The triaxial compression and direct shear tests are the most common tests for determining the ϕ' values in the laboratory. The triaxial test is more suitable for clayey soils. The direct shear test is commonly used for sandy soils and requires a simpler test procedure than the triaxial test. The tests employed in the field include the vane shear test or any other indirect method (Mousavi et al., 2012; Murthy, 2008). However, experimental determination of the strength parameters is extensive, cumbersome, and costly. Also, it is not always possible to conduct the tests in every new situation. In order to cope with such problems, numerical solutions have been developed to estimate the ϕ' values. The fact that most of the available empirical models are based on limited experimental data raises doubts about their generality. On the other hand, despite the multivariable dependency of soils such correlations are developed on the basis of only one soil index property (Kayadelen et al., 2009). Incorporating simplifying assumptions into the development of the statistical and numerical methods may also lead to very large errors (Shahin et al., 2001).

In this chapter, the TGP and LGP techniques are used to obtain generalized relationships between ϕ' and the physical properties of the clayey and sandy soils. The proposed correlations are developed based on the consolidated-drained (CD) triaxial test results obtained from the literature. The most important factors representing the ϕ' behavior are selected based on the literature review (Kayadelen et al., 2009; Korayem et al., 1996; Mousavi et al., 2012; Murthy, 2008) and after a trial study. Consequently, the ϕ' (°) formulation is considered to be as follows:

$$\phi' = f(\text{FC, LL}, \gamma) \tag{12.1}$$

where

FC (%): fine-grained content,
LL (%): liquid limit,
γ (g/cm^3): soil bulk density.

The significant influence of these parameters in determining ϕ' is well understood. FC and LL represent the intrinsic soil properties and γ carries information on the state of the soil and its compressibility and previous history. The best TGP and LGP correlations are chosen on the basis of a multiobjective strategy as follows:

1. The simplicity of the model, although this is not a predominant factor.
2. The best fitness value on the learning set of data.
3. The best fitness value on a validation set of data.

Performance Measures

The correlation coefficient (R), root mean square error (RMSE), and mean absolute percent error (MAPE) are used to evaluate the performance of the proposed correlations. R, RMSE, and MAPE are given in the form of equations as follows:

$$R = \frac{\sum_{i=1}^{n} (h_i - \overline{h}_i)(t_i - \overline{t}_i)}{\sqrt{\sum_{i=1}^{n} (h_i - \overline{h}_i)^2 \sum_{i=1}^{n} (t_i - \overline{t}_i)^2}} \tag{12.2}$$

$$\text{RMSE} = \sqrt{\frac{\sum_{i=1}^{n} (h_i - t_i)^2}{n}} \times 100 \tag{12.3}$$

$$\text{MAPE} = \frac{1}{n} \sum_{i=1}^{n} \left[\frac{|h_i - t_i|}{h_i} \right] \times 100 \tag{12.4}$$

where h_i and t_i are, respectively, the actual and predicted output values for the ith output; $\overline{h}_i$ and $\overline{t}$ are, respectively, the average of the actual and predicted outputs; and n is the number of samples.

Experimental Database

A comprehensive database containing the results of 135 CD triaxial tests presented by Kayadelen et al. (2009) and Mousavi et al. (2012) is used for the model development. For the TGP and LGP analyses, the available data sets are randomly divided into learning, validation, and testing subsets. The learning data are used for training (genetic evolution). The validation data are used for model selection. In other words, the learning and validation data sets are used to select the best-evolved programs and included in the training process. Thus, they are categorized into one group, referred to as *training data*. The testing data are used to measure the performance of the models obtained by TGP and LGP on data that play no role in building the models. A trial study is conducted to find a consistent data division. The selection is such that the statistical properties (e.g., mean and standard deviation) of the training and testing subsets are similar. Out of the 135 data sets, 108 are used as the training data (96 sets as the learning data and 12 sets as the validation data). The remaining 27 data sets are taken for testing of the generalization capability of the models. Although normalization is not strictly necessary in the

Table 12.1 The Variables Used in Model Development

Parameters	Inputs			Output
	FC (%)	LL (%)	γ (g/cm³)	ϕ' (°)
Mean	61.39	41.00	1.81	26.41
Standard deviation	20.51	11.19	0.14	3.39
Sample variance	420.72	125.16	0.02	11.46
Range	84.00	76	0.84	22
Minimum	15.00	22	1.43	18
Maximum	99.00	98	2.27	40
Normalized form	$FC_n = FC/100$	$LL_n = LL/100$	$\gamma_n = \gamma/2.5$	$\phi'_n = \phi'/50$

GP-based analyses, better results are usually reached after normalizing the variables. This is mainly due to influence of unification of the variables, no matter their range of variation. Thus, both input and output variables are normalized between 0 and 1. The ranges, normalized values, and statistics of different input and output parameters involved in the model development are given in Table 12.1.

TGP-Based Formulation of the Angle of Shearing Resistance

A TGP analysis is performed to compare LGP with a classical GP approach. Various parameters involved in the TGP predictive algorithm are shown in Table 12.2. It is worth mentioning that a notable limitation of GP and its variants is that these methods are parameter sensitive, especially when difficult experimental training data sets are employed. Using any form of optimally controlling the parameters of the run (e.g., GAs) can improve the performance of the TGP and LGP algorithms. In this study, several runs are conducted considering different values for the TGP parameters. The parameters are selected based on some previously suggested values (Javadi et al., 2006; Johari et al., 2006) and after a trial-and-error approach. Basic arithmetic operators and mathematical functions are used to get the optimum TGP models. Three levels are set for the population size and two levels are considered for the crossover and mutation rates. There are $3 \times 2 \times 2 = 12$ different combinations of the parameters. All of these combinations are tested and 10 replications for each combination are performed. This makes 120 runs for the TGP algorithm. A fairly large number of generations are tested on each run to find models with minimum error. The program is run until the runs terminated automatically. A TGP software, GPLAB (Silva, 2007), is used in this study in conjunction with subroutines coded in MATLAB.

The prediction equation for ϕ', for the best result by the TGP algorithm, is as follows:

$$\phi'_{TGP}(°) = 8\gamma^2 + \frac{LL(FC^2 - 100FC)(8\gamma^2 - 25)}{250,000} \tag{12.5}$$

where FC, LL, and γ are the predictor variables shown in Table 12.1. A comparison of the experimental versus predicted ϕ' values is shown in Figure 12.5.

Table 12.2 Parameter Settings for the TGP Algorithm

Parameter	Settings
Function set	$+, -, \times, /$, sin, cos
	500, 1500, 3000
Maximum tree depth	10
Total generations	4000
Initial population	Ramped half-and-half
Sampling	Tournament
Expected number of offspring method	Rank 89
Fitness function error type	Linear error function
Termination	Generation 40
Crossover rate (%)	50, 95
Mutation rate %)	50, 95
Real max level	30
Survival mechanism	Keep best

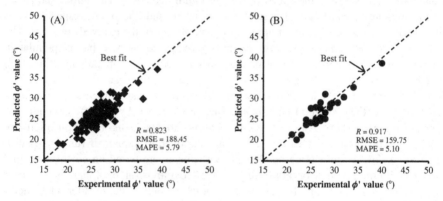

Figure 12.5 Experimental versus predicted ϕ' values using the TGP model: (A) training data, (B) testing data.

LGP-Based Formulation of the Angle of Shearing Resistance

The available database is used for generating an LGP prediction model relating ϕ' to FC (%), LL (%), and γ (g/cm^3). Various parameters involved in the LGP algorithm are shown in Table 12.3. The parameter selection will affect the model generalization capability of LGP. Several runs are conducted to come up with a parameterization of LGP that provided enough robustness and generalization to solve the problem. The parameters are selected based on previously suggested values (Alavi and Gandomi, 2011a, b, 2012; Alavi et al., 2008, 2011) and also after a trial study. Three levels are set for the population size and two levels are considered for the crossover and mutation rates. The success of the LGP algorithm usually increases as the initial and maximum program size parameters increase. In this case, the complexity of the evolved

Table 12.3 Parameter Settings for the LGP Algorithm

Parameter	Settings
Function set	$+, -, \times, /, \sqrt{}$, sin, cos
Population size	500, 1500, 3000
Maximum number of tournaments	900,000
Maximum program size	256
Initial program size	80
Crossover rate (%)	50, 95
Homologous crossover (%)	95
Mutation rate (%)	50, 95
Block mutation rate (%)	30
Instruction mutation rate (%)	30
Data mutation rate (%)	40
Number of demes	20

functions increases, and the speed of the algorithm decreases. The initial and maximum program sizes are set to optimal values of 80 and 256 bytes, respectively, as trade-offs between the running time and the complexity of the evolved solutions. The number of demes is set to 20. This parameter is related to the way that the population of programs is divided. Note that demes are semi-isolated subpopulations in which evolution proceeds faster than in a single population of equal size (Brameier and Banzhaf, 2001). In this study, basic arithmetic operators and mathematical functions are used to get the optimum LGP models. There are $3 \times 2 \times 2 = 12$ different combinations of the parameters. All of these combinations are tested and five replications for each combination are carried out. Therefore, the overall number of runs is equal to $12 \times 10 = 120$. A fairly large number of tournaments are tested on each run to find models with minimum error. For each case, the program is run until there is no longer significant improvement in the performance of the models or the runs terminate automatically. Each run is observed for overfitting while in progress. In checking for overfitting, situations are examined in which the fitness of the samples for the learning of LGP is negatively correlated with the fitness on the validation data sets. For the runs showing signs of overfitting, the LGP parameters are progressively changed to reduce the computational power available to the LGP algorithm until observed overfitting is minimized. The resulting run is then accepted as the production run. For the LGP-based analysis, the Discipulus software (Conrads et al., 2004) is used, which works on the basis of the AIMGP platform.

The LGP-based formulation of ϕ' is as follows:

$$\phi'_{LGP}(°) = 8\gamma^2 - \frac{(FC^2 - 100FC)LL(-2 + 0.01FC)((1 + 0.01FC)(0.4\gamma - 0.01FC + 1) + (\gamma - 5)/\gamma)}{20,000}$$
$$- \frac{50(FC^2 - 100FC)LL^2(0.01FC - 0.4\gamma)}{1,000,000FC - 200,000,000}$$

$$(12.6)$$

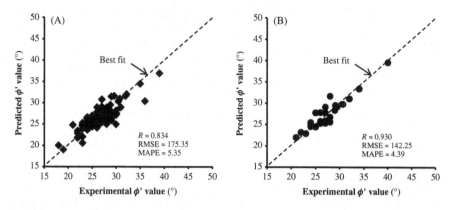

Figure 12.6 Experimental versus predicted ϕ' values using the LGP model: (A) training data, (B) testing data.

where FC, LL, and γ, respectively, denote the fine-grained content, liquid limit, and soil bulk density. Figure 12.6 shows a comparison of the experimental versus predicted ϕ' values.

Performance Analysis

According to Smith (1986), if a model gives a correlation coefficient (R) > 0.8, and the error (e.g., RMSE and MAPE) values are at minimum, there is a strong correlation between the predicted and the measured values. The model can therefore be judged as very good. It can be observed from Figures 12.5 and 12.6 that the TGP and LGP models with high R and low RMSE and MAPE values predict the target values with a high degree of accuracy. The performance of the models on the testing data is better than that on the training (learning and validation) data. The LGP-based correlation has produced better results on the training and testing data than the TGP correlation.

It is known that the models derived using soft computing techniques have a predictive capability within the data range used for their calibration in most cases. Thus, the amount of data used in the training process is an important issue, as it bears heavily on the reliability of the final models. To cope with this limitation, Frank and Todeschini (1994) argue that the minimum ratio of the number of objects over the number of selected variables for model acceptability is 3. It is also suggested that considering a higher ratio, perhaps 5, is safer. In the present study, this ratio is much higher and is equal to $135/3 = 45$. Furthermore, new criteria recommended by Golbraikh and Tropsha (2002) are checked for the external validation of the TGP and LGP models on the testing data sets. It is suggested that at least one slope of regression lines (k or k') through the origin should be close to 1. Also, the performance indexes of m and n should be lower than 0.1. Recently, Roy and Roy (2008) introduced a confirm indicator (R_m) of the external predictability

Table 12.4 Statistical Parameters of the TGP and LGP Models for the External Validation

Item	Formula	Condition	TGP	LGP
1	R	$0.8 < R$	0.917	0.930
2	$k = \dfrac{\sum_{i=1}^{n}(h_i \times t_i)}{h_i^2}$	$0.85 < K < 1.15$	1.011	0.993
3	$k' = \dfrac{\sum_{i=1}^{n}(h_i \times t_i)}{t_i^2}$	$0.85 < K' < 1.15$	0.986	1.004
4	$m = \dfrac{R^2 - \mathrm{Ro}^2}{R^2}$	$m < 0.1$	-0.181	-0.153
5	$n = \dfrac{R^2 - \mathrm{Ro}'^2}{R^2}$	$n < 0.1$	-0.176	-0.154
6	$R_{\mathrm{m}} = R^2 \left(1 - \sqrt{|R^2 - \mathrm{Ro}^2|}\right)$	$0.8 < R$	0.513	0.551
where	$\mathrm{Ro}^2 = 1 - \dfrac{\sum_{i=1}^{n}(t_i - h_i^o)^2}{\sum_{i=1}^{n}(t_i - \bar{t_i})^2}, \quad h_i^o = k \times t_i$		0.994	0.997
	$\mathrm{Ro}'^2 = 1 - \dfrac{\sum_{i=1}^{n}(h_i - t_i^o)^2}{\sum_{i=1}^{n}(h_i - \bar{h_i})^2}, \quad t_i^o = k' \times h_i$		0.990	0.999

of models. For $R_{\mathrm{m}} > 0.5$, the condition is satisfied. Either the squared correlation coefficient (through the origin) between predicted and experimental values (Ro^2), or the coefficient between experimental and predicted values (Ro'^2) should be close to 1. The validation criteria and the relevant results obtained by the models are presented in Table 12.4. As is seen, the models satisfy the required conditions. These facts ensure that the derived models are strongly valid, have good prediction power, and are not chance correlations.

Comparative Study

Kayadelen et al. (2009) employed a new variant of GP, namely genetic expression programming (GEP), to predict the ϕ' of soils. Recently, Mousavi et al. (2012) presented a novel hybrid method coupling GP and orthogonal least squares algorithm (OLS), called GP/OLS, and least squares regression technique (LSR) to formulate ϕ'. The results obtained by these methods are included in the comparative study and are shown in Table 12.5. A comparison of the predictions made by different methods for the entire database is displayed in Figure 12.7. It can be observed from Table 12.5 and Figure 12.7 that TGP and LGP have remarkably better generalization capabilities than GEP, GP/OLS, and LSR.

Sensitivity Analysis

Sensitivity analysis is of utmost concern for selecting the important input variables. The contributions of the predictor variables to the prediction of ϕ' are evaluated

Table 12.5 Performance Statistics of the ϕ' Prediction Models for the Testing Data

Model	Performance		
	R	RMSE	MAPE
TGP	0.917	159.75	5.10
LGP	0.930	142.25	4.39
GEP (Kayadelen et al., 2009)	0.879	216.21	5.80
GP/OLS (Mousavi et al., 2012)	0.909	160.61	5.12
LSR (Mousavi et al., 2012)	0.874	194.54	5.90

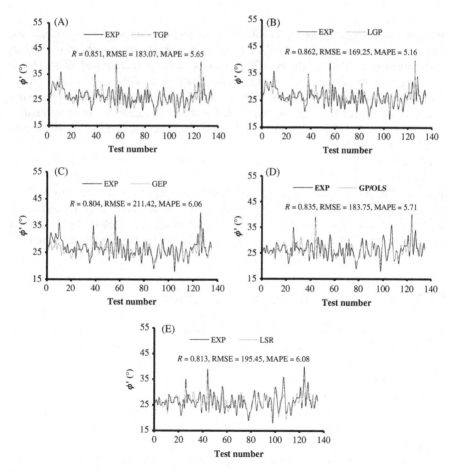

Figure 12.7 A comparison of the ϕ' predictions made by different models for the entire database: (A) LGP, (B) TGP, (C) GEP, (D) GP/OLS, and (E) LSR.

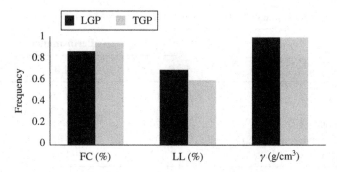

Figure 12.8 Contributions of the predictor variables in the TGP and LGP models.

through a sensitivity analysis. To this aim, frequency values of the variables are obtained. A frequency value equal to 1 for an input indicates that this variable has been appeared in 100% of the best 30 programs evolved by TGP and LGP. This is a common methodology for the sensitivity analysis in GP-based studies (Alavi et al., 2011; Gandomi et al., 2010, 2011d). The frequency values of the input parameters of the correlations are presented in Figure 12.8. According to these results, it can be found that among the three influencing parameters, ϕ' is more sensitive to γ and FC than is LL.

Besides, Figure 12.9 shows the variations of the experimental/predicted ϕ' values with FC, LL, and γ. As the scattering increases in these figures, the accuracy of the model consequently decreases. It can be observed from these figures that the predictions made by the proposed models have a very good accuracy with no significant trend with respect to the design parameters. In the case of LL (see Figure 12.9C), the scattering slightly decreases as this parameter increases.

12.5   Discussion and Future Directions

TGP and LGP introduce completely new characteristics and traits. One of the major advantages of the TGP and LGP approaches over the traditional regression analyses is their ability to derive explicit relationships without assuming prior forms of the existing relationships. The best solutions (equations) evolved by these techniques are determined after controlling numerous preliminary models, even millions of linear and nonlinear models. For instance, the proposed LGP model for the estimation of ϕ' is selected among approximately 413,116,600 programs. This is the sum of the programs evolved and evaluated during the conducted 120 runs.

It is worth mentioning that the GP- and ANN-based approaches are well suited to modeling the complex behavior of most geotechnical engineering problems with extreme variability in their nature (Gandomi and Alavi, 2012; Shahin et al., 2009). Despite some similarities, there are some important differences between GP

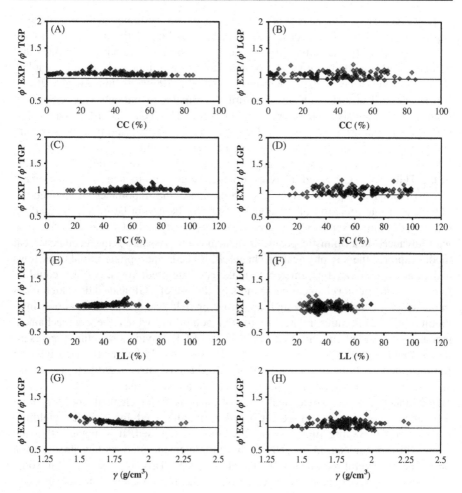

Figure 12.9 The ratio between the experimental and the predicted ϕ' values with respect to the design parameters.

and ANN. ANNs suffer from a few shortcomings, including lack of transparency and knowledge extraction. The main advantage of GP over ANN is that GP generates a transparent and structured representation of the system being studied. An additional advantage of GP over ANN is that determining the ANN architecture is a difficult task. In GP, the number and combination of terms are automatically evolved during model calibration (Gandomi and Alavi, 2012; Shahin et al., 2009).

It is notable that the underlying assumption that the input parameters are reliable is not always the case. Since fuzzy logic can provide a systematic method to deal with imprecise and incomplete information, the process of developing hybrid fuzzy and GP-based models for such problems can be a suitable topic for further studies.

However, one of the goals of introducing expert systems, such as GP-based approaches, into the design processes is better handling of the information in the predesign phase. In the initial steps of design, information about the features and properties of targeted output or process are often imprecise and incomplete (Gandomi and Alavi, 2012; Kraslawski et al., 1999; Shahin et al., 2009). Nevertheless, it is ideal to have some initial estimates of the outcome before performing any extensive laboratory or field work. The TGP and LGP approaches employed in this research are based on the data alone to determine the structure and parameters of the models. Thus, the derived constitutive models can particularly be valuable in the preliminary design stages. For more reliability, the results of the TGP- and LGP-based analyses are suggested to be treated as a complement to conventional computing techniques. In any case, the importance of engineering judgment in interpretation of the obtained results should not be underestimated. In order to develop a sophisticated prediction tool, TGP and LGP can be combined with advanced deterministic geomechanical models. Assuming the geomechanical model captures the key physical mechanisms, it needs appropriate initial conditions and carefully calibrated parameters to make accurate predictions. An idea could be to calibrate the geomechanical parameters by the use of TGP and LGP, which takes into account historic data sets and laboratory or field test results. This allows integrating the uncertainties related to *in situ* conditions which the geomechanical model does not explicitly account for. TGP and LGP provide a structured representation for the constitutive material model that can readily be incorporated into the finite element or finite difference analyses. In this case, it is possible to use a suitably trained GP-based material model instead of a conventional (analytical) constitutive model in a numerical analysis tool such as finite element code or finite difference software (like FLAC). Consequently, the need for complex yielding/plastic potential/failure functions or flow rules is avoided. It is notable that the numerical implementation of ANN in the finite element analyses has already been presented by several researchers (Javadi et al., 2005). This strategy has led to some qualitative improvement in the application of the finite element method in engineering (Javadi and Rezania, 2009).

12.6   Conclusions

In this chapter, the TGP and LGP paradigms were introduced for the behavioral modeling of geotechnical engineering systems. The viabilities of these techniques to model the behavior of the geotechnical phenomena were demonstrated through solving a complex, prediction-type example. The representative problem considered was the assessment of the effective angle of shearing resistance. The results indicated that TGP and LGP are effectively capable of simulating the nonlinear behavior of the investigated systems. LGP had better overall behavior for the analysis of the considered problem than TGP. For more validity verification, the models created by TGP and LGP were applied to a part of the experimental results that were not included in the training process. New criteria were also checked for the external validation of TGP and LGP. The proposed methods provided more accurate predictions of the angle of shearing resistance than the existing methods (i.e., GEP, GP/ OLS, and LSR). TGP and LGP have an advantage that once the evolved models are trained, they can be used as quick and accurate tools for prediction purposes. The verification phases confirmed the effectiveness and robustness of these methods for their future applications to geotechnical problems. A major advantage of utilizing the TGP and LGP methods is that the geotechnical design parameters can be estimated directly from the available experimental data. Thus, there is no need to go through sophisticated and time-consuming field experiments. TGP- and LGP-based models are expected to be very useful for the evaluation of design parameters in the preplanning and predesign stages. TGP and LGP are especially practical for cases where the conventional methods are unable to describe various aspects of the behavior effectively.

References

Alavi, A.H., Gandomi, A.H., 2011a. A robust data mining approach for formulation of geotechnical engineering systems. Eng. Comput. 28 (3), 242–274.

Alavi, A.H., Gandomi, A.H., 2011b. Prediction of principal ground-motion parameters using a hybrid method coupling artificial neural networks and simulated annealing. Comput. Struct. 89 (23–24), 2176–2194.

Alavi, A.H., Gandomi, A.H., 2012. Energy–based models for assessment of soil liquefaction. Geosci. Front 3 (4), 541–555.

Alavi, A.H., Heshmati, A.A.R., Gandomi, A.H., Askarinejad, A., Mirjalili, M., 2008. Utilisation of computational intelligence techniques for stabilised soil. In: Papadrakakis, M., Topping, B.H.V. (Eds.), Engineering Computational Technology. Civil-Comp Press, Edinburgh, UK, paper 175.

Alavi, A.H., Gandomi, A.H., Gandomi, M., Sadat Hosseini, S.S., 2009. Prediction of maximum dry density and optimum moisture content of stabilized soil using RBF neural networks. IES J. A Civ. Struct. Eng. 2 (2), 98–106.

Alavi, A.H., Gandomi, A.H., Mollahasani, A., Heshmati, A.A.R., Rashed, A., 2010a. Modeling of maximum dry density and optimum moisture content of stabilized soil using artificial neural networks. J. Plant Nutr. Soil Sci. 173, 368–379.

Alavi, A.H., Gandomi, A.H., Mousavi, M., Mollahasani, A., 2010b. High-precision modeling of uplift capacity of suction caissons using a hybrid computational method. Geomech. Eng. 2 (4), 253–280.

Alavi, A.H., Gandomi, A.H., Sahab, M.G., Gandomi, M., 2010c. Multiexpression programming: a new approach to formulation of soil classification. Eng. Comput. 26, 111–118.

Alavi, A.H., Aminian, P., Gandomi, A.H., Arab Esmaeili, M., 2011. Genetic-based modeling of uplift capacity of suction caissons. Exp. Syst. Appl. 38 (10), 12608–12618.

Banzhaf, W., Nordin, P., Keller, R., Francone, F.D., 1998. Genetic Programming—An Introduction. On the Automatic Evolution of Computer Programs and Its Application. Dpunkt/Morgan Kaufmann, Heidelberg/San Francisco.

Brameier, M., Banzhaf, W., 2001. A comparison of linear genetic programming and neural networks in medical data mining. IEEE Trans. Evol. Comput. 5 (1), 17–26.

Brameier, M., Banzhaf, W., 2007. Linear Genetic Programming. Springer Science + Business Media, New York, NY.

Cabalar, A.F., Cevik, A., 2009a. Genetic programming—based attenuation relationships: an application of recent earthquakes in Turkey. Comput. Geosci. 35, 1884—1896.

Cabalar, A.F., Cevik, A., 2009b. Modelling damping ratio and shear modulus of sand—mica mixtures using neural networks. Eng. Geol. 104, 31—40.

Cabalar, A.F., Cevik, A., Guzelbey, I.H., 2009. Constitutive modeling of Leighton Buzzard sands using genetic programming. Neural Comput. Appl. 19 (5), 657—665.

Cevik, A., Cabalar, A.F., 2009. Modelling damping ratio and shear modulus of sand—mica mixtures using genetic programming. Exp. Syst. Appl. 36 (4), 7749—7757.

Conrads, M., Dolezal, O., Francone, F.D., Nordin, P., 2004. Discipulus—Fast Genetic Programming Based on AIM Learning Technology. Register Machine Learning Technologies Inc., Littleton, CO.

Francone, F.D., Deschaine, L.M., 2004. Extending the boundaries of design optimization by integrating fast optimization techniques with machine-code-based, linear genetic programming. Inf. Sci. 161, 99—120.

Frank, I.E., Todeschini, R., 1994. The Data Analysis Handbook. Elsevier, Amsterdam, The Netherland.

Gandomi, A.H., Alavi, A.H., 2011. Multi-stage genetic programming: a new strategy to nonlinear system modeling. Inf. Sci. 181 (23), 5227—5239.

Gandomi, A.H., Alavi, A.H., 2012. A new multi-gene genetic programming approach to nonlinear system modeling. Part II: geotechnical and earthquake engineering problems. Neur. Comput. Appl. 21, 189—201.

Gandomi, A.H., Alavi, A.H., Sadat Hosseini, S.S., 2008. Discussion on genetic programming for retrieving missing information in wave records along the west coast of India. Appl. Ocean Res. 30 (4), 338—339.

Gandomi, A.H., Alavi, A.H., Sahab, M.G., 2010. New formulation for compressive strength of CFRP confined concrete cylinders using linear genetic programming. Mater. Struct. 43 (7), 963—983.

Gandomi, A.H., Alavi, A.H., Mirzahosseini, M.R., Moqhadas Nejad, F., 2011a. Nonlinear genetic-based models for prediction of flow number of asphalt mixtures. J. Mater. Civ. Eng. 23 (3), 248—263.

Gandomi, A.H., Alavi, A.H., Mousavi, M., Tabatabaei, S.M., 2011b. A hybrid computational approach to derive new ground-motion prediction equations. Eng. Appl. Artif. Intell. 24 (4), 717—732.

Gandomi, A.H., Tabatabaie, S.M., Moradian, M.H., Radfar, A., Alavi, A.H., 2011c. A new prediction model for load capacity of castellated steel beams. J. Constr. Steel Res. 67 (7), 1096—1105.

Gandomi, A.H., Alavi, A.H., Yun, G.J., 2011d. Nonlinear modeling of shear strength of SFRC beams using linear genetic programming. Struct. Eng. Mech. 38 (1), 1—25.

Goh, A.T.C., 1994. Seismic liquefaction potential assessed by neural networks. J. Geotech. Eng. 120 (9), 1467—1480.

Golbraikh, A., Tropsha, A., 2002. Beware of q^2. J. Mol. Graph. Model. 20, 269—276.

Guven, A., Azamathulla, H.M., Zakaria, N.A., 2009. Linear genetic programming for prediction of circular pile scour. Ocean Eng. 36 (12—13), 985—991.

Jafarian, Y., Kermani, E., Baziar, M.H., 2010. Empirical predictive model for the vmax/amax ratio of strong ground motions using genetic programming. Comput. Geosci. 36 (12), 1523—1531.

Javadi, A.A., Rezania, M., 2009. Applications of artificial intelligence and data mining techniques in soil modeling. Geomech. Eng. 1 (1), 53—74.

Javadi, A.A., Tan, T.P., Elkassas, A.S.I., 2005. Intelligent finite element method. In: Bathe K.J. (Ed.), The Third MIT Conference on Computational Fluid and Solid Mechanics. Cambridge, Massachusetts, USA.

Javadi, A.A., Rezania, M., Mousavi Nezhad, M., 2006. Evaluation of liquefaction induced lateral displacements using genetic programming. Comput. Geotech. 33 (4—5), 222—233.

Johari, A., Habibagahi, G., Ghahramani, A., 2006. Prediction of soil—water characteristic curve using genetic programming. J. Geotech. Geoenviron. Eng. 132 (5), 661—665.

Juang, C.H., Jiang, T., Christopher, R.A., 2001. Three-dimensional site characterisation: neural network approach. Geotechnique. 51, 799—809.

Kayadelen, C., Günaydin, O., Fener, M., Demir, A., Özvan, A., 2009. Modeling of the angle of shearing resistance of soils using soft computing systems. Exp. Syst. Appl. 36, 11814—11826.

Kim, Y.S., Kim, B.T., 2008. Prediction of relative crest settlement of concrete-faced rockfill dams analyzed using an artificial neural network model. Comput. Geotech. 35, 313—322.

Korayem, A.Y., Ismail, K.M., Sehari, S.Q., 1996. Prediction of soil shear strength and penetration resistance using some soil properties. Missouri J. Agri. Res. 13 (4), 119—140.

Koza, J., 1992. Genetic Programming, on the Programming of Computers by Means of Natural Selection. MIT Press, Cambridge, MA.

Kraslawski, A., Pedrycz, W., Nyström, L., 1999. Fuzzy neural network as instance generator for case-based reasoning system: an example of selection of heat exchange equipment in mixing. Neur. Comput. Appl. 8 (2), 106—113.

Levasseur, S., Malécot, Y., Boulon, M., Flavigny, E., 2007. Soil parameter identification using a genetic algorithm. Int. J. Numer. Anal. Methods Geomech. 32, 189—213.

Levasseur, S., Malécot, Y., Boulon, M., Flavigny, E., 2009. Statistical inverse analysis based on genetic algorithm and principal component analysis: method and developments using synthetic data. Int. J. Numer. Anal. Methods Geomech. 33, 1485—1511.

McCombie, P., Wilkinson, P., 2002. The use of the simple genetic algorithm in finding the critical factor of safety in slope stability analysis. Comput. Geotech. 29, 699—714.

Metenidis, M.F., Witczak, M., Korbicz, J., 2004. A novel genetic programming approach to nonlinear system modelling: application to the DAMADICS benchmark problem. Eng. Appl. Artif. Intell. 17, 363—370.

Mousavi, S.M., Alavi, A.H., Gandomi, A.H., Mollahasani, A., 2011a. A hybrid computational approach to formulate soil deformation moduli obtained from PLT. Eng. Geol. 123, 324—332.

Mousavi, S.M., Alavi, A.H., Gandomi, A.H., Mollahasani, A., 2011b. Nonlinear Genetic-Based Simulation of Soil Shear Strength Parameters. J. Earth Syst. Sci. 120 (6), 1001—1022.

Mousavi, S.M., Alavi, A.H., Gandomi, A.H., Mollahasani, A., 2012. Formulation of soil angle of shearing resistance using a hybrid GP and OLS method. Eng. Comput. doi: 10.1007/s00366—011—0242—x (in press).

Murthy, S., 2008. Geotechnical Engineering: Principles and Practices of Soil Mechanics. second ed. CRC Press, Taylor and Francis, UK.

Narendra, B.S., Sivapullaiah, P.V., Suresh, S., Omkar, S.N., 2006. Prediction of unconfined compressive strength of soft grounds using computational intelligence techniques: a comparative study. Comput. Geotech. 33, 196—208.

Oltean, M., Grossan, C., 2003. A comparison of several linear genetic programming techniques. Adv. Complex Syst. 14 (4), 1−29.

Pal, S., Wije Wathugala, G., Kundu, S., 1996. Calibration of a constitutive model using genetic algorithms. Comput. Geotech. 19, 325−348.

Rashed, A., Bolouri Bazaz, J., Alavi, A.H., 2012. Nonlinear modeling of soil deformation modulus through LGP-based interpretation of pressure meter test results. Eng. Appl. Artif. Intell. doi: 10.1016/j.engappai.2011.11.008 (in press).

Rezania, M., Javadi, A.A., 2007. A new genetic programming model for predicting settlement of shallow foundations. Can. Geotech. J. 44 (12), 1462−1473.

Roy, P.P., Roy, K., 2008. On some aspects of variable selection for partial least squares regression models. QSAR Comb. Sci. 27, 302−313.

Shahin, M.A., Maier, H.R., Jaksa, M.B., 2001. Artificial neural network applications in geotechnical engineering. Aust. Geomech. 36 (1), 49−62.

Shahin, M.A., Jaksa, M.B., Maier, H.R., 2008. State of the art of artificial neural networks in geotechnical engineering. Electron. J. Geotech. Eng. 8, 1−26. <http://www.ejge.com/Bouquet08/Shahin>

Shahin, M.A., Jaksa, M.B., Maier, H.R., 2009. Recent advances and future challenges for artificial neural systems in geotechnical engineering applications. Adv. Artif. Neural Syst. vol. 2009, Article ID 308239, 9 pages. doi:10.1155/2009/308239.

Shahnazari, H., Dehnavi, Y., Alavi, A.H., 2010. Numerical modeling of stress−strain behavior of sand under cyclic loading. Eng. Geol. 116 (1−2), 53−72.

Silva, S., 2007. GPLAB, a genetic programming toolbox for MATLAB. Available from <http://gplab.sourceforge.net> Last accessed date: January 2010.

Simpson, A.R., Priest, S.D., 1993. The application of genetic algorithms to optimisation problems in geotechnics. Comput. Geotech. 15, 1−19.

Smith, G.N., 1986. Probability and Statistics in Civil Engineering. Collins, London.

Torres, R.S., Falcão, A.X., Gonçalves, M.A., Papa, J.P., Zhang, B., Fan, W., et al., 2009. A genetic programming framework for content-based image retrieval. Pattern Recognit. 42, 283−292.

Yang, C.X., Feng, X.T., 2005. Evolutionary self-organizing identification of nonlinear dynamics of landslides. Chin. J. Rock Mech. Eng. 24 (6), 911−914.

Yang, C.X., Tham, L.G., Feng, X.T., Wang, Y.J., Lee, P.K.K., 2004. Two-stepped evolutionary algorithm and its application to stability analysis of slopes. J. Comput. Civ. Eng. 18, 145−153.

13 An EPR Approach to the Modeling of Civil and Geotechnical Engineering Systems

Akbar A. Javadi[1], Alireza Ahangar-Asr[1], Asaad Faramarzi[2] and Nasim Mottaghifard[1]

[1]Computational Geomechanics Group, College of Engineering, Mathematics and Physical Sciences, University of Exeter, UK
[2]Department of Civil Engineering, School of Engineering, University of Greenwich, UK

13.1 Introduction

Some engineering problems lack precise analytical theories or models for their solutions. This is usually because of inadequate understanding of the phenomena involved and the factors affecting them, as well as a limited quantity and poor quality of available information. In order to cope with the complexity of engineering problems, traditional forms of engineering design solutions have been widely developed. The information has been usually collected, synthesized, and presented in the form of design charts, tables, or empirical formulas (Rezania et al., 2008).

In recent years, computers have become an inseparable integral part of everyday engineering computations and design activities. Through rapid developments in computer software and hardware in the past few decades, several alternative computer-aided data classification and pattern recognition methods have been developed. The main idea behind a pattern recognition system (e.g., neural networks or fuzzy logic) is that it learns adaptively from experience and extracts various discriminants, each appropriate for its purpose. Genetic programming (GP) has been used in modeling different engineering problems (Alavi and Gandomi, 2011a, b; Gandomi and Alavi, 2012a, b; Gandomi et al., 2010); however, artificial neural networks (ANNs) are the most widely used pattern recognition system to capture nonlinear interactions between various parameters in complex systems. So far, ANNs have been used for a wide range of civil engineering disciplines, such as in geotechnical engineering (Abu-Kiefa, 1998; Alavi and Gandomi, 2011a, b; Javadi, 2006; Juang, et al., 2001), structural engineering

Metaheuristics in Water, Geotechnical and Transport Engineering. DOI: http://dx.doi.org/10.1016/B978-0-12-398296-4.00013-1

(Ankireddi and Yang, 1999; Feng and Bahng, 1999; Huang and Loh, 2001), con-struction engineering (Adeli and Karim, 1997; Adeli and Wu, 1998; Arditi et al., 1998), environmental and water resources engineering (Coulibaly et al., 2000; Liu and James, 2000; Thirumalaiah and Deo, 1998), and transportation engineering (Celikoglu and Cigizoglu, 2007; Gagarin et al., 1994), among many others (Alavi et al., 2010; Javadi and Rezania, 2009).

A neural network consists of a large number of interconnected processing ele-ments, commonly referred to as *neurons*. The neurons are arranged into two or more layers and interact with each other via weighted connections. The data are presented to the neural network using an input layer; and an output layer holds the response of the network to the input. The input–output relationship is captured by repeatedly presenting examples of the input–output data sets to the ANN and adjusting the model coefficients (i.e., connection weights) in an attempt to mini-mize an error function between the desired outputs and the outputs predicted by the model (Javadi et al., 2005).

Although it has been shown by many researchers that ANNs offer great advan-tages in the analysis of many engineering applications, they are also known to suf-fer from a number of drawbacks. One of these is that the optimum structure of the network (such as number of inputs, hidden layers, and transfer functions) must be identified *a priori*, which is usually done through a time-consuming trial-and-error procedure. Furthermore, the main disadvantage of the neural network–based mod-els is the large complexity of the network structure, as it represents the knowledge in terms of a weight matrix that is not accessible to the user.

In this research, a new data mining technique is used to model some engineering systems. This new technique is called *evolutionary polynomial regression* (EPR), and it uses evolutionary searching to find polynomial expressions that represent the behavior of a system. Previous applications of EPR have proved to be effective in the fields of environmental modeling (Giustolisi et al., 2007) and water system management (Savic et al., 2006). The capabilities of the EPR technique will be demonstrated here by applying it to a number of practical examples.

An important application of material modeling is the numerical analysis of boundary value problems, and recently, it has been shown that neural net-work–based constitutive models can be practically incorporated in a finite element code as a material model (Hashash et al., 2004, 2011; Osouli et al., 2010; Savic et al., 2006). Javadi and his colleagues carried out extensive research on the appli-cation of neural networks in constitutive modeling of complex materials in general and soils in particular. They have developed an ANN-based finite element model (NeuroFE code) based on the integration of a back-propagation neural network in finite element analysis. The ANN-based finite element model has been applied to a wide range of boundary value problems, including several geotechnical applica-tions (Javadi et al., 2003, 2004a, b, 2009) and has shown that neural networks can be very effective in learning and generalizing the constitutive behavior of complex materials. The third example in this research will be presented to demonstrate the capabilities of the EPR-based models in constitutive modeling of materials in finite element analysis.

13.2 Evolutionary Polynomial Regression

EPR is a data-driven method based on evolutionary computing, aimed to search for polynomial structures representing a system. It integrates numerical and symbolic regression to perform EPR. The strategy uses polynomial structures to take advantage of their favorable mathematical properties. The key idea behind the EPR is to use evolutionary search for exponents of polynomial expressions by means of a genetic algorithm (GA) engine. This allows (i) easy computational implementation of the algorithm, (ii) efficient search for an explicit expression, and (iii) improved control of the complexity of the expression generated (Giustolisi and Savic, 2006). A physical system, having an output y, dependent on a set of inputs $\mathbf{X}$, and parameters θ, can be mathematically formulated as

$$y = F(\mathbf{X}, \theta) \tag{13.1}$$

where F is a function in an m-dimensional space and m is the number of inputs. To avoid the problem of the length of mathematical expressions growing rapidly with time, in EPR, the evolutionary procedure is conducted in a way that it searches for the exponents of a polynomial function with a fixed maximum number of terms. During one execution, EPR returns a number of expressions with increasing numbers of terms up to a limit set by the user, to allow the optimum number of terms to be selected. The general form of expression used in EPR can be presented as (Giustolisi and Savic, 2006):

$$y = \sum_{j=1}^{m} F(\mathbf{X}, f(\mathbf{X}), a_j) + a_0 \tag{13.2}$$

where y is the estimated vector of output of the process, a_j is a constant, F is a function constructed by the process, $\mathbf{X}$ is the matrix of input variables, f is a function defined by the user, and m is the number of terms of the target expression. The first step in identification of the model structure is to transfer Eq. (13.2) into the following vector form:

$$Y_{N \times 1}(\theta, Z) = [I_{N \times 1} Z_{N \times m}^{j}] \times [a_0 \, a_1 \ldots a_m]^{\mathrm{T}} = Z_{N \times d} \times \theta_{d \times 1}^{\mathrm{T}} \tag{13.3}$$

where $\mathbf{Y}_{N \times 1}(\theta, \mathbf{Z})$ is the least squares (LS) estimate vector of the N target values; $\theta_{1 \times d}$ is the vector of $d = m + 1$ parameters a_j and a_0 (θ^{T} is the transposed vector); and $\mathbf{Z}_{N \times d}$ is a matrix formed by $\mathbf{I}$ (unitary vector) for bias a_0 and m vectors of variables $\mathbf{Z}^{j}$. For a fixed j, the variables $\mathbf{Z}^{j}$ are a product of the independent predictor vectors of inputs, $\mathbf{X} = \langle \mathbf{X}_1 \mathbf{X}_2 \cdots \mathbf{X}_k \rangle$.

In general, EPR follows a two-stage procedure for constructing symbolic models. Initially, using a standard GA, it searches for the best form of the function structure (i.e., a combination of vectors of independent inputs, $X_{s=1:k}$); second, it performs an LS regression to find the adjustable parameters θ for each combination of inputs. In this way, a global search algorithm is implemented for both the best set of input combinations and related exponents simultaneously, according to the user-defined cost function (Giustolisi and Savic, 2006). The adjustable parameters a_j are evaluated by means of the linear LS method based on minimization of the sum of squared errors (SSE) as the cost function. The SSE function, which is used to guide the search process toward the best-fit model, is as follows:

$$SSE = \frac{\sum_{i=1}^{N}(y_a - y_p)^2}{N} \tag{13.4}$$

where y_a is the target values in the training data set and y_p is the model predictions. The global search for the best form of the EPR equation is performed by means of a standard GA over the values in the user-defined vector of exponents. The GA operates based on Darwinian evolution, which begins with the random creation of an initial population of solutions. Each parameter set in the population represents the individual's chromosomes, and each individual is assigned a fitness based on how well it performs in its environment. Through crossover and mutation operations, with the probabilities P_c and P_m, respectively, the next generation is created. Fit individuals are selected for mating, whereas weak individuals die off. The mated parents create a child (offspring) with a chromosome set that is a mix of the parents' chromosomes. In EPR, integer GA coding with a single-point crossover is used to determine the location of the candidate exponents.

The EPR process stops when the termination criterion, which can be either the maximum number of generations, the maximum number of terms in the target mathematical expression, or a particular allowable error, is satisfied. A typical flow diagram for the EPR procedure is illustrated in Figure 13.1.

In the evolutionary process of building EPR models, a number of constraints can be implemented to control the output models in terms of the type of functions used, number of terms, range of exponents, number of generations, etc. In this process, there is the potential to achieve different models for a particular problem, which enables the user to gain additional information for different scenarios (Rezania et al., 2008). Applying the EPR procedure, the evolutionary process starts from a constant mean of output values. By increasing the number of evolutions, it gradually picks up different participating parameters in order to form equations describing the relationship between the parameters of the system. Each proposed model is trained using the training data and tested using the testing data provided. The level of accuracy at each stage is evaluated based on the coefficient of determination (COD); i.e., the fitness function is

$$COD = 1 - \frac{\sum_N (Y_a - Y_p)^2}{\sum_N (Y_a - (1/N)\sum_N Y_a)} \tag{13.5}$$

where Y_a is the actual output value, Y_p is the EPR-predicted value, and N is the number of data on which COD is computed. If the model fitness is not acceptable or the other termination criteria (in terms of maximum number of generations and maximum number of terms) are not satisfied, the current model goes through another evolution in order to obtain a new model.

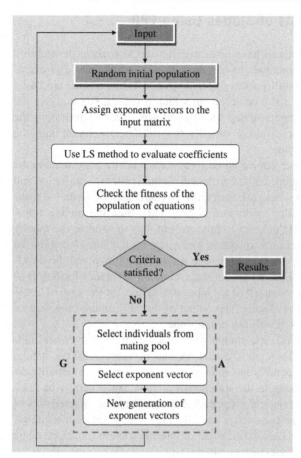

Figure 13.1 General representation of the EPR procedure.

13.3   Data Preparation

In every case, to select the most robust representation, a statistical analysis was performed on the input and output parameters on several randomly selected training and validation data sets. The aim of this analysis was to ensure that the statistical properties of the data in each of the subsets were as close to the others as possible and thus represented similar statistical populations (Rezania and Javadi, 2008). The testing sets of data were chosen in a way that all parameters were in the range between the maximum and the minimum values in the training data set. The minimum, maximum, mean, and standard deviation values were calculated for all contributing parameters for different random combinations of training and testing data sets. Among these cases, the most statistically consistent ones were chosen to be used in training and validation of the EPR models.

13.4   Stability Analysis of Slopes Using EPR

Although the conventional methods have been widely used to analyze the stability of soil and rock slopes, these methods have a number of shortcomings. For example, the existing methods of stability analysis for slopes on cohesive soils are based on (i) assuming a slip surface and a center about which it rotates, (ii) studying the equilibrium of the forces/moments acting on this surface, and (iii) repeating the analysis on several different trial failure surfaces from different centers, until the most critical slip surface is found.

A number of researchers have implemented some metaheuristic algorithms in the stability analysis of slopes (Samui, 2008; Samui and Kothari, 2011). Sakellariou and Ferentinou (2005) used a neural network to acquire a relationship between the parameters involved in analyzing the stability of slopes. They used the models introduced by Hoek and Bray (1981) in order to produce test data to validate the quality of training of the ANN model. In this research, a new approach is introduced for the analysis of the stability of slopes using EPR. The methodology involved the development and verification of an EPR model for determination of factor of safety (FS) for soil slopes. The input data consisted of six input parameters for the case of circular failure mechanism for cohesive soils. The output of the EPR model presented an FS that demonstrated the status of stability of the slope.

Two data sets consisting of 67 case studies of slopes with a circular critical failure mechanism were used in this study (Sakellariou and Ferentinou, 2005). A total of 25 cases involved dry soil conditions, and the other 42 were in wet conditions. The main parameters contributing to the stability of a slope can generally be categorized in two classes of geotechnical properties and geometrical characteristics of the slope. More specifically, the parameters used for the circular failure mechanism in soils were unit weight (γ), cohesion (c), angle of internal friction (ϕ), slope angle (β), height (H), and pore water pressure parameter (r_u).

The data was divided into two sets: the training set (57 out of 67 cases) used for developing the EPR model and the testing set (10 out of 67 cases) kept for validation and evaluation of the generalization capabilities of the developed EPR model. Among the resulting equations developed by the EPR process, the one with the highest COD was selected:

$$F_s = -\frac{1.49H}{\gamma^2} - 1.8 \cdot r_u^2 + \tan(\varphi)[2.59 - 2.18\tan(\beta)]$$
$$+ 0.014 \cdot c - 5.19 \times 10^{-5}c^2 + 0.817 \tag{13.6}$$

Figure 13.2 shows the comparison of the results in terms of FS predicted by the EPR model together with the ones from ANN analysis (Sakellariou and Ferentinou, 2005) and the field data for the training cases. The results of the EPR model are in close agreement with the field data, as well as those predicted by the ANN model.

After training, the performance of the trained EPR model was validated based on the validation data that was not used during the model development process. Equation (13.6) was used to predict the FS for the unseen data cases, and the results are shown

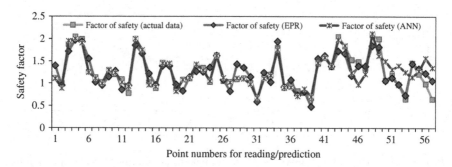

Figure 13.2 EPR training results versus ANN and field measurements for soil slope analysis.

in Figure 13.3. A very good agreement can be seen between the model results and the field data demonstrating the excellent capability of the EPR-based model in generalizing the relationship with unseen cases. The COD values for the developed EPR model, as well as the ANN model, are shown in Table 13.1. It is shown that the EPR model outperforms the ANN model in terms of both finding the COD values for the training and testing and providing a transparent and easy-to-implement expression.

A parametric study was carried out to evaluate the prediction capabilities of the proposed EPR model, the extent to which it represents the physical relationships between different parameters, and the effects of different input parameters on the output model. This was done through a basic approach to sensitivity analysis that is to set all but one input variable to their mean values and vary the remaining one within the range of its maximum and minimum values. This procedure was repeated consecutively for all input parameters, and the results are shown in Figure 13.4. These results indicate that the developed EPR model has been able to capture, with very high accuracy, the important physical patterns of behavior of slopes, and the relationship between the slope stability and its contributing factors.

13.5   EPR Modeling of the Behavior of Rubber Concrete

In recent years, much research has been carried out to investigate the possibilities for the reuse of abandoned tires by grinding them into small particles (crumb rubber or tire chips) and using in asphalts, sealants, and rubber sheets. Of particular interest has been the use of waste tires as aggregate in Portland cement concrete (Sukontasukkul and Chaikaew, 2006). Studies revealed that the addition of rubber aggregates leads to the reduction in the basic engineering properties of concrete, and the reduction in strength appears to be more remarkable as the rubber content in the composite increases (Benazzouk et al., 2003; Eldin and Senouci, 1993; Khatip and Bayomy, 1999; Topcu, 1995). In this research, EPR is introduced as a new approach to model the compressive strength of rubber concrete.

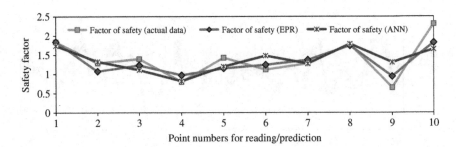

Figure 13.3 EPR testing results versus ANN and field measurements for soil slope analysis.

Table 13.1 COD Values (Soil Slope Analysis Example)

Model	COD Training (%)	COD Testing (%)
ANN	97.6	93.7
EPR	98.3	97.1

Data from an experimental study (Guneyisi et al., 2004) was used to develop an EPR model. Out of the 70 experimental data cases, 56 were used to train the EPR model while the remaining data were used to validate the developed model. The same training and testing data sets as those used by Guneyisi et al. (2004) for developing ANN and GP models were considered, so that direct comparison between the results of the EPR model with those of ANN and GP models became possible.

The computational time for the model development with an Intel Core 2 Quad CPU Q6600 @ 2.40 GHz processor was about 4 min. Among the resultant models developed using EPR, the one with the highest value of COD was selected to represent the compressive strength (f_c) of rubber concrete:

$$f_c = -\frac{986.15FA^3}{SP^2 \cdot CA^3 \cdot (W/C)^3} + 6.59 \times 10^{-3} \left(SP \cdot CR \left(\frac{W}{C} \right) \right)^{0.5} \cdot \left(\frac{CA}{FA} \right)^3 TC$$
$$- \frac{379.13FA^3}{CA^3 \cdot SP^2} \left(\frac{W}{C} \right)^2 \cdot \left(\frac{SF \cdot CR}{TC} \right)^{0.5}$$
$$- 1.45 \left(SP \left(\frac{W}{C} \right) \right)^{0.5} \cdot \left(\frac{FA}{CA} \right)^3 TC + 100.21 \tag{13.7}$$

where C, SF, W, SP, CA, FA, CR, and TC are cement, silica fume, water, superplasticizer, coarse aggregate, fine aggregate, crumb rubber, and tire chip contents, respectively.

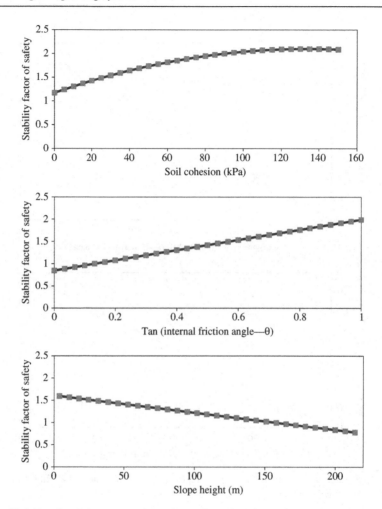

Figure 13.4 Results of the parametric study conducted on the EPR model developed for soil slope analysis.

Figure 13.5 shows the comparison between the results of the EPR model predictions with the experimental data for the training and testing data cases, and a very close agreement between the EPR model predictions and the experimental data can be seen. The CODs for EPR, linear regression, ANN, and GP (Gesoglu et al., 2009) techniques are presented in Table 13.2. The results indicate that EPR is able to model the compressive strength of rubber concrete with high accuracy.

To verify the proposed model, a sensitivity analysis was conducted in a way similar to what was described in the case of the slope stability model. The results show that increasing the amount of fine-grained aggregate and tire chips decreases the compressive strength of the mixture, but any increase in the coarse-grained aggregate content improves the compressive strength of the rubber concrete

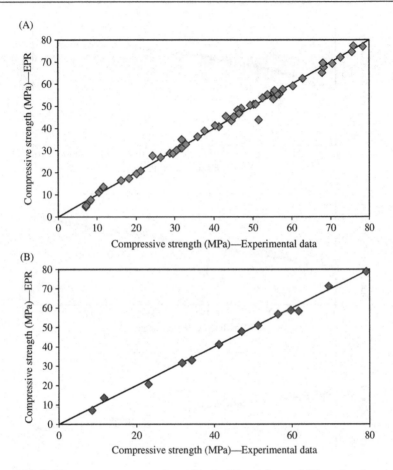

Figure 13.5 Rubber concrete compressive strength: (A) training and (B) testing data cases compared to the EPR model.

Table 13.2 COD Values (%) for Different Models

Different Models	COD Value (f_c)—Testing
LR	86.89
GP	98.18
ANN	99.94
EPR	99.5

(Figure 13.6), which was consistent with the expected behavior. It can be seen that the EPR model, with the advantage of being directly developed from experimental data, is capable of capturing and representing the complex mechanical behavior of rubber concrete.

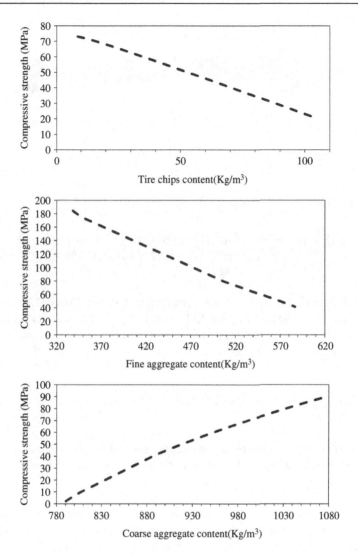

Figure 13.6 Sensitivity analysis results for EPR compressive strength model (rubber concrete).

13.6   Application of EPR in Constitutive Modeling of Materials

A plane stress beam (Figure 13.7) under a uniformly distributed load is considered here. ABAQUS (a commercial finite element software) was used to create data corresponding to a material with elastoplastic behavior, and the data was used to develop and validate the EPR-based material model. After training, the

UDL

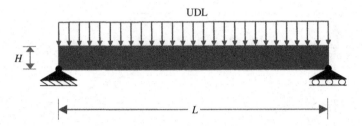

Figure 13.7 Beam under uniformly distributed load (hinge supports).

following equation was selected as the best EPR model based on the COD value:

$$\sigma = 2.17 \times 10^{11}\varepsilon - 6.3 \times 10^{15}\varepsilon^3 + 1.03 \times 10^{18}\varepsilon^4 - 8.03 \times 10^{19}\varepsilon^5 + 3.44$$
$$\times 10^{21}\varepsilon^6 - 8.02 \times 10^{22}\varepsilon^7 + 8.77 \times 10^{23}\varepsilon^8 - 2.54 \times 10^{24}\varepsilon^9 - 6.5 \times 10^6$$

$$(13.8)$$

where ε is the strain and σ is the corresponding stress. Figure 13.8 shows the stress–strain curve predicted by the EPR model (Eq. (13.8)), together with the data used in the model development and validation. It is seen that the EPR model has captured the nonlinear stress–strain behavior of the material with very high accuracy.

The material model is used to provide the material stiffness matrix. For infinitesimal strain increments ($d\varepsilon$), J is the Jacobian continuum:

$$J = \frac{\partial(d\sigma)}{\partial(d\varepsilon)} \tag{13.9}$$

Equation (13.9) was employed to build the material stiffness matrix. The constitutive relationships are generally given in the form of (Owen and Hinton, 1980):

$$\Delta\sigma = \mathbf{D}\Delta\varepsilon \tag{13.10}$$

where $\mathbf{D}$ is the material stiffness matrix. For an elastic and isotropic material matrix, $\mathbf{D}$ is represented in terms of Young's modulus (E) and Poisson's ratio (v) (Stasa, 1986). The developed EPR constitutive model (Eq. (13.8)) was used to describe the material behavior in the finite element analysis. Linear behavior was assumed for small load increments in the nonlinear finite element analysis, and the following equation was used to calculate the tangential elastic modulus of the material:

$$E_t = \frac{d\sigma}{d\varepsilon} = 2.17 \times 10^{11} - 1.89 \times 10^{16}\varepsilon^2 + 4.14 \times 10^{18}\varepsilon^3 - 4.02 \times 10^{20}\varepsilon^4$$

$$+ 2.06 \times 10^{22}\varepsilon^5 - 5.62 \times 10^{23}\varepsilon^6 + 7.01 \times 10^{24}\varepsilon^7 - 2.27 \times 10^{25}\varepsilon^8$$

$$(13.11)$$

The stiffness matrix was developed using the elastic Young's modulus (Eq. (13.11)) and used to conduct the finite element analysis. To evaluate the proposed methodology,

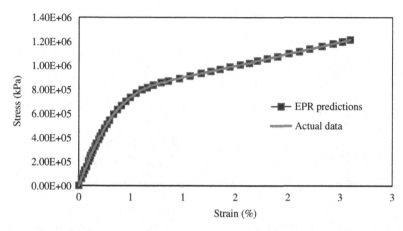

Figure 13.8 EPR model predictions compared to the original data.

displacement of the midspan of the beam predicted using the EPR-based finite element analysis was compared with that of conventional finite element analysis results. Figure 13.9 shows the load–displacement curves in the middle point of the beam obtained using both the elastoplastic finite element analysis and the proposed EPR-based finite element model. It is shown that the results of the EPR-based finite element model closely agree with the ones from elastoplastic finite element analysis.

13.7 Summary and Conclusion

Most of the current analysis and design processes in civil engineering involve the use of conventional/empirical techniques to evaluate the field or experimental data that involve complex relationships between various parameters. The traditional methods usually suffer from the lack of physical understanding, and the simplifying assumptions that are usually made to develop the traditional methods may lead to large errors in some cases. A number of alternative pattern recognition techniques like ANNs have recently begun to be implemented in the analysis of engineering problems. These methods have the advantage that they do not require any simplifying assumptions in developing the model. However, the neural network–based models also suffer from a number of shortcomings, including (i) their inability to present an explicit relationship between the input and the output parameters, (ii) the fact that they require the structure of the neural network (e.g., number of inputs, kernel type, transfer functions, and number of hidden layers) to be identified *a priori*, and (iii) that the optimum structure and the parameters of the network are obtained by trial and error.

In this contribution, a new approach was introduced for the analysis of complex civil engineering problems using EPR. The capabilities of the EPR methodology were illustrated by applying it to two practical problems involving the prediction of the stability status of slopes and the compressive strength of rubber concrete.

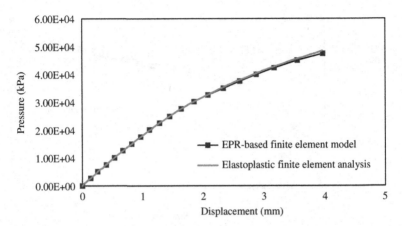

Figure 13.9 EPR-based and elastoplastic finite element analysis results of the plane stress supported beam.

The proposed EPR models generate a transparent and structured representation of the system allowing physical interpretation of the problem, which gives the user an insight into the relationship between input and output data. In the EPR approach, no preprocessing of the data is required and there is no need for normalization or scaling of the data. Another major advantage of the EPR approach is that as more data becomes available, the quality of the prediction can be improved by retraining the EPR model using the new data. However, it should be noted that the EPR models should not be used for extrapolation and if used, the predicted results should be taken with caution and allowance should be made for the uncertainty.

Implementation of the EPR-based model in the finite element analysis was also presented. EPR model was used to create the stiffness matrix for implementation in finite element analysis. The results of the analysis were compared to those obtained from an elastoplastic finite element analysis. Comparison of the results showed very good agreement between the conventional elastoplastic finite element analysis and the proposed EPR-based finite element approach. The results also showed the capability of the EPR-based models in representing the material constitutive behavior in the numerical analysis of boundary value problems.

References

Abu-Kiefa, M.A., 1998. General regression neural networks for driven piles in cohesionless soils. ASCE J. Geotech. Geoenviron. Eng. 124 (12), 1177−1185.

Adeli, H., Karim, A., 1997. Scheduling/cost optimization and neural dynamics model for construction. ASCE J. Constr. Eng. Manage. 123 (4), 450−458.

Adeli, H., Wu, M., 1998. Regularization neural network for construction cost estimation. ASCE J. Constr. Eng. Manage. 124 (1), 18−24.

Alavi, A.H., Gandomi, A.H., 2011a. A robust data mining approach for formulation of geo-technical engineering systems. Eng. Comput. 28 (3), 242−274.

Alavi, A.H., Gandomi, A.H., 2011b. Prediction of principal ground-motion parameters using a hybrid method coupling artificial neural networks and simulated annealing. Comput. Struct. 89 (23−24), 2176−2194.

Alavi, A.H., Gandomi, A.H., Mollahasani, A., Heshmati, A.A.R., 2010. Modeling of maximum dry density and optimum moisture content of stabilized soil using artificial neural networks. J. Plant Nutr. Soil Sci. 173 (3), 368−379.

Ankireddi, S., Yang, H.T.Y., 1999. Neural networks for sensor fault correction in structural control. ASCE J. Struct. Eng. 125 (9), 1056−1064.

Arditi, D., Oksay, F.E., Tokdemir, O.B., 1998. Predicting the outcome of construction litigation using neural networks. Comput. Aided Civ. Infrastruct. Eng. 13 (2), 75−81.

Benazzouk, A., Douzane, O., Queneudec, M., 2003. Effect of rubber aggregates on physico-mechanical behaviour of cement−rubber composites-influence of the alveolar texture of rubber aggregates. Cem. Concr. Compos. 25, 711−720.

Celikoglu, H.B., Cigizoglu, H.K., 2007. Public transportation trip flow modeling with generalized regression neural networks. J. Adv. Eng. Softw. 38 (2), 71−79.

Coulibaly, P., Anctil, F., Bobee, B., 2000. Neural network based long term hydropower forecasting system. Comput. Aided Civ. Infrastruct. Eng. 15 (5), 355−364.

Eldin, N.N., Senouci, A.B., 1993. Rubber tyre particles as concrete aggregate. J. Mater. Civ. Eng. ASCE. 5 (4), 478−496.

Feng, M.Q., Bahng, E.Y., 1999. Damage assessment of jacketed RC columns using vibration tests. ASCE J. Struct. Eng. 125 (3), 256−271.

Gagarin, N., Flood, I., Albrecht, P., 1994. Computing truck attributes with artificial neural networks. ASCE J. Comput. Civ. Eng. 8 (2), 179−200.

Gandomi, A.H., Alavi, A.H., 2012a. A new multi-gene genetic programming approach to nonlinear system modeling. Part I: materials and structural engineering problems. Neural Comput. Appl. 21, 171−187.

Gandomi, A.H., Alavi, A.H., 2012b. A new multi-gene genetic programming approach to non-linear system modeling. Part II: geotechnical and earthquake engineering problems. Neural Comput. Appl. 21, 189−201.

Gandomi, A.H., Alavi, A.H., Sahab, M.G., Arjmandi, P., 2010. Formulation of elastic modulus of concrete using linear genetic programming. J. Mech. Sci. Tech. 24 (6), 1011−1017.

Gesoglu, M., Guneyisi, E., Ozturan, T., Ozbay, E., 2009. Modeling the mechanical behavior of rubberized concretes by neural networks and genetic programming. Mater. Struct. 43, 31−45, online publication.

Giustolisi, O., Savic, D., 2006. A symbolic data-driven technique based on evolutionary polynomial regression. J. Hydroinf. 8 (3), 207−222.

Giustolisi, O., Doglioni, A., Savic, D.A., Webb, B., 2007. A multi-model approach to analysis of environmental phenomena. Environ. Model. Softw. 22 (5), 674−682.

Guneyisi, E., Gesoglu, M., Ozturan, T., 2004. Properties of rubberized concretes containing silica fume. Cem. Concr. Res. 34 (12), 2309−2317.

Hashash, Y.M., Jung, S., Ghaboussi, J., 2004. Numerical implementation of a neural network based material model in finite element analysis. Int. J. Numer. Methods Eng. 5, 989−1005.

Hashash, Y.M.A., Song, H., Osouli, A., 2011. Three-dimensional inverse analyses of a deep excavation in Chicago clays. Int. J. Numer. Anal. Methods Geomech. 35 (9), 1059−1075.

Hoek, E., Bray, J.W., 1981. Rock Slope Engineering. Institution of Mining and Metallurgy, London.

Huang, C.C., Loh, C.H., 2001. Nonlinear identification of dynamic systems using neural networks. Comput. Aided Civ. Infrastruct. Eng. 16 (1), 28−41.

Javadi, A.A., 2006. Estimation of air losses in compressed air tunneling using neural network. J. Tunnell. Undergr. Space Technol. 21 (1), 9−20.

Javadi, A.A., Rezania, M., 2009. Applications of artificial intelligence and data mining techniques in soil modelling. Geomech. Eng. 1 (1), 53−74.

Javadi, A.A., Tan, T.P., Zhang, M.X., 2003. Neural network for constitutive modelling in finite element analysis. Comput. Assisted Mech. Eng. Sci. 10, 523−529.

Javadi, A.A., Tan, T.P., Elkassas, A.S.I., 2004a. An Intelligent Finite Element Method. Weimar, s.n.

Javadi, A.A., Tan, T.P., Elkassas, A.S.I., Zhang, M., 2004b. Intelligent Finite Element Method: Development of the Algorithm. Beijing, s.n.

Javadi, A.A., Farmani, R., Tan, T.P., 2005. An intelligent self-learning genetic algorithm; development and engineering applications. J. Adv. Eng. Inf. 19, 255−362.

Javadi, A.A., Tan, T.P., Elkassas, A.S.I., 2009. Intelligent finite element method and application to simulation of behavior of soils under cyclic loading. Foundations of Computational Intelligence. Springer Verlag, Germany, pp. 317−338.

Juang, C.H., Jiang, T., Christopher, R.A., 2001. Three-dimensional site characterisation: neural network approach. Geotechnique. 51 (9), 799−809.

Khatip, Z.K., Bayomy, F.M., 1999. Rubberized Portland cement concrete. J. Mater. Civ. Eng. ASCE. 11 (3), 206−213.

Liu, W., James, C.S., 2000. Estimation of discharge capacity in meandering compound channels using artificial neural networks. Can. J. Civ. Eng. 27, 297−308.

Osouli, A., Hashash, Y.M.A., Song, H., 2010. Interplay between field measurements and soil behavior for capturing supported excavation response. ASCE J. Geotech. Geoenviron. Eng. 36 (1), 69−84.

Owen, D.R.J., Hinton, E., 1980. Finite Elements in Plasticity: Theory and Practice. Pineridge Press, Swansea.

Rezania, M., Javadi, A.A., 2008. Settlement Prediction of Shallow Foundations: A New Approach. Dundee, s.n.

Rezania, M., Javadi, A., Giustolisi, O., 2008. An evolutionary-based data mining technique for assessment of civil engineering systems. J. Eng. Comput. 25 (6), 500−517.

Sakellariou, M.G., Ferentinou, M.D., 2005. A study of slope stability prediction using neural networks. Geotech. Geol. Eng. 23, 419−445.

Samui, P., 2008. Slope stability analysis: a support vector machine approach. Environ. Geol. 56 (2), 255−267.

Samui, P., Kothari, D.P., 2011. Utilization of a least square support vector machine (LSSVM) for slope stability analysis. Sci. Iran. 18 (1), 53−58.

Savic, D.A., et al., 2006. Modeling sewers failure using evolutionary computing. Proc. ICE Water Manage. 159 (2), 111−118.

Stasa, F.L., 1986. Applied Finite Element Analysis for Engineers. CBS College Publishing, New York, NY.

Sukontasukkul, P., Chaikaew, C., 2006. Properties of concrete pedestrian block mixed with crumb rubber. Construct. Build. Mater. 20, 450−457.

Thirumalaiah, K., Deo, M.C., 1998. Real-time flood forecasting using neural networks. Comput. Aided Civ. Infrastruct. Eng. 13 (2), 101−111.

Topcu, I.B., 1995. The properties of rubberized concretes. Cem. Concr. Res. 25 (2), 304−310.

14 Slope Stability Analysis Using Multivariate Adaptive Regression Spline

Pijush Samui

Centre for Disaster Mitigation and Management, VIT University, Vellore, Tamil Nadu, India

14.1 Introduction

In civil engineering, a slope is an unsupported, inclined surface of soil and/or rock. Slopes are formed for railway formations, highway embankments, earth dams, canal banks, levees, and many other locations. The failure of slope is a major concern in civil engineering, so the determination of the stability of slope is a major task. The design of slope is often carried out with the use of stability numbers, as originally introduced by Taylor (1948). The charts, providing the variation of stability numbers, are available in the literature for homogeneous soil slopes (Chen, 1975; Michalowski, 1994, 2002; Taylor, 1948). Geotechnical engineers use limit equilibrium to determine slope stability (Bishop, 1955; Bishop and Morgenstern, 1960; Fellenius, 1936; Morgenstern and Price, 1965). However, a major disadvantage of this method is that it does not address the issue of kinematics (Kumar and Samui, 2006). A number of investigations have been performed recently that deal with the stability of slopes using upper-bound limit analysis (Chen and Liu, 1990; Chen et al., 1969; Karal, 1977a,b; Kumar, 2000, 2004; Michalowski, 1994, 1995, 2002). Fellenius (1936) used the method of slices to assess the stability of slopes. In order to solve the stability problem, Fellenius assumed that the result of inter-slice forces acts in a direction parallel to the base of each slice. It was seen that this method generally provides a conservative estimate of the factor of safety (FOS). Taylor (1948) used the friction circle method to obtain the stability numbers (N_s) for homogeneous soil slopes; the stability number (N_s) was defined by the expression $N_s = (\gamma H_c/c)$, where H_c is the critical height of the slope (on the verge of failure) that is associated with the critical failure surface. Taylor provided the charts indicating the variation of stability number (N_s) for homogenous slopes with changes in slope angle (β) for various values of soil friction angle ϕ. It was indicated that for slope angle (β) greater than $53°$, with $r_u = k_h = 0$, toe failure invariably occurs. If the value of β is less than $53°$, there were found to be two

Metaheuristics in Water, Geotechnical and Transport Engineering. DOI: http://dx.doi.org/10.1016/B978-0-12-398296-4.00014-3

possibilities: slope failure and the base failure. Bishop (1955) used the method of slices in obtaining stability of slopes. In order to solve the problem, Bishop assumed that the result of interslice forces acts in the horizontal direction. This method is being used very widely in the literature, and the results obtained from it compare very closely to more rigorous approaches such as the finite element method. Janbu (1957) solved the problem by assuming the point of application of the interslice forces. Janbu not only used his method to obtain the stability of slopes but also extended it to deal with the determination of the bearing capacity of foundations. Morgenstern and Price (1965) attempted to satisfy all the equations of statical equilibrium in obtaining the solution of the stability problem using the method of slices. They assumed different distributions of interslice forces to obtain the solution.

It should be mentioned that the previously available methods of slices (Bishop, 1955; Fellenius, 1936; Janbu, 1957) do not satisfy all the conditions of statical equilibrium. Chen (1975) used the upper-bound theorem of the limit analysis to obtain the critical heights for homogenous soil slopes. A rotational discontinuity mechanism was assumed in this analysis; it was indicated that in order for the rupture surface to remain kinematically admissible, its shape should become an arc of the logarithmic spiral. However, Chen (1975) did not incorporate in his analysis either the effect of pseudostatic earthquake body forces or the pore water pressure. Michalowski (1994) also used the upper-bound theorem of limit analysis in order to obtain the stability numbers for homogenous soil slopes. Similar to Chen (1975), a rigid body rotation of the soil mass, bounded by an arc of log-spiral failure surface, was assumed in his analysis. Michalowski (1994) also incorporated the effect of pore water pressure in his work; it was taken into account by using the pore water pressure coefficient $r_u = u/\gamma z$, where u is the pore water pressure at any point along the failure surface, γ is the total average bulk unit weight of the soil mass vertically above the failure surface, and z is the depth of the point below the soil surface.

Using upper-bound limit analysis, Michalowski (1995) presented a stability analysis of slopes based on a translational mechanism failure mechanism. A collapse mechanism was selected in the form of rigid vertical blocks similar to the traditional methods of slices. This allows one to relate the proposed analysis to the traditional method of slices and to assess the consequences of the statical assumptions made in them. The effect of the pore water pressure was also included in this work. Michalowski (2002) also used the upper-bound theorem of limit analysis in order to obtain the stability numbers for homogeneous slopes in the presence of pore water pressures as well as pseudostatic horizontal earthquake body forces. This work was an extension of the previous work of Michalowski (1994). Researchers use different metaheuristic models in geotechnical engineering (Alavi and Gandomi, 2011a,b; Gandomi and Alavi, 2011, 2012). Sah et al. (1994) used the maximum likelihood method, gave an equation for prediction of the FOS, and observed that the value matches well with that obtained using the limit equilibrium method. FOS, derived from the ratio of the resistance to the disturbance force, can be considered more as an index of stability than as a physical parameter. A FOS > 1.0 is considered a stable slope; otherwise, it is a failed slope.

Yang et al. (2004) used genetic programming and also presented an equation for the FOS. Gao (2009) successfully adopted ant colony clustering algorithm for slope stability analysis. Hwang et al. (2009) applied decision trees for slope stability analysis. Recently, the artificial neural network (ANN) has been successfully used in the slope stability problem (Chen and Yang, 2005; Fu et al., 2003; Lan et al., 2009; Li and Liu, 2005; Lu and Rosenbaum, 2003; Samui and Kumar, 2006; Wang et al., 2005). Chen et al. (2011) successfully used adaptive neuro-fuzzy inference system for prediction of stability of slope. However, ANN has some limitations:

- Unlike other statistical models, ANN does not provide information about the relative importance of the various parameters (Park and Rilett, 1999).
- The knowledge acquired during the training of the model is stored in an implicit manner; hence, it is very difficult to come up with a reasonable interpretation of the overall structure of the network (Kecman, 2001).
- In addition, ANN has some inherent drawbacks, such as slow convergence speed, less generalized performance, arriving at the local minimum, and overfitting problems.

Researchers have used the support vector machine (SVM), least squares support vector machine (LSSVM), and relevance vector machine (RVM) to overcome these limitations of ANN (Samui, 2008; Samui and Kothari, 2011; Samui et al., 2010). SVM was developed based on statistical learning theory (Vapnik, 1998), and it is a very powerful classification and regression tool (Ying, 2012). RVM was proposed by Tipping (2000), and it is a probabilistic version of SVM. It is highly insensitive to dimensionality. LSSVM is a modified version of SVM that was introduced by Suykens and Vandewalle (1999). The main difference between SVM and LSSVM is that LSSVM uses a set of linear equations for training, while SVM uses a quadratic optimization problem (Tsujinishi and Abe, 2003).

This chapter examines the capability of multivariate adaptive regression spline (MARS) for predicting the FOS of slope. MARS was developed by Friedman (1990). It creates an explicit model. Researchers have successfully used MARS to solve various problems in engineering (Ekman and Kubin, 1999; Jin et al., 2000; Ko and Osei-Bryson, 2004; Okine et al., 2003, 2009; Prasad and Iverson, 2000; MacLean and Mix, 1991; Sharada et al., 2008; Veaux et al., 1993). This chapter has taken the data set from the work of Sakellatiou and Ferentinou (2005). The data set contains information about unit weight (γ), cohesion (c), angle of internal friction (ϕ), slope angle (β), height (H), pore pressure ratio (r_u), and FOS. It has the following aims:

- to investigate the feasibility of MARS for predicting FOS of slope,
- to develop an equation for prediction of FOS of slope based on MARS,
- to make a comparative study between the developed MARS and the other metaheuristic models,
- to do sensitivity analysis to determine the effect of each input parameter.

14.2   Method

14.2.1   Details of MARS

This section will describe the details of MARS for predicting FOS of slope. It is a procedure for adaptive nonparametric regression (Friedman, 1990). The general expression of nonparametric regression is given as follows:

$$y_i = f(x_{i1}, x_{i2}, \ldots, x_{ik}) + \varepsilon_i \tag{14.1}$$

The main goal of the nonparametric regression is to estimate the regression function $f(x_{i1}, x_{i2}, \ldots, x_{ik})$ directly, rather than to estimate parameters. In nonparametric regression, it is assumed that $f(x_{i1}, x_{i2}, \ldots, x_{ik})$ is a smooth, continuous function, except for wavelet regression (Fox, 2002).

MARS uses the following expression for predicting output (y):

$$y = c_0 + \sum_{m=1}^{M} c_m B_m(x) \tag{14.2}$$

where x is the input variable, c_0 is a constant, $B_m(x)$ is basis function (Figure 14.1), and c_m is the coefficient of $B_m(x)$. In this study, the input variables are γ, c, ϕ, ψ, H, and r_u. The output of MARS is FOS. So, $x = [\gamma, c, \phi, \psi, H, r_u]$ and $y = [\text{FOS}]$.

The spline function consists of two segments, i.e., truncated functions of the left side of Eq. (14.3) and the right side of Eq. (14.4) separated from each other by a so-called knot location (Friedman, 1990), as follows:

$$b_q^-(x - t) = [-(x-t)]_+^q = \begin{cases} (t-x)^q & \text{if } x > t \\ 0 & \text{otherwise} \end{cases} \tag{14.3}$$

$$b_q^+(x - t) = [+(x-t)]_+^q = \begin{cases} (x-t)^q & \text{if } x > t \\ 0 & \text{otherwise} \end{cases} \tag{14.4}$$

where t is the knot location and $b_q^-(x - t)$ and $b_q^+(x - t)$ are the spline functions.

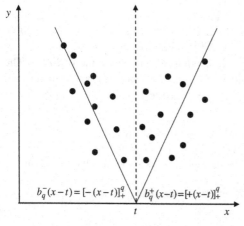

Figure 14.1 The basis function and knot location.

In general, MARS contains the following three steps:

1. The constructive phase
2. The pruning phase
3. Selection of optimum MARS.

In the constructive phase, the basis functions are introduced to define Eq. (14.2), and the selection of these basis functions is carried out using the generalized cross-validation (GCV) statistic. The value of GCV is determined by the following equation:

$$GCV(M) = \left(\frac{1}{n}\right) \frac{\sum_{m=1}^{M} (y_i - \hat{y}_i)^2}{(1 - C(M)/n)^2} \tag{14.5}$$

where n is the number of data objects, y_i is the response value for object i, $\hat{y}_i$ is the predicted response value for object i, and $C(M)$ is a penalty factor. The value of $C(M)$ is determined from the following expression:

$$C(M) = M + dM \tag{14.6}$$

where d is a cost penalty factor for each basis function optimization. Overfitting can occur due to many basis functions, and to prevent this, some basis functions are deleted in the pruning phase. In the third step, the optimum MARS model is selected.

Analysis of variance (ANOVA) decomposition of the MARS model is given by the following expression:

$$f(x) = \beta_0 + \sum_{B=1} f_i(x_i) + \sum_{B=2} f_{ij}(x_i, x_j) + \sum_{B=3} f_{ijk}(x_i, x_j, x_k) + \cdots \tag{14.7}$$

$\sum_{B=1} f_i x_i$ is overall basis functions that involve only a single variable, $\sum_{B=2} f_{ij}(x_i, x_j)$ is overall basis functions that involve exactly two variables, and $\sum_{B=3} f_{ij}(x_i, x_j, x_k)$ represents the contributions from three variable interactions (if present).

14.3 Application of MARS to Slope Stability Analysis

The above MARS has been adopted for the prediction of FOS. The data are normalized against their maximum values. To develop MARS, the data (Table 14.1) were divided into the following two groups:

- Training data set: This was required to develop the model. This chapter uses 32 out of 46 data sets as the training data set (Table 14.2).
- Testing data set: This was required to verify the developed model. The remaining 14 data sets were used as the testing data set (Table 14.3).

The statistical parameters of all the data are given in Table 14.4. Tables 14.5 and 14.6 show the statistical parameters of the training and testing data sets, respectively. The program of MARS was developed by using MATLAB.

Table 14.1 Data Set Used in This Study

γ (kN/m³)	c (kPa)	ϕ (°)	β (°)	H (m)	r_u	FOS
18.68	26.34	15	35	8.23	0	1.11
16.5	11.49	0	30	3.66	0	1
18.84	14.36	25	20	30.5	0	1.875
18.84	57.46	20	20	30.5	0	2.045
28.44	29.42	35	35	100	0	1.78
28.44	39.23	38	35	100	0	1.99
20.6	16.28	26.5	30	40	0	1.25
14.8	0	17	20	50	0	1.13
14	11.97	26	30	88	0	1.02
25	120	45	53	120	0	1.3
26	150.05	45	50	200	0	1.2
18.5	25	0	30	6	0	1.09
18.5	12	0	30	6	0	0.78
22.4	10	35	30	10	0	2
21.1	10	30.34	30	20	0	1.7
22	20	36	45	50	0	1.02
22	0	36	45	50	0	0.89
12	0	30	35	4	0	1.46
12	0	30	45	8	0	0.8
12	0	30	35	4	0	1.44
12	0	30	45	8	0	0.86
23.47	0	32	37	214	0	1.08
16	70	20	40	115	0	1.11
20.41	24.9	13	22	10.67	0.35	1.4
19.63	11.97	20	22	12.19	0.405	1.35
21.82	8.62	32	28	12.8	0.49	1.03
20.41	33.52	11	16	45.72	0.2	1.28
18.84	15.32	30	25	10.67	0.38	1.63
18.84	0	20	20	7.62	0.45	1.05
21.43	0	20	20	61	0.5	1.03
19.06	11.71	28	35	21	0.11	1.09
18.84	14.36	25	20	30.5	0.45	1.11
21.51	6.94	30	31	76.81	0.38	1.01
14	11.97	26	30	88	0.45	0.625
18	24	30.15	45	20	0.12	1.12
23	0	20	20	100	0.3	1.2
22.4	100	45	45	15	0.25	1.8
22.4	10	35	45	10	0.4	0.9
20	20	36	45	50	0.25	0.96
20	20	36	45	50	0.5	0.83
20	0	36	45	50	0.25	0.79
20	0	36	45	50	0.5	0.67
22	0	40	33	8	0.35	1.45

(*Continued*)

Table 14.1 (Continued)

γ (kN/m^3)	c (kPa)	ϕ (°)	β (°)	H (m)	r_u	FOS
24	0	40	33	8	0.3	1.58
20	0	24.5	20	8	0.35	1.37
18	5	30	20	8	0.3	2.05

Table 14.2 Training Data Set for Developing the MARS Model

γ (kN/m^3)	c (kPa)	ϕ (°)	β (°)	H (m)	r_u	Actual FOS	Predicted FOS by MARS
18.68	26.34	15	35	8.23	0	1.11	1.09
18.84	14.36	25	20	30.5	0	1.875	1.90
28.44	29.42	35	35	100	0	1.78	1.81
28.44	39.23	38	35	100	0	1.99	1.96
14.8	0	17	20	50	0	1.13	1.10
14	11.97	26	30	88	0	1.02	1.01
25	120	45	53	120	0	1.3	1.29
26	150.05	45	50	200	0	1.2	1.19
18.5	25	0	30	6	0	1.09	0.98
18.5	12	0	30	6	0	0.78	0.90
21.1	10	30.34	30	20	0	1.7	1.68
22	20	36	45	50	0	1.02	0.99
12	0	30	35	4	0	1.46	1.37
12	0	30	45	8	0	0.8	0.90
12	0	30	45	8	0	0.86	0.90
23.47	0	32	37	214	0	1.08	1.07
19.63	11.97	20	22	12.19	0.405	1.35	1.28
20.41	33.52	11	16	45.72	0.2	1.28	1.28
18.84	15.32	30	25	10.67	0.38	1.63	1.64
21.43	0	20	20	61	0.5	1.03	1.00
19.06	11.71	28	35	21	0.11	1.09	1.15
18.84	14.36	25	20	30.5	0.45	1.11	1.20
14	11.97	26	30	88	0.45	0.625	0.59
18	24	30.15	45	20	0.12	1.12	1.08
23	0	20	20	100	0.3	1.2	1.19
22.4	100	45	45	15	0.25	1.8	1.80
22.4	10	35	45	10	0.4	0.9	0.87
20	20	36	45	50	0.5	0.83	0.81
20	0	36	45	50	0.25	0.79	0.77
20	0	36	45	50	0.5	0.67	0.70
24	0	40	33	8	0.3	1.58	1.58
18	5	30	20	8	0.3	2.05	2.04

Table 14.3 Testing Data Set for Developing the MARS Model

γ (kN/m³)	c (kPa)	ϕ (°)	β (°)	H (m)	r_u	Actual FOS	Predicted FOS by MARS
16.5	57.46	0	30	3.66	0	1	0.97
18.84	16.28	20	20	30.5	0	2.045	2.00
20.6	10	26.5	30	40	0	1.25	1.43
22.4	0	35	30	10	0	2	1.92
22	0	36	45	50	0	0.89	0.82
12	70	30	35	4	0	1.44	1.40
16	24.9	20	40	115	0	1.11	1.09
20.41	8.62	13	22	10.67	0.35	1.4	1.40
21.82	0	32	28	12.8	0.49	1.03	1.01
18.84	6.94	20	20	7.62	0.45	1.05	0.95
21.51	20	30	31	76.81	0.38	1.01	0.98
20	0	36	45	50	0.25	0.96	0.93
22	0	40	33	8	0.35	1.45	1.44
20	0	24.5	20	8	0.35	1.37	1.40

Table 14.4 Statistical Parameters of All the Data

Input Variable	Mean	Standard Deviation	Kurtosis	Skewness
γ (kN/m³)	19.71	3.88	3.18	− 0.15
c (kPa)	20.47	31.71	9.66	2.60
ϕ (°)	27.51	10.98	3.55	− 0.83
β (°)	32.93	10.08	1.86	0.11
H (m)	43.91	48.68	6.31	1.81
r_u	0.17	0.19	1.51	0.42
FOS	1.24	0.38	2.55	0.66

Table 14.5 Statistical Parameters of the Training Data

Input Variable	Mean	Standard Deviation	Kurtosis	Skewness
γ (kN/m³)	19.80	4.28	2.78	− 0.03
c (kPa)	22.38	35.32	8.45	2.49
ϕ (°)	28.20	11.18	3.62	− 0.84
β (°)	33.93	10.66	1.77	− 0.03
H (m)	49.77	53.52	5.36	1.65
r_u	0.16	0.19	1.64	0.52
FOS	1.22	0.39	2.34	0.56

Table 14.6 Statistical Parameters of the Testing Data

Input Variable	Mean	Standard Deviation	Kurtosis	Skewness
γ (kN/m^3)	19.49	2.92	4.05	− 1.31
c (kPa)	15.30	22.24	4.21	1.57
ϕ (°)	25.92	10.74	3.43	− 0.91
β (°)	30.64	8.53	2.14	0.32
H (m)	60.50	33.07	4.02	1.37
r_u	0.18	0.20	1.27	0.20
FOS	1.28	0.36	3.03	1.04

14.4 Results and Discussion

The performance of the developed MARS was assessed by the coefficient of correlation (R). The value of R was determined using the following equation:

$$R = \frac{\sum_{i=1}^{n}(\text{FOS}_{ai} - \overline{\text{FOS}}_a)(\text{FOS}_{pi} - \overline{\text{FOS}}_p)}{\sqrt{\sum_{i=1}^{n}(\text{FOS}_{ai} - \overline{\text{FOS}}_a)}\sqrt{\sum_{i=1}^{n}(\text{FOS}_{pi} - \overline{\text{FOS}}_p)}} \tag{14.8}$$

where FOS_{ai} and FOS_{pi} are the actual and predicted FOS values, respectively; $\overline{\text{FOS}}_a$ and $\overline{\text{FOS}}_p$ are the mean of the actual and predicted FOS values corresponding to n patterns. Figure 14.2 shows the flowchart of the MARS. To develop the MARS model, a different number of basis functions was introduced in the constructive phase. Figure 14.3 shows the effect of the number of basis functions on testing performance (R). It is clear from Figure 14.3 that 12 basis functions gave the best performance. In the pruning phase, 4 basis functions were deleted. So, the final MARS contained 8 basis functions. The expression of the final MARS is given as follows:

$$\text{FOS} = 0.057 + \sum_{m=1}^{8} c_m B_m(x) \tag{14.9}$$

This is achieved by putting $y = \text{FOS}$, $M = 8$, and $c_0 = 0.057$ in Eq. (14.2).

Table 14.7 shows the details of c_m and $B_m(x)$.

Equation (14.8) was used to determine the performance of the training and testing data sets. Figures 14.4 and 14.5 depict the performance of training and testing data sets, respectively.

It can be seen from Figures 14.4 and 14.5 that the value of R is close to 1 for training as well as testing the data set. So, the developed MARS is a robust model for predicting FOS of slope. The performances of the training and testing data sets were almost the same, so there is no overtraining in the MARS and it has good generalization capability. The developed MARS has been applied in several studies (Hoek and Bray, 1981; Hudson, 1992b; Lin et al., 1988; Madzie, 1988). Table 14.8 shows the data in the literature. The developed MARS has been compared with ANN (Sakellatiou and

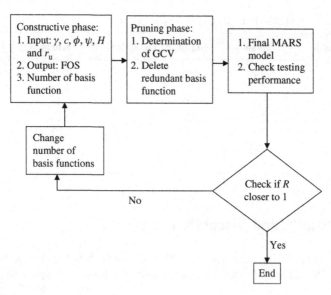

Figure 14.2 Flowchart of MARS.

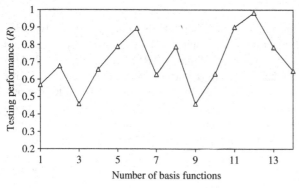

Figure 14.3 Variation of testing performance (R) with number of basis functions.

Ferentinou, 2005), SVM (Samui, 2008), RVM (Samui et al., 2010), and LSSVM (Samui and Kothari, 2011) for the data sets in the literature. The comparison was carried out in terms of root mean square error (RMSE) and mean absolute error (MAE). The values of RMSE and MAE were determined by using the following equations:

$$RMSE = \sqrt{\frac{\sum_{i=1}^{n} (FOS_{ai} - FOS_{pi})^2}{n}} \qquad (14.10)$$

$$MAE = \frac{\sum_{i=1}^{n} |FOS_{ai} - FOS_{pi}|}{n} \qquad (14.11)$$

where FOS_a and FOS_p are the actual and predicted FOS values, respectively; n is the number of data. Table 14.9 depicts the values of MAE and RMSE for the ANN,

Table 14.7 Values of $B_m(x)$ and c_m

Basis Function $\{B_m(x)\}$	Equation	Coefficient (c_m)
$B_1(x)$	$\max(0,\gamma-0.010)\times\max(0,H-0.316)$	2.402
$B_2(x)$	$\max(0,\gamma-0.010)\times\max(0,\phi-0.074)$	1.257
$B_3(x)$	$\max(0,\gamma-0.010)\times\max(0,0.074-\phi)$	-1.246
$B_4(x)$	$B_1(x)\times\max(0,0.592-r_u)$	35.346
$B_5(x)$	$\max(0,\psi-0.453)$	0.763
$B_6(x)$	$B_5(x)\times\max(0,c-0.275)$	2.528
$B_7(x)$	$B_5(x)\times\max(0,0.275-0.275)$	-3.988
$B_8(x)$	$\max(0,\gamma-0.010)\times\max(0,0.316-H)\times$ $\max(0,c-0.052)$	2.293

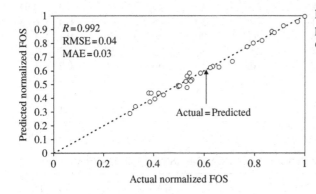

Figure 14.4 The performance of the training data set.

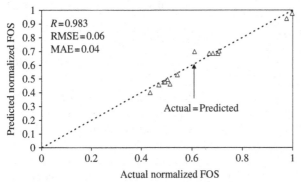

Figure 14.5 The performance of the testing data set.

SVM, LSSVM, RVM, and MARS models. It is clear from Table 14.9 that the developed MARS outperforms the ANN model, and the performance of MARS is comparable to the SVM, RVM, and LSSVM models. SVM and LSSVM use two tuning parameters (for SVM: capacity factor (C) and kernel parameter; for LSSVM:

Table 14.8 Data from Different Case Histories

Reference	γ (kN/m^3)	c (kPa)	ϕ (°)	β (°)	H (m)	r_u	FOS
Hoek and Bray (1981)	21	20	40	40	12	0	1.84
	21	45	25	49	12	0.3	1.53
	21	30	35	40	12	0.4	1.49
	21	35	28	40	12	0.5	1.43
	20	10	29	34	6	0.3	1.34
	20	40	30	30	15	0.3	1.84
	18	45	25	25	14	0.3	2.09
	19	30	35	35	11	0.2	2
	20	40	40	40	10	0.2	2.3
Hudson (1992b)	18.85	24.8	21.3	29.2	37	0.5	1.07
	18.85	10.34	21.3	34	37	0.3	1.29
Lin et al. (1988)	18.8	30	10	25	50	0.1	1.4
	18.8	25	10	25	50	0.2	1.18
	18.8	20	10	25	50	0.3	0.97
	19.1	10	10	25	50	0.4	0.65
	18.8	30	20	30	50	0.1	1.46
	18.8	25	20	30	50	0.2	1.21
	18.8	20	20	30	50	0.3	1
	19.1	10	20	30	50	0.4	0.65
Madzie (1988)	22	20	22	20	180	0	1.12
	22	20	22	20	180	0.1	0.99

Table 14.9 Comparison Between the ANN, SVM, LSSVM, and MARS Models

Model	RMSE	MAE	R
ANN	0.374	0.313	0.523
SVM	0.281	0.241	0.745
LSSVM	0.284	0.232	0.769
RVM	0.249	0.211	0.789
MARS	0.272	0.267	0.800

regularization parameter (γ) and kernel parameter). Meanwhile, RVM and MARS use only one tuning parameter (for RVM, kernel parameter; for MARS, number of basis functions). In ANN, there is a larger number of tuning parameters, including the number of hidden layers, number of hidden nodes, learning rate, momentum term, number of training epochs, transfer functions, and weight initialization methods. The developed RVM model has some limitations, such as a highly nonlinear optimization process and difficulties in finding the optimum solution for the large data set. Table 14.10 shows the result of ANOVA decomposition.

The value of GCV is comparable for all the functions. However, the value of GCV is at its maximum for γ, H, and r_u. Therefore, γ, H, and r_u have the maximum effect on the predicted FOS.

Table 14.10 Results of ANOVA Decomposition

Function	Standard Deviation	GCV	Basis Function	Variable
1	0.073	0.018	1	ψ
2	0.096	0.022	2	γ, c
3	0.072	0.016	1	γ, H
4	0.077	0.017	2	c, ψ
5	0.054	0.012	1	γ, c, H
6	0.099	0.023	1	γ, H, r_u

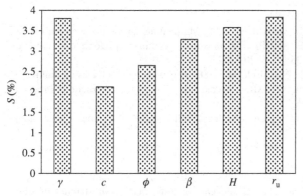

Figure 14.6 Sensitivity analysis of the input parameters.

A sensitivity analysis has been done to determine the effect of each input parameter. The basic idea was that each input of the model is offset slightly, and the corresponding change in the output was reported. The procedure has been taken from the work of Liong et al. (2000), in which the sensitivity (S) of each input parameter was calculated by the following formula:

$$S(\%) = \frac{1}{N} \sum_{j=1}^{N} \left(\frac{\% \text{ change in ouput}}{\% \text{ change in input}} \right)_j \times 100 \tag{14.12}$$

where N is the number of data points. The analysis was carried out on the trained model by varying each input parameter, one at a time, at a constant rate of 20%. Figure 14.6 shows the result of the sensitivity analysis.

It is clear from Figure 14.6 that γ, H, and r_u have the maximum effect on the predicted FOS.

14.5 Conclusion

This chapter successfully applied MARS for the prediction of FOS of slope. The performance of the developed MARS is encouraging, in that it was better than the performance of ANN. The developed equation can be used to predict FOS of slope.

The developed MARS proves the generalization capability to handle the data cited in the literature. ANOVA and sensitivity analysis indicate that the effect of γ, H, and r_u is at its maximum on FOS. In summary, it can be concluded that MARS can be used as an efficient tool for solving various problems in civil engineering.

References

Alavi, A.H., Gandomi, A.H., 2011a. A robust data mining approach for formulation of geotechnical engineering systems. Eng. Comput. 28 (3), 242−274.

Alavi, A.H., Gandomi, A.H., 2011b. Prediction of principal ground-motion parameters using a hybrid method coupling artificial neural networks and simulated annealing. Comput. Struct. 89 (23−24), 2176−2194.

Attoh-Okine, N.O., Cooger, K., Mensah, S., 2009. Multivariate adaptive regression (MARS) and hinged hyperplanes (HHP) for doweled pavement performance modeling. Construct. Build. Mater. 23 (9), 3020−3023.

Attoh-Okine, N.O., Mensah, S., Nawaiseh, M., 2003. A new technique for using multivariate adaptive regression splines (MARS) in pavement roughness prediction. Proc. ICE Transport. 156 (1), 51−55.

Bishop, A.W., 1955. The use of slip circle in the stability analysis of slopes. Geotechnique. 5 (1), 7−17.

Bishop, A.W., Morgenstern, N.R., 1960. Stability coefficients for earth slopes. Geotechnique. 10 (4), 129−150.

Chen, W.F., 1975. Limit Analysis and Soil Plasticity. Elsevier, Amsterdam.

Chen, W.F., Giger, M.W., Fang, H.Y., 1969. On the limit analysis of stability of slopes. Soils Found. 9 (4), 23−32.

Chen, W.F., Liu, X.L., 1990. Limit Analysis in Soil Mechanics. Elsevier, Amsterdam.

Chen, Ch.F., Yang, Y., 2005. Fuzzy reasoning system driven by HGA−ANN for estimation of slope stability. Chin. J. Rock Mech. Eng. 24 (19), 3459−3464.

Chen, Ch.-F., Xiao, Zh.-Y., Zhang, G.B., 2011. Stability assessment model for epimetamorphic rock slopes based on adaptive neuro-fuzzy inference system. Electron. J. Geotech. Eng. 16, 93−107.

De Veaux, R.D., Psichogios, D.C., Ungar, L.H., 1993. A comparison of two nonparametric estimation schemes: MARS and neural networks. Comput. Chem. Eng. 17 (8), 819−837.

Ekman, T., Kubin, G., 1999. Nonlinear prediction of mobile radio channels: measurements and MARS model designs. IEEE Int. Conf. Acoust., Speech Signal Process. 2667−2670.

Fellenius, W., 1936. Calculation of stability of earth dams. In: Transactions of the Second Congress on Large Dams. Washington, DC, vol. 4, pp. 445−463.

Fox, J., 2002. Nonparametric Regression. Appendix to an R and S-PLUS Companion to Applied Regression. Sage Publications, Thousand Oaks, CA.

Friedman, J.H., 1990. Multivariate adaptive regression splines. Ann. Stat. 19, 1−67.

Fu, Y., Liu, S., Liu, D., 2003. Predicting models to estimate stability of rock slope based on RBF neural network. J. Wuhan Univ. Technol. 27 (2), 170−173.

Gandomi, A.H., Alavi, A.H., 2011. Multi-stage genetic programming: a new strategy to nonlinear system modeling. Inform. Sci. 181 (23), 5227−5239.

Gandomi, A.H., Alavi, A.H., 2012. A new multi-gene genetic programming approach to nonlinear system modeling. Part II: geotechnical and earthquake engineering problems. Neural Comput. Appl. 21 (1), 189−201.

Gao, W., 2009. Analysis of stability of rock slope based on ant colony clustering algorithm. Rock Soil Mech. 30 (11), 3476−3480.

Hoek, E., Bray, J.W., 1981. Rock Slope Engineering. third ed. Institution of Mining and Metallurgy, London.

Hudson, J.A., 1992b. Rock Engineering—Theory and Practice. Ellis Horwood Limited, West Sussex.

Hwang, S.G., Guevarra, I.F., Yu, B.O., 2009. Slope failure prediction using a decision tree: a case of engineered slopes in South Korea. Eng. Geol. 104 (1−2), 126−134.

Janbu, N., 1957. Earth pressures and bearing capacity calculation by generalized procedure by slices. In: Proceedings of the Fourth International Conference on Soil Mechanics. London, vol. 2, pp. 207−212.

Jin, R., Chen, W., Simpson, T.W., 2000. Comparative Studies of Metamodeling Techniques under Multiple Modeling Criteria. American Institute of Aeronautics and Astronautics, Reston, VA, AIAA-2000-4801.

Karal, K., 1977a. Application of energy method. J. Geotech. Eng. ASCE. 103 (5), 381−399.

Karal, K., 1977b. Energy method for soil stability analyses. J. Geotech. Eng. ASCE. 103 (5), 431−447.

Kecman, V., 2001. Learning and Soft Computing: Support Vector Machines, Neural Networks, and Fuzzy Logic Models. MIT Press, Cambridge, MA.

Ko, M., Osei-Bryson, K.M., 2004. Using regression splines to assess the impact of information technology investments on productivity in the health care industry. Inform. Syst. 14, 43−63.

Kumar, J., 2000. Slope Stability Calculations Using Limit Analysis, 101. ASCE Special Publication, New York, NY, pp. 239−249.

Kumar, J., 2004. Stability factors for slopes with non-associated flow rule using energy consideration. Int. J. Geomech. ASCE. 4 (4), 264−272.

Kumar, J., Samui, P., 2006. Determination for layered soil slopes using the upper bound limit analysis. Geotech. Geol. Eng. 24 (6), 1803−1819.

Lan, H., Q., Li, Chunyu, H., 2009. Slope stability evaluation based on generalized regression neural network. Rock Soil Mech. 30, 3460−3463.

Li, S., Liu, Y., 2005. Slope stability estimation with probabilistic neural networks. J. Comput. Inform. Syst. 1 (4), 679−685.

Lin, P.S., Lin, M.H., Lee, T.M., 1988. An investigation on the failure of a building constructed on hillslope. In: Bonnard (Ed.), Landslides, 1. Balkema, pp. 445−449.

Liong, S.Y., Lim, W.H., Paudyal, G.N., 2000. River stage forecasting in Bangladesh: neural network approach. J. Comput. Civ. Eng. 14 (1), 1−8.

Lu, P., Rosenbaum, M.S., 2003. Artificial neural networks and grey systems for the prediction of slope stability. Nat. Hazards. 30, 383−398.

MacLean, M., Mix, P., 1991. Measuring hospital productivity and output: the omission of outpatient services. Health Rep. 3, 229−244.

Madzie, E., 1988. Stability of unstable final slope in deep open iron mine. In: Bonnard (Ed.), Landslides, 1. Balkema, pp. 455−458.

Michalowski, R.L., 1994. Limit analysis for slopes subjected to pore pressure. In: Siriwardane, H.J, Zaman, M.M. (Eds.), Proceedings of the Conference on Computing and Advances in Geomechanics. Balkema, Rotterdam, The Netherlands.

Michalowski, R.L., 1995. Slope stability analysis: a kinematical approach. Geotechnique. 45 (2), 283−293.

Michalowski, R.L., 2002. Stability charts for uniform slopes. J. Geotech. Geoenviron. Eng. ASCE. 128 (4), 351−355.

Morgenstern, N.R., Price, V.E., 1965. The analysis of the stability of general slip surfaces. Geotechnique. 15 (1), 79−93.

Park, D., Rilett, L.R., 1999. Forecasting freeway link ravel times with a multi-layer feed forward neural network. Comput. Aided Civ. Infrastruct. Eng. 14, 358−367.

Prasad, A.M., Iverson, L.R., 2000. Predictive vegetation mapping using a custom built model-chooser: comparison of regression tree analysis and multivariate adaptive regression splines. In: Fourth International Conference on Integrating GIS and Environmental Modeling (GIS/EM4). Problems, Prospects and Research Needs. Banff, Alberta, Canada.

Sah, N.K., Sheorey, P.R., Upadhyaya, L.N., 1994. Maximum likelihood estimation of slope stability. Int. J. Rock Mech. Mining Sci. Geomech. 31 (1), 47−53.

Sakellatiou, M.G., Ferentinou, M.D., 2005. A study of slope stability prediction using neural networks. Int. J. Geotech. Geol. Eng. 23, 419−445.

Samui, P., 2008. Slope stability analysis: a support vector machine approach. Environ. Geol. 56 (2), 255−267.

Samui, P., Das, S., Sitaram, T.G., 2010. Soft Computing: In Geotechnical Engineering. VDM Verlag Dr. Müller, Germany.

Samui, P., Kothari, D.P., 2011. Utilization of a least square support vector machine (LSSVM) for slope stability analysis. Sci. Iranica. 18 (1), 53−58.

Samui, P., Kumar, B., 2006. Artificial neural network prediction of stability numbers for two-layered slopes with associated flow rule. Electron. J. Geotech. Eng.

Sharada, V.N., Prasher, S.O., Patel, R.M., Ojasvi, P.R., Prakash, C., 2008. Performance of multivariate adaptive regression splines (MARS) in predicting runoff in mid-Himalayan microwatersheds with limited data. Hydrol. Sci. J. Sci. Hydrol. 53 (6), 1165−1175.

Suykens, J.A.K., Vandewalle, J., 1999. Least squares support vector machine classifiers. Neural Process. Lett. 9 (3), 293−300.

Taylor, D.W., 1948. Fundamental of Soil Mechanics. Wiley, New York, NY.

Tipping, M.E., 2000. The Relevance Vector Machine. Advances in Neural Information Proceeding Systems, 12. MIT Press, Cambridge, MA, pp. 652−658.

Tsujinishi, D., Abe, S., 2003. Fuzzy least squares support vector machines for multi-class problems. Neural Networks. 16, 785−792.

Vapnik, V.N., 1998. Statistical Learning Theory. Wiley, New York, NY.

Wang, H.B., Xu, W.Y., Xu, R.C., 2005. Slope stability evaluation using back propagation neural networks. Eng. Geol. 80, 302−315.

Yang, C.X., Tham, L.G., Feng, X.T., Wang, Y.J., Lee, P.K.K., 2004. Two-stepped evolutionary algorithm and its application to stability analysis of slopes. J. Comp. Civ. Eng. 18 (2), 145−153.

Ying, W., 2012. An improved discriminative common vectors and support vector machine based face recognition approach. Expert Syst. Appl. 39, 4628−4632.

Part Three

Transport Engineering

15 Scheduling Transportation Networks and Reliability Analysis of Geostructures Using Metaheuristics

Nikos Kallioras[1], George Piliounis[2], Matthew G. Karlaftis[1] and Nikos D. Lagaros[2]

[1]Department of Transportation Planning and Engineering, National Technical University of Athens, Athens, Greece, [2]Institute of Structural Analysis and Seismic Research, National Technical University of Athens, Athens, Greece

15.1 Introduction

Heuristic and metaheuristic algorithms are nature- or bio-inspired, as they were developed based on the successful evolutionary behavior of natural systems. Nature has been solving various problems over millions or even billions of years. Only the best and robust solutions remain, based on the principle of the survival of the fittest. Similarly, heuristic algorithms use trial and error, learning, and adaptation to solve problems. Modern metaheuristic algorithms are almost guaranteed to perform efficiently to solve a wide range of optimization problems. The main aim of research in optimization and algorithm development is to design and/or choose the most suitable and efficient algorithms for a given optimization problem. Loosely speaking, modern metaheuristic algorithms for engineering optimization include the genetic algorithms (Goldberg, 1989), simulated annealing (Kirkpatrick et al., 1983), particle swarm optimization (Kennedy and Eberhart, 1995), ant colony algorithm (ACO; Dorigo and Stützle, 2004), artificial bee colony algorithm (Bozorg et al., 2005), harmony search (HS; Geem et al., 2001), and firefly algorithm (Yang, 2008), among many others.

Natural hazards such as earthquakes, floods, tsunamis, and hurricanes can cause enormous damage to both social and infrastructure networks. Following such disasters, local communities and search and rescue crews are faced with rapidly degrading infrastructure networks, which may result in much slower response times, delays in population evacuation, and significant complications in infrastructure repair. Because of the obvious importance of minimizing the adverse impacts from

Metaheuristics in Water, Geotechnical and Transport Engineering. DOI: http://dx.doi.org/10.1016/B978-0-12-398296-4.00015-5

natural disasters, the literature has systematically dealt with what are considered the four main steps in disaster response (Altay and Greene, 2006). First, there is mitigation, which includes assessing seismic hazards (Dong et al., 1987), probabilistic damage projection (Peizhuangm et al., 1986; Tamura et al., 2000), and decision support systems for integrating the emergency process (Mendonca et al., 2001, 2006). Second is preparedness, which has mainly focused on preparing infrastructure networks for dealing with potential disasters and for accommodating evacuation needs (Nicholson and Du, 1997; Sakakibara et al., 2004; Sohn, 2006; Song et al., 2003; Verter and Lapierre, 2002; Viswanath and Peeta, 2003). Third is response, which has evolved around two main research paths: (i) planning the response—relief logistics operations (Barbarosoglou and Arda, 2004; Barbarosoglou et al., 2002; Fiedrich et al., 2000; Ozdamar et al., 2004) and (ii) assessing the performance of the infrastructure system following a natural disaster (Bell, 2000; Chang and Nojima, 2001; Karaouchi et al., 2001; Li and Tsukaguchi, 2001). Fourth is recovery operations, which have attracted limited attention despite their importance in practice; for example, work has concentrated on infrastructure element protection (Cret et al., 1993), general assessment of relief performance (Song et al., 1996), and fund allocation for infrastructure repairs following disasters (Karlaftis et al., 2007; Lagaros and Karlaftis, 2011; Plevris et al., 2010). Recovery operations are very important in all natural disasters, particularly for the speedy revitalization of community activities. However, the mathematical complexity of organizing postdisaster operations has hampered research efforts. Frequently, following a disaster, civil infrastructure elements must be inspected and evaluated, and repairs prioritized. In order to deal with these problems efficiently, it is required to formulate and solve complex districting and routing problems. For example, the affected area must be partitioned into districts of responsibility for repair crews and, then, inspection sequences—infrastructure elements to be inspected first, second, and so on—must be determined.

Reliability analysis can be performed either with simulation methods, such as the Monte Carlo simulation (MCS) method, or with other approximation methods. First- and second-order reliability methods (FORM and SORM, respectively) require prior knowledge of the mean and the variance of each random variable. Furthermore, these methods require a differentiable failure function. On the other hand, although the major advantage of MCS is that accurate solutions can be obtained for almost every problem, yet it requires excessive computational cost in many cases. Variance reduction techniques, such as importance sampling, directional simulation, antithetic variates, and adaptive sampling, have been proposed in order to reduce the computational effort of MCS. The disadvantage of these methods is that they require prior knowledge of the behavior of the system and the characteristics of the problem at hand in order to determine the most effective sampling region, which for many practical problems is not clearly identifiable. Recent results (Koutsourelakis et al., 2004) reveal that variance reduction techniques still require a significant number of system response evaluations to estimate probabilities of the order less than 10^{-3}. Other recently proposed simulation methods, such as line sampling (Koutsourelakis et al., 2004) and subset simulation (Au and Beck, 2001), proved to be very efficient in reducing the required sample size and the computational cost; however, their performance and ergodicity were sensitive to the values of certain parameters that were not known *a priori*. For the application of a computationally efficient MCS method to complex models, it would be necessary to have an approximate knowledge of the limit-state function $g(x)$. On the other hand, since complex reliability problems are characterized by the implicit nature of the limit-state functions, the implementation of FORM or SORM requires an explicit approximation of either the entire limit-state function $g(x)$ or its limit-state surface $g(x) = 0$ in the space of the random variable x (Hurtado, 2004; Rajashekhar and Ellingwood, 1993).

In this chapter, two distinct optimization problems are considered for assessing the performance of metaheuristics. In particular, the problem of infrastructure network restoration and minimization of the adverse impacts of natural hazards on civil infrastructure and the problem of reliability analysis of geostructures, which is formulated as an optimization problem, are considered. In the first problem, an important issue within the scope of postnatural disaster actions is taken into account; we formulate the districting and routing problems for scheduling infrastructure inspection crews following a natural disaster in urban areas and a combined treatment of these two separate problems is performed. In the second problem, based on the Lasofer–Lind reliability index, the problem of reliability analysis is formulated as an optimization problem and it is implemented in a real-life piled foundation test case. In both problems, the HS algorithm is implemented for solving optimization problems, while the ACO algorithm is also considered for solving the routing problem.

15.2 Problem Statement and Research Impact

Natural disasters have always affected human life and activities, but it is some of the major transformations of the modern world and society that have increased the impacts and associated risks of natural catastrophes. Factors such as the increased size and density of human communities and the introduction of critical infrastructures in areas prone to natural catastrophes have enhanced the impact of catastrophic events. National and local economies depend on efficient and reliable civil infrastructures that provide added value and competitive advantage to an area's social and economic growth. Inevitably, a catastrophic event will severely affect the structural integrity and operation of civil infrastructures and will lead to both immediate and long-term economic losses for surrounding communities and even the national and international economies. The recent earthquake and tsunami events in Japan yielded an estimated immediate cost of over €100 billion due to the destruction of the country's infrastructure, while the impact on the international economy is still undocumented. The impacts and the associated risks of natural disasters can be mitigated through careful planning; disaster management is a multistage process which begins with predisaster planning and system improvement, and extends to postdisaster system response, recovery, and reconstruction. The postdisaster stage involves tactical and operational decision making for providing critical emergency, recovery, and reconstruction services to support society.

The theory and methods of structural reliability analysis have been developed significantly during the last 20 years and have been documented in an increasing number of publications. These advancements in structural reliability theory and the attainment of more accurate quantification of the uncertainties associated with structural loads and resistances have stimulated the interest in the probabilistic treatment of structures. The reliability of a system or its probability of failure under various loading scenarios is an important factor in the design, construction monitoring, and maintenance procedures of a structure or geostructure since it investigates the probability of the structure or geostructure to complete its design requirements successfully. Therefore, reliability analysis is at the heart of risk analysis methodologies that have been developed for very important structures or geostructures and subsidies (essentially the decision-making procedures), leading to safety measures that the owners and the engineers have to take into account due to the aforementioned uncertainties. Although from a theoretical point of view, the field has reached a stage where the developed methodologies are becoming widespread, from a computational point of view, serious obstacles have been encountered in practical implementations.

The problems considered in this study are inherently complex because of their stochastic characteristics and combinatorial nature. Indeed, as can be seen from the literature, existing modeling efforts exploit advanced mathematical programming formulations and optimization methods for obtaining realistic results while keeping these models computationally tractable. In this context, this work is focused on developing novel, computational intelligence (CI) methodologies for addressing the two optimization problems. CI methodologies were introduced 30 years ago as a new family of computational methods that are based on heuristic approaches rather than on rigorous closed-form mathematics. Optimization is a field where extensive research has been conducted over the last several decades, where many intricate problems have been addressed and many algorithms have been developed. Engineers are constantly challenged with the desire to search for optimal solutions in complex system analysis and designs. The ever-increasing advances in computational power have fueled this temptation, and the oft-used brute force design methodologies are systematically replaced by state-of-the-art, nature-inspired techniques.

Many optimization problems in diverse fields have been solved using different optimization algorithms. Traditional optimization techniques such as linear programming, nonlinear programming (NLP), and dynamic programming (DP) have had major roles in solving these problems. However, their drawbacks generate demand for other types of algorithms, such as heuristic optimization approaches; the most promising of these are algorithms based on the analogies between natural and artificial phenomena (and even the development of hybrid approaches). These algorithms are frequently referred to as *nature-inspired optimization techniques* and have been proved to be quantitatively appealing in that they converge, in general, to satisfactory solutions, in an effective and efficient way even for the most complex problems examined.

15.3 Metaheuristic Algorithms

The two metaheuristic optimization algorithms tested in this chapter appear to be very promising, as they have been implemented in various challenging problems with success. We present here a short description of the algorithms.

15.3.1 Harmony Search

The HS algorithm was originally inspired by the improvisation process of jazz musicians (Geem et al., 2001). According to the analogy between improvisation and optimization, each musician (saxophonists, bassists, guitarists, etc.) corresponds to each decision variable; each musical instrument's pitch range corresponds to a decision variable's value range. Musical harmony at certain times corresponds to a solution vector at certain iterations, and the audience's esthetics corresponds to the objective function. According to this algorithmic concept, the HS algorithm consists of the following five steps: parameter initialization, harmony memory initialization, new harmony improvisation, harmony memory update, and termination criterion check.

Parameter initialization: In the first step, the optimization problem is specified where n is the number of decision variables (equivalent to the number of music instruments), while $s_i^L \leq s_i \leq s_i^U$, $i = 1, 2, \ldots, n$ determines the range of the ith decision variable's value. The HS algorithm parameters are also specified in this step: HMS is the harmony memory size that corresponds to the number of simultaneous solution vectors stored in harmony memory, HMCR defines the harmony memory considering rate, and PAR is the pitch-adjusting rate.

Harmony memory initialization: In the second step, the harmony memory (HM) is initialized with HMS randomly generated solution vectors defining the musician's harmony memory matrix:

$$\text{HM} = \begin{bmatrix} s_1^1 & s_2^1 & s_3^1 & \cdots & s_n^1 \\ s_1^2 & s_2^2 & s_3^2 & \cdots & s_n^2 \\ \vdots & \vdots & \vdots & \ddots & \vdots \\ s_1^{\text{HMS}} & s_2^{\text{HMS}} & s_3^{\text{HMS}} & \cdots & s_n^{\text{HMS}} \end{bmatrix} \tag{15.1}$$

New harmony improvisation: In the third step, a new harmony vector is improvised following three rules: memory consideration, memory consideration, and pitch adjustment. According to the memory consideration, the value of the decision variable s_i is chosen randomly from the pitches stored in $\text{HM} = [s_i^1, s_i^2, \ldots, s_i^{\text{HMS}}]$ with probability of HMCR ($0 \leq \text{HMCR} \leq 1$) or according to random selection it is randomly chosen with a probability of $(1 - \text{HMCR})$ within its value range, as a musician plays any pitch within the instrument's pitch range:

$$s_i = \begin{cases} s_i \in [s_i^1, s_i^2, \ldots, s_i^{\text{HMS}}] \text{ with probability HMCR} \\ s_i^L \leq s_i \leq s_i^U \text{with probability } (1 - \text{HMCR}) \end{cases} \tag{15.2}$$

After the value s_i is randomly picked according to this memory consideration process, it can be further adjusted into neighboring values by adding certain amount to the value, with probability of HMCR × PAR ($0 \leq PAR \leq 1$), while the original pitch obtained in HM consideration is just kept with a probability of HMCR × $(1-PAR)$:

$$s_i = \begin{cases} s_i(k+m) \text{ with probability HMCR} \times \text{PAR} \\ s_i \text{ with probability HMCR} \times (1 - \text{PAR}) \end{cases} \quad (15.3)$$

Harmony memory update: If the new generated harmony vector is better than the worst harmony vector of the HM, with reference to the objective function value, the worst harmony is replaced by the new harmony vector.

15.3.2 Ant Colony Algorithm

The ACO algorithm (Dorigo, 1992; Dorigo and Stützle, 2004) is a population-based probabilistic optimization method inspired by the behavior of real ants in nature and implemented mainly for finding optimal paths through graphs. In ACO, a set of software agents called *artificial ants* search for good solutions to the optimization problem of finding the best path on a weighted graph. The ants incrementally build solutions by moving on the graph. Consider a population of m ants where, at each iteration, every ant defines a "route" by visiting every node sequentially. Initially, ants are set on randomly chosen nodes. At each construction step during an iteration, ant k applies a random action choice rule, called the *random proportional rule*, to decide which node to visit next. While defining the route, an ant k currently at node i maintains a memory $\mathbf{M}^k$, which contains the nodes already visited, in the order they were visited. This memory is used to define the feasible neighborhood $\mathbf{N}^k_i$ that is the set of nodes that have not yet been visited by ant k. In particular, the probability with which ant k, currently at node i, chooses to go to node j is

$$p^k_{i,j} = \frac{(\tau_{i,j})^\alpha \cdot (\eta_{i,j})^\beta}{\sum_{\ell \in \mathbf{N}^k_i}((\tau_{i,\ell})^\alpha \cdot (\eta_{i,\ell})^\beta)}, \quad \text{if } j \in \mathbf{N}^k_i \quad (15.4)$$

where $\tau_{i,j}$ is the amount of pheromone on connection between i and j nodes, α is a parameter to control the influence of $\tau_{i,j}$, β is a parameter to control the influence of $\eta_{i,j}$, and $\eta_{i,j}$ is heuristic information that is available *a priori*, denoting the desirability of connection i,j, given by

$$\eta_{i,j} = \frac{1}{d_{i,j}} \quad (15.5)$$

According to Eq. (15.4), the heuristic desirability of going from node i to node j is inversely proportional to the distance between cities i and j. By definition, the probability of choosing a city outside $\mathbf{N}^k_i$ is zero. By this probabilistic rule, the probability of choosing a particular connection i,j increases with the value of the associated pheromone trail $\tau_{i,j}$ and of the heuristic information value $\eta_{i,j}$. The selection of parameters α and β is very important. After all, ants have defined their routes, the amount of pheromone for each connection between i and j nodes is updated for the next iteration $t + 1$ as follows:

$$\tau_{i,j}(t + 1) = (1 - \rho) \cdot \tau_{i,j}(t) + \sum_{k=1}^{m} \Delta\tau_{i,j}^{k}(t), \quad \forall(i,j) \in A \tag{15.6}$$

where ρ is the rate of pheromone evaporation, a constant parameter of the method, A is the set of arcs (edges or connections) that fully connect the set of nodes, and $\Delta\tau_{i,j}^{k}(t)$ is the amount of pheromone ant k deposits on the connections that it has visited through its tour T^k, typically given by

$$\Delta\tau_{i,j}^{k} = \begin{cases} \dfrac{Q}{L(T^k)} & \text{if connection } (i,j) \text{ belongs to } T^k \\ 0 & \text{otherwise} \end{cases} \tag{15.7}$$

The coefficient ρ must be set to a value <1 to avoid unlimited accumulation of trail (Dorigo and Stützle, 2004), while Q is a constant. In general, connections that are used by many ants and are parts of short tours receive more pheromone, and therefore they are more likely to be chosen by ants in future iterations of the algorithm.

15.4 Scheduling Transportation Networks

This part of the study deals with developing an innovative postdisaster management methodology. Given the catastrophe, the postdisaster phase includes actions for emergency relief and evacuation. Combining the initial conditions with information regarding the natural phenomenon, as well as additional data by other sources, the authorities will be able to prioritize their interventions effectively and assist the population immediately following the disaster. Therefore, postdisaster management is vital in immediate and effective restoration of urban activities. Following a seismic event, all structures must be inspected in the shortest time possible. Scheduling inspection crews is a combinatorial problem on which metaheuristics can be applied, while it is formulated as a two-step problem. In the first step, the structural blocks to be inspected are optimally assigned into a number of inspection areas (a districting problem), while in the second step, the scheduling problem (inspection prioritization) is solved for each of the areas obtained from the first step. In formulating the optimization problems, the area examined is composed of N_{SB} structural blocks, while N_{IG} is the number of crews available for inspecting the condition of the area's infrastructure system.

15.4.1   Step 1: The Optimal Districting Problem

The optimal districting problem is defined as an NLP optimization problem as follows:

$$\min\sum_{i=1}^{N_{\text{IG}}}\sum_{k=1}^{n_{\text{SB}}^{(i)}}\left[\frac{D(k)}{U_{\text{in}}}+\frac{d(\text{SB}_k,C_i)}{U_{\text{tr}}}\right],\quad D(k)=A(k)\cdot\text{BP}(k) \tag{15.8}$$

where $n_{\text{SB}}^{(i)}$ is the number of structural blocks assigned to the ith district (frequently referred to in the infrastructure literature as an *inspection group*), $d(\text{SB}_k,C_i)$ is the distance between the SB_k building block from the starting block of the crew responsible for the ith group of structural blocks, U_{in} is the inspection speed of crews, and U_{tr} is the traveling speed of crews, while $D(k)$ is the "demand" for inspection for the kth building block defined as the product of the building block total area $A(k)$ times the built-up percentage $\text{BP}(k)$ (i.e., percentage of the area with structures). This is a discrete optimization problem since the design variables **s** are integers and denote the inspection groups to which each built-up block has been assigned. Therefore, the total number of the design variables is equal to the number of structural blocks and the range of the design variables is $[1, N_{\text{IG}}]$.

15.4.2   Step 2: The Inspection Prioritization Problem

The inspection prioritization problem is a typical traveling salesman problem (TSP), also defined as an integer optimization problem. In TSP, the task is to find a Hamiltonian tour of minimal length; i.e., to find a closed tour of minimal length that visits each node of a network once. For an N number of cities TSP, there are $(N-1)!$ different tours; the TSP can be represented by a complete weighted graph $G=(N,A)$, with N the set of nodes and A the set of arcs (edges or connections) that fully connects the components of N. A cost function is assigned to every connection between two nodes i and j, represented by the distance between the two nodes $d_{i,j}$ $(i\neq j)$. A solution to the TSP is a permutation $\mathbf{p}=\{p(1),\ldots,p(N)\}$ of the node indices $\{1,\ldots,N\}$, as every node must appear only once in a solution. The optimal solution is the one that minimizes the total length $L(\mathbf{p})$ given by

$$L(\mathbf{p})=\sum_{i=1}^{N-1}(d_{p(i),\,p(i+1)})+d_{p(N),\,p(1)} \tag{15.9}$$

Thus, the corresponding scheduling problem is defined as follows:

$$\min\left[\sum_{k=1}^{n_{\text{SB}}^{(i)}-1}d(\text{SB}_k,\text{SB}_{k+1})+d(\text{SB}_{n_{\text{SB}}^{(i)}},\text{SB}_1)\right],\quad i=1,\ldots,N_{IG} \tag{15.10}$$

where $d(SB_k, SB_{k+1})$ is the distance between the kth and kth $+ 1$ building blocks. The main objective is to define the shortest possible route between the structural blocks that have been assigned to each inspection group in Step 1.

15.4.3 Case Study

In order to assess the performance of the metaheuristic formulation discussed in the districting and TSP frameworks, we consider the city of Patras, in Greece. The city is composed of $N_{SB} = 112$ grouped structural blocks with different areas and built-up percentages. We assume that a total of 10 crews are available for the inspection of the city, divided into two 8 h shifts. This means that for 16 h/day, 5 inspection crews are available. A reduction of computational loads can be achieved by grouping several small building blocks into bigger ones. This can be applied in neighboring building blocks with equal building factors which are surrounded by major road arteries. These grouping parameters ensure the quality of the solution.

In order to identify the performance of the algorithm, a sensitivity analysis regarding all parameters is performed. The sensitivity analysis is performed with reference to the following parameters: (i) the HM size is defined in the range of [5, 20], (ii) the first action probability of the HS algorithm is defined in the range of (0.0, 0.7], (iii) the second action probability of the HS algorithm is defined in the range of (0.0, 1.0), (iv) the number of ants in the colony are defined in the range of [1, 100], (v) the pheromone influence rate is defined in the range of [0.0, 2.0], (vi) the desirability influence rate is defined in the range of $[-1.0, 1.0]$, (vii) the pheromone evaporating rate is defined in the range of [0.0, 1.0), while (viii) the pheromone depositing rate is defined in the range of [0.0, 1.0]. On the other hand, the following parameters remain unchanged: (i) the harmony maximum iterations are equal to 2×10^5, (ii) the ant colony maximum iterations are equal to 400, (iii) the inspection speed is equal to 50 m^2/min, and (iv) the traveling speed is equal to 10 km/h. In order to identify the best combination of the parameters for each metaheuristic algorithm, 32 combinations of the parameters are generated by means of the Latin hypercube sampling method, while for each combination, 100 optimization runs are performed to calculate the mean and the coefficient of variation with reference to the objective function value (Kallioras, 2011). The optimal parameters for the problem are defined based on the minimum time required for the inspection of the city and on the minimum variance of the working hours among the crews. The completion time is defined as the working time of the last finishing crew and minimum difference is defined by the standard deviation of the working time between the crews. The optimal parameters are obtained for the 12th test run, while its parameters are HMS $= 10$, HMCR $= 0.1726$, PAR $= 0.658$, $m = 63$, $\alpha = 0.910$, $\beta = 0.970$, $p = 0.210$, and $q = 0.306$. A detailed description of the sensitivity analysis performed with reference to the algorithm parameters can be found in Kallioras (2011). The objective function value for optimal parameters is 6193.56 h, which represents the sum of working time required for each crew. The working time for each crew is 1062.906, 1193.225, 1100.856, 1397.140, and 1436.904 h for the first to the fifth crew, respectively. As can be seen, the HS method is efficient, resulting in equal distribution of the demand to the five inspection crews.

In the second part of this study, the best combinations of the parameters, found in the previous part of the investigation, are used for solving the problem defined in Eq. (15.8). Figure 15.1 depicts the solution obtained for the optimal allocation problem for the five inspection groups considered. As it can be seen from this figure, a balanced decomposition of the city of Patras is obtained with the implementation of the HS algorithm. In order to compare the resulted optimum designs, the scheduling problem has to be solved for each inspection group by means of the ACO method. Therefore, the inspection prioritization problem defined in Eq. (15.10) is solved by means of the ACO algorithm. Figure 15.2 depicts the optimal routes achieved when the five inspection groups are employed. These solutions correspond to the least time-consuming route required for each inspection crew departing from their base.

15.5   Reliability Analysis of Geostructures

The advancements in reliability theory during the last 20 years and the attainment of more accurate quantification of the uncertainties associated with loads and resistances have stimulated the interest in the probabilistic treatment of the systems (Schuëller, 2006). The reliability of a system or its probability of failure is an important factor in the design procedure since it investigates the probability of the system to accomplish its design requirements successfully. Reliability analysis leads to safety measures that a design engineer has to take into account due to the aforementioned uncertainties. Although from a theoretical point of view, the field has reached a stage where the developed methodologies are becoming widespread,

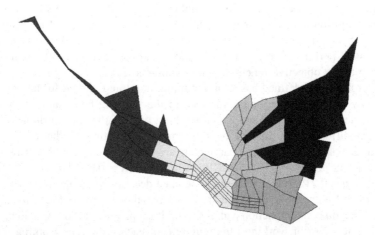

Figure 15.1 Subdivision into structural blocks.

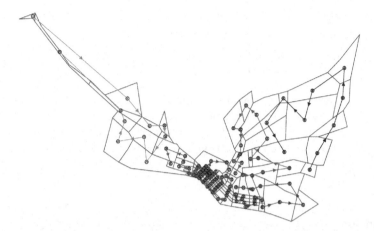

Figure 15.2 Best route for each inspection group.

from a computational point of view, serious obstacles have been encountered in practical implementations. First- and second-order reliability methods that have been developed to perform system reliability, although they lead to elegant formulations, they require prior knowledge of the means and variances of the component random variables and the definition of a differentiable limit-state function. On the other hand, the MCS method is not restricted by the form and the knowledge of the limit-state function but is characterized by high computational cost.

15.5.1 Monte Carlo Simulation

The MCS method is applied in stochastic mechanics when an analytical expression of the limit-state function is not attainable. This is mainly the case in complex problems with a large number of random variables, where all other stochastic analysis methods are not applicable. In stochastic analysis problems, the probability of violation of the behavioral constraints can be written as

$$p_{\text{viol}} = \int\limits_{g(\mathbf{x}) \geq 0} f_x(\mathbf{x}) d\mathbf{x} \tag{15.11}$$

where $f_x(\mathbf{x})$ denotes the joint probability of violation for the random variables, the limit-state function $g(\mathbf{x}) < 0$ defines the safe region and $\mathbf{x}$ is the vector of the m random variables. Considering that MCS is based on the theory of large numbers (N_∞), an unbiased estimator of the probability of violation is given by

$$p_{\text{viol}} = \frac{1}{N_\infty} \sum_{j=1}^{N_\infty} I(\mathbf{x}_j) \tag{15.12}$$

where $\mathbf{x_j}$ is the jth vector of the random parameters and $I(\mathbf{x_j})$ is an indicator for successful and unsuccessful simulations, defined as

$$I(\mathbf{x_j}) = \begin{cases} 1 & \text{if} \quad g(\mathbf{x_j}) \geq 0 \\ 0 & \text{if} \quad g(\mathbf{x_j}) < 0 \end{cases} \tag{15.13}$$

In order to estimate p_{viol}, an adequate number of N_{sim} independent random samples is produced using a specific, uniform probability density function of the vector $\mathbf{x_j}$. The value of the violation function is computed for each random sample $\mathbf{x_j}$ and the Monte Carlo estimation of p_{viol} is given in terms of sample mean by

$$p_{\text{viol}} \cong \frac{N_H}{N_{sim}} \tag{15.14}$$

where N_H is the number of successful simulations and N_{sim} is the total number of simulations.

The basic MCS is simple to use and has the ability to handle practically every possible case regardless of its complexity. However, for typical reliability problems, the computational effort involved becomes excessive due to the enormous sample size required. To reduce the computational effort, more elaborate simulation methods, called *variance reduction techniques*, have been developed. Their efficiency, though, is limited for larger probability values. Moreover, despite the improvements achieved on the efficiency of computational methods, they still require disproportional computational effort for reliability analysis of realistic problems.

15.5.2 First-Order Reliability Method

In the general case of a nonlinear limit-state function, the main objective of the first-order reliability method is to calculate the reliability index β. The Hasofer—Lind reliability index β (Jiang et al., 2007) is calculated by a process of minimization, and the probability of violation is approximated by

$$p_{viol} = \Phi(-\beta) \tag{15.15}$$

where Φ is the standard normal cumulative distribution function. This equation is exact when the failure criterion is linear and all random variables have normal distributions. Given a vector of basic variables $\mathbf{x}$, a failure surface $\partial\omega$, on which the failure criterion $g(\mathbf{x}) = 0$ is satisfied, and a safe region denoted by $g(\mathbf{x}) > 0$, the vector of the reduced variables $\mathbf{z}$ is defined as follows:

$$\mathbf{z} \equiv \mathbf{S}_x^{-1} \cdot (\mathbf{x} - \mathbf{\mu}_x) \tag{15.16}$$

where $\mathbf{S}_x$ is a diagonal matrix of the standard deviations and $\mathbf{\mu}_x$ is the vector of mean values. Then, the Hasofer—Lind reliability index β is defined as

$$\beta \equiv \min_{\mathbf{z} \in \partial\omega} \sqrt{\mathbf{z}^T \mathbf{z}} \tag{15.17}$$

The point on the failure surface $g(\mathbf{x}) = 0$, where its transformation to the $\mathbf{z}$ space satisfies Eq. (15.16), is called the *design* or *most probable point* and will be denoted as $\mathbf{z}_D$. The design point $\mathbf{z}_D$ is located on the limit-state surface, $g(\mathbf{x}) = 0$ and has a minimum distance from the origin in the standard normal space. For applying either first- or second-order methods to complex structural models, it is necessary to have an explicit expression or an approximation of either the entire limit-state function $g(\mathbf{x})$ or its limit-state surface $g(\mathbf{x}) = 0$ in the space of the random variable $\mathbf{x}$. This is because these methods require knowledge not only of the function but also of its gradient in the vicinity of its limit-state surface. In the case of unknown expressions, the limit-state function is usually approximated by the response surface method.

15.5.3 Case Study

In the design of structural or geostructural systems, limiting uncertainties and increasing safety is an important issue to consider. Probabilistic analysis of structures or geostructures, which is defined as the probability that the system meets some specified demands for a specified time period under specified environmental conditions, is used as a probabilistic measure to evaluate the reliability of the system. The performance function of a structural or geostructural system must be determined to describe the system's behavior and to identify the relationship between the basic parameters in the system. It should be noted that in the earthquake-loading environment, the uncertainties related to seismic demand and structure or the geostructure's capacity are strongly coupled. The main scope of the present study is to compare the MCS-based reliability analysis procedure with that of the FORM-based one implemented by the HS method. For this purpose, a real-life pile-group design is considered for a particular soil type and for a given axial load corresponding to the weight of the superstructure and reliability analysis is performed. For clay soil conditions, an elastic−plastic material exhibiting plasticity in the deviatoric stress−strain response only is employed. The volumetric stress−strain response is linear-elastic and is independent of the deviatoric response. This material law can simulate monotonic or cyclic response of materials whose shear behavior is insensitive to the confinement change, such as organic soils or clay under undrained loading conditions. During the application of gravity load, material behavior is linear-elastic. In the subsequent dynamic loading phase(s), the stress−strain response is considered elastic−plastic. Plasticity is formulated based on the multisurface (nested surfaces) concept, with an associative flow rule, while the yield surfaces are of the von Mises type.

Nonlinear static or dynamic analysis needs a detailed simulation of the pile foundation in the regions where inelastic deformations are expected to develop. In order to consider the inelastic behavior of the piles either the plastic-hinge or the fiber approach can be adopted. The plastic-hinge approach has limitations in terms of accuracy, particularly in cyclic loading, and therefore, the fiber beam-column elements are preferred (Fragiadakis and Papadrakakis, 2008). According to the fiber approach, each structural element is discretized into a number of integration sections restrained to the beam kinematics, and each section is divided into a number of fibers with specific material properties. Every fiber in the section can be assigned to different material properties, e.g., concrete, structural steel, or reinforcing bar material properties (Figure 15.3). The sections are located at the Gaussian integration points of the elements. The main advantage of the fiber approach is that every fiber has a simple uniaxial material model allowing an easy and efficient implementation of the inelastic behavior. This approach is considered to be suitable for inelastic beam-column elements under dynamic loading and provides a reliable solution compared to other formulations. In the numerical test examples section that follows, all analyses were performed using the OpenSees platform (McKenna and Fenves, 2001). A bilinear material model with pure kinematic hardening is adopted for the steel reinforcement of the piles. For the simulation of the concrete, the modified Kent and Park (1971) model, as extended by Scott et al. (1982), is employed. This model was chosen because it allows an accurate prediction of the demand for flexure-dominated reinforced concrete (RC) members despite its relatively simple formulation. The transient behavior of the reinforcing bars was simulated with the Menegotto and Pinto (1973) model. Spring elements are implemented for modeling the interaction between piles and the surrounding soil in order to simulate the soil—pile interface. Without the use of springs, the soil and pile elements move together when subjected to any loading or ground motion. With the use of these springs, a more realistic model is achieved, and the relative displacements between the soil and each pile can be simulated. T_z springs were used for the vertical components of the pile interface and P_y springs for the horizontal components (Figure 15.4; Sherif and Elgamal, 2003). All the nodes of the ground base are fully constrained in both x (horizontal) and y (vertical) directions, while the side boundaries are constrained in the x (horizontal) direction.

One real-world building was considered: The 31-story building of the Hyde Park Cavalry Barracks founded on clay (Figure 15.5; Tomlinson and Woodward, 2008), in London. The building is 90 m tall and its weight was calculated to be 228 MN. It is estimated that at the end of construction, 60% (0.60 MN × 228 MN = 136.80 MN) of the building load is carried by the piles and 40% by the raft. In this study, various sources of uncertainty are considered: on the ground motion excitation, which influences the level of seismic demand, and on the modeling and the material properties, which affects the structural capacity. The characteristics of the random variables are provided in Table 15.1. The reliability problem is defined as follows:

$$P_{viol} = P(d_f > 2.0) \qquad\qquad (15.18)$$

where d_f is the horizontal displacement of the foundation at the superstructure level and the objective is to define the probability of exceeding the limit state of 2.0 cm

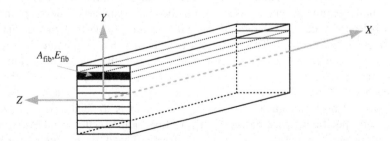

Figure 15.3 Modeling of the inelastic behavior—the fiber approach.

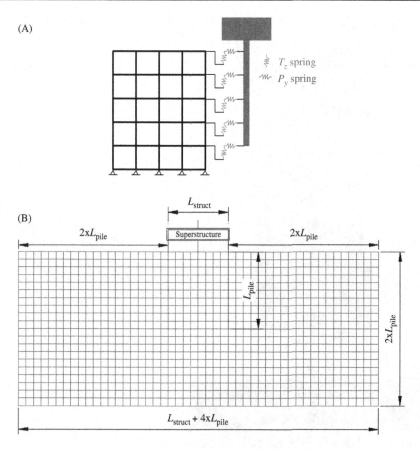

Figure 15.4 (A) Components of the soil−pile interface (T_z and P_y springs) and (B) mesh dimensions.

for the 10% in the 50-year hazard level. The probability that horizontal displacement exceeds the value of 2.0 cm calculated by means of MCS and FORM is given in Table 15.2, along with the number of finite element (FE) analyses required. It can be seen that the FE analyses required by FORM is almost three orders of magnitude less than those required by MCS, while a very good approximation of the probability is achieved. It is worth mentioning that a parametric study was performed with reference to the FE analyses required by MCS in order to define the less required (Piliounis, 2011).

15.6 Conclusions

In this chapter, we presented two successful implementations of metaheuristics. In the first one, a crew-scheduling problem in densely populated metropolitan regions

Figure 15.5 The Hyde Park Cavalry Barracks.

Table 15.1 Random Variables

Random Variable	Distribution	Mean	COV (%)
Rho	Uniform	1.8	15
Shear modulus	Uniform	1.50×10^5	15
Reference bulk modulus	Uniform	7.50×10^5	15
Cohesion	Uniform	75.0	15
Maximum shear strength	Uniform	0.10	15

for critical infrastructure inspection following an earthquake was solved; while in the second, a problem of reliability analysis of a piled foundation was formulated as an optimization problem. Both problems were dealt with efficiently with the HS optimization method, and for the first one, the ACO algorithm was also used successfully. In both problems, a large number of solutions needed to be found and evaluated in search of the optimum one. The two metaheuristics employed in this

Table 15.2 Efficiency of the Procedures

Procedure	P_{viol} (%)	Iterations
MCS	3.27×10^{-4}	1.00×10^6
FORM–HS	3.35×10^{-4}	2.49×10^3

study were found to be efficient in finding an optimized solution, overcoming excessive computational effort, local optima while they are capable of dealing with discrete variables when needed. Metaheuristics offer a broad field for further research and development of sophisticated methods in structural seismic design, reliability analysis, and postseismic infrastructure network restoration. Excessive use of these methods will inevitably lead to further improved metaheuristics and its implementation fields.

References

Altay, N., Greene, W.G., 2006. OR/MS research in disaster operations management. Eur. J. Oper. Res. 175, 475–493.

Au, S-K., Beck, J.L., 2001. Estimation of small failure probabilities in high dimensions by subset simulation. Probab. Eng. Mech. 16 (4), 263–277.

Barbarosoglou, G., Arda, Y., 2004. A two-stage stochastic programming framework for transportation planning in disaster response. J. Oper. Res. Soc. 55 (1), 43–53.

Barbarosoglou, G., Ozdamar, L., Cevik, A., 2002. An interactive approach for hierarchical analysis of helicopter logistics in disaster relief operations. Eur. J. Oper. Res. Soc. 140 (1), 118–133.

Bell, M.G.H., 2000. A game theory approach to measuring the performance reliability of transportation networks. Transport. Res. 34 (6), 533–545.

Bozorg Haddad, O., Afshar, A., Mariño, M.A., 2005. Honey bees mating optimization algorithm: a new heuristic approach for engineering optimization. In: Abdalla J., Sadek I. (Eds.), Proceeding of the First International Conference on Modelling, Simulation and Applied Optimization (ICMSA0/05). Sharjah, UAE, pp. 1–3.

Chang, S.E., Nojima, N., 2001. Measuring post-disaster transportation system performance: the 1995 Kobe earthquake in comparative perspective. Transport. Res. 35 (6), 475–494.

Cret, L., Yamakazi, F., Nagata, S., Katayama, T., 1993. Earthquake damage estimation and decision-analysis for emergency shutoff of city gas networks using fuzzy set theory. Struct. Saf. 12 (1), 1–19.

Dong, W.M., Chiang, W.L., Shah, H.C., 1987. Fuzzy information processing in seismic hazard analysis and decision making. Soil Dynam. Earthquake Eng. 6 (4), 2202–2226.

Dorigo, M., 1992. Optimization, Learning and Natural Algorithms. Politecnico di Milano, Milano, Italy.

Dorigo, M., Stützle, T., 2004. Ant Colony Optimization. The MIT Press, Massachusetts, USA.

Fiedrich, F., Gehbauer, F., Rickers, U., 2000. Optimized resource allocation for emergency response after earthquake disasters. Saf. Sci. 35 (1-3), 41–57.

Fragiadakis, M., Papadrakakis, M., 2008. Modelling, analysis and reliability of seismically excited structures: computational issues. Int. J. Comput. Methods. 5 (4), 483–511.

Geem, Z.W., Kim, J.H., Loganathan, G.V., 2001. A new heuristic optimization algorithm: harmony search. Simulation. 76, 60−68.

Goldberg, D.E., 1989. Genetic Algorithms in Search Optimization and Machine Learning. Addison Wesley, Boston, MA, USA.

Hurtado, JE., 2004. An examination of methods for approximating implicit limit state functions from the viewpoint of statistical learning theory. Struct. Saf. 26 (3), 271−293.

Jiang, C., Han, X., Liu, G.R., 2007. Optimization of structures with uncertain constraints based on convex model and satisfaction degree of interval. Comput. Methods Appl. Mech. Eng. 196 (49−52), 4791−4800.

Kallioras, N.K., 2011. Optimal Inspection of a Building Area Following a Seismic Event with the use of Metaheuristic Methods. National Technical University of Athens, School of Civil Engineering, Diploma Thesis.

Karaouchi, F., Lida, Y., Shimada, H., 2001. Evaluation of road network reliability considering traffic regulation after a disaster. In: Bell, M.G.H., Lida, Y. (Eds.), The Network Reliability of Transport: Proceedings of the First International Symposium on Transportation Network Reliability (INSTR). Elsevier, Oxford, UK.

Karlaftis, M.G., Kepaptsoglou, K.L., Lampropoulos, S., 2007. Fund allocation for transportation network recovery following natural disasters. J. Urban Plann. Dev. 133 (1), 82−89.

Kennedy, J., Eberhart, R., 1995. Particle swarm optimization. In: Proceedings of IEEE International Conference on Neural Networks IV, Piscataway, NJ, pp. 1942−1948.

Kent, DC, Park, R., 1971. Flexural members with confined concrete. J. Struct. Div. 97 (7), 1969−1990.

Kirkpatrick, S., Gelatt Jr., C.D., Vecchi, M.P., 1983. Optimization by simulated annealing. Science. 220, 671−680.

Koutsourelakis, P.S., Pradlwarter, H.J., Schuëller, G.I., 2004. Reliability of structures in high dimensions, part I: algorithms and applications. Probab. Eng. Mech. 19 (4), 409−417.

Lagaros, N.D., Karlaftis, M.G., 2011. A critical assessment of metaheuristics for scheduling emergency infrastructure inspections. Swarm Evol. Comput. 1 (3), 147−163.

Li, Y., Tsukaguchi, H., 2001. Improving the reliability of street networks in highly densely populated urban areas. In: Bell, M.G.H., Lida, Y. (Eds.), The Network Reliability of Transport: Proceedings of the First International Symposium on Transportation Network Reliability (INSTR). Elsevier, Oxford, UK.

McKenna, F., Fenves, G.L., 2001. The OpenSees Command Language Manual—Version 1.2. Pacific Earthquake Engineering Research Centre, University of California, Berkeley, CA.

Mendonca, D., Beroggi, G.E.G., Wallace, W.A., 2001. Decision support for improvisation during emergency response operations. Int. J. Emerg. Manage. 1 (1), 30−38.

Mendonca, D., Beroggi, G.E.G., van Gent, D., 2006. Designing gaming simulations for the assessment of group decision support systems in emergency response. In: Safety Science. vol. 44. Elsevier, Wallace, WA, pp. 523−535.

Menegotto, M., Pinto, P.E., 1973. Method of analysis for cyclically loaded reinforced concrete plane frames including changes in geometry and non-elastic behaviour of elements under combined normal force and bending. In: Proceedings, IABSE Symposium on Resistance and Ultimate Deformability of Structures Acted on by Well Defined Repeated Load. Zurich, Switzerland, Secretariat of IABSE. pp. 15−22.

Nicholson, A., Du, Z-P., 1997. Degradable transportation networks systems: an integrated equilibrium model. Transport. Res. 31 (3), 209−223.

Ozdamar, L., Ekinci, E., Kucukyazici, B., 2004. Emergency logistics planning in natural disasters. Ann. Oper. Res. 129 (1−4), 217−245.

Peizhuangm, W., Xihui, L., Sanchez, E., 1986. Set-valued statistics and its application to earth-quake engineering. Fuzzy Set. Syst. 18 (3), 347–356.

Piliounis, G., 2011. Reliability Analysis of Geostructures using Metaheuristics. National Technical University of Athens, School of Civil Engineering, Diploma Thesis.

Plevris, V., Karlaftis, M.G., Lagaros, N.D., 2010. A swarm intelligence approach for emergency infrastructure inspection scheduling. In: Gopalakrishnan, K., Peeta, S. (Eds.), Sustainable Infrastructure Systems: Simulation, Imaging, and Intelligent Engineering. Springer, pp. 201–230.

Rajashekhar, M.R., Ellingwood, B.R., 1993. A new look at the response surface approach for reliability analysis. Struct. Saf. 12 (3), 205–220.

Sakakibara, H., Kajitani, Y., Okada, N., 2004. Road network robustness for avoiding functional isolation in disasters. J. Transport. Eng. 130 (5), 560–567.

Schuëller, G.I., 2006. Developments in stochastic structural mechanics. Arch. Appl. Mech. 75 (10–12), 755–773.

Scott, B.D., Park, R., Priestley, M.J.N., 1982. Stress–strain behaviour of concrete confined by overlapping hoops at low and high strain rates. ACI J. 79, 13–27.

Sherif, K., Elgamal, A., 2003. Modelling of the Humboldt–Bay Bridge. UCSD, Cal Poly, San Luis Obispo.

Sohn, J., 2006. Evaluating the significance of highway network links under the flood damage: An accessibility approach. Transport. Res. 40 (6), 491–506.

Song, B., Hao, S., Murakami, S., Sadohara, S., 1996. Comprehensive evaluation method on earthquake damage using fuzzy theory. J. Urban Plann. Dev. 122 (1), 1–17.

Song, J., Kim, T.J., Hewings, G.J.D., Lee, J.S., Jang, S.-G., 2003. Retrofit priority of transport network links under an earthquake. J. Urban Plann. Dev. 129 (4), 195–210.

Tamura, H., Yamamoto, K., Tomiyama, S., Hatono, I., 2000. Modelling and analysis of decision making problem for mitigating natural disaster risks. Eur. J. Oper. Res. 122 (2), 461–468.

Tomlinson, M., Woodward, J., 2008. Pile Design and Construction Practice. fifth ed. Taylor and Francis, New York, Chapter 5, pp. 284–287.

Verter, V., Lapierre, S., 2002. Location of preventive healthcare facilities. Ann. Oper. Res. 110, 123–132.

Viswanath, K., Peeta, S., 2003. Multicommodity maximal covering network design problem for planning critical routes for earthquake response. Transport. Res. Rec. 1857, 1–10.

Yang, X.S., 2008. Nature-Inspired Metaheuristic Algorithms. Luniver Press, Frome.

16 Metaheuristic Applications in Highway and Rail Infrastructure Planning and Design: Implications to Energy and Environmental Sustainability

Manoj K. Jha

Center for Advanced Transportation and Infrastructure Engineering Research, Department of Civil Engineering, Morgan State University, Baltimore, MD

16.1   Introduction

Highway and rail infrastructure planning and design present a complex combinatorial optimization problem since many conflicting objectives have to be considered simultaneously in the optimization process. Many metaheuristic applications have been proposed to solve the highway and rail infrastructure planning and design problem, among which genetic algorithms (GAs) are most popular. This is primarily due to GA's ability to perform exhaustive searches and avoid local optima.

The traditional optimization models for highway and rail infrastructure planning do not necessarily feature the minimization of environmental impact and energy (power) consumption. Due to a greater recognition for building green infrastructure and rising gasoline prices, there is a need for considering environmental and energy aspects in sustainable highway and rail infrastructure planning and design.

This chapter presents an overview of two metaheuristic applications, namely, GAs and ant colony optimization (ACO) in highway and rail infrastructure planning and design. A brief overview of the problems dealing with highway and rail infrastructure planning and design is presented, followed by a description of GA and ACO. Finally, a methodological framework for GA and ACO application in highway and rail infrastructure design is presented, followed by a discussion of energy and environmental aspects in sustainable transportation infrastructure planning and design.

Metaheuristics in Water, Geotechnical and Transport Engineering. DOI: http://dx.doi.org/10.1016/B978-0-12-398296-4.00016-7

16.2   Highway Infrastructure Planning and Design

Highway infrastructure planning and design involves the selection of the best economical route, subject to design and operational constraints. In highway design,
points of intersection (PIs) are treated as decision variables. Appropriate curves are
fitted to represent a three-dimensional highway alignment. The traditional method
of highway design involves laying out the horizontal alignment first and then fitting
appropriate vertical alignment (Jha et al., 2006; Jong et al., 2000), to come up with
the full scope of the three-dimensional alignment. In recent years, newer methods
have been introduced to obtain a three-dimensional course of alignment using
spline functions and Beazer curves (Jha et al., 2011a, b; Karri et al., 2012; Kuhn
and Jha, 2011; Kuhn et al., 2011). These methods eliminate the need for designing
horizontal and vertical alignments in separate stages, which may prevent design
inconsistencies when combining horizontal and vertical alignments.

    As noted earlier, among the available metaheuristic applications for highway
infrastructure planning and design, GAs have been extensively applied to obtain an
optimized alignment, subject to design and operational constraints. Other search
methods are generally not appropriate due to the indirect relationship between the
decision variables (which are the PIs) and the objective function (which is a sum of
alignment costs). This issue is discussed in detail in Jong et al. (2000).

16.3   Rail Infrastructure Planning and Design

Similar to highway infrastructure planning and design, rail infrastructure planning
and designing an optimal route for railway track is desired (Jha et al., 2007).
However, the design and operational conditions for railway track design are different than that for highways since the geometric curves of the railway track should
accommodate the safe passage of the entire length of the train. For subway, metro,
and light rail systems designed to serve an urban area, identifying locations of stations become the most important criteria. The station locations can be identified
from the analysis of an origin—destination trip matrix and prevailing demand of the
analysis area. GAs have been used in recent works (Samanta and Jha, 2008, 2011)
to optimally locate station points under variable demand conditions along railway
track routes.

16.4   Discussion of Metaheuristics Commonly Applied in
         Highway and Rail Infrastructure Planning and Design

16.4.1   Genetic Algorithms

The GA is a powerful search algorithm that has been used in many fields for optimization. It is a technique for solving optimization problems, but not all problems

can be solved in its default format. For different systems, one may have to develop different solution procedures based on the philosophies and principles of GAs and the nature of the problem. There is no rigorous proof to show why a GA will converge toward the global optimum. Nevertheless, several theorems and hypotheses have been developed to give a theoretical explanation to the effectiveness of GAs. It has been shown that GAs work quite successfully in many practical applications (Jha et al., 2006; Jong and Schonfeld, 2003).

The theoretical foundations of GAs rely on a binary string representation of solutions (called *chromosomes*) and the notation of a schema. A schema is a string of symbols taken from the alphabet {0,1,*}, where * is a "don't care" symbol, which can be 0 or 1. The number of 0s and 1s in a chromosome is called the *order*. The distance between the positions of the first and the last non-* symbol in a chromosome is called the *defining length.*

The schema theorem and the building block hypothesis (BBH) explain the power of GAs in terms of how schemata are processed. The schema theorem (Goldberg, 1989; Holland, 1975) states that schemata with short defining length, low-order, and better fitness (called *building blocks*) receive exponentially increasing trials in subsequent generations of a GA. The BBH states that a GA seeks near-optimal performance through the juxtaposition of building blocks and genetic operators.

Limitations of GAs

The schema theorem is widely taken to be the foundation for explanations of the power of GAs. Yet some disagreement (Jha et al., 2004; Lee, 1995; Lovell and Jha, 2005; Poli, 2001; Poli et al., 1998; Shakya, 2003; Stephens and Waelbroeck, 1997, 1998, 1999a, b; Yang and Li, 2003) has been expressed in recent years as to its implications. Interpretations of the schema theorem have implicitly assumed that a correlation exists between parent and offspring fitnesses. However, this assumption may be misleading. According to Holland (1975), a particular schema grows as the ratio of the average fitness of the schema to the average fitness of population, i.e., the selection process allocates an increasing number of samples to above-average fit schemata. However, the selection process alone does nothing to promote exploration of new regions of the search space, i.e., it selects only the chromosomes that are already present in the current generation. To avoid such a case, crossover and mutation operators are needed. But crossover and mutation both can create schemata as well as destroy them. The schema theorem considers only the destructive effect of crossover and mutation, i.e., the effect that decreases the number of instances of the schema that occur in subsequent generations.

We can conclude that the short, low-order, above-average fit schemata receive increasing samples in subsequent generations. These short, low-order, above-average fit schemata are known as *building blocks*. The BBH states that the GA seeks the near-optimal performance through the juxtaposition of these building blocks (Goldberg, 1989).

It is important to understand that GAs depend upon the recombination of building blocks to seek the best point. However, if the building blocks are misleading due to the coding used or the function itself, the problem may require long waiting times to arrive at the near-optimal solution. The schema theorem, by itself, addresses the positive effect of selection allocating increasing samples of schemas with observed high performance, but only the negative aspect of crossover (Mitchell, 1998), i.e., it takes into account only the lower bound of survival of schema after crossover takes place. However, many researchers have seen crossover as the major source of the search power of GAs. The schema theorem does not address the question of how crossover works to recombine highly fit schemas. The BBH states that crossover combines short, low-order, highly fit schemas into increasingly fit candidate solutions but does not describe how this combination takes place (Mitchell, 1998).

In schema theory, the search space is partitioned into subspaces of varying levels of generality, and mathematical models are constructed that estimate how the number of individuals in the population belonging to certain schema can be expected to grow in the next generation. The BBH attempts to explain how a GA solves a problem by positing that near-optimal solutions were forged from small, low-order, fitter-than-average schemata. Although long-term solutions can be obtained from certain schema theorems, the mathematics of it becomes quite difficult without the inclusion of infinite population assumptions and the use of Markov modeling methods (Davis and Jose, 1993; Nix and Vose, 1992; Spears and De Jong, 1997). Moreover, the vast majority of schema theoretic results have concentrated on what happens from one generation to the next. For this reason, schema theory can be considered a local analysis method. Much controversy has surrounded the schema theory. The main contention has been its apparent lack of utility. Opponents of schema theory argue that it tells us very little about what is really going on inside the optimal search. Moreover, the traditional Holland/Goldberg schema theorem is pessimistic in the sense that it provides only a lower bound on expected schema growth. Further, it was traditionally developed for GAs with fixed-length, binary representation using standard GA genetic operators.

16.4.2 Ant Colony Optimization

ACO is inspired by the concept of self-organization of swarms and is derived from swarm intelligence. The fundamental idea is that ants organize themselves to travel to the food source and have the ability to follow each other. The two important properties of ACO that basically simulate the real ant system are as follows (Bonabeau et al., 1999):

- *Stigmergy*: This is a property that plays an important role in developing a collective behavior of the social insects. The stimulatory factor pheromone trail is secreted from an ant, the amount of which decides the preference for the next ant to choose a path. This basically depicts the property called *self-organization*.
- *Autocatalysis*: According to this property, the shorter the path, the sooner the pheromone is deposited by the ants, and the more ants use the shorter path. This ensures the fact that the algorithm introduces the chance of *rapid convergence* while heading toward the optimal solution. The important property of this algorithm is the decaying of pheromone,

which influences the convergence by governing the amount of accumulation of pheromone in the paths. It ensures that the search process does not get stuck in the local optima.

The principle of ACO can be explained by the following simple example (Figure 16.1). Suppose that there are two different reversible paths, $\overleftrightarrow{ECADF}$ and $\overrightarrow{ECBDF}$ available to a group of ants. The ants can travel in either direction, with the objective of deriving the shortest path. The $\overleftrightarrow{CAD}$ leg is twice as long as the $\overleftrightarrow{CBD}$ leg. The underlying concept is that ants lay pheromone along the traveled path, which evaporates over time. Thus, the shorter a traveled leg, the longer the pheromone lasts. An ant traveling in the $\overrightarrow{ECDF}$ direction is faced with two options, $\overrightarrow{CBD}$ or $\overrightarrow{CAD}$ at point C. The decision of choosing a path over the other at point C is purely arbitrary and has equal probability. But the probability of choosing the shorter path $\overrightarrow{CBD}$ grows in time for the follower ants, as the pheromone trail left by preceding ants lasts longer on the shorter path. After sufficient time intervals, all ants converge to the shortest path. A three-time-interval scenario with a 30-ant sample is shown in Figure 16.1.

So far, ACO has been widely and successfully implemented for solving discrete optimization problems. It has been tried on both static and dynamic combinatorial optimization problems (Dorigo et al., 1999). A few examples of static optimization problem are given as follows:

Traveling salesman problem: In this problem, n cities are traveled in such a way that the total traveling cost is minimized. ACO has shown better performance than the GA for a small problem (30-city problem), but not for a larger problem (Dorigo et al., 1997).

Quadratic assignment problem: It is the problem of assigning n facilities in n locations so that the total cost of assignment is minimized. The results obtained for this problem is shown in Dorigo et al. (1999).

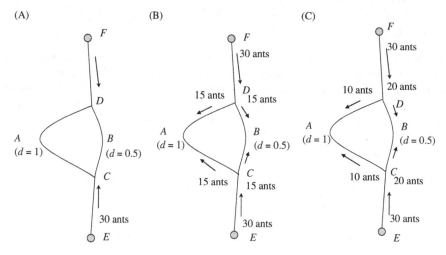

Figure 16.1 An example of ACO: (A) $t = 0$, (B) $t = 1$, and (C) $t = 2$.

Job-shop scheduling problem: A set *J* of jobs are to be assigned on *M* machines in such a fashion that the total completion time is minimal, satisfying the constraint that no two jobs can be processed on the same machine at the same time. ACO was able to find 10% of the optimal value of the results for a 15-machine, 15-job problem (Dorigo et al., 1999).

Vehicle routing problem: This problem is about obtaining minimum-cost vehicle routes for a fleet of vehicles starting from a depot or more than one depot. Dorigo et al. (1999) applied ACO on this problem and obtained reasonable results.

Some of the dynamic optimization problems where ACO is applied are the connection-oriented network routing problem and the connectionless network routing problem.

ACO for Searching in a Continuous Space

Although ACO has been applied to many discrete optimization problems, not many applications (Bilchev and Parmee, 1995; Kaveh and Talatahari, 2010) to continuous optimization problems have been observed. Bilchev and Parmee (1995) tried to solve a flight trajectory for an air-launched winged rocket that will achieve orbit before returning to atmosphere for a conventional landing by discretizing the continuous search space.

This approach says that the continuous nest neighborhood is divided in a finite number of directions represented by vectors (Figure 16.2). Feasible regions are first randomly placed in the search space, or they may correspond to regularly sampled directions from the nest. The agent then chooses a random direction and moves a short distance from the region's center in that direction with a probability proportional to the virtual pheromone concentration of the path that goes from the nest to the region. Agents reinforce their paths according to their performance, depending on the diffusion, evaporation, and recombination of the trails.

Limitations of ACO

Essentially, then, it can be concluded that ACO fails to exploit a continuous search space the way that GA does. The continuous space must be discretized for ACO application, one of the inherent impediments in ACO application that is often ignored by researchers. Using the discretization concept for ACO application described in Bilchev and Parmee (1995), many real-world problems can be solved

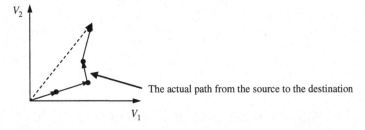

Figure 16.2 Discretization of a continuous search space.

and results compared with those obtained with GAs. We have performed a comparative assessment of GA and ACO in highway and rail-alignment optimization in some of our recent works (Jha and Samanta 2006; Samanta and Jha, 2012). Using the discretization concept, we attempted to reformulate the highway alignment optimization (HAO) problem in order to seek its solution with ACO. Preliminary analyses were presented in some of our earlier works (Jha, 2001, 2002).

16.5   GA Application in Highway and Rail Infrastructure Planning and Design

In highway infrastructure planning and design, an optimal highway alignment needs to be obtained that connects any two given end points in the feasible solution space. Since numerous possibilities exist to connect two points in space, one cannot start with a given set of alignments and then implement the search algorithm, as it will not necessarily ensure a global optimum. Our research team (Jha et al., 2006), therefore, devised a procedure based on what is called "orthogonal cutting planes" in our previous studies to generate a highway alignment randomly.

It was assumed that a highway alignment can be sufficiently described by a set of intermediate points $P_i's$ between given start and end points. In order to describe an alignment, first a straight line is drawn, connecting the given start and end points. Next, orthogonal planes (lines for two-dimensional alignments) at random intervals are cut across that line (Figure 16.3). $P_i's$ are randomly placed along these lines. If a P_i falls along the straight line, then a tangent section is obtained; otherwise, a curved section is obtained. Appropriate curves can be fitted using American Association of State Highway and Transportation Officials design criteria. Thus, an alignment can be sufficiently described by (i) random location of cutting planes and (ii) random location of $P_i's$ along the planes. Once an alignment is described, its associated costs (such as right-of-way, pavement, construction, environmental

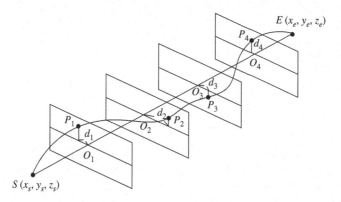

Figure 16.3 Representation of a three-dimensional alignment for optimization formulation (Jha et al., 2006).

impact, accident, travel-time delay, and vehicle operating costs) can be calculated by developing cost functions in sufficient detail (Jong and Schonfeld, 1999). Note that in Figure 16.3, each intersection point in the three-dimensional space can be determined by only two decision variables (the abscissa and ordinate on the vertical cutting plane). This helps to reduce the dimension of the search space. Then, by trigonometry, we can transform these intersection points further into the Cartesian coordinate system. Let O_i be the point at which the line segment $\overline{SE}$ intersects the ith cutting plane, where S and E are the start and end points of the alignment. Then the X and Y coordinates of O_i can be obtained by simple trigonometric transformation. On each vertical cutting plane, the abscissa, denoted by d_i, is defined as the axis that passes through O_i and parallels the XY plane, with O_i as its origin. The ordinate on the ith cutting plane is simply defined as the Z coordinate in the Cartesian coordinate system to reduce coordinate transformation requirements.

Let P_i be the intersection point on the ith cutting plane, whose coordinates are (d_i, z_i). Then, the Cartesian coordinates of P_i, denoted by $(x_{P_i}, y_{P_i}, z_{P_i})$ can be obtained by:

$$\begin{bmatrix} x_{P_i} \\ y_{P_i} \\ z_{P_i} \end{bmatrix} = \begin{bmatrix} x_{O_i} \\ y_{O_i} \\ 0 \end{bmatrix} + \begin{bmatrix} d_i \cos\theta \\ d_i \sin\theta \\ z_i \end{bmatrix} \tag{16.1}$$

where (x_{O_i}, y_{O_i}) are the coordinates of the origin of the abscissa on the ith cutting plane and θ is the angle of cutting planes on the XY plane.

16.5.1 Optimization Formulation

The single-objective optimization included formulation of a single-objective function and a set of constraints. The objective function consists of alignment-sensitive costs, such as user cost (C_U), right-of-way cost (C_R), pavement cost (C_P), earthwork cost (C_R), and structure cost (C_S), as shown in Eq. (16.2a). These costs are formulated as functions of decision variables. Additional cost functions can be formulated as desired.

$$\underset{x_{P_1}, y_{P_1}, z_{P_1}, \dots, x_{P_n}, y_{P_n}, z_{P_n}}{\text{Minimize}} \quad C_T = C_U + C_R + C_P + C_E + C_S \tag{16.2a}$$

$$\text{subject to } x_O \leq x_{P_i} \leq x_{\max}, \quad \forall\, i = 1, \dots, n \tag{16.2b}$$

$$y_O \leq y_{P_i} \leq y_{\max}, \quad \forall\, i = 1, \dots, n \tag{16.2c}$$

$$z_O \leq z_{P_i} \leq z_{\max}, \quad \forall\, i = 1, \dots, n \tag{16.2d}$$

where (x_O, y_O) = the X, Y coordinates of the bottom-left corner of the study region (shown in Figure 16.4):

(x_{P_i}, y_{P_i}) = the X, Y coordinates of PIs, P_i;
$(x_{\max}, y_{\max})$ = the X, Y coordinates of the top-right corner of the study region.

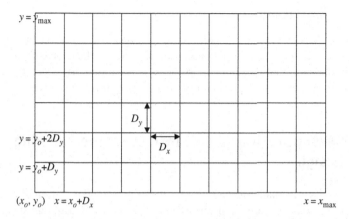

Figure 16.4 An example of study area for alignment optimization.

The user cost consists of travel-time cost, vehicle operating cost, and accident cost. The right-of-way cost consists of the land area taken by the alignment and damage to the properties (Jha and Schonfeld, 2000) and is calculated directly from a Geographic Information System (GIS) and transmitted to GAs during optimal search. The detailed formulations of alignment-sensitive costs shown in Eq. (16.2a) are provided in previously published works (Jha and Schonfeld, 2004; Jha et al., 2006) and have been omitted here for brevity. To eliminate any confusion, it is pointed out here that effects of traffic congestion, travel-time delay, and topography are all considered in the developed cost functions. There are also many design and operational constraints to be met in alignment optimization. Among those the minimum length of vertical curves, gradient constraint, sight-distance constraint, and environmental constraints are important ones, which have been sufficiently formulated in previous works.

16.5.2 Genetic Encoding of Alignment Alternatives

In GAs, the decision variables are encoded as binary or real numbers, called *chromosomes*. For an alignment represented by n PIs, the encoded chromosome is composed of $3n$ genes. Thus, the chromosome is defined as (Jong and Schonfeld, 2003):

$$\Lambda = [\lambda_1, \lambda_2, \lambda_3, \ldots, \lambda_{3n-2}, \lambda_{3n-1}, \lambda_{3n}] = [x_{P_1}, y_{P_1}, z_{P_1}, \ldots, x_{P_n}, y_{P_n}, z_{P_n}] \quad (16.3)$$

where

Λ = chromosome
λ_i = the ith gene, for all $i = 1, \ldots, 3n$
$(x_{P_i}, y_{P_i}, z_{P_i})$ = the coordinates of the ith point of intersection, for all $i = 1, \ldots, n$.

For the HAO problem, the chromosomes are encoded by real number representation.

16.5.3   Genetic Operators

Eight problem-specific genetic operators are designed to work on the encoded PI rather than on individual genes (Jong and Schonfeld, 2003). Extensive tests are conducted to ensure that these operators assist in obtaining efficient solutions while exploiting the entire search space and without getting trapped in local optima.

16.6   GA Application to Rail Infrastructure Planning and Design

In rail infrastructure planning and design, locating potential station sites based on the prevailing demand pattern of an urban region is a key problem to be investigated for the met heuristics application (Figure 16.5). In this problem, any point along the given rail line is a feasible station location so long as the minimum station spacing constraint is satisfied. Recall from the HAO formulation that the decision variables were set to points along the orthogonal cutting planes. Thus, it is observed that the station location optimization (SLO) problem is a condensed version of the HAO problem with the following differences: (i) the rail line is available *a priori* and (ii) curve fitting is not required.

16.6.1   The Genetic Station Location Optimization Algorithm

The genetic station location optimization algorithm (GSLOA) to the proposed research problem can be described as follows (Jha and Oluokun, 2004; Jha and Samanta, 2006).

Let there be N possible intermediate stations that can be accommodated between the starting station (S) and destination (CBD). The total rail line length is L. Let the minimum distance to be maintained between station pairs for acceleration and deceleration be ΔS_{min}. Therefore, $N \leq \{(L/\Delta S_{min}) - 1\}$, where N is a positive integer. Further, let us assume that once the stations $Z = \{S, S_1, S_2, \ldots, S_k, \ldots, S_N,$ CBD$\}$ are

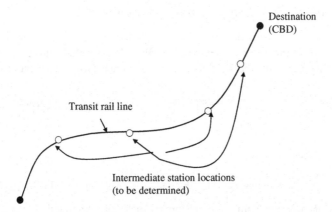

Figure 16.5 The transit rail station location optimization problem.

specified, the approximate travel times from a demand node P_i ($i = 1,2,\ldots,R$) to a station (or if driving directly to the CBD is attractive, the user can so choose) can be obtained from a GIS database.

Steps of the GSLO Algorithm

Step 1: Generate a random number (positive integer) between 1 and N: $r_d(1,N) = k$, where k is an integer. This is the number of intermediate stations.

Step 2: Generate k random numbers $(r_{d1},r_{d2},\ldots,r_{dk}) = [\lambda_1,\lambda_2,\ldots,\lambda_k]$ in the interval $(\Delta S_{min},\ L - \Delta S_{min})$, which represent the interstation spacings. This also represents the initial population. $\lambda_1,\lambda_2,\ldots,\lambda_n$ represent the distances of the possible intermediate stations from the starting station.

Step 3: Calculate the fitness (total cost function) of the population members.

Step 4: Apply mutation and crossover operators. Mutation operator is developed by randomly selecting a gene and replacing it by a randomly selected real number within the limiting values. The mutated chromosome becomes $[\lambda_1,\lambda_2,\ldots,\lambda_k',\ldots,\lambda_n]$, where λ_k' is the replaced value (the changed distance from the starting station).

Step 5a: Develop the crossover operator by selecting two solutions from the population randomly. A part containing one or more than one station is identified from each of these two solutions, and they are exchanged maintaining the feasibility conditions to produce two new solutions.

Step 5b: Develop a selection–replacement scheme to ensure efficient convergence toward the global optimum.

Step 6: Iterate through the specified number of generations by repeating steps 1–5.

Step 7: Stop when a specified number of generations have been searched or improvement in the objective function value is negligible (within 1%). Obtain the optimal station sequence and associated optimal cost.

The total cost function (or the fitness function) for the SLO problem may consist of user, operator, and construction costs. Let TC be the total cost incurred for a station location, and the objective function is defined as

$$\text{Minimize TC} = \text{UC} + \text{OC} + \text{CC} \tag{16.4}$$

where

UC = user cost = unit travel cost (UTC) $\times$ total travel time,
total travel time = access time (t_a) + waiting time (t_w) + in-vehicle travel time (t_i)
OC = operator cost = unit operator cost (UOC) $\times$ [vehicle travel time (t_v) + standing time (t_s) + time loss in acceleration–deceleration (t_{loss})].

These costs have to be formulated as a function of the decision variables, which are the coordinates of the station points.

16.7 The Ant Highway Alignment Optimization Algorithm

Recall that the orthogonal cutting plane principle of alignment representation in the preceding section. In this study, we limit our analysis to two-dimensional HAO and leave the three-dimensional coverage for future works. A set of possible PI points are generated randomly along each orthogonal line. We assume that the number of possible PIs, which is equal to the number of orthogonal cuts, is fixed in the search space. We then assume that each artificial ant always starts at the "start point" and ends its tour at the "end point." Let $k = 1,\ldots,m$ be the number of ants; $i = 1,\ldots,n$ be the number of orthogonal lines (0 and $n + 1$ represent start and end points, respectively); and $j(i) = 1,\ldots,p(i)$ be the number of PIs along the ith orthogonal line (Figure 16.6). Now, m ants will select the best point as the PI among the set of possible points along each orthogonal line depending on the minimum cost. The pheromone intensities are updated locally and globally. We start with the first orthogonal line. The steps are repeated for jth iterations to obtain the best PI on the first orthogonal line. The same procedure is performed for the other orthogonal lines. There may be different number of PIs along each vertical cut $\forall i = 1,\ldots,n$. After we obtain n PIs on n orthogonal lines, the total cost is calculated. Further, let $|\Delta h_{j(i)l(i+1)}|$, $|\Delta r_{j(i)l(i+1)}|$, and $|\Delta d_{j(i)l(i+1)}|$ be the absolute values of elevation differences (for three-dimensional alignment), unit land cost differences, and the distance between $j(i)$ and $j(i + 1)$, respectively. Now, let us define the ant visibility as

$$\eta_{j(i)l(i+1)} = \alpha\frac{\min|\Delta h_{j(i)\xi_h(i+1)}|}{|\Delta h_{j(i)l(i+1)}|} + \beta\frac{\min|\Delta r_{j(i)\xi_r(i+1)}|}{|\Delta r_{j(i)l(i+1)}|} + \gamma\frac{\min|\Delta d_{j(i)\xi_d(i+1)}|}{|\Delta d_{j(i)l(i+1)}|}$$

(16.5)

where $\xi_h(i + 1)$ is the point whose height difference with that at $j(i)$ is the minimum among all points along the vertical cut $(i + 1)$. Similarly, $\xi_r(i + 1)$ is the point whose unit land cost difference with that at $j(i)$ is the minimum and $\xi_d(i + 1)$ is the

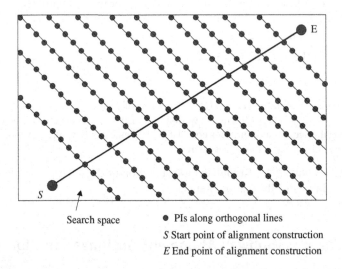

Search space   • PIs along orthogonal lines
                     S Start point of alignment construction
                     E End point of alignment construction

Figure 16.6 Configuration of the search space.

point that is nearest to $j(i)$. α, β, and γ are user-defined parameters such that $\alpha + \beta + \gamma = 1$. The redefinition of the visibility in this problem is based on the fact that the less differences in terrain elevations and unit land costs for any pair of points between successive vertical cuts, and the closer those points are to each other, the more desirable the connecting path becomes. This is because a lower value of terrain elevation difference will result in a smaller earthwork cost. Similarly, a low value of unit land cost difference will result in a smaller earthwork cost, and shorter links will result in smaller user costs (i.e., travel time, vehicle operating, and accident costs). Thus, the total cost, which is the objective function, will be smaller, and the resulting alignment more desirable.

Let $l_k(x)$ be the tour length of the kth ant between time t and $t + x$ and $Q_{l_k}(x)$ be the objective function value of the alignment corresponding to $l_k(x)$. It is noted that using the intermediate PIs that form $l_k(x)$, a smooth alignment (Figure 16.7) is constructed by fitting appropriate curves using the procedure of Jong and Schonfeld (2003). The total cost (objective function value) of the alignments is computed using the cost functions developed by Jha et al. (2006). The pheromone laid at link $\{j(i)l(i + 1)\}$ by the kth ant between time t and $t + x$ is specified as

$$\Delta\tau_{j(i)l(i+1)}^{k} = \begin{cases} \dfrac{\min\{Q_{l_g}(x)\}}{Q_{l_k}(x)} & \text{if the } k\text{th ant uses edge} \\ & \{j(i)l(i + 1)\} \text{ in its tour between time } t \text{ and } t + x \\ 0 & \text{otherwise} \end{cases}$$

$$(16.6)$$

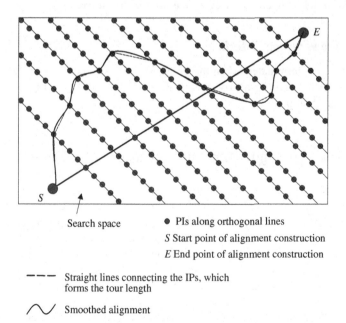

Figure 16.7 Tour length and smoothed alignment.

where min $\{Q_{l_g}(x)\}$ is the minimum value of the objective function corresponding to the tour lengths formed by all ants between time t and $t + x$. This equation implies that the lower the value of $Q_{l_k}(x)$, the higher the amount of pheromone laid. The trial intensity, $\tau_{j(i)l(i+1)}\{t + x\}$ on edge $j(i)l(i + 1)$ at time $t + x$, is specified as

$$\tau_{j(i)l(i+1)}\{t + x\} = \rho \tau_{j(i)l(i+1)}\{t\} + \Delta \tau_{j(i)l(i+1)} \tag{16.7}$$

where ρ is the coefficient such that $(1 - \rho)$ represents the evaporation of trail between time t and $t + x$. $\Delta \tau_{j(i)l(i+1)}$ is expressed as

$$\Delta \tau_{j(i)l(i+1)} = \sum_{k=1}^{m} \Delta \tau_{j(i)l(i+1)}^{k} \tag{16.8}$$

The transition probability $P_{j(i)l(i+1)}^{k}$ for ant k from point $j(i)$ to $l(i + 1)$ is specified as

$$P_{j(i)l(i+1)}^{k} = \begin{cases} \dfrac{[\tau_{j(i)l(i+1)}\{t\}]^{\alpha_1} [\eta_{j(i)l(i+1)}]^{\beta_1}}{\sum_{v(i+1) \in \Omega_{k(i+1)}} [\tau_{j(i)v(i+1)}\{t\}]^{\alpha_1} [\eta_{j(i)v(i+1)}]^{\beta_1}} & \text{if } l(i + 1) \in \Omega_{k(i+1)} \\ 0 & \text{otherwise} \end{cases} \tag{16.9}$$

where $\Omega_{k(i+1)}$ is the set of all feasible nodes to be visited by ant k along the vertical cut $(i + 1)$, which is updated for every ant after every move.

16.7.1 Convergence Criteria

The algorithm is terminated when the objective functions corresponding to the tour lengths of every ant is nearly identical, i.e., if $Q_{l_1}(t + x) = Q_{l_2}(t + x) = \cdots = Q_{l_m}(t + x)$, then stop.

16.8 The Ant Algorithm Applied to the SLO Problem

It is worth noting that the Ant Algorithm (AA) concept is based on "ant movement," and therefore, a suitable path must be designed along which artificial ant movement can occur. Using this concept for the SLO problem, we assume that m number of ants reside at r demand nodes (population centers) at $t = 0$. Here, it is assumed that transit demand of the catchment area is concentrated at r different population centers. Further, it is assumed that the demand pattern is many to one, i. e., every transit rider from the demand nodes is destined to the final station along the rail line (typically, the CBD).

For the SLO problem, n sets of stations are generated first. Each of these sets contains k intermediate stations. At $t = 1$, the first ant from each of the demand nodes selects a set of feasible transit stations along the rail line with equal probability. At this stage, the problem reduces to the classic AA, with the exception that the objective is to minimize the total cost rather than the travel distance. As time elapses, the follower ants will select the set where some pheromone trail exists. Following this trend after sufficient time intervals, the ants will choose the set that results in the minimum total cost. The total cost corresponding to the selected number of stations at every iteration (the movement of the m ants between $t = 0$ to $t = t + m$ for every station set is called an *iteration*) is calculated for every iteration of randomly generated station sets at the conclusion of $t + m$ ant moves. After a sufficient number of iterations, the solutions converge to an optimum, i.e., an optimum number of stations and their positions along the transit line are obtained.

Let $\tau_{ij}(t)$ be the intensity of pheromone trail on the path connecting the ith demand node to the jth set of stations at time t. After every ant move, the next ant experiences a pheromone trail, given as

$$\tau_{ij}(\Delta t) = \rho \tau_{ij}(t) + \Delta \tau_{ij} \tag{16.10}$$

where ρ is a coefficient such that $(1 - \rho)$ represents the *evaporation* of trail during the Δt time interval. $\Delta \tau_{ij}$ is the quantity per unit cost of pheromone laid on the path ij in the Δt time interval and is given by:

$$\Delta \tau_{ij} = \sum_{k=1}^{m} \Delta \tau_{ij}^{k} \tag{16.11}$$

where $\Delta \tau_{ij}^{k}$ is the pheromone quantity per unit cost laid on path ij in the Δt time interval by the kth ant. It is given as

$$\Delta \tau_{ij}^{k} = \left\{ \begin{array}{ll} \dfrac{Q}{C_{ij}^{k}} & \text{if the } k\text{th ant travels link } (i,j) \text{ in time } \Delta t \\ 0 & \text{otherwise} \end{array} \right\} \tag{16.12}$$

Each ant chooses a set of stations with a probability that is an inverse function of the total cost and of the amount of pheromone trail present on the connecting path. It chooses the set of stations to go to with a probability given by:

$$p_{ij}^{k} = \left\{ \begin{array}{ll} \dfrac{[\tau_{ij}(t)]^{\alpha}[\eta_{ij}]^{\beta}}{\sum_{k \in \text{allowed}_k}[\tau_{ik}(t)]^{\alpha}[\eta_{ik}]^{\beta}} & \text{if } j \text{ belongs to allowed}_k \\ 0 & \text{otherwise} \end{array} \right. \tag{16.13}$$

where allowed$_k = \{m - \text{tabu}_k\}$ and α and β are parameters that control the relative importance of trail versus visibility. Therefore, the movement probability is a trade-off between visibility (which implies that economical stations should be chosen with high probability, thus implementing a greedy constructive heuristic) and trail intensity at time t (which implies that if many ants travel along link (i,j), then it is highly desirable, thus implementing the autocatalytic process). The ant visibility η_{ij} is given as

$$\eta_{ij} = \frac{1}{TC_{ij}} \tag{16.14}$$

where TC_{ij} = total cost incurred to travel to the jth set of stations from the demand node i.

16.8.1   The Ant Station Location Optimization Algorithm

The steps of the ant station location optimization algorithm are given here:

Step 1: Generate a random number (an integer) between 1 and N: $r_d(1,N) = k$, where k is an integer. This is the number of intermediate stations.
Step 2: Generate k random numbers $(r_{d1}, r_{d2}, \ldots, r_{dk}) = [\lambda_1, \lambda_2, \ldots, \lambda_k]$ in the interval $(\Delta S_{min}, L - \Delta S_{min})$, which represent the interstation spacings. $\lambda_1, \lambda_2, \ldots, \lambda_n$ represent the distances of the possible intermediate stations from the starting station. The decision variables are the coordinates of the points representing the intermediate stations.
Step 3: Generate n such sets of points, each representing a set of intermediate stations such as $S_1 = [\lambda_1, \lambda_2, \ldots, \lambda_k]$, $S_2 = [\lambda'_1, \lambda'_2, \ldots, \lambda'_k] \ldots S_n$.
Step 4a: Perform the first ant move to any of the sets probabilistically at $t = t + 1$.
Step 4b: Update pheromone trail intensity depending on the associated total cost.
Step 5: Make the next ant move to any of the sets. The movement occurs probabilistically depending on the intensity of pheromone present on the corresponding link.
Step 6: Go to step 4a and continue with m ants. The option with the least cost involved will be the best option for that particular iteration.
Step 7: Update the pheromone intensities globally.
Step 8: Perform step 4a−7 for j iterations.
Step 9: Stop when the solution converges and obtain the optimal station sequence and associated optimal cost.

16.9   Implications to Environment and Energy Sustainability

In the highway and rail infrastructure planning procedures described in the preceding sections, the factors contributing to environmental and energy sustainability are not comprehensively formulated. Some of the environmental factors, such as impacts to floodplains, wetlands, and other environmentally sensitive regions, are considered in the highway and rail-alignment optimization modeling framework by imposing appropriate penalties (Davis and Jha, 2011; Kang et al., 2012; Samanta and Jha, 2011). In addition, the travel-time delay and vehicle operating costs (including fuel consumption costs) are also formulated in previous works for HAO (Jong and Schonfeld, 1999).

16.9.1 Air Pollution due to Vehicular Traffic

Air pollution is a phenomenon by which particles (solid or liquid) and gases contaminate the environment. Such contamination can result in health effects on the population, which might be either chronic (arising from long-term exposure) or acute (due to accidents). The vehicular emission is the highest contributor (as high as 60%) to the total pollution in the air; therefore, measures should be taken to reduce the traffic volume, especially during rush hours. An assessment of the potential air pollution impact of a proposed road development relies on the following information:

> *Traffic volume*: The key factor in air emissions is the traffic volume (measured as vehicle–kilometers per hour by vehicle type). Often, an understanding of traffic peaks and their duration will be required in order to make meaningful projections of emission levels.
> *Traffic composition*: A percentage breakdown of the number of vehicles by type. Heavy trucks and buses are distinguished from light passenger vehicles, newer vehicles from older ones, and diesel-powered vehicles from those that are gasoline-powered.
> *Speed of traffic*: Average speed of vehicles, with some indication of the consistency of speed (degree of traffic congestion).
> *Dispersion dynamics*: Dispersion of pollutants is dictated by prevailing wind direction, weather conditions, roadside vegetation, topography, and distance from road.
> *Vehicle emission levels, by major pollution*: Useful indicators might be mean annual emissions, hourly concentration peaks, and daily value exceeded once a year.
> *Road surface*: Whether the road is paved or not makes a difference to the amount of dust generated. Once the current and projected pollution levels have been determined, comparisons can be made with industrial, regional, and national standards for air quality.

Generally, there is no penalty to the road user due to the vehicular emission that each vehicle contributes toward the atmospheric pollution. In future works, a modified user equilibrium formulation can be developed to penalize the road user by imposing a toll for elevated levels of vehicular emission (Jha et al., 2011a, b).

16.9.2 Suggested Approaches to Considering Environmental and Energy Sustainability in Highway and Rail Infrastructure Planning

An alternate approach to considering environmental sustainability in the highway infrastructure planning process is by developing a modified user equilibrium formulation under the assumption that users consider the impact of vehicular pollution in addition to travel time minimization in choosing their routes between specified Origin-Destination (O-D) pairs (Jha et al., 2011a, b). For rail infrastructure planning, noise seems to be the main contributor to environmental impact. Therefore, various competing corridors for railway route planning can be analyzed based on the objective of noise level reduction. A formulation for perceived noise level due to railway movement can be provided and considered in the rail transit optimization models (Samanta and Jha, 2009, 2011).

For considering energy sustainability in highway infrastructure planning, highway alignments should be analyzed based on the criteria of fuel cost minimization. Generally, congestion- and obstruction-free, flatter, and straighter roads will require less frequent acceleration and deceleration, resulting in savings in fuel costs. Similarly, rail tracks along flatter grades and express trains with less frequent stoppages will reduce energy costs related to train motion. A comprehensive formulation of factors contributing to environmental and energy sustainability will be developed in future works.

16.10 Conclusions and Future Works

In this chapter, two metaheuristic applications (namely, GA and ACO) have been extensively discussed for highway and rail infrastructure planning. In the past 15 years, the problem of HAO has been extensively studied and solved by GAs. However, ACO are not fully explored for both highway and rail infrastructure planning. In general, ACO works well for discrete optimization problems (Samanta and Jha, 2012). But only a few studies have been done to test ACO in continuous search spaces by developing specialized procedures to discretize the search space. The comparison of GAs and ACO is done only on limited problem sets. Therefore, additional study should be done in future works.

In this study, we also discussed the implication of environmental and energy sustainability, which seems to be a critical element for future infrastructure planning and design problems. We provided a brief framework to incorporate environmental and energy sustainability in highway and rail infrastructure planning. Additional work, including comprehensive formulation of the associated factors, need to be done in future studies.

Acknowledgments

This work was completed at the Center for Advanced Transportation and Infrastructure Engineering Research at Morgan State University.

References

Bilchev, G., Parmee, C., 1995. The ant colony metaphor for searching continuous design spaces. In: Fogarty, T.C. (eds.): Proceedings of the AISB Workshop on Evolutionary Computation. LNCS Vol. 993, Springer-Verlag, Berlin, Germany, 25—39.
Bonabeau, E., Dorigo, M., Theraulaz, G., 1999. From Natural to Artificial Swarm Intelligence. Oxford University Press, London.
Davis, C., Jha, M.K., 2011. A dynamic modeling approach to stimulating socioeconomic measures in highway planning and sustainability. Transport. Res., Part A. 45 (7), 598—610.

Davis, T.E., Jose, C., 1993. A Markov chain framework for the simple genetic algorithm. Evol. Comput. 1 (3), 269–288.

Dorigo, M., Caro, G.D., Gambardella, L.M., 1997. Ant colonies for the traveling salesman problem. Biosystems. 43, 73–81.

Dorigo, M., Caro, G.D., Gambardella, L.M., 1999. Ant algorithms for discrete optimization. Artif. Life. 5 (3), 137–172.

Goldberg, D.E., 1989. Genetic Algorithms in Search, Optimization, and Machine Learning. Addison-Wesley Publishing Company, Inc., Boston, MA.

Holland, J.H., 1975. Adaptation in Natural and Artificial Systems. The University of Michigan Press, Ann Arbor, MI.

Jha, M.K., 2001. Geographic information systems and artificial intelligence integration in transportation: an overview. In: Proceedings of the Joint Meeting of the Fifth World Multiconference on Systematics, Cybernatics and Informatics and Seventh International Conference on Information Systems Analysis and Synthesis. Vol. VII, 11–16, Orlando, FL.

Jha, M.K., 2002. Optimizing highway networks: a genetic algorithm and swarm intelligence based approach. In: Songer, A., Miles, J. (Eds.), Computing in Civil Engineering. ASCE Press, Reston, VA, pp. 76–89.

Jha, M.K., Schonfeld, P., 2000. Geographic information system-based analysis of right-of-way cost for highway optimization. Transport. Res. Rec. 1719, 241–249.

Jha, M.K., Oluokun, C., 2004. Optimizing station locations along transit rail lines with geographic information systems and artificial intelligence. In: Allan, J., Brebbia, C.A., Hill, R.J., Sciutto, G., Sone, S. (Eds.), Computers in Railways IX (COMPRAIL 2004). WIT Press, Southampton, UK.

Jha, M.K., Schonfeld, P., 2004. A highway alignment optimization model using geographic information systems. Transport. Res., Part A. 38 (6), 455–481.

Jha, M.K., Samanta, S., 2006. A comparative study of GA and AA for transportation location optimization problems. In: Transportation Research Board, Proceedings of the 85th Transportation Research Board Annual Meeting, Washington DC, Paper # 06-1582.

Jha, M.K., Lovell, D.J., Kim, E., 2004. Discussion of the Paper "Genetic-Algorithm-Based Approach for Optimal Location of Transit Repair Vehicles on Complex Network, by M. Karlaftis, K. Kepaptsoglou, and A. Stathopoulos. Transport. Res. Rec. 1879, 41–50.

Jha, M.K., Schonfeld, P., Jong, J.-C., Kim, E., 2006. Intelligent Road Design. WIT Press, Southampton, UK.

Jha, M.K., Schonfeld, P., Samanta, S., 2007. Optimizing transit rail routes with genetic algorithms and GIS. J. Urban Plann. Dev. 133 (3), 161–171.

Jha, M.K., Karri, G.A.K., Kuhn, W., 2011a. A new 3-dimensional highway design methodology for sight distance measurement. Transport. Res. Rec. J. Transport. Res. Board. 2262, 74–82.

Jha, M.K., Namdeo, A., Shiva Nagendra, S.M., Li, J., 2011b. Investigating the relative impacts of travel-time delay and vehicular emission on road user choice behavior: case studies from US, UK, and India. In: *proceedings of the 1st Conference of Transportation Research Group of India,* Bangalore, India, Dec. 7–10.

Jong, J.-C., Schonfeld, P., 1999. Cost functions for optimizing highway alignments. Transport. Res. Rec. 1659, 58–67.

Jong, J.-C., Schonfeld, P., 2003. An evolutionary model for simultaneously optimizing 3-dimensional highway alignments. Transport. Res., Part B. 37 (2), 107–128.

Jong, J.-C., Jha, M.K., Schonfeld, P., 2000. Preliminary highway design with genetic highway alignments. *Computer-Aided Civil and Infrastructure Engineering.* 15 (4), 261–271.

Kang, M.-W., Jha, M.K., Schonfeld, P., 2012. Applicability of highway alignment optimization models. Transport. Res., Part C. 21 (1), 257–286.

Karri, G., Maji, A., Jha, M.K., 2012. Optimizing geometric elements of a three-dimensional alignment in a single-stage highway design process. In: Proceedings of the 2012 Annual Transportation Research Board (TRB) Meeting, Washington, DC, Paper Number 12-4470.

Kaveh, A., Talatahari, S., 2010. An improved ant colony optimization for constrained engineering design problems. Eng. Comput. 27 (1), 155–182.

Kuhn, W., Jha, M.K., 2011. Methodology for checking shortcomings in the three-dimensional alignment. Transport. Res. Rec. J. Transport. Res. Board. 2262, 13–21.

Kuhn, W., Volker, H., Kubik, R., 2011. Workplace simulator for geometric design of rural roads. Transport. Res. Rec. J. Transport. Eng. 2241, 109–117.

Lee, A., 1995. The schema theorem and price's theorem. In: Whitley, D., Vose, M.D. (Eds.), Foundations of Genetic Algorithms 3. Morgan Kaufmann, San Francisco, CA, pp. 23–49.

Lovell, D.J., Jha, M.K., 2005. Issues with schema theorem in genetic algorithms: a new upper bound for the number of matching schemata. In: Proceedings of the *First International Conference on Simulation and Applied Optimization*, Sharjah, United Arab Emirates, February, Paper # 250.

Mitchell, M., 1998. An Introduction to Genetic Algorithm. MIT Press, Cambridge, MA.

Nix, A.E., Vose, M.D., 1992. Modeling genetic algorithms with Markov chains. Ann. Math. Artif. Intell. 5, 79–88.

Poli, R., 2001. Exact schema theory for genetic programming and variable-length genetic algorithms with one-point crossover. Genetic Programming and Evolvable Machines. Kluwer Academic Publishers, Dordrecht, The Netherlands, pp. 123–163.

Poli, R., Langdon, W.B., O'Reilly, U.-M., 1998. Analysis of schema variance and short term extinction likelihoods. Genetic Programming. Morgan Kaufmann, San Francisco, CA, pp. 284–292.

Samanta, S., Jha, M.K., 2008. Identifying feasible locations for rail transit stations: two-stage analytical model. Transport. Res. Rec. 2063, J. Transport. Res. Board. 81–88.

Samanta, S., Jha, M.K., 2011. Modeling a rail transit alignment considering different objectives. Transport. Res., Part A. 45 (1), 31–45.

Samanta, S., Jha, M.K., 2012. Applicability of genetic and ant algorithms in highway alignment and rail transit station location optimization. Int. J. Oper. Res. Inf. Syst. 3 (1), 13–36.

Shakya, S.K., 2003. A Simple Analysis of Schema Theorem and Royal Road Functions. School of Computing, The Robert Gordon University, Aberdeen, Scotland, UK (http://sidshakya.com/Articals/SchemataAnalysisandRRfunctions.htm).

Spears, W.M., De Jong, K.A., 1997. Analyzing GAs using Markov models with semantically ordered and lumped States. Foundations of GAs IV. Morgan Kaufmann, San Francisco, CA, pp. 85–100.

Stephens, C.R., Waelbroeck, H., 1997. Effective degree of freedom in genetic algorithms and the block hypothesis. In: Back, T. (Ed.), Proceedings of the Seventh International Conference on Genetic Algorithms. Morgan Kaufmann, San Mateo, pp. 34–40.

Stephens, C., Waelbroeck, H., 1998. Analysis of the effective degrees of freedom in genetic algorithms. Phys. Rev. E57, 3251–3264.

Stephens, C., Waelbroeck, H., 1999a. Schemata evolution and building blocks. Evol. Comput. 7 (2), 109–124.

Stephens, C., Waelbroeck, H., 1999b. Schemata as building blocks: does size matter? In: Banzhaf, W., Reeves, C. (Eds.), Foundations of Genetic Algorithms 5. Morgan Kaufmann, San Francisco, CA, pp. 117–135.

Yang, H., Li, M., 2003. Form invariance of schema and exact schema theorem. Sci. Chin. Ser. 46 (6), 475–484.

17 Multiobjective Optimization of Delay and Stops in Traffic Signal Networks

Khewal Bhupendra Kesur

School of Statistics and Actuarial Science, University of the
Witwatersrand, Johannesburg, South Africa

17.1 Introduction

When optimizing a network of traffic signals, the traffic engineer aims to set signal timings so as to minimize or maximize one or more objective criteria. The most commonly used criteria are delay and the number of stops. These measures of quality of service have been cited as the most important factors in user perception of level of service (Flannery et al., 2005; Pecheux et al., 2004; Sutaria and Haynes, 1977). It is not possible in general to obtain a utopian signal timing plan that simultaneously minimizes both these objectives. Instead, a compromise between these two important but competing objectives has to be made. The trade-off between delay and the number of stops has been demonstrated in several studies (Berg and Do, 1981; Jovanis and May, 1978; Leonard and Recker, 1997; Leonard and Rodegerdts, 1998).

In practice, delay and the number of stops are usually combined into a single measure and a single-objective optimization methodology is applied to this composite measure. For example, the commercial signal timing optimization package TRANSYT-7F (Hale, 2005) minimizes a weighted sum of delay and the number of stops called the *disutility index*. Weights can be chosen to reflect excess fuel consumption or operating cost of vehicles. Often, a pure delay-minimization strategy is used, as it has been found to provide a reasonable compromise with respect to the number of stops and other competing objectives (Berg and Do, 1981; Jovanis and May, 1978; Park et al., 2000). Ultimately, the weightings applied to these noncommensurable quantities are subjective, and optimization will result in a single-compromise solution. Furthermore, the weightings must be chosen without any knowledge of the set of possible solutions.

A more informed decision could be made if the entire set of trade-off solutions could be identified. This can be accomplished by performing multiobjective optimization. Using the concept of Pareto-optimality, multiobjective optimization can be used to identify the set of solutions that offer the best possible compromise between delay and the number of stops. Multiobjective optimization also allows

Metaheuristics in Water, Geotechnical and Transport Engineering. DOI: http://dx.doi.org/10.1016/B978-0-12-398296-4.00017-9

the traffic engineer to evaluate and understand the compromise between these competing objectives.

Genetic algorithms (GAs), which are heuristic optimizers based on the evolutionary concepts, have emerged as a proven and widely accepted technique for optimizing traffic signal timing, with many successful applications in the case of a single objective[1]. GAs have also been found to be most successful in multiobjective optimization (Deb, 2001). The Nondominated Sorting Genetic Algorithm II (NSGA-II) is one of the most popular and best performing multiobjective GAs (Deb et al., 2002).

In this chapter, we investigate the feasibility of applying NSGA-II to the multiobjective problem of minimizing delay and the number of stops in large traffic networks under fixed-time signal control. Under- and oversaturated traffic conditions will be considered. This is a challenging optimization problem due to the presence of a large number of decision variables. Furthermore, detailed microscopic stochastic traffic simulation models are required to perform realistic modeling in signalized networks. These models are computationally expensive, placing a limit on the number of signal timing policies that can be examined in the GA search. Furthermore, these simulation models can only provide estimates of mean delay and the number of stops.

Furthermore, simple modifications to NSGA-II will be tested to obtain the optimal algorithm design. After identifying the optimal design for NSGA-II, multiobjective optimization will be compared to single-objective optimization as follows: The minimum delay solutions from the set of compromise solutions produced by multiobjective optimization will be contrasted with those arising from a pure delay-minimization strategy and evaluated in terms of delay and the associated number of stops. A similar comparison will be made for the solution with the minimal number of stops. This comparison will serve to highlight any advantages and disadvantages offered by multiobjective optimization.

The trade-off between delay and the number of stops will be evaluated by examining the set of optimal compromise solutions produced by multiobjective optimization. Finally, the role of signal timings in this trade-off is examined to provide further insight.

The material in the remainder of this chapter is organized as follows:

- Section 17.2 provides background information, covering fundamental concepts in multiobjective optimization, a description of the NSGA-II multiobjective optimization algorithm, and a review of applications of GAs in traffic signal optimization.
- Section 17.3 covers the potential modifications to the design of NSGA-II that will be tested in order to identify the most efficient algorithm design for multiobjective optimization of delay and the number of stops.
- Section 17.4 is an exposition of the study methodology. The Microscopic Stochastic Traffic Network Simulator (MSTRANS), the traffic simulation model used for estimating delay and the number of stops, is covered first. The encoding scheme used to transform signal timing variables into genetic material amenable to manipulation by genetic operators follows. Details of the study test cases are given next. The approach for ranking and comparing optimization results from the execution of NSGA-II with different designs is

---

[1] A thorough review will be provided in a later section of this chapter.

also described. Finally, further details on the implementation of the multi- and single-objective optimizers are given.

- Section 17.5 presents and interprets the results from the experiments performed. The alternative designs of NSGA-II are examined and the optimal design is identified based on the results. After identifying the optimal design for NSGA-II, results from single- and multiobjective optimization runs are compared. The trade-off between delay and the number of stops are examined next. Finally, the role of cycle length, green splits, and signal phasing in this trade-off are examined.
- Section 17.6 is the conclusion, which summarizes the findings, recommendations, and ideas for further research on the topic considered in this chapter.
- References are provided at the end of the chapter.

17.2 Background

17.2.1 Pareto-optimality

Suppose that we have three separate signal timing plans, A, B, and C, with associated delay and number of stops, as illustrated in Figure 17.1.

Clearly, signal timing plan B is preferable to A, as it produces less delay for the same number of stops. A solution is said to "Pareto-dominate" another if it is no worse in all objectives and provides an improvement in at least one objective. Thus, points B and C dominate A. However, B and C do not dominate each other. Comparing B and C, we find that B provides a reduction in delay at the cost of an increased number of stops. The set of feasible solutions that are not dominated by any other solutions is called the Pareto-optimal set. The set of objective vectors associated with the Pareto-optimal set is called the *Pareto front* and is illustrated in Figure 17.1. Solutions in the Pareto-optimal set represent the best possible compromises with respect to the competing objectives. In multiobjective optimization, the aim is to obtain or approximate this set.

17.2.2 Nondominated Sorting Genetic Algorithm II

NSGA-II (Deb et al., 2002) is a heuristic multiobjective optimizer based on the GA optimization approach. Unlike most conventional search algorithms, GAs search from a population of points, producing an entire set of solutions as the optimization

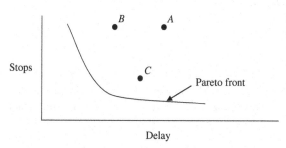

Figure 17.1 Pareto-dominance and Pareto-optimality.

outcome. The multiobjective GA can thus produce a set of solutions that approximates the Pareto-optimal set in a single run. The search progresses by manipulating a population of N solutions using operators that mimic the evolutionary process in biology. The individual solutions are analogous to the individuals in an evolving population. A generational approach is used in that the entire population of points is evolved from one generation to the next. The algorithm details are discussed under the following headings:

- Problem encoding
- Initialization
- Selection
- Reproduction
- Replacement
- Stopping criteria.

Problem Encoding

The decision variables of each individual are encoded into a form of genetic material that is acted upon by genetic operators. The most common encoding scheme is binary encoding, where the decision variables are transformed into a binary string. For the optimization of a problem in a single, real-valued variable $x \in [a,b]$, a binary string of length l can be used. If

$$y = y_l y_{l-1} \ldots y_2 y_1, \quad y_k \in \{0, 1\} \quad \forall \; k \in \{1, 2, \ldots, l\} \tag{17.1}$$

denotes a particular point in the search space in the binary coding, then the corresponding value in the problem space, x, can be computed by converting the binary value to a decimal value in the range $\{0,1,2,\ldots,2^l - 1\}$ and then converting this into a value in the range $[a,b]$ using a linear mapping, i.e.,

$$x = a + \frac{b-a}{2^l - 1} \sum\nolimits_{k=1}^{l} y_k 2^{k-1} \tag{17.2}$$

For a multiparameter optimization problem, each decision variable x_j may have its own search domain $[a_j, b_j]$ and binary string z_j of length l_j. The vector of decision variables $\underline{x} = (x_1, x_2, \ldots)$ can then be represented by a single binary string of length $L = \sum_{j=1} l_j$ which is the concatenation of the binary strings $z_1, z_2, \ldots$.

Initialization

The initial population or generation of N individuals is allocated to random points in the search domain. This can be performed by independently setting the bit values in the binary strings for each individual to either 0 or 1 with equal probability.

Selection

Selection is employed to form a mating pool of individuals. The selection process is biased so as to guide the search toward higher-performance and less-represented regions in the objective space. First, a ranking of the points based on dominance is obtained. The set of nondominated individuals in the population are assigned a domination rank of 1. The nondominated individuals in the remaining set of points are assigned a domination rank of 2 and so on. A set of solutions with the same domination rank is referred to as a *nondominated front.*

After ranking by dominance is performed, a crowding distance measure is computed for each individual in each nondominated front. For the case of the bivariate traffic signal optimization problem with m individuals in the nondominated front illustrated in Figure 17.2, the crowding distance for individual i is obtained as the perimeter of the rectangle formed by using the nearest neighbors as the vertices. In order to allow for the different scales along each dimension in the objective space, the length of the rectangle along the dimension of delay is standardized by dividing it by $\text{Delay}^{max} - \text{Delay}^{min}$. A similar standardization is performed for the length of the rectangle along the dimension of stops. The crowding distance for individuals 1 and m at the edge of the nondominated front are assigned a crowding distance of ∞.

Binary tournament selection is then applied to insert individuals into the mating pool by randomly selecting two individuals and choosing the individual with the lower domination rank. If the two individuals have the same domination rank, the one with the larger crowding distance measure is chosen. This procedure is repeated until a mating pool of the same size as the population is obtained. Discrimination between individuals based on dominance and crowding distance encourages the discovery of a diverse approximation to the Pareto set. Individuals in the mating pool are then paired for reproduction.

Reproduction

Crossover and mutation are employed to generate two children from each pair in the mating pool. These operators act on the encoded genetic material. Crossover randomly combines or blends genetic material of the two parents. For the case of a binary problem encoding, uniform crossover is recommended (Eshelman et al., 1989; Syswerda, 1989). With uniform crossover, individual bits in the binary

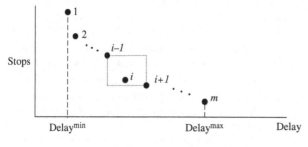

Figure 17.2 Illustration of crowding distance measure.

strings of two parents are swapped independently with probability 0.5 forming two children. Crossover allows for the combination of useful traits. For example, two solutions, one with a favorable cycle length and another with a favorable phase sequence may be combined to form an offspring inheriting positive traits from the parents. Mutation induces random alterations to the decision variables to allow the examination of new search points, as well as the restoration of lost genetic material. Mutation is performed by flipping each bit in the children independently based on a specified probability.

Replacement

An elitist replacement scheme is used to determine the constituents of the subsequent generation. The parent and child populations are combined, forming a population of $2N$ individuals. The individuals in the combined population are sorted by increasing order of domination rank first and then by decreasing order based on the crowding distance measure. After sorting, the first N individuals are retained to form the constituents of the subsequent generation.

Stopping Criteria

Initialization is performed and the remaining steps are repeated until a stopping criterion is met. The most common stopping condition is the completion of a prespecified number of objective function evaluations. The individuals with a domination rank of 1 in the final generation are taken as the approximation of the Pareto-optimal set. A flowchart encompassing the various steps of NSGA-II is shown in Figure 17.3.

17.2.3   GAs in Traffic Signal Optimization

Single-Objective Optimization

Foy et al. (1992) were the first to apply GAs to the traffic signal timing problem, considering delay minimization for a hypothetical four-signal network. Memon and Bullen (1996) compared the optimization efficiency of a quasi-Newton search and a GA on a five-signal traffic network and noted superior performance of the GA for more complicated optimization scenarios. Oda et al. (1996) found that GA optimization outperformed both hill-climbing and random searching on two large-grid networks. Park (1998) also found GA optimization to provide an improvement over hill-climbing for hypothetical two- and four-signal networks. For a simplified two-signal network, the GA found a solution with delay only 1% larger than that of a full enumerative search. Other successful applications of GAs in single-objective optimization of traffic signals are Park et al. (2000, 2001), Stevanovic et al. (2007), and Kesur (2009, 2012). Commercial signal timing software packages such as TRANSYT-7F (Hale, 2005) and PASSER V (Chaudhary et al., 2002) now have support for signal timing optimization by GAs.

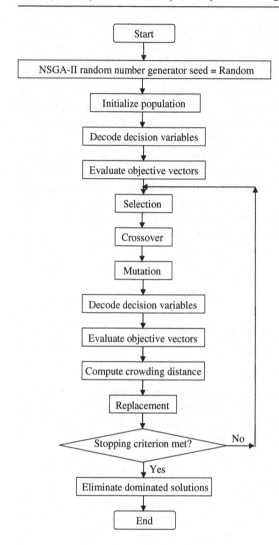

Figure 17.3 NSGA-II flowchart.

Multiobjective Optimization

Sun et al. (2003) applied NSGA-II to the minimization of average control delay and the average number of stops for an isolated intersection under two-phase control. Delay and the number of stops were computed via an analytical formula. Branke et al. (2007) also applied NSGA-II to an isolated intersection, but they considered a traffic-actuated controller and emphasized the minimization of travel time and the number of stops.

Although Abbas et al. (2007) already considered multiobjective optimization of delay and the number of stops for a small, three-signal network, NSGA-II chose signal settings from a predetermined set obtained from a single-objective optimization at different cycle lengths. Thus, the multiobjective optimizer only considered variations in the cycle length, simplifying the exercise considerably. In this study, the full range of signal timings will be examined in the optimization.

The author has already applied some of the findings from this chapter, which were unpublished at the time, for the minimization of both overall delay and the inequitable distribution of delay along various routes through a network using NSGA-II (Kesur, 2010). The multiobjective optimizer was successfully applied and was found to offer considerable advantages over single-objective optimization. The work in this chapter serves as a precursor for this application of multiobjective optimization.

Stevanovic et al. (2011) have applied a multiobjective GA to the optimization of both throughput and a surrogate safety measure in a 12-signal arterial network. The multiobjective GA used was identical to NSGA-II, but without a crowding distance mechanism for discriminating between solutions with identical domination rank. However, the quality of the approximate Pareto-optimal front was not examined, and a comparison with single-objective optimization was not made.

17.3 Modifications to NSGA-II Design

From an extensive review of literature on multiobjective GAs, several alternatives to the NSGA-II design have been identified as potential methods for improving the efficiency of the optimization.

GAs traditionally use a binary encoding of the decision variables. Most applications in the optimization of traffic signal timings have used binary encoding. Real or integer representations have become more popular over time. With these representations, encoding and problem spaces correspond. Genetic operators thus manipulate genetic material directly in the problem space. A real encoding of signal timing variables has been found to improve optimization in the case of a single objective (Kesur, 2009). When applying a real-problem encoding, crossover and mutation operators need to be modified. Blend crossover (BLX-α), which is a generalization of a number of proposed real crossover operators will be tested (Deb, 2001). BLX-α crossover is applied independently to each decision variable and requires a parameter of $\alpha \geq 0$. For the ith decision variable, let x_i and y_i denote the values of the decision variables in the two parents and without loss of generality, assume that $x_i \leq y_i$. The value of this same decision variable in an offspring is computed by simulating a uniform random variable on the interval $(x_i - \varepsilon_i, y_i + \varepsilon_i)$, where

$$\varepsilon_i = \alpha(y_i - x_i) \qquad\qquad (17.3)$$

A graphical illustration of the interval above is seen in Figure 17.4.

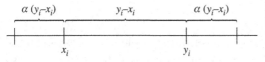

Figure 17.4 Graphical illustration of blend crossover.

The value of $\alpha = 0.5$ has been shown to be optimal for the traffic signal timing problem (Kesur, 2009). Mutation with real encoding can be achieved by replacing the value of a decision variable with a uniformly distributed value in the allowable search domain (Deb, 2001). The potential improvement in multiobjective optimization using the real encoding is investigated in this chapter.

A further avenue for improvement is the selection mechanism. NSGA-II uses tournament selection, which favors better-performing individuals. This may lead to predominant dissemination of genetic material from nondominated individuals, leading to premature convergence. In single-objective optimization, the selective pressure induced by the elitist replacement mechanism alone has been found to be sufficient in guiding the search toward better-performing regions of the search space (Eshelman, 1991). Selective pressure can be reduced using the following uniform selection mechanism: The mating pool is obtained by copying the individuals from the current population in random order. This ensures that each individual receives exactly one reproductive opportunity. This is the selection mechanism used in the Cross-generational elitist selection, Heterogeneous recombination, and Cataclysmic mutation single-objective optimization algorithm (CHC) (Eshelman, 1991) which has shown excellent performance. There is some limited evidence demonstrating the effectiveness of uniform selection methods in multiobjective optimization (Kim et al., 2004; Mumford, 2004; Toffolo and Benini, 2003). The uniform selection method will be evaluated on the multiobjective traffic signal optimization problem.

The uniform selection method has been recommended for application in conjunction with reproduction without mutation (Eshelman, 1991). A highly explorative binary crossover operator (HUX) is recommended in this case. HUX randomly swaps half the differing bits from the two parents to form the two children. Uniform selection and HUX slow convergence, allowing for reproduction to take place without the need for the disruptive effect of mutation to reintroduce diversity. The augmentation of uniform selection with HUX and the removal of mutation will be tested.

The sensitivity of NSGA-II to the choice of the population size will also be examined. GAs are known to be sensitive to parameter settings, the population size being of particular importance. Although parameter tuning for traffic signal optimization has been performed (Kesur, 2009), it has been done for the single-objective case.

17.4 Methodology

17.4.1 Microscopic Stochastic Traffic Network Simulation

The MSTRANS model is a stochastic microscopic traffic simulation model developed by Kesur (2011). The logic and parameters are based on findings from the literature on driver behavior and vehicle characteristics. The model applies a fixed increment time step to advance the simulation. Vehicle status and kinematics are updated each second, along with the traffic signal indications. New vehicles are generated at the network boundary with exponentially distributed headways. Vehicle routing through the network is generated stochastically based on the specified expected turning proportions. Linear acceleration and constant deceleration models are applied to lead vehicles, and the behavior of following vehicles is based on the Gipps microscopic car-following model (Gipps, 1981). Left-turning vehicles receiving an unprotected green signal use a gap-acceptance model, where the probability of accepting a gap increases with the gap size according to a logistic function. Stopping decisions on amber signals are also made using a logistic model. Response delays are modeled and lane changes are governed by pragmatic rules. The delay measure considered in this study is control delay, which is computed as the difference between actual and uninterrupted travel time in MSTRANS. The number of stops is computed using the approach of Rakha et al. (2001). According to their approach, the partial stops for a vehicle at time step t are given by:

$$s_t = \frac{\max(v_{t-1} - v_t, 0)}{v_f} \tag{17.4}$$

where

v_t = instantaneous speed of vehicle at time step t and
v_f = free-flow speed.

The number of stops experienced by a vehicle is computed by summing the partial stops over all time steps. By accounting for partial stops, the measurement for the number of stops is appropriate in both under- and oversaturated conditions.

An initialization period is completed before results are recorded. Delay and the number of stops are averaged for all vehicles that complete their trips through the network during the specified simulation period. However, to account for the delay and stops being experienced by residual queues remaining at the end of the simulation period, the run length is extended until all remaining vehicles have cleared the network.

Several validation exercises have been performed, and MSTRANS has been found to give results comparable to the Corridor Simulation Model (CORSIM) (FHWA, 2003; Kesur, 2011), a commerically available and widely used microscopic traffic simulator. A thorough account of the functional details of the model and a review of the literature justifying the logic is given in Kesur (2011). A full code listing is provided in Kesur (2007).

17.4.2 Problem Encoding

The encoding of the signal timing variables is performed using the fraction-based encoding scheme of Kesur (2009). This encoding allows for all signal timing variables to be considered in the optimization; namely, cycle length, green times, signal offsets, and phase structure and sequence. The encoding of phase structure and sequence is discussed first, followed by the specification of the encoding of all other signal timing variables.

Encoding of Phase Structure and Sequence

For the case of a signal controlling conflicting flow from two perpendicular two-way streets, there are four approaches, which can be labeled North (N), East (E), South (S), and West (W). It is assumed that green indications are given to vehicles on the N and S approaches (the N/S green phase), followed by green indications for vehicles on the E and W approaches (the E/W green phase).

For approach E, green indications can be given for all movements for the entire duration of the E/W green phase. Alternatively, a green indication can be given for only part of the E/W green phase. If this is the case, then the remainder of the phase can be used to allow for protected left-turn movements approach W. Let δ_E = green staging for traffic movements on approach E.

The three staging possibilities of green indications for approach E are:

1. $\delta_E = 0$, i.e., no protected stage for opposing left-turning movements
2. $\delta_E = 1$, i.e., protected stage for opposing left-turning movements at the end of the green stage (lagging stage)
3. $\delta_E = 2$, i.e., protected stage for opposing left-turning movements at the beginning of the green stage (leading stage).

Let

ρ = the duration of E/W green phase(s)
g_E^1 = the duration of stage with green indications for all traffic movements on approach E(s), and
g_E^2 = the duration of stage allowing protected left-turn movements for left-turning vehicles on approach W(s).

The staging possibilities are illustrated in Figure 17.5.

Similarly, three staging possibilities can be defined for green indications for approach W. Taking the cross-product of the staging possibilities for the E and W approaches, we obtain nine phasing possibilities for the E/W green phase. Green indications applied in the N/S green phase can also be allotted in nine possible ways. Taking the cross-product, we obtain 81 phasing possibilities at each traffic signal.

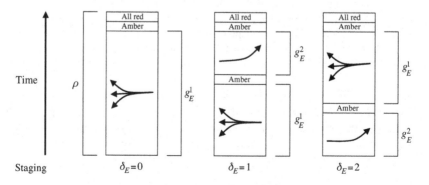

Figure 17.5 Staging of green time for traffic movements on approach E.

Genetic Encoding of all Signal Timing Variables

Let

N_s = the number of traffic signals in the network under study
i = a particular traffic signal $\in \{1, 2, \ldots, N_s\}$
j = green phase $\in \{N/S, E/W\}$
k = the approach at signal $\in \{N, E, S, W\}$
$G(k)$ = the green phase corresponding to a particular approach

$$= \begin{cases} N/S \text{ for } k \in \{N, S\} \\ E/W \text{ for } k \in \{E, W\} \end{cases} \qquad (17.5)$$

A = the duration of the amber interval (s)
R = the duration of the all-red interval (s)
C_{min} = the minimum cycle length to search (s)
C_{max} = the maximum cycle length to search (s)
$g_{min,stage}$ = the minimum duration of any green stage (s), and
$g_{min,phase}$ = the minimum duration of the green phase, including signal change period (s)[2]:

$$= 2(g_{min,stage} + A) + R \qquad (17.6)$$

Now, for a particular individual in the genetic search, let

C = the cycle length (s)
ϕ_i = the offset of traffic signal i with respect to the beginning of its N/S green phase relative to the start of the N/S green phase at signal 1 (s)
$\rho_{i,j}$ = the duration of green phase j at traffic signal i, including the amber and all-red transition periods (s)
$\delta_{i,k}$ = the sequence of green stages for approach k at traffic signal i, ($\delta_{i,k} \in \{0,1,2\}$)
$g_{i,k}^1$ = the duration of the stage with green indications for all traffic movements on approach k at traffic signal i (s), and
$g_{i,k}^2$ = the duration of stage allowing for protected left-turn movements on the approach with vehicle flow opposing that on approach k at traffic signal i (s).

For a particular decision variable, let

x = the value of the decision variable in the problem space
l = the number of bits in the binary encoding representing x
$y = y_l y_{l-1} \ldots y_1$ = the binary value mapping to x, and
$D(x)$ = the decoded value of x expressed as a fraction in the range [0,1]:

$$= \frac{1}{2^l - 1} \sum_{m=1}^{l} y_m 2^{m-1} \qquad (17.7)$$

Two bits are used for the encoding of each $\delta_{i,k}$. Let $y_{i,k,1}$ and $y_{i,k,2}$ denote the values of these two bits. The binary genetic material of a particular individual can be decoded into a traffic signal timing scheme using the following formulas:

---

[2] An allowance must be made for the possibility of two green stages during each green phase.

$$C = C_{\min} + \left\lfloor (C_{\max} - C_{\min})D(C) \right\rfloor \tag{17.8}$$

$$\phi_i = \begin{cases} 0 & \text{for} \quad i = 1 \\ \left\lfloor (C - 1)D(\phi_i) \right\rfloor & \text{for} \quad i \in \{2, 3, \ldots, N_s\} \end{cases} \tag{17.9}$$

$$\rho_{i,N/S} = g_{\min,\text{phase}} + \left\lfloor (C - 2g_{\min,\text{phase}})D(\rho_{i,N/S}) \right\rfloor \tag{17.10}$$

$$\rho_{i,E/W} = C - \rho_{i,N/S} \tag{17.11}$$

$$\delta_{i,k} = \begin{cases} 0 & \text{for} \quad (y_{i,k,1} = 0) \\ 1 & \text{for} \quad (y_{i,k,1} = 1, y_{i,k,2} = 0) \\ 2 & \text{for} \quad (y_{i,k,1} = 1, y_{i,k,2} = 1) \end{cases} \tag{17.12}$$

$$g^1_{i,k} = \begin{cases} \rho_{i,G(k)} - A - R & \text{for} \quad \delta_{i,k} = 0 \\ g_{\min,\text{stage}} + \left\lfloor (\rho_{i,G(k)} - g_{\min,\text{phase}})D(g^1_{i,k}) \right\rfloor & \text{for} \quad \delta_{i,k} \neq 0 \end{cases} \tag{17.13}$$

$$g^2_{i,k} = \begin{cases} 0 & \text{for} \quad \delta_{i,k} = 0 \\ \rho_{i,G(k)} - g^1_{i,k} - 2A - R & \text{for} \quad \delta_{i,k} \neq 0 \end{cases} \tag{17.14}$$

When a real problem encoding is applied, the blend crossover and real mutation operators are applied to the normalized values of the decision variables (i.e., $\{D(x)\}$). The decision variables related to the structure and sequence of signal phases (i.e., $\{\delta_{i,k}\}$) are binary in nature and are always subject to binary crossover and mutation operators.

17.4.3 Test Networks

The study objectives are achieved by analyzing the results from the optimization of a 9-signal arterial network and a 14-signal grid network. Applying the problem encoding from the previous section produces optimization problems with 90 and 140 decision variables for the arterial network and the grid network, respectively. The specifications for both networks are based on real work data sets. The arterial network is Canal Street in New Orleans, LA (Gartner et al., 1990), and the grid network is based on a data set for downtown Ann Arbor, MI (Stamatiadis and Gartner, 1996). The structure and spacing for each network are given in Figure 17.6.

Two through lanes, exclusive turning lanes, and a free-flow speed of 54 ft/s (37 mph) have been assumed for each link. The average total flows into the arterial and grid networks are 8191 and 6546 vehicles/h, respectively. A breakdown of the average flows by individual entry nodes and the expected turning proportions at each approach are given in Gartner et al. (1990) and Kesur (2011) for the arterial and grid networks, respectively. A 15 min initialization period is used, followed by a 15 min analysis period for each network when evaluating the quality of signal

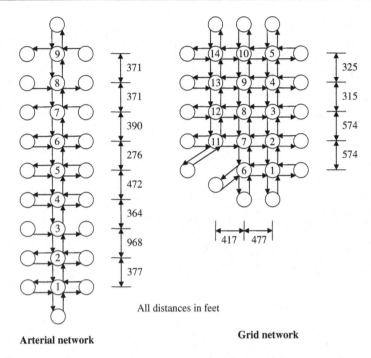

All distances in feet

Arterial network   **Grid network**

Figure 17.6 Test networks.

timing plans in MSTRANS. The flow rates for the arterial and grid network as given constitute undersaturated conditions. The average flows were increased by 50% and 60%, respectively, to construct oversaturated conditions. The following four test networks are thus considered:

1. Arterial undersaturated
2. Arterial oversaturated
3. Grid undersaturated
4. Grid oversaturated

A minimum green stage time of 5 s is implemented (i.e., $g_{min,stage} = 5$). An amber period of 3 s and an all-red interval of 2 s are assumed for all phase transitions (i.e., $A = 3$ and $R = 2$). The domains of cycle lengths examined in the optimization are given in Table 17.1.

17.4.4 Evaluating Multiobjective Optimizers

In this chapter, we evaluate several modifications to the NSGA-II search process to determine the algorithm design that best approximates the Pareto set when optimizing traffic signal timings. Evaluating the closeness of an approximate Pareto front to the true Pareto front becomes difficult when the true Pareto front is unknown, as is the case in this study. The hypervolume metric (Deb, 2001), which

Table 17.1 Domain of Cycle Lengths Examined in Optimization

Network	Cycle Length (s)	
	C_{min}	C_{max}
Arterial undersaturated	50	120
Arterial oversaturated	100	200
Grid undersaturated	50	120
Grid oversaturated	60	180

can be applied when the true Pareto front is unknown, has been selected as the measure of the quality of the approximate Pareto-optimal set.

Hypervolume

The hypervolume measure is explained using an example. Suppose that a multi-objective optimizer provides the three solutions, labeled $A1$, $A2$, and $A3$ in Figure 17.7, as an approximation of the Pareto front.

The area of the region dominated by these points up to a reference point W is given by the area shaded in gray. For more than three objective dimensions, this region becomes a hypervolume. Clearly, the hypervolume is increased as the following occurs:

• The approximate set moves closer to the Pareto front.
• More points are added to the approximate nondominated set.
• The extent of the approximate set increases.

Thus, hypervolume provides a composite measure of the quality of the approximate Pareto set and allows the different approximate Pareto sets to be evaluated and compared.

Multiple Runs and the Attainment Surface

NSGA-II is a stochastic search algorithm with random elements in the initialization, selection, and reproductive stages. Thus, the outcome of an optimization run is not always the same. When comparing alternative designs of NSGA-II, it is important to perform several independent runs of the search algorithm to produce more stable results and allow for statistical comparisons to be made. The sample values of a hypervolume can be used to test for differences in the population mean of a hypervolume using a two-sample t-test.

However, a graphical illustration of the approximate Pareto front becomes difficult to create with multiple runs. Superimposing the approximate Pareto front from each independent run may clutter the display. A summary of the approximate Pareto sets can be made using the attainment surface approach (Deb, 2001) which is explained with the aid of an example. Suppose that three independent runs of a

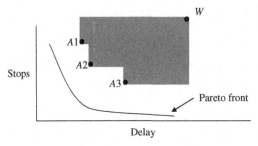

Figure 17.7 Illustration of the hypervolume measure.

fixed design of NSGA-II are performed on a particular traffic signal network, and they produce the three approximate sets shown in Figure 17.8.

For a delay value of D, NSGA-II produces solutions with stops that dominate $S1$ in all three runs. A solution with stops that dominates $S2$ is obtained in two out of the three runs. A solution that dominates $S3$ is obtained in the other run. We can take the median of $S1$, $S2$, and $S3$ to obtain a measure of centrality. We do this for all D and obtain a curve called the *attainment surface*. Solutions that dominate points to the right and above the curve are obtained in half the NSGA-II runs. The curve can be interpreted as the "average" Pareto front obtained by the multiobjective optimizer.

17.4.5 Multiobjective Optimizer Specification

For this study, NSGA-II has been executed with a population size of 100, crossover probability of 1.0, and mutation probability of 0.01. For each problem, NSGA-II is run for 200 generations, which for a population size of 100 amounts to the examination of $F = 20{,}000$ signal timing policies. Intermediate results after 100 generations are also examined to allow comparison of alternative algorithms for a smaller level of computational resources. For each test network, 20 independent runs of each search algorithm are performed. Within each run, a signal replication of MSTRANS is performed to estimate delay and the number of stops. The same random number seed is used in MSTRANS when evaluating each individual in a particular run of a GA, thus subjecting each signal timing policy to identical arrival and routing patterns. This reduces the variability in the estimated delay and stops, allowing the GA to identify improved solutions with higher accuracy (Kesur, 2009). A different random number seed is applied in MSTRANS for each independent run of NSGA-II. The nondominated solutions in the final generation are taken as the outcome of each run. Each of these points is re-evaluated using 100 replications of MSTRANS to obtain an unbiased and accurate measure of delay and stops. Although these points are nondominated based on evaluations performed during the optimization run, this is not necessarily true after re-evaluation. Using the revised estimates of mean delay and the number of stops from the re-evaluations, some points may now dominate others. The dominated points are removed to provide the final nondominated set based on precise estimates of delay and the number of stops. A flowchart illustrating the process discussed here is given in Figure 17.9.

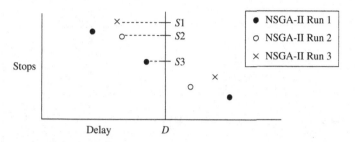

Figure 17.8 Construction of the attainment surface.

17.4.6 Single-Objective Optimizer Specification

In this study, results from the multiobjective optimization of delay and the number of stops are compared to those arising from optimization with a single objective. Single-objective minimization of delay or the number of stops is performed using a modification of the CHC GA (Eshelman, 1991) called Real CHC (Kesur, 2009), with a population size of 50. This GA includes an incest prevention mechanism to prevent similar individuals from mating and cataclysmic mutation to reintroduce diversity. This has been found to be the most efficient GA for single-objective optimization of traffic signals (Kesur, 2009). To ensure fair comparisons, a total of 20,000 signal timing policies are examined by the single-objective optimizer. We perform 20 independent runs of Real CHC on each test network. As with the multiobjective optimizer, a single replication of MSTRANS with the same random number seeds is performed to evaluate delay in a particular run of the GA. Unbiased and precise measures of delay and the number of stops of the best solution from each run are obtained by performing 100 re-evaluations of the solution in MSTRANS.

17.5 Results

17.5.1 Identifying the Most Efficient Design of NSGA-II

Optimization with NSGA-II using binary encoding with uniform crossover and bit-flip mutation is compared to optimization using real encoding with BLX-0.5 crossover and real mutation in Table 17.2.

With a real encoding, the sample mean of hypervolume is increased in each instance, with highly statistically significant differences measured for a number of cases. We can thus conclude that multiobjective optimization is enhanced using the real encoding. Subsequent optimization runs were all performed using real encoding with BLX-0.5 crossover and real mutation.

A comparison of standard tournament selection with the uniform selection procedure is given in Table 17.3.

With uniform selection, the sample mean of a hypervolume is reduced in most cases, with the reductions being statistically significant in two instances. The

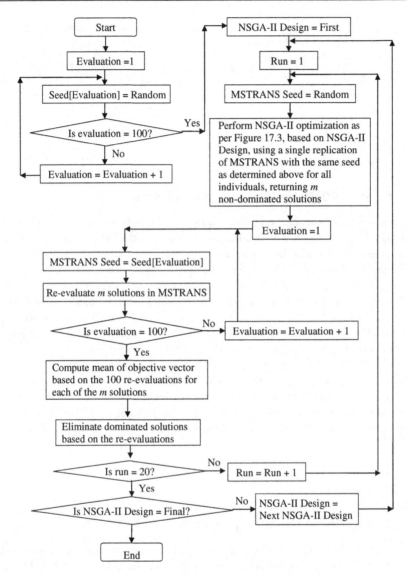

Figure 17.9 Process flowchart for performing multiple runs of NSGA-II.

uniform selection procedure was then combined with the HUX crossover operator and real mutation was removed. The results of this process are given in Table 17.4.

With uniform selection, HUX, and no mutation, the sample mean of a hypervolume is increased in all cases except two, where statistically insignificant reductions are observed. The increase in hypervolume for the grid network in undersaturated conditions for $F = 10,000$ is statistically significant. Thus, uniform selection in conjunction with HUX and reproduction without mutation provides equivalent or improved

Table 17.2 Comparison of NSGA-II Using Binary and Real Encoding

Network	F	Sample Mean of Hypervolume		p Value (%)
		Binary Encoding	Real Encoding	
Arterial undersaturated	10,000	59.95	61.55	27.44
Arterial undersaturated	20,000	62.54	63.99	31.51
Arterial oversaturated	10,000	886.38	998.41	0.92
Arterial oversaturated	20,000	993.72	1042.09	21.57
Grid undersaturated	10,000	11.80	15.34	0.00
Grid undersaturated	20,000	14.08	16.73	0.00
Grid oversaturated	10,000	115.00	137.67	0.00
Grid oversaturated	20,000	138.54	146.89	3.84

Table 17.3 Comparison of NSGA-II Using Tournament and Uniform Selection

Network	F	Sample Mean of Hypervolume		p Value (%)
		Tournament Selection	Uniform Selection	
Arterial undersaturated	10,000	61.55	57.54	0.41
Arterial undersaturated	20,000	63.99	61.81	11.38
Arterial oversaturated	10,000	998.41	955.35	8.49
Arterial oversaturated	20,000	1042.09	1014.19	29.84
Grid undersaturated	10,000	15.34	15.22	62.77
Grid undersaturated	20,000	16.73	16.77	76.34
Grid oversaturated	10,000	137.67	129.69	1.62
Grid oversaturated	20,000	146.89	150.64	24.61

multiobjective optimization relative to tournament selection. Subsequent optimization runs were all performed using uniform selection with HUX and no mutation.

Other variants to the search algorithm, such as use of the neighborhood selection mechanism (Kim et al., 2004), the inclusion of genetic diversity as an additional objective (Toffolo and Benini, 2003), and the use of more complex mechanisms for discrimination between individuals with identical domination ranks (Deb et al., 2005), were tested. However, these were not found to improve optimization, and those results are not discussed here.

Finally, the effect of the population size, N, was investigated. Optimization runs at population sizes of 60, 80, 100, 120, and 140 were compared. For each population size, intermediate output after 10,000 and 20,000 signal timing policies were examined. The number of generations of evolution before considering output was adjusted for each population size, allowing for comparisons to be made for the same level of computation resources (e.g., 125 generations of evolution with a

Table 17.4 Comparison of NSGA-II Using Tournament Selection Against Uniform Selection with HUX and No Mutation

Network	F	Sample Mean of Hypervolume		p Value (%)
		Tournament Selection	Uniform Selection with HUX and No Mutation	
Arterial undersaturated	10,000	61.55	58.40	7.89
Arterial undersaturated	20,000	63.99	61.20	10.58
Arterial oversaturated	10,000	998.41	1006.99	63.76
Arterial oversaturated	20,000	1042.09	1046.66	84.78
Grid undersaturated	10,000	15.34	16.23	0.04
Grid undersaturated	20,000	16.73	17.09	8.27
Grid oversaturated	10,000	137.67	138.35	83.03
Grid oversaturated	20,000	146.89	150.03	37.63

population size of 80 correspond to the examination of 10,000 signal timing plans). Summary statistics are given in Table 17.5.

A one-factor ANOVA to test for statistically significant differences in the mean hypervolume due to the choice of population size could not be applied, as Bartlett's test (Snedcor and Cochran, 1989) revealed that the required assumption of equal variances was violated for the oversaturated grid network at both $F = 10,000$ and $F = 20,000$ (p value less than 0.01% in both cases). Instead, Welch test (Welch, 1951), which allows for unequal variance when testing for equality of means, was applied. This test was applied consistently across all cases, even those for which there was no evidence of heteroscedasticity. The p values from the Welch test are given in Table 17.5 and reveal that choice of population size has a statistically significant effect on the mean optimization outcome for the larger and more complex grid network.

The Games–Howell multiple-comparison test (Games and Howell, 1976) was then used to identify which particular choices for the population sizes degrade the quality of the mean optimization outcome in the grid network. All comparisons were performed assuming a 5% significance level. For $F = 10,000$, the mean optimization outcome using $N = 140$ is significantly inferior to all other choices of population size in undersaturated conditions and significantly inferior to $N \in \{100,120\}$ in oversaturated conditions. In addition, for $F = 10,000$, $N = 120$ is significantly inferior to $N = 100$ in undersaturated conditions. For $F = 20,000$, $N = 60$ is significantly inferior to $N \in \{100,120,140\}$ in undersaturated conditions and significantly inferior to $N = 120$ in oversaturated conditions. Thus, for the grid network, optimization is less efficient if the population is set too large for a less extensive search. For a more thorough search, optimization performance is degraded if the population size is too small. Overall, a population size of 100 is a robust choice for the multiobjective optimization.

Evaluation of the multiobjective optimization approach in the subsequent sections consider only the most efficient design; namely, NSGA-II with real encoding, uniform selection, HUX, no mutation, and a population size of 100.

17.5.2 Comparison to Single-Objective Optimization

Multiobjective optimization is compared to single-objective optimization by considering solutions at the edge of the approximate Pareto front. For instance, the solution with minimum delay from the Pareto front represents the traffic signal timing plan with minimum delay and the best possible compromise with regard to the number of stops. The quality of this solution is compared to that from a pure delay-minimization strategy. The mean of the solution with minimum delay from each Pareto front over the 20 optimization runs of NSGA-II is given in Table 17.6. The sample mean of the minimum delay solution from the 20 optimization runs of Real CHC are also given. In addition, the corresponding number of stops associated with these minimum delay solutions are averaged over each of the 20 optimization runs and are also presented. The relative improvement/degradation in delay and the

Table 17.5 Comparison of NSGA-II with Different Choices of Population Size N

Network	F	Sample Mean of Hypervolume					p Value (%)
		$N = 60$	$N = 80$	$N = 100$	$N = 120$	$N = 140$	
Arterial undersaturated	10,000	59.14	60.15	58.40	59.04	58.72	84.92
Arterial undersaturated	20,000	60.19	61.32	61.20	62.00	63.35	40.84
Arterial oversaturated	10,000	980.16	993.41	1006.99	965.51	974.23	12.00
Arterial oversaturated	20,000	994.89	1014.42	1046.66	1019.81	1025.61	24.98
Grid undersaturated	10,000	15.99	16.32	16.23	15.77	15.06	0.00
Grid undersaturated	20,000	16.27	16.84	17.09	17.23	16.99	0.03
Grid oversaturated	10,000	137.73	139.07	138.35	137.44	129.16	0.17
Grid oversaturated	20,000	141.66	146.06	150.03	153.61	153.32	1.49

Table 17.6 Comparison of NSGA-II with Real CHC Using Delay Minimization

Network	F	Sample Mean of Delay (s/vehicle)		Sample Mean of Corresponding Number of Stops (per vehicle)		Relative Improvement Using NSGA- II	
		Real CHC	Minimum Delay Solution from NSGA-II	Real CHC	Minimum Delay Solution from NSGA-II	Delay (%)	Stops (%)
Arterial undersaturated	10,000	37.65	40.39	1.29	1.21	−7	6
Arterial undersaturated	20,000	37.20	39.81	1.28	1.18	−7	8
Arterial oversaturated	10,000	133.58	136.33	1.29	1.15	−2	11
Arterial oversaturated	20,000	133.36	134.43	1.29	1.09	−1	16
Grid undersaturated	10,000	35.76	36.21	1.36	1.24	−1	9
Grid undersaturated	20,000	35.48	35.61	1.35	1.22	0	9
Grid oversaturated	10,000	106.16	108.92	1.42	1.21	−3	15
Grid oversaturated	20,000	103.39	106.84	1.39	1.19	−3	14

number of stops using the multiobjective optimization relative to single-objective optimization is also given.

We find that in all cases, the sample mean of delay is marginally higher with multiobjective optimization. Applying a two-sample t-test, we find that the difference is statistically insignificant for the arterial network in oversaturated conditions when $F = 20,000$ (p value = 32.81%) and for the grid network in undersaturated conditions for both $F = 10,000$ (p value = 9.68%) and $F = 20,000$ (p value = 59.7%). The degradation in the minimum delay solution is only practically significant for the arterial network in undersaturated conditions. However, in all cases, the multiobjective optimizer produces substantial improvements in the associated number of stops. These improvements are highly statistically significant in all cases (p value <0.02%) and improvements of up to 16% are achieved. The improvement in the associated number of stops is larger in oversaturated conditions.

A similar exercise is performed by comparing the solution with the minimal number of stops from the Pareto front with single-objective optimization that targets the minimization of the number of stops. Results are given in Table 17.7. The mean delays associated with the solutions with a minimal number of stops are also given, along with the relative improvement/degradation in delay and the number of stops using the multiobjective optimization.

Comparing Table 17.6 and Table 17.7, we find that the minimization of the number of stops produces a relatively poor compromise with respect to delay, particularly when single-objective optimization is used. Furthermore, since the measurement of the number of stops is based on partial stops, the number of stops can be lower in oversaturated conditions due to vehicles traveling at lower speeds.

From Table 17.7, we find that the sample mean of the number of stops with multiobjective optimization is marginally higher in all cases. The differences are statistically insignificant for the arterial network in oversaturated conditions. The multiobjective optimizer produces improvements of up to 82% in the associated delay when minimizing stops relative to single-objective optimization. The improvements in the associated value of delay are highly statistically significant in all cases and larger in oversaturated conditions.

Therefore, multiobjective optimization is able to produce solutions of a similar quality to those from single-objective optimization, while simultaneously improving the quality of the competing objective.

17.5.3   Trade-Off Between Delay and Stops

The attainment surfaces from the 20 optimization runs of NSGA-II at $F = 20,000$ are given in Figure 17.10.

Starting from a minimum delay solution, the number of stops can be reduced at the cost of increasing delay. Table 17.8 presents a numerical quantification of this trade-off, where reductions in the number of stops (relative to the number of stops associated with the minimum delay solution) by increasing delay (relative to minimum delay) are given.

Table 17.7 Comparison of NSGA-II with Real CHC When Minimizing the Number of Stops

Network	F	Sample Mean of Number of Stops (per vehicle)		Sample Mean of Delay (s/vehicle)		Relative Improvement Using NSGA-II	
		Real CHC	Minimum Delay Solution from NSGA-II	Real CHC	Minimum Delay Solution from NSGA-II	Stops (%)	Delay (%)
Arterial undersaturated	10,000	1.02	1.09	169.24	51.16	−7	70
Arterial undersaturated	20,000	1.01	1.08	178.22	50.28	−7	72
Arterial oversaturated	10,000	0.87	0.91	1188.91	222.43	−4	81
Arterial oversaturated	20,000	0.86	0.88	1177.63	209.17	−2	82
Grid undersaturated	10,000	1.13	1.17	55.06	43.22	−3	22
Grid undersaturated	20,000	1.12	1.15	55.32	43.43	−2	21
Grid oversaturated	10,000	0.96	1.09	320.03	132.54	−13	59
Grid oversaturated	20,000	0.95	1.05	312.07	128.28	−10	59

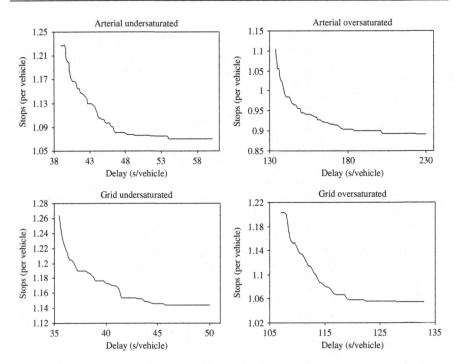

Figure 17.10 Attainment surfaces from 20 optimization runs of best variant of NSGA-II.

From this table, we find that increases in delay provide progressively smaller reductions in stops. Thus, for a signal timing exercise where delay minimization is a primary objective, it may be feasible to further reduce the number of stops by only a small margin. On the other hand, for applications where the number of stops is of greater importance, it is possible to obtain large reductions in delay at the cost of a small increase in the number of stops. Furthermore, sacrifices in delay produce larger relative reductions in stops in oversaturated conditions.

17.5.4 Role of Signal Timing Variables in the Trade-Off Between Delay and Stops

In this section, we look at how signal timing plans are adjusted by the multiobjective optimizer to generate solutions along the Pareto front. This is achieved by plotting the signal timing variables against delay for each Pareto-optimal point in the 20 multiobjective optimization runs performed for $F = 20,000$. Cycle lengths, green splits, and signal phasing are considered in turn.

Cycle Length

Cycle length is plotted against delay in Figure 17.11.

Table 17.8 Trade-Off Between Delay and Stops

Relative Increase in Delay (%)	Relative Reduction in the Number of Stops			
	Test Network			
	Arterial Undersaturated (%)	Arterial Oversaturated (%)	Grid Undersaturated (%)	Grid Oversaturated (%)
0	0.00	0.00	0.00	0.00
1	0.00	4.45	3.04	1.63
2	1.97	4.96	4.01	4.16
3	2.32	7.58	4.74	5.05
4	4.87	10.01	5.28	6.25
5	4.97	10.97	5.81	7.31
6	5.94	10.97	5.81	8.80
7	6.50	11.51	5.81	9.79
8	6.89	12.51	6.03	10.26
9	7.30	12.76	6.34	11.15
10	7.99	12.93	6.86	11.34
11	8.13	13.58	6.86	11.66
12	8.23	14.49	6.88	12.05
13	9.77	14.49	7.17	12.05
14	9.91	14.93	7.38	12.05
15	10.14	14.93	7.47	12.24
16	10.61	14.97	7.51	12.24
17	10.64	14.97	8.65	12.24
18	11.12	15.53	8.75	12.27
19	11.87	15.53	8.75	12.27
20	11.87	16.14	8.75	12.35

We find that lower delay solutions use the shorter cycle lengths. Since we are considering signal timing plans on the Pareto front, increases in delay are accompanied by reductions in the number of stops. Thus, solutions with a smaller number of stops (and larger delay) are obtained by increasing the cycle length. This is a consequence of the well-known result that the cycle length that minimizes the number of stops is larger than the cycle length that minimizes delay. Within the set of Pareto-optimal solutions, the relationship between delay and cycle length is approximately linear.

Green Splits

The proportion of green time allocated to cross-street movements, averaged over the nine signals in the arterial network, is plotted against delay in Figure 17.12.

With the majority of traffic flowing along the arterial network, the average proportion of cross-street green time allocated is less than 50% for all points in the Pareto-optimal set. From the figure, we find that stops are reduced by reducing the

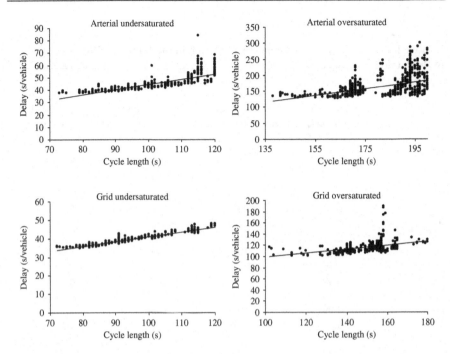

Figure 17.11 Delay versus cycle length.

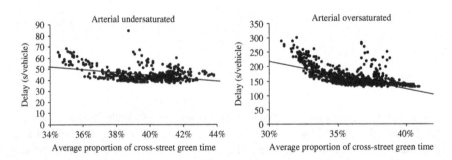

Figure 17.12 Delay versus average proportion of cross-street green time in the arterial network.

proportion of cross-street green time and increasing the proportion of green time to arterial movements at the cost of increasing delay. Given that the bulk of traffic eventually flows on the arterial network, it seems sensible that an increase in the proportion of green time for arterial movements will reduce the incidence of repeated stops. Even though the reduction in cross-street green time will increase the number of stops for vehicles originating at cross-streets, many of these vehicles turn into the arterial network and subsequently benefit from the reduced incidence of repeated stops along the network. Thus, reductions in cross-street green time result in a reduction in the overall number of stops. However, it appears that the resulting increase in delay for cross-street movements at their first encounter of a traffic signal outweighs the reduction in delay experienced along arterial links due to an increase in proportion of green time to arterial movements. The net effect of reducing the proportion of cross-street green time is an increase in network wide average delay.

Phasing

A composite measure of the degree of multiple phase operation is obtained by summing the number of phases for all signals and subtracting two for each signal. This gives the number of additional phases over and above two-phase control. For example, if one signal is three phase and all others are two phase, the number of additional phases is 1. The number of additional phases is plotted against delay in Figure 17.13.

Aside from the grid network in undersaturated conditions, lower delay solutions tend to use two-phase signal operation at all signals. In order to decrease the number of stops, multiple phases are required (at the cost of increased delay).

17.6   Conclusion

This chapter has examined minimization of delay and the number of stops in fixed-time traffic signal networks using the heuristic multiobjective optimizer NSGA-II.

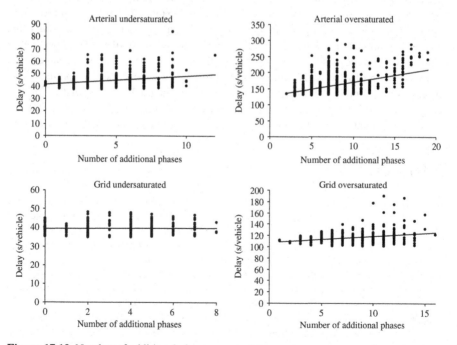

Figure 17.13 Number of additional phases versus delay.

On large signalized networks, an adequate approximation of the Pareto-optimal trade-off surface can be obtained by examining 20,000 signal timing plans. NSGA-II is found to be most efficient when using a real encoding of the signal timing variables with blend crossover. Uniform selection using reproduction without mutation is preferable to tournament selection with mutation. Optimization performance is robust for a population size of 100.

When compared to a single-objective delay-minimization strategy, the multiobjective optimizer is able to produce solutions with similar delay and a smaller associated number of stops. The reductions in the number of stops using the multiobjective optimizer are between 6% and 16% on the test networks considered relative to a pure delay-minimization strategy. The benefits with multiobjective optimization over single-objective optimization are greater in oversaturated conditions. The set of Pareto-optimal solutions generated by multiobjective optimization provide the traffic engineer with an entire set of trade-off solutions. Trading delay in favor of reducing the number of stops is found to become increasingly more costly as more delay is sacrificed. Sacrificing delay so as to reduce the number of stops is more beneficial in oversaturated conditions.

The cycle length plays an important role in this trade-off, with longer cycle lengths contributing to fewer stops and larger delays. In arterial networks, shifting green time from cross-street movements to arterial movements reduces the number of stops at the cost of increasing delay. The addition of protected turning phases also reduces the number of stops while increasing delay.

Commercial signal timing software packages such as TRANSYT-7F (Hale, 2005) and PASSER V (Chaudhary et al., 2002) have support for signal timing optimization by GAs. These single-objective GAs can easily be extended to perform multiobjective optimization according to the approach suggested in this chapter. Although this chapter has focused on delay and the number of stops, many of the findings can be generalized for the case where other objectives are considered. For instance, the optimal design of NSGA-II identified in this chapter was successfully applied to the minimization of both overall delay and the extent of inequality in the distribution of delay along different routes through the network (Kesur, 2010). Extension of the work in this chapter for actuated traffic signal controllers should also be considered.

References

Abbas, M.M., Rakha, H.A., Li, P., 2007. Multi-objective strategies for timing signal systems under oversaturated conditions. In: Proceedings of the 18th IASTED International Conference. ACTA Press, Anaheim, CA, pp. 580–585.

Berg, W.D., Do, C.-U., 1981. Evaluation of network traffic performance measures by use of computer simulation models. Transport. Res. Rec. 819, 43–49.

Branke, J., Goldate, P., Prothmann, H., 2007. Actuated traffic signal optimization using evolutionary algorithms. In: 6th European Congress and Exhibition on Intelligent Transport Systems and Services (ITS07), Aalborg, Denmark, 4 June 2007 – 6 June 2007.

Chaudhary, N.A., Kovvali, V.G., Mahabubul Alam, S.M., 2002. Guidelines for Selecting Signal Timing Software Report FHWA/TX-03/0-4020-P2. Texas Transportation Institute, College Station, TX.

Deb, K., 2001. *Multi-Objective Optimization Using Evolutionary Algorithms.* John Wiley & Sons, Chichester, England.

Deb, K., Pratap, A., Argawal, S., Meyarivan, T., 2002. A fast and elitist multiobjective genetic algorithm: NSGA-II. IEEE Trans. Evol. Comput. 6, 182–197.

Deb, K., Mohan, M., Mishra, S., 2005. Evaluating the ε-domination based multi-objective evolutionary algorithm for quick computation of Pareto-optimal solutions. Evol. Comput. 13, 501–525.

Eshelman, L.J., 1991. The CHC adaptive search algorithm: how to have safe search when engaging in nontraditional genetic recombination. Foundations of Genetic Algorithms. Morgan Kaufmann, San Mateo, CA, pp. 265–283.

Eshelman, L.J., Caruana R.A., Schaffer, J.D., 1989. Biases in the crossover landscape. In: Schaffer, J.D., (Ed.), *Proceedings of the Third International Conference in Genetic Algorithms.* Morgan Kaufmann, San Mateo, CA, pp. 10–19.

Federal Highway Administration, 2003. Traffic Software Integrated System (TSIS) Version 5.1 User Guide. Federal Highway Administration Office of Operations Research, Development and Technology, McLean, VA.

Flannery, A., Wochinger, K., Martin, A., 2005. Driver assessment of service quality on urban streets. Transport. Res. Rec. 1920, 25–31.

Foy, M.D., Benekohal, R.F., Goldberg, D.E., 1992. Signal timing determination using genetic algorithms. Transport. Res. Rec. 1365, 108–115.

Games, P.A., Howell, J.F., 1976. Pairwise multiple comparison procedures with unequal N's and/or variance: a Monte Carlo study. J. Educ. Stat. 1, 113–125.

Gartner, N.H., Assmann, S.G., Lasaga, F., Hou, D.L., 1990. Multiband—A variable-bandwidth arterial progression scheme. Transport. Res. Rec. 1287, 212–222.

Gipps, P.G., 1981. A behavioural car-following model for computer simulation. Transport. Res. B. 15, 105–111.

Hale, B., 2005. *Traffic Network Study Tool, TRANSYT-7F, United States Version 10.* McTrans Center, University of Florida, Gainesville, Florida.

Jovanis, P.P., May, A.D., 1978. Alternative objectives in arterial-traffic management. Transport. Res. Rec. 682, 1–8.

Kesur, K.B., 2007. Advances in Genetic Algorithm Optimization of Traffic Signals. M. Sc. Dissertation, University of the Witwatersrand, South Africa, <http://wiredspace.wits.ac. za/bitstream/handle/10539/4900/TrafficSignalOptimization.pdf?sequence=1>.

Kesur, K.B., 2009. Advances in genetic algorithm optimization of traffic signals. J. Transport. Eng. 135, 160–173.

Kesur, K.B., 2010. Generating more equitable signal timing plans. Transport. Res. Rec. 135, 108–115.

Kesur, K.B., 2011. *Traffic Signals: Advances in Genetic Algorithm Optimization.* VDM Verlag, Saarbrucken, Germany.

Kesur, K.B., 2012. Optimization of mixed cycle length traffic signals. J Adv. Transport. (in press, early preview avaiable at <http://onlinelibrary.wiley.com/doi/10.1002/atr.1190/ abstract>).

Kim, M., Hiroyasu, T., Miki M., Watanabe S., 2004. SPEA2+: improving the performance of the strength Pareto evolutionary algorithm 2. In: Yao, X., Burke, E., Lozano, J.A., Smith, J., Merelo-Guervós, J.J., Bullinaria, J.A., et al., (Eds.), Parallel Problem Solving From Nature (PPSN VIII). Springer, Berlin, pp. 742–751.

Leonard, J.D., Recker, W.W., 1997. A streamlined methodology for application of TRANSYT-7F. ITE J. 67, 26–32.

Leonard, J.D., Rodegerdts, L.A., 1998. Comparison of alternate signal timing policies. J. Transport. Eng. 124, 510−520.

Memon, G.Q., Bullen, A.G.R., 1996. Multivariate optimization strategies for real-time traffic control signals. Transport. Res. Rec. 1554, 36−42.

Mumford C.L., 2004. Simple population replacement strategies for a steady-state multi-objective evolutionary algorithm. In: Deb, K., Poli, R., Banzhaf, W., Beyer, H.G., Burke, E., Darwen, P., et al., (Eds.), Genetic and Evolutionary Computation (GECCO 2004). Springer, Berlin, pp. 1389−1400.

Oda, T., Otokita, T., Tsugui, T., Kohno, M., Mashiyama, Y., 1996. Optimization of signal control parameters using a genetic algorithm. In: Third Annual World Congress on Intelligent Transportation Systems, Orlando, FL, 14 October 1996 − 18 October 1996.

Park, B., 1998. Development of Genetic Algorithm Based Signal Optimization Program for Oversaturated Intersections, Ph. D. Thesis, Texas A&M University.

Park, B., Messer, C.J., Urbanik II, T., 2000. Enhanced genetic algorithm for signal-timing optimization of oversaturated intersections. Transport. Res. Rec. 1727, 32−41.

Park, B., Rouphail, N.M., Sacks, J., 2001. Assessment of stochastic signal optimization method using microsimulation. Transport. Res. Rec. 1748, 40−45.

Pecheux, K.K, Flannery, A., Wochinger, K., Rephlo, J., Lappin, J., 2004. Automobile driver's perception of service quality on urban streets. Transport. Res. Rec. 1883, 167−175.

Rakha, H., Kang, Y.-S., Dion, F., 2001. Estimating vehicle stops at under-saturated and over-saturated fixed-time signalized intersections. Transport. Res. Rec. 1776, 128−137.

Snedcor, G.W., Cochran, W.G., 1989. Statistical Methods. eighth ed. Iowa State University Press, Ames, Iowa.

Stamatiadis, C., Gartner, N.H., 1996. MULTIBAND-96: a program for variable bandwidth progression optimization of multiarterial traffic networks. Transport. Res. Rec. 1554, 9−17.

Stevanovic, A., Martin, P.T., Stevanovic, J., 2007. VisSim-based genetic algorithm optimization of signal timings. Transport. Res. Rec. 2035, 59−68.

Stevanovic, A., Stevanovic, J., Kergaye, C., 2011. Optimizing signal timings to improve safety of signalized arterials. In: 3rd International Conference on Road Safety and Simulation, Indianapolis, USA, 14 September 2011 − 16 September 2011.

Sun, D., Benekohal, R.F., Waller, S.T., 2003. Multi-objective traffic signal optimization using non-dominated sorting genetic algorithm. In: Proceedings of the IEEE Intelligent Vehicles IEEE, IEEE Xplore Digital Library, Symposium, Piscataway, NJ, pp. 198−203.

Sutaria, T.C., Haynes, J.J., 1977. Level of service at signalized intersections. Transport. Res. Rec. 644, 107−113.

Syswerda, G. 1989. Uniform crossover in genetic algorithms In: Schaffer, J.D., (Ed.), Proceedings of the Third International Conference in Genetic Algorithms. Morgan Kaufmann, San Mateo, CA, pp. 2−9.

Toffolo, A., Benini, E., 2003. Genetic diversity as an objective in multi-objective evolutionary algorithms. Evol. Comput. 11, 151−167.

Welch, B.L., 1951. On the comparison of several mean values: an alternative approach. Biometrika. 38, 330−336.

18 An Improved Hybrid Algorithm for Stochastic Bus-Network Design

Ana Carolina Olivera[1], Mariano Frutos[2] and Jessica Andrea Carballido[1]

[1]Departamento de Ciencias e Ingeniería de la Computación, Universidad Nacional del Sur, Blanca, Argentina, [2]Departamento de Ingeniería, Universidad Nacional del Sur, Blanca, Argentina

18.1 Introduction

These days, the pollution increment and parking problems in urban areas are becoming increasingly important around the world, having a direct relation with both private and public transits. In this context, the bus-network design problem (BNDP) deals with the task of finding a bus network that fulfills various conflicting objectives. Therefore, for a proper design, it is necessary to consider the interests of several entities: the user, who is the passenger of the buses; the authorities that impose the system regulations; and the operator of the lines (Ceder and Israeli, 1998; Deb et al., 2002).

These entities usually have different expectations that are generally confronted. The BNDP, therefore, involves the determination of many aspects: routes, frequencies, time schedules, fleet size, and number of employees (Bielli et al., 2002; Bleuler et al., 2003). The whole process can be decomposed into several activities (Deb et al., 2002). In particular, the design of a route for the determination of the number of lines and paths, as well as the specification of line frequencies, comprise the most important ones because their results are directly used to perform the subsequent activities. In this context, the initial crucial activity is to find the route's design as an arrangement of lines and routes. The second one is to define the frequencies that vary according to the time of day (midday, midnight, rush hour, etc.) and the synchronicity of transfers, fleet size, resources, and employees, which should be assigned for each line. Together with these requirements, the BNDP involves the optimization of many objectives, such as the maximization of the quality of the service and the maximization of the benefit for the transport operator. The main challenging edges of this problem lie on its NP-completeness (Deb, 2002) regarding computational issues, economic and social interests, and technical difficulties.

Metaheuristics in Water, Geotechnical and Transport Engineering. DOI: http://dx.doi.org/10.1016/B978-0-12-398296-4.00018-0

Furthermore, an extra hindrance is constituted by the need to consider temporal features in order to build a realistic model of the problem.

In this work, we have focused on undertaking the BNDPs activities of line scheduling and route design. Also, it is important to mention that the user and operator entities were given the same level of importance. In this context, the main contribution of this work is constituted by the technique called elastic hybrid algorithm (Elastic HA), which was carefully designed and implemented as a combination of a Floyd−Warshall (WFI) algorithm (Fonseca and Fleming, 1993) and a multiobjective evolutionary algorithm (MOEA) (Gruttner et al., 2002; Zitzler et al., 2001) with a simulation tool. The main idea behind the core procedure of the method is that the individuals in the population of the EA are bus networks, and their fitness values are calculated by simulating their behavior during a working day. This last component, the simulation for the calculus of the fitness value of each bus network, was included in order to proportion representative time-related elements, which are necessary to carry off the dynamics of real scenarios with precision.

This chapter has been divided into five sections. Section 18.1 lays out a review of the literature and background. In Section 18.2, the equations that are used to model the entities that participate in the bus-network optimization problem are introduced. Then, in Section 18.3, each stage of the hybrid procedure is presented: the WFI algorithm and the hybrid algorithm (HA), which uses simulation to calculate fitness values, are described, culminating with the presentation of the Elastic HA as an integration of them. Section 18.4 describes experimental studies based on a real-life case study, with the analysis of the results and of the influence of WFI in the HA. Finally, Section 18.5 discusses the conclusions and outlines some speculations about future work on this issue.

18.1.1 Literature Review

A reasonable design of a bus network entails the optimization of several conflicting objectives under complex constraints. This feature characterizes our problem instance as a multiobjective optimization problem (MOP). Different methods have been proposed to solve some of the stages of the BNDP: mathematical optimization, heuristics (Bleuler et al., 2003; Ceder and Israeli, 1998), and several metaheuristics like genetic algorithms (GAs), ant colonies, simulated annealing (SA), and combinations of them (Bielli et al., 2002; Bleuler et al., 2003; Olivera et al., 2008; Pattnaik et al., 1998). The approaches based on mathematical optimization usually have rigorous problem statements and a complete solution search space. However, these strategies may be too sensitive to the different settings of certain design parameters (Bleuler et al., 2003). Moreover, they also have drawbacks such as rapid convergence to a unique solution, which is clearly not desired in the case of a MOP. The same happens with traditional heuristic approaches like greedy algorithms, rollout algorithms, tabu search, and SA (Zitzler and Künzli, 2004). Even more, these techniques usually obtain a local optimal, a suboptimal, or a unique solution that satisfies a number of requirements, thus their results are not global in most cases, or, worse, they are not even locally optimal (Bielli et al., 2002). Finally, the biologically inspired metaheuristics like GAs are also used for solving the BNDP (Bleuler et al., 2003; Uur, 2008). The main advantage of EA-based methods is that they can be used naturally to tackle MOPs, since they work with a population of individuals; therefore, they yield a set of feasible solutions instead of a single one. Also, it is easy to incorporate constraints into a solution-searching EA process.

However, EAs used as stand-alone strategies have their shortcomings too. They cannot guarantee that the optimal solution will be found, and in large-scale problems, they need to search huge solution spaces with an associated high cost. However, far too little attention has been paid to the stochastic process of arrival of the passengers. Even more, whenever this issue is considered, the studies are carried out in a very simplified manner and the formulation of the problem is extremely elementary. Therefore, in order to attain better and more realistic results, the tendency in the last decade has been to combine two or more metaheuristics with the objective of dealing with the limitations of the existing studies (Zhao and Zeng, 2006). In this context, we have integrated several techniques, with an MOEA as the core of the whole procedure, to solve the BNDP, considering several issues related to the user. This algorithm was constructed after several implementation stages. In the first stage, a prototype tested on a simple academic case study was built on the base of the elitist nondominated sorting genetic algorithm version II (NSGA-II) (Deb et al., 2002; Olivera et al., 2008; Olivera et al., 2009). Later, a complete study based on the Platform and Programming Language Independent Interface for Search Algorithms (PISA) platform (Bleuler et al., 2003) was carried out in order to decide whether the NSGA-II, strength Pareto evolutionary algorithm 2 (SPEA2; Zitzler et al., 2001) or indicator-based evolutionary algorithm (IBEA; Zitzler and Künzli, 2004) was the best MOEA (Olivera et al., 2008) to use. As a result of this study, SPEA2 turned out to be the most suitable method, and therefore, the whole hybrid procedure presented in this chapter was constructed on the base of this evolutionary algorithm.

18.1.2   Background

In this section, relevant concepts about the bus-network design and the HA used for its treatment are introduced.

Bus-Network Design

Ceder and Wilson (1986) established the general planning of the transit network scheduling problem into five stages: network design, frequency setting, timetable development, bus scheduling, and driver scheduling. In this chapter, we focus on network design and bus scheduling, considering both entities (user and operators) to be at the same level of importance. The goal in the network design is to define a set of bus routes in a particular area, each route being determined by a sequence of bus stops. The purpose of bus scheduling is to obtain a feasible sequence of line runs (also called *bus service*), thereby determining the number of buses required for the considered period, usually one day.

Every day, thousands of people cross cities for different reasons. A proper scheduling of public urban transportation (and in particular, of the bus network) gives the user an interesting alternative to private transportation. For the treatment of the bus-network design, it is necessary to consider two entities with conflicting objectives: the **user** of the bus and the **operator** of the network. A *bus* has a capacity and a frequency, it has a route (or *line*) to follow, and this route is shown in a *map* of a city. The points of the route where the bus must stop are the *bus stops* of a given line. If two lines have a common bus stop, it is called a *transfer point*. All the routes in a map create the *bus network*.

Multiobjective Optimization

The multiobjective optimization refers to the optimization of more than one objective at the same time. Moreover, these objectives can be in conflict, i.e., the improvement of an objective can provoke the detriment of the others. There exist several strategies to deal with multiobjective problems. In this chapter, we use "inside" our hybrid algorithm the SPEA2 proposes by Zitzler et al. (2001). The idea of this technique is to find an approximation of the Pareto optimal set.

Pareto Optimal Set

When we talk about several objectives, the notion of optimum changes because in multiobjective problems, the aim is to find good compromises rather than a single solution as in global optimization. In the case of bus-network design, both the user of the bus network and the operator of the service have different, conflicting objectives. The users want to arrive to their destinations quickly, as well as having the option of several lines and a high frequency of bus arrivals. On the other hand, the operators need to reduce the cost of maintaining the lines and buses. A solution to a problem like this is said to "Pareto-dominate" another one if it is no worse in all objectives and provides an improvement in at least one objective. The set of all solutions that are not dominated is called the Pareto optimal set.

Strength Pareto Evolutionary Algorithm 2

SPEA2 (Ziztler et al., 2002) is a revised version of SPEA that incorporates three new features: First is a fine-grained fitness assignment strategy, which takes into account for each individual the number of individuals that dominate it and the number of individuals that it dominates. Second, it uses a nearest-neighbor density estimation technique to guide the search; and third, it has an enhanced archive truncation method that guarantees the preservation of the boundary solutions. The SPEA2 uses this external archive to preserve nondominated solutions that were previously found. At each generation, nondominated individuals are copied to the external set.

The Pareto dominance is used to ensure that solutions are in the Pareto optimal set. The whole procedure is explained in Section 18.3.2.

18.2 The Main Entities of the BNDP: The Operator and the User

In order to optimize the BNDP, the objectives associated to the main entities of the problem are based on the model proposed by Gruttner et al. (2002). The equations associated with the operator can be expressed as follows:

$$\sum_{L=1}^{M} FO_L = \sum_{L=1}^{M} (IO_L - CO_L) \tag{18.1}$$

$$IO_L = AF_L T_L \tag{18.2}$$

$$CO_L = D_L K_L \tag{18.3}$$

where AF_L is the total client influx for line L, T_L is the price for a journey in line L, D_L is the travel distance for each line L, and K_L is the unitary operative cost of operation per kilometer for line L. In Eq. (18.1), FO_L corresponds to the economic benefit of the operator. Equation (18.2) represents the total income for line L (IO_L), taking into account the total influx for L (AF_L) and the price per trip in line L (T_L). In contrast, Eq. (18.3) represents the cost (CO_L) that the operator has to assume, considering traveling distance (D_L) and the unitary operative cost per kilometer (K_L). In order to model the dynamic influence of clients, access, wait, and journey times associated with the bus-network services are defined. The access time (t_{ij}^A) is defined as the time used by the user to get into the bus once the bus has arrived. The wait time (t_{ij}^W) is calculated as the period of time between the user's arrival to the bus stop and the bus's arrival. The period of time that the user is traveling inside the bus is called the journey time (t_{ij}^J), calculated as follows:

$$\sum_{L=1}^{M} FU_L = \sum_{L=1}^{M} \sum_{i=1}^{N} \sum_{j=1}^{N} (\delta t_{ijL}^A + t_{ijL}^W + \eta t_{ijL}^J) VST_{ijL} V_{ijL} \tag{18.4}$$

where VST_{ijL} is the subjective value for the time that employs line L to travel between each origin−destiny (OD) (i,j) pair, V_{ijL} is the number of journeys for each pair (i,j) that employs line L, t_{ijL}^A is the access time for line L, t_{ijL}^J is the journey time for line L, t_{ijL}^W is the waiting time for line L, and δ and η are the relative weights between access time and waiting time with respect to the journey time. In Eq. (18.4), FU_L represents the cost related to transporting a client in line L. The customer's cost FU_L considers the access time and the queue for line L.

From the formulation of these equations arises the need to model the attribute of time. For this reason, as several variables are time-dependent, their estimation will be explained in the context of the simulation procedure performed during the evaluation of the fitness function in the EA that constitutes the core of the new strategy.

18.3   Hybrid Method for Stochastic Bus-Network Design

In this section, the complete architecture of a method designed to solve the BNDP is presented. The main stages of the method comprise the initialization, which constitutes the estimation of the paths between each pair of bus stops and the corresponding distance; and the core, which yields several entire bus networks. The routes and distances from any pair of bus stops are defined by means of the Floyd–Warshall method. Then, it is the turn of the multiobjective EA (Zitzler et al., 2001), which is based on the SPEA2 and performs the most important task, assisted by the simulation tool. Under a broad point of view, the individuals of the EA are bus networks and their fitness values are calculated by simulating a working day for each of them, considering the equations presented in the previous section. The general layout of the whole procedure is depicted in Figure 18.1.

18.3.1   The Initialization: WFI Algorithm

The WFI algorithm (Floyd, 1962) calculates the different paths and distances between every pair of bus stops, using a directed graph that corresponds to the map of the city, where the weights represent the meters between each pair of nodes. The WFI also maintains a matrix such that at iteration k, d_{ij} contains the shortest path from bus stop i to bus stop j using nodes $1,\ldots,k$ as intermediate nodes. After the algorithm terminates, the shortest path between i and j is d_{ij} (Floyd, 1962). Therefore, the results of this stage are the distances and the routes between every bus stop in the map. The Elastic HA takes this information and uses it for the calculus of the fitness function during the evolutionary stage, considering the topology of the lines.

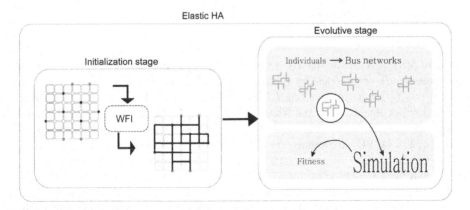

Figure 18.1 A general view of the Elastic HA.

Algorithm 18.1. Floyd−Warshall
Require: M: Map with n Nodes
  1: **for** $\forall$ $(i,j) \in M, i \neq j$ **do**
  2:     $d_{ij} \leftarrow c_{ij}$
  3: **end for**
  4: **for** $i = 1$ **to** n **do**
  5:     $d_{ij} \leftarrow \infty$
  6: **end for**
  7: **for** $j = 1$ **to** n **do**
  8:     **for** $i = 1$ **to** n **do**
  9:        **for** $k = 1$ **to** n **do**
10:          $d_{ik} \leftarrow \min \{d_{ik}, d_{ij} + d_{jk}\}$
11:          **if** $d_{ik} > d_{ij} + d_{jk} \wedge i \neq j \neq k$ **then**
12:            $e_{ik} \leftarrow j$
13:          **end if**
14:        **end for**
15:     **end for**
16: **end for**
17: **return** $[d_{ij}]$ distances between i to j
18: **return** $[e_{ij}]$ routes between i to j

Algorithm 18.1 gives some details of the implementation of the WFI. It is important to note that this algorithm was chosen to be the first step of the Elastic HA since after iteration k, d_{ij} is the shortest distance from i to j involving a subset of nodes in $\{1,\ldots,k\}$ as intermediate nodes. The accomplishment of this property is imperative in order to minimize the cost of operation of the lines and the time of journey of the users in the BNDP.

18.3.2 The Core: EA

Several authors have shown that EAs succeed in obtaining suboptimal solutions for NP-hard problems (Deb et al., 2002). The multiobjective EA proposed in this work is based on the SPEA2. It starts from a randomly generated initial population, the distances between the bus stops obtained through the WFI, and a set of parameters. The parameters are the following: (i) number of bus stops with concentrated demand; (ii) information of each bus stop (position, time between arrivals); (iii) number of lines (M); (iv) fee for traveling with each line L (T_L); (v) unitary cost of operation per kilometer for each line L (K_L); (vi) starting nodes (B_S); (vii) return nodes (B_F); and (viii) maximum capacity of buses (C_{max}). As will be explained in detail later in this chapter, the individuals of the population represent bus networks. Then, the EA evolves the population until several satisfactory bus networks are achieved in the last generation.

The Individuals: Bus Networks

An individual represents a bus network by means of a set of lines. Each route of a line is modeled by an ordered list of integers, starting from an initial bus stop B_S and ending with a final bus stop B_F. It is important to note that not all the integer strings constitute feasible solutions. In this regard, restrictions related to the feasibility of individuals will be defined according to set theory. Consider that $L_1,...,L_M$ are the lines represented by the individual x. A feasible solution for the BNDP should fulfill the following constraints:

1. The solution should contain the same number of routes as the amount of lines (M) that were defined as input parameter of the algorithm:

$$x = \{L_1,...,L_M\} \qquad (18.5)$$

2. The network must be connected (L_i and L_j are lines of x):

$$\forall L_i \forall L_j, i \neq j \; \exists \; \text{a path between } i \text{ and } j \qquad (18.6)$$

3. Every bus stop should be present on at least one of the lines:

$$\{B_S\} \cup \{B_I\} \cup \{B_F\} = \{L_1\} \cup \cdots \cup \{L_M\} \qquad (18.7)$$

For the creation of the population, there are two types of maps that should be considered: connected and disconnected. In the case of a connected map, the initialization is performed as follows. The bus stops on each line are determined from those that constitute the sets of initial (B_S), intermediate (B_I), and final (B_F) bus stops. Each bus stop in each set has a probability of being chosen that is equal to 1 divided by the amount of elements in the corresponding set. Then, for every line in the network, a random number between 1 and the amount of possible bus stops in the set is generated. This first number, namely #B_I, corresponds to the quantity of bus stops that will be selected for that line. Then, #B_I random numbers are generated in order to select the corresponding bus stops for the given line. This task is repeated for each line. When this job has finished, Eqs. (18.5)–(18.7) are evaluated. If some of them are not satisfied, then the individual is modified according to some feasibility rules. For example, in the case where a line is not connected with any other line (Eq. (18.6)), an intermediate bus stop is randomly selected from another line and inserted into the disconnected one.

Whenever the map is disconnected, a matrix that contains information about the directly connected bus stops is created and the following procedure is performed: the bus stops in the set B_F are ordered, a random bus stop b_i from set B_S is selected, and a random number a is generated between 1 and #B_I. The intermediate bus stop b_a is put behind b_i. Then a random number between 1 and #B_I and a random number L between 1 and #B_I are generated. The matrix with the information about the bus stops directly connected to b_a is generated, the k bus stops in the matrix that have the lowest cost per kilometer between b_a and b_k are found, and a probability of acceptance is calculated that considers whether b_k already is in another line (the probability is 40%) or not (the probability is 60%). If b_k is chosen, it is inserted into the line and b_k is the new b_a. These steps are repeated until a bus stop does not have another bus stop to be chosen, or until the number of intermediate bus stop in the line reaches L. Finally, a bus stop of B_F is selected and inserted into the line L.

Design of the Fitness Function: FO_L and FU_L

From the modeling of the BNDP described earlier, two objectives arise: FO_L and FU_L. Then, it is clear that two contradicting objectives have to be optimized; therefore, there is not a single solution to the problem. Moreover, both objectives (FO_L and FU_L) are equally important. For these reasons, the treatment of the BNDP has to be tackled as an MOP from a multiobjective perspective. In this context, the multiobjective problem can be formalized as the problem of finding a decision variable x that minimizes the objective function presented in Eq. (18.8). Hence, the solution set of the BNDP consists of all the decision vectors whose objective values cannot be improved in one dimension without degrading the other:

$$f(x) = \begin{bmatrix} FO \\ FU \end{bmatrix} \tag{18.8}$$

$$FO(x) = \frac{1}{1 + \sum_{L=1}^{M} FO_L} \tag{18.9}$$

$$FU(x) = \sum_{L=1}^{M} FU_L \tag{18.10}$$

Estimation of the Fitness Function: The Simulation

First, it is important to bear in mind that the simulation procedure that will be explained in this section is performed for each individual of the population (i.e., each bus network) in order to obtain its fitness value. The manner in which the simulation is faced depends to a large extent on the optimization model built for the problem at hand. For the formal BNDP treatment, it was necessary to apply simulation techniques related to queue theory and access to resources. The static structure of a bus network is basically composed of different routes, the operator's fleets, the network users, and the transfer points. It should be taken into account that a bus moves from the initial stop to the final one, and then it travels back to the initial point. During the simulation, each entity is associated with some information that should be obtained, so that this information is later used to calculate the fitness function of the EA. For each user modeled in the bus network, we are interested in its own waiting time (t_{ijL}^{W}), trip time (t_{ijL}^{J}), access time (t_{ijL}^{A}), destination node, and (in case the user is traveling) in which bus is in. Each node keeps two lists: a list of clients who are queuing (i.e., waiting to get into a mobile (there is a list for each line crossing that node)) and a list of the users who will arrive at that stop. It also stores information on the number of users that arrived to this center (V_{ijL}). The line's attributes include a list of the nodes the line goes through, its fleet, and the total influx of trips (AF_L). Finally, each mobile has information about the point it is on, the next node, the current capacity (C_A), the maximum capacity (C_M), and the present clock (t^{Now}).

The simulation time progresses in a discrete and synchronic way. For our problem instance, the planned event that advances the clock is the arrival of a mobile to a node. Before effectively beginning the simulation for a mobile's working day, the arrival of each client to the respective transfer centers is generated, and the arrivals are put on the customers' list that will potentially arrive to the stop.

The simulation begins by generating each mobile's arrival to its route's initial point. The simulation clock advances to the first arrival of the first transport of the fleet. Two lists are updated: the list of clients that arrived at the node and the list of clients that will arrive later. Besides, the transport arrival to the next node is planned considering the present clock and the distance that has been traveled, which was obtained through the WFI algorithm. The following attributes are updated: those that belong to the lines (AF_L, V_{ijL}), to the user ($t_{ijL}^W, t_{ijL}^J, t_{ijL}^A$), and to the transport (C_A, t^{Now}). In this way, the clock moves forward until the simulation is over.

18.3.3 Evolutionary Operators

Crossover

A variation of the two-point crossover operator (OX) was adopted for this problem. It is interesting to point out that the one-point approach yielded unsatisfactory results since the children's fitness value became lower than their parents'. The crossover method implemented for our technique works in the following manner: The operator selects two edges for each route of each parent. By using the constraints defined in Eqs. (18.5)–(18.7), the algorithm detects unfeasible individuals in order to keep feasible individuals in the population. When the crossover produces unfeasible children, they are discarded and new cut points for their parents are selected.

For example, suppose that the first parent's routes are as follows: Line 1: 1−5−14−15−24−26−34−35−36; Line 2: 2−6−13−16−23−25−33−37; Line 3: 2−7−12−11−17−23−27−32−38; Line 4: 3−8−11−18−22−28−29−30−40; Line 5: 4−9−10−20−19−21−29−31−39 (Figure 18.2). The second parent's routes are as follows: Line 1: 1−5−14−16−15−24−26−34−35−37; Line 2: 2−7−6−13−16−25−33−36; Line 3: 3−12−17−23−27−33−32−38; Line 4: 3−8−11−18−23−22−28−30−40; Line 5: 4−8−9−10−20−19−21−29−31−39 (Figure 18.3). For the first child (Figure 18.4), the algorithm picks the first half of each line of the first parent (Line 1: 15−24; Line 2: 23−25−; Line 3: 11−17; Line 4: 18−22−28; Line 5: 20−19−21−29), and they are connected to the final part of the second parent (Line 1: 26−34−35−37; Line 2: 36; Line 3: 33−32−38; Line 4: 30−40; Line 5: 21−29−31−39). Likewise, the first portion of the second parent (Line 1: 1−5−14; Line 2: 2−7−6−13; Line 3: 3−12; Line 4: 3−8−11; Line 5: 4−8−9−10) is connected to the final part of the first parent (Line 1: 25; Line 2:

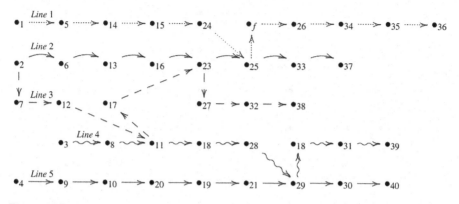

Figure 18.2 Parent 1.

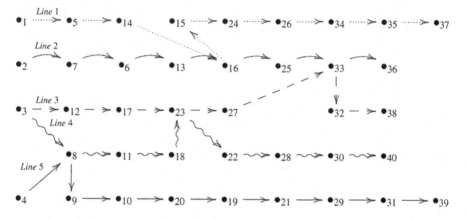

Figure 18.3 Parent 2.

16–23; Line 3: 23–27; Line 4: 29; Line 5: −). This operator was inspired by the OX operator designed for the traveling salesman problem (Uur, 2008).

Mutation

As far as mutation is concerned, the children have two options: edge mutation (Figure 18.5) or bus stop mutation (Figure 18.6). Edge mutation randomly chooses two edges from two different routes and inverts them (Figure 18.7). On the other hand, with bus stop mutation, there are two alternatives: the insertion of a randomly selected node into another route or the inclusion of a randomly selected node into the same route (Figure 18.8). In both cases, if the resulting individual is unfeasible, it is discarded. Then, the original individual is mutated again until a feasible individual is obtained. In the worst case, the mutated child has the same structure

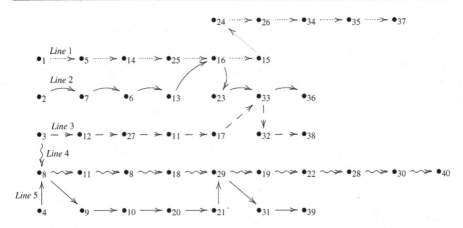

Figure 18.4 Child.

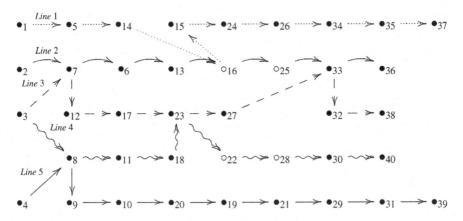

Figure 18.5 Selection of edges for edge mutation (o).

of the original child. It is important to note that eventually, the mutation always finds a feasible individual.

18.3.4   Floyd−Warshall + SPEA2 × SIMULATION = Elastic HA

In this section, we will explain how all the aforementioned pieces are assembled in the Elastic HA. First, starting from the map of the city, the WFI algorithm calculates the different paths and distances between every pair of bus stops. Then, based on the new map yielded by the Floyd−Warshall stage, a random population of feasible bus networks is built. Later, in order to evaluate the fitness of the individuals in the population,

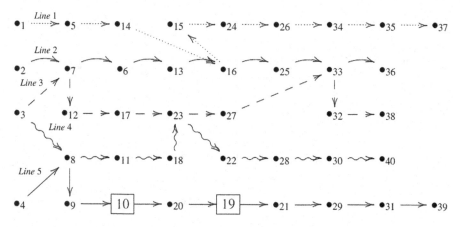

Figure 18.6 Random selection of bus stops in an individual instance.

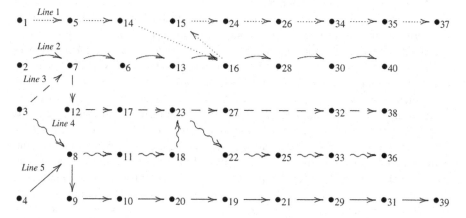

Figure 18.7 Edge mutation.

it is necessary to calculate the value of each of the objectives, FO_L and FU_L. For this aim, the algorithm simulates a day of work of the lines represented by each of the individuals, and returns the values of all of the variables involved in those equations shown in Mutation section. With regard to the other features of the evolutionary process, it is well known that SPEA2 is a revised version of SPEA. SPEA2 uses an external archive containing nondominated solutions previously found, the so-called external nondominated set. At each generation, nondominated individuals are copied to the external set. For each individual in this external set, a strength value is computed. SPEA2 has a fine-grained fitness assignment strategy that takes into account, for each individual, the number of individuals that dominate it and the number of individuals that it dominates. Then, it uses a nearest density estimation technique for guiding

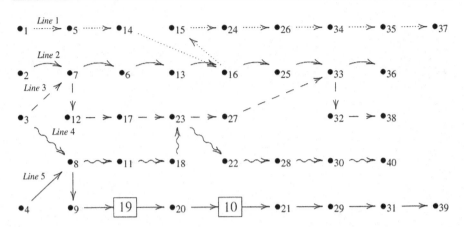

Figure 18.8 Bus stop mutation.

the search. Finally, it also has an enhanced archive truncation method that guarantees the preservation of boundary solutions.

In Algorithm 18.2, we show a simplified pseudocode of the whole procedure. Furthermore, in this section, a short complexity analysis of the algorithm will be presented. The initialization algorithm is executed only once at the beginning of the Elastic HA, with an execution time in the $O(\text{nodes}^3)$ (Floyd, 1962; Olivera et al., 2008) being equal to the number of nodes of the map. Concerning the fitness assignment scheme of SPEA2, each individual is assigned a fitness that is calculated on the basis of the number of solutions that dominate it. In order to build the next generation, SPEA2 combines the offspring with the current population. Subsequently, the best individuals are chosen and the archive incorporates the non-dominated solutions. For diversity preservation, SPEA2 uses a truncation procedure that requires a computational complexity of $O((N + N')^2 \log(N + N'))$, where N is the population size and N' is the archive population size. The simulation takes $O(NM)$ for a population with a size equal to N, with each individual representing a bus network with M lines.

In g generations, the entire time of the EA is $O(g(N + N')^2)$. The whole procedure takes $O(\text{nodes}^3)$, considering that, in a real city, nodes is always larger than N.

18.4 Practical Experience

In this section, the experimental assessment of the new method will be presented. In the first case, the BNDP was studied in the context of a hypothetical city consisting of 100 nodes. The results of this experimental phase were compared with those reported in Olivera et al. (2008), which constituted the first prototype of the technique developed for a simplified version of the BNDP. The second case study, a traditional benchmark, is a Swiss city described by Mandl in 1980 with 15 bus stops and, in this case, the comparison was performed against two other methods: the technique presented by Mandl (1980) and the one introduced in Baaj and Mahmassani (1995).

Algorithm 18.2. Elastic HA

Require: M: Map; B_S, B_F, B_I: set of bus stops

  Require: g: integer (g is the number of generations)

  Require: N : integer (N is the number of individuals in the population)

  1:   Floyd$-$Warshall(M, B_S, B_F, B_I)

  2:   Generate random population P_0

  3:   Create empty external set N'

  4:   **for** $x \in P_0$ **do**

  5:   **for** $\exists$ line L **do**

  6:   simulate(AF$_L$, T$_L$, D$_L$, K$_L$, t_{ijL}^A, t_{ijL}^J, t_{ijL}^W)

  7:   **end for**

  8:   Evaluate Objective FO and FU for each individual x in Q_i

  9:   **end for**

  10: Copy all individual evaluating to nondominated vector P_0 and N' to N'

  11: Use the truncation operator to remove elements from E' when capacity of the file has been extended

  12: **if** capacity (N') $>$ limitCapacity **then**

  13:   Use dominated individuals in P_0 to fill N'

  14: **end if**

  15: Parents Binary Tournament selection

  16: $Q_0 \leftarrow$ Crossover (Parents)

  17: Mutation (Q_0)

  18: **for** $i = 0$ to $g - 1$ **do**

  19:   **for** $x \in Q_i$ **do**

  20:   **for** $\exists$ line L **do**

  21:   simulate(AF$_L$, T$_L$, D$_L$, K$_L$, t_{ijL}^A, t_{ijL}^J, t_{ijL}^W)

  22:   **end for**

  23:   Evaluate Objective FO and FU for each individual x in Q_i

  24:   **end for**

  25:   Copy all individual evaluating to nondominated vector Q_i and N' to N'

  26:   Use the truncation operator to remove elements from N' when capacity of the file has been extended

  27:   **if** capacity(N') $>$ limitCapacity **then**

  28:   Use dominated individuals in Q_i to fill N'

  29:   **end if**

  30:   Parents Binary Tournament selection($Q_i + N'$)

  31: $Q_i \leftarrow$ Crossover(Parents)

  32:   Mutation(Q_i)

  33: **end for**

  34: **return** [AF$_L$, T$_L$, D$_L$, K$_L$]: set of real

  35: **return** P_g

18.4.1 Hypothetical City

The map of the hypothetical city used during the first experimental stage is represented with 100 nodes, 2 initial stops (B_S), 6 intermediate stops (B_I), and 2 final stops (B_F). Figure 18.9 shows the node layout and the positions of the stops on the map. As it can be observed in this figure, B_S = (61,21) (circle), B_F = (70,40) (triangle), and B_I = (17,42,48,65,83,89) (square). For the generation of the initial population, the connected bus stop procedure was used. The main goal of this experimental phase was to design the routes for two bus lines L_0 and L_1 with the Elastic HA, and compare the results with those obtained by the algorithm Time Dependent Hybrid Algorithm (TDHA), presented in (Olivera et al., 2008).

Some rules were established with regard to the BNDP. There is room for 25 people in each bus, so that is the maximum amount to be admitted. Only one time frequency (the morning) was modeled, which is set at the beginning of the run of the EA. The client arrivals and the corresponding destinations are arranged on a table that constitutes the entry to the simulation phase. The bus arrivals at each line's initial stop are also stored on a table that is used during the simulation. Figure 18.9 represents the nodes and bus stops with numbers, and the roads' directions are shown with arrows. Table 18.1 shows the OD matrix between 6 a.m.

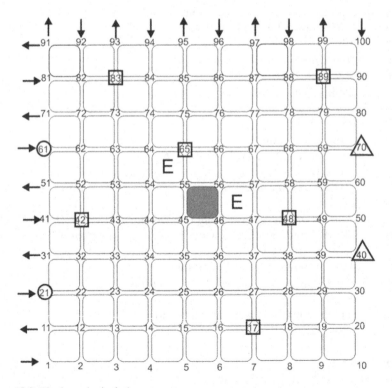

Figure 18.9 The hypothetical city.

and 10 a.m., and Table 18.2 illustrates the parameters of the exponential distribution used for the arrival of passengers to the bus stops. The parameters that were defined after several preliminary runs and remained fixed during the experimentation were the following. For the evolutionary stage, pop size = 100, PC = 0.9, PM = 0.2; and for the simulation procedure, the bus frequency was 20 min.

A total of 20 runs of the TDHA and 20 runs of the Elastic HA were performed, and the quality of the results regarding the operator's and user's perspectives was analyzed. In Table 18.3, the outcomes yielded by the Elastic HA were compared with those reported in Olivera et al. (2008). As can be seen mainly in the last rows of this table, the waiting times and operative costs are represented, and the nondominated individuals obtained by the new technique exhibit a better behavior regarding both of the objectives being studied. The comparison is based on the variables that characterize the quality of the transit-route network configurations (Desaulniers and Hickman, 2007), such as percentages of trips with zero, one, or more than one transfers, total travel time in a nondominated solution, and operation cost. These differences are shown in the operation cost of services and average length routes, and they reflect the positive influences of the Floyd–Warshall method on the whole method. Furthermore, this information assesses the importance of an accurate initialization that also influences the price of the bus's tax.

Table 18.1 Matrix OD—6 a.m. to 10 a.m.

Bus Stop	17	42	48	65	83	89
17	200	200	80	200	80	
42	0	0	400	200	80	200
48	100	800	0	80	200	80
65	0	0	0	0	80	120
83	800	0	0	100	0	400
89	100	400	100	0	0	0

Table 18.2 α Parameter Arrival Distribution Probability—6 a.m. to 10 a.m.

Bus Stop	17	42	48	65	83	89
17	0.000000	0.833333	0.833333	0.333333	0.833333	0.333333
42	0.000000	0.000000	1.666667	0.833333	0.333333	0.833333
48	0.416667	3.333333	0.000000	0.333333	0.833333	0.333333
65	0.000000	0.000000	0.000000	0.000000	0.333333	0.500000
83	3.333333	0.000000	0.000000	0.416667	0.000000	1.666667
89	0.416667	1.666667	0.416667	0.000000	0.000000	0.000000

Table 18.3 The Hypothetical City's Outcomes

Statistical Results	THDA	Elastic HA
% demand 0-transfer	80.36	80.14
% demand 1-transfer	19.64	19.86
% demand +1-transfer	0	0
% unsatisfied demand	0	0
Average total route length	Line 0: 5230	Line 0: 3450
	Line 1: 4360	Line 1: 3690
Average wait time for a user	4' 35"	3' 36"
Average operation cost	3572 m.u.	2745 m.u.

Table 18.4 Matrix OD for Mandl Instance

Bus Stops	0	1	2	3	4	5	6	7	8	9	10	11	12	13	14
0	0	400	200	60	80	150	75	75	30	160	30	25	35	0	0
1	400	0	50	120	20	180	90	90	15	130	20	10	10	5	0
2	200	50	0	40	60	180	90	90	15	45	20	10	10	5	0
3	60	120	40	0	50	100	50	50	15	240	40	25	10	5	0
4	80	20	60	50	0	50	25	25	10	120	20	15	5	0	0
5	150	180	180	100	50	0	100	100	30	880	60	15	15	10	0
6	75	90	90	50	25	100	0	50	15	440	35	10	10	5	0
7	75	90	90	50	25	100	50	0	15	440	35	10	10	5	0
8	30	15	15	15	10	30	15	15	0	140	20	5	0	0	0
9	160	130	45	240	120	880	440	440	140	0	600	250	500	200	0
10	30	20	20	40	20	60	35	35	20	600	0	75	95	15	0
11	25	10	10	25	15	15	10	10	5	250	75	0	70	0	0
12	35	10	10	10	5	15	10	10	0	500	95	70	0	45	0
13	0	5	5	5	0	10	5	5	0	200	15	0	45	0	0
14	0	0	0	0	0	0	0	0	0	0	0	0	0	0	0

18.4.2   The Swiss City

Mandl's case study (Mandl, 1980) is presented in this section, and the results achieved by the Elastic HA are reported and analyzed. The experiments were carried out considering the original map provided by Mandl, which constitutes a traditional benchmark in this research area. At this point, it is important to bear in mind the limitation of the study performed in Mandl's work, in the sense that they do not consider the dynamics of the arrivals of the passengers to the bus stops. Therefore, for this experiment, the arrivals were modeled with an exponential distribution in the context of the OD matrix of Mandl's case (Table 18.4). Moreover, an estimation of the capacity of the buses was included for this experience. Finally, for the generation of the initial population, the disconnected bus stop procedure was used. The map of the Swiss city

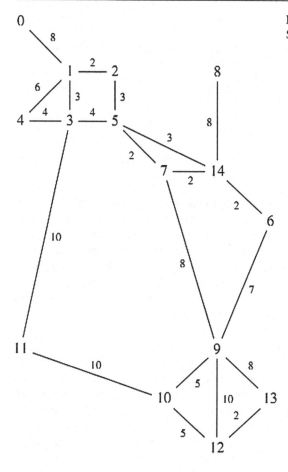

Figure 18.10 Mandl's map of a Swiss city.

provided by Mandl (1980) is represented with 15 bus stops (Figure 18.10). This map shows the position of the stops and the travel times between the nodes.

The main goal of this experiment is to design the routes for three bus lines: L_0, L_1, and L_2. This number of lines was set on the basis that each bus has room for 20 people, which is the maximum amount to be admitted. Also, a day of trips with one time frequency that is set at the beginning of the run was modeled. The client's arrivals and the corresponding destinations are arranged on a table that constitutes the input of the simulation phase. The bus arrivals at each line's initial stop are also stored on a table that is used during the simulation.

During the experimentation, 20 runs of the Elastic HA with 200 generations for the EA were performed. At this point, it is important to note that an average of 20 nondominated solutions was obtained after each execution of the EA. The parameters of each technique were as follows: for the EA, pop size = 100, PC = 0.9, and PM = 0.2; and for the simulation, the bus frequency was 30 min.

Table 18.5 shows a comparison between Mandl's method (Mandl, 1980), the procedure in Baaj and Mahmassani (1995), and the Elastic HA. Mandl (1980) and

Table 18.5 Mandl's Outcomes

	Elastic HA Results Mean	Baaj and Mahmassani's Solution	Mandl's Solution
% 0-transfers	83.1	80.99	69.94
% 1-transfers	16.9	19.01	29.93
% 2-transfers	0	0	0.13
% demand unsatisfied	0	0	0

Baaj and Mahmassani (1995) was taken from Zhao and Zeng (2006). The comparison is based on the variables that traditionally characterize the quality of the transit-route network configurations, such as percentages of trips with zero, one, or more than one transfers, total travel time in a nondominated solution, etc. These variables are the same as those chosen for their examples by the referents in the field (Desaulniers and Hickman, 2007). For each user that gets on the bus at stop i and wants to get off at stop j, the number of transfers he or she has to make, in order to go from i to j, must be reported. For this reason, all the simulated users were classified in different rows of the table. If a user needs no transfer between lines, the journey is called a 0-transfer trip. Similarly, when a user needs q transfers between lines, the journey is called a q-transfer trip. Then, the transfer percentage is calculated from the total amount of journeys that took place (see the first three rows in Table 18.5). The value associated to the unsatisfied demand percentage (shown in row 4 of Table 18.5) corresponds to the number of users who could not reach their destinations as the simulation concluded. Therefore, from the information provided by the table, it can be inferred that the best results were obtained with the Elastic HA. Moreover, this method not only yields the best outcomes but it is also the only one, to the best of our knowledge, that considers several time-related issues that are relevant to the problem.

18.5 Conclusions and Future Research Work

In this chapter, we have introduced the Elastic HA to solve the BNDP, which integrates three important procedures in an efficient manner: a WFI algorithm, an EA, and a simulation strategy. The core of the algorithm is constituted by a MOEA, and the simulation features are used in the calculus of the fitness function in order to model the time-related aspects of the problem. The WFI algorithm works as the initialization of the entire method and plays a very important role in producing the final results of the technique. The performance of the Elastic HA was assessed on the basis of two different scenarios. In the first scenario, it was established that the magnitude of the initialization stage and the influence of the distance of the travels in the user's and operator's cost. In the second experiment, the Elastic HA was compared to two

other salient techniques in a real-life case study that is especially accepted in the literature about BNDP. The evidence of the first stage of the experimentation suggests that
a high-quality initialization of the distances between the bus stops directly contributes
in a positive manner to the final line scheduling cost and user satisfaction, and this is
successfully achieved by the Elastic HA. Moreover, regarding the second experimental phase, the findings for the study of Mandl's city clearly suggest that the new
method achieves better results compared to those obtained by two other significant
methods. In addition, these well-known methods do not consider the dynamic aspects
of the problem, which constitutes a major issue in the novel technique presented here.

This research has raised many questions that need further investigation. Mainly,
further work needs to be done with regard to the objectives of the problem. We propose analyzing the inclusion of a new objective related to the environmental influence
of the bus network in the microcenters of cities. Finally, it would be interesting to
assess the method in a city in Argentina, namely Bahía Blanca, which presents several
problems with regard to the BNDP.

Acknowledgments

The authors acknowledge CONICET, the ANPCyT for Grant N11-1652, and SeCyT (UNS)
for Grant PGI 24/N026. Mariano Frutos is supported by post-doctoral fellowships from the
Consejo Nacional de Investigaciones Científicas y Técnicas (www.conicet.gov.ar).

References

Baaj, H.M., Mahmassani, H.S., 1995. Hybrid route generation heuristic algorithm for the
    design of transit networks. Transport. Res. C. 3 (1), 31−50.
Bielli, M., Caramia, M., Carotenuto, P., 2002. Genetic algorithms in bus network optimiza-
    tion. Transport. Res. C. 10 (1), 19−34.
Bleuler, S., Laumanns, M., Thiele, L., Zitzler, E., 2003. PISA—A Platform and
    Programming Language Independent Interface for Search Algorithms. Springer,
    Conference on Evolutionary Multi-Criterion Optimization (EMO 2003). LNCS,
    pp. 494−508.
Ceder, A., Wilson, N.H.M., 1986. Bus network design. Transport. Res. B. 20 (4), 331−344.
Ceder, A., Israeli, Y., 1998. User and operator perspectives in transit network design.
    Transport. Res. Rec. J. Transport. Res. Board. 1623, 3−7.
Deb, K., Agrawal, S., Pratap, A., Meyarivan, T., 2002. A Fast Elitist Non-Dominated Sorting
    Genetic Algorithm for Multi-Objective Optimization: NSGA-II. IEEE Transactions on
    Evolutionary Computation, 6, pp. 182−197.
Desaulniers, G., Hickman, M.D., 2007. Public Transit. In: Barnhart, C., Laporte G., (Eds.),
    Transportation, Volume 14 of Handbooks in Operations Research and Management
    Science, Elsevier, Amsterdam, pp. 69−127 (Chapter 2).
Floyd, R.W., 1962. Algorithm 97: shortest path. Commun. ACM. 5 (6), 345.
Fonseca, C.M., Fleming, P.J., 1993. Genetic algorithms for multiobjective optimization: for-
    mulation, discussion and generalization. ICGA. 416−423.

Gruttner, E., Pinninghoff, M.A., Tudela, A., Díaz, H., 2002. Recorridos Óptimos de líneas de transporte público usando algoritmos genéticos. In: Jornadas Chilenas de Computación, Chile.

Mandl, C.E., 1980. Evaluation and optimization of urban public transportation networks. Eur. J. Oper. Res. 5 (6), 396−404.

Olivera, A.C., Frutos, M., Carballido, J.A., Brignole, N.B., 2008. Bus network optimization with a time-dependent hybrid algorithm. J. Universal Comput. Sci. 14 (15), 2512−2531.

Olivera, A.C., Frutos, M., Carballido, J.A., Ponzoni, I., Brignole, N.B., 2009. Bus network scheduling problem: grasp + eas with pisa * simulation. In: Joan, C., Francisco S., Alberto P., Juan M.C., (Eds.), Bio-Inspired Systems: Computational and Ambient Intelligence, IWANN (1), vol. 5517, Lecture Notes in Computer Science, pp. 1272−1279.

Pattnaik, S.B., Mohan, S., Tom, V.M., 1998. Urban bus transit route network design using genetic algorithm. J. Transport. Eng. 124 (4), 368−375.

Uur, A., 2008. Path planning on a cuboid using genetic algorithms. Inf. Sci. 178 (16), 3275−3287.

Zhao, F., Zeng, X., 2006. Simulated annealing−genetic algorithm for transit network optimization. J. Comput. Civ. Eng. 20 (1), 57−68.

Zitzler, E., Künzli, S., 2004. Indicator-based selection in multiobjective search. In: Proceedings of the Eighth International Conference on Parallel Problem Solving from Nature (PPSN VIII). Springer, pp. 832−842.

Zitzler, E., Laumanns, M., Thiele, L., 2001. Spea2: improving the strength pareto evolutionary algorithm. Technical Report 103, Gloriastrasse 35, CH-8092 Zurich, Switzerland.

19 The Hybrid Method and its Application to Smart Pavement Management

Fereidoon Moghadas Nejad and Hamzeh Zakeri

Department of Civil and Environmental Engineering,
Amirkabir University of Technology, Tehran, Iran

19.1 Introduction

A pavement management system (PMS) is a crucial system that aggregates inputs such as distress, condition, and properties gathered from road surface inventory and used for several goals, such as optimum network and project strategy selection, enhancement level of serviceability and budget activities, and scheduling maintenance with optimum benefits for entire networks and for projects specifically. General tasks of PMS include inventorying and evaluating pavement conditions, classification of networks, segmentation of roads, and maintenance scheduling. However, the final decision made by humans and PMS propose several alternative policies (Ismail et al., 2009). Data gathered based on visual or automatic methods need special software to prepare and enhance to use in the mother PMS engine. A golden PMS engine plays a crucial role in pavement maintenance and management performance selection and analysis. Artificial intelligence (AI) is a branch of computer science that attempts to replicate human intelligence and make it possible to perceive, reason, and act on rules by autonomous engines (Luger and Stubbelfield, 1993; Winston, 1992). AI methods cover a wide range of engineering applications, such as game playing, speech recognition, computer vision, expert systems (ES), and civil engineering. The AI methods are used with increasing frequency to solve problems in the civil domain. Based on computer models, AI methods implemented and replaced our classic models of human reasoning. It uses frames, rules, cases, and expectations (Alavi and Gandomi, 2011a, b; Gandomi and Alavi, 2011; Gandomi et al., 2011a, b). Several studies have been done in the field of pavement engineering based on AI methods (Alavi et al., 2011; Gandomi et al., 2011a, b; Nejad and Zakeri, 2011a, b, c). Many researchers have looked into the application of AI in the PMS (Flintsch, 2003).

In the area of pavement engineering, the domain of input of PMS inherently used expert knowledge, and it has been straightforward to crystallize ambiguous and

Metaheuristics in Water, Geotechnical and Transport Engineering. DOI: http://dx.doi.org/10.1016/B978-0-12-398296-4.00019-2

uncertainties (incomplete data). Uncertain, ambiguous, and incomplete inputs in knowledge base call on all the AI methods to overcome these problems in subsystems of PMS. These methods are used for condition assessment, evaluation, prediction (performance, distress, and budget), need analysis, prioritization, and optimization treatments. Because selection and furniture architects of AI methods are very important for such unusual inputs to handle these types of tasks, the aggregating and design of agents play a crucial role in the general performance of PMS. Hybrid method (HM) in PMS is a complex collection of methods in the fields of pavement management, pavement quantification, and distress detection and classification.

Several hybrid strategies have been proposed for promoting PMS to Smart PMS (SPMS). Flexible agents which are provided by Hybrid AI systems can be used for solving the problem. Generally, hybrid systems (HSs) made with intelligent methods will produce more robust agents in subsystems than traditional agents in PMS. As in other fields of engineering, the hybrid agent is considered to be an effective tool to solve complex problems, and indeed it is providing intelligence for huge system. A PMS generally consists of several agents by different tasks, which can separate into several subsystems. Every subsystem has a special task, and autonomous ability is a first-limited condition for this kind of subsystem.

Recently, in the field of PMS, and especially in evaluation and quantification, multiresolution analysis such as wavelet analysis provides effective tools for variable scales of pavement surface analysis and distress classification. Automatic pavement diagnosis systems in the nondestructive testing world of detection and classification of pavement distress (cracking) based on multiresolution methods is a very good example of the performance of HSs in the heart of a subsystem.

Chen (2007) has stated that several fields are particularly attractive for platform of AI individually and complex. These characteristics are uncertain, with ambiguous and incomplete information, objective (numerical) and subjective (linguistic) knowledge and information, verity of expert knowledge, large amounts of assets, and several feasible treatments (Chen, 2007).

In this chapter, we first concentrate on the application of elements and architect of HS which are mostly used for fault diagnosing and treatment selection. Complex AI techniques have been discussed because they allow us to handle incomplete, ambiguous, and huge data. Recently, many ES, fuzzy, and NN techniques have been used in pavement domain for distress detection and classification. These algorithms show various degrees of success. The image processing techniques, AI methods and HS, will be introduced. This chapter presents basic theory of HM elements and several of its application toward smart decision making in PMS. A quick overview of the image processing techniques will be discussed in the multiresolution section.

19.2 Methodology

The HM is a combination of several intelligent methods which promotes the whole system to a new smart one, HS.

The main goal of HM is combining the advantages of different AI methods within a single system. HM and its application in smart management in PMS is a new vision of aggregation AI methods in the fields of automatic pavement distress detection and classification. Several architect models are predicted for HSs. The preliminary HSs usually integrate two intelligent methods, such as (FL + NN), (NN + ES), (NN + NN), (ES + ES), (ES + FL), and (FL + FL) (Figure 19.1).

However, recent applications tend toward hybrid integration containing two or more intelligent technologies (Hatzilygeroudis and Prentzas, 2001, 2004; Sahina et al., 2012). Sahina et al. (2012) offers a survey of current approaches and applications that contains a review from 1988 to 2010; also, Liao (2005) provides a review of ES methodologies and applications from 1995 to 2004 (Liao, 2003, 2005; Sahina et al., 2012). Robots and fast systems often need a synergistic integration of the complementary agents into HSs. Some benefits of hybridized agents in pavement distress detection and classification in smart pavement management are tractability, robustness, low solution cost, and better rapport with reality than other HSs. For example, while assessing the feasibility of using the Fuzzy Type II method for threshold selection in automatic pavement distress detection, because of the high degree of uncertainty in surface cracking, it can be easily concluded that the ES had potential only in a hybridized version of Fuzzy Type II, not as a stand-alone solution. A combined HS makes it possible to use for more complex problem such as pavement quantification (Figure 19.2).

User-friendly and decision-support tools play important roles in the increasing robustness of intelligent techniques in combined HSs. Generally, this complex agent can handle uncertain, subjective, incomplete, ambiguous information; generate knowledge by learning from examples and/or experts in self-healing learning; and promote their performance as it is designed in the center of the PMS. Wavelet, ridgelet, and curvelet fuzzy NNs and ES hybridization are possible modes for pavement distress detection and classification (Figure 19.3).

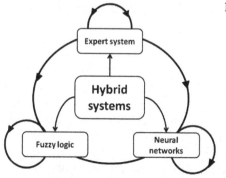

Figure 19.1 General architect of HS.

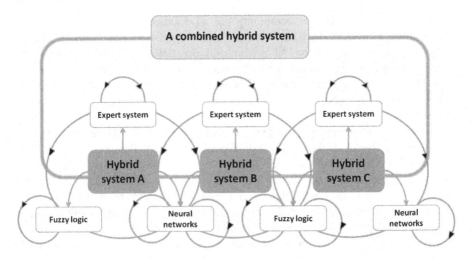

Figure 19.2 General architect of a combined HS.

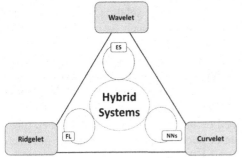

Figure 19.3 Multiresolution HS.

It is predictable that these kinds of tools can predominate smart pavement management in the near future. This section is organized as follows: In Section 19.2.1, the theory and role of image processing methods in PMS are briefly reviewed. In Section 19.2.2, the roles of AI methods in PMS and their combinations are explained.

19.2.1   Image Processing and PMS

Inspection of pavement surface conditions is considered to be a foundation of the highway PMS because decisions about pavement maintenance are made based on inspection results. Pavement surface conditions are mostly determined by manual or visual inspection. The simplest method is for human experts to visually examine and evaluate the pavement. This approach, however, involves high labor costs and produces unreliable and inconsistent results. Furthermore, it exposes inspectors to dangerous working conditions on highways. To overcome the limitations of the subjective visual evaluation process, several attempts have been made to develop an automatic procedure to replace it. Most current systems use computer vision and image processing technologies to do this. Surveys have revealed that operating costs decrease dramatically between manual and automated methods for collecting data. In recent years, many efforts have been made to develop a more automated pavement inspection system in both the pavement image acquisition and the pavement image processing (Chambon and Moliard, 2011; Nejad and Zakeri, 2011a, b, c).

Generally speaking, most automated pavement inspection systems lies in automated analysis and evaluation of pavement distress information (Chambon and Moliard, 2011; Nejad and Zakeri, 2011a, b, c; Zhou, 2004; Zhou et al., 2006). The currently existing systems are robust for with collecting pavement images with distress. In this kind of management system, image processing is a controversial issue. It includes distress detection and isolation, distress classification, pavement condition quantification, image compression and noise reduction, distress segmentation, and maintenance (Nejad and Zakeri, 2011b; Zhang et al., 2006). Accurate and real-time information on their condition is necessary for pavement evaluation and management. With a high-speed Charge-Coupled Device (CCD) camera-based image-processing device, we can automatically extract distress from an image and develop the information to characterize it. Because a huge amount of data is expected to be collected, it is desirable that initial screening of data is done to detect the existence of distress and isolate the bad frames from the good ones. Due to the nonuniform background of pavement distress, it is more difficult to find a certain threshold to segment distress according to the traditional edge detection method (Brown, 2000; Chua and Xu, 1994; Nejad and Zakeri, 2011c).

Imaging techniques tend to result in images with poor contrast and relatively high noise level. Image enhancement is crucial because it can help to improve the quantity and quality of information on asphalt pavement distress. Conventional image enhancement methods such as histogram equalization tend to amplify noise and at the same time enhance the visibility of an object's characteristics. Considerable success has been achieved in the development of wavelet-transform image enhancement algorithms with noise suppression (Brown, 2000; Candès, 1998; Candès and Donoho, 2000; Candès et al., 2006; Cheng et al., 2001; Claudio and Jacob, 2003; Do and Vetterli, 2003a, b; Donoho, 2001; Donoho and Duncan, 2000; Farook and Gao, 2003; Laine et al., 1994; Nejad and Zakeri, 2011a, b, c; Sattar et al., 1997; Shan et al., 2005; Starck et al., 1999; Wang et al., 2007; Zhang et al., 2009a, b; Zong et al., 1996). Wavelets perform very well for objects with point singularities, but their performance is not good for representing one-dimensional singularities. As extension of the wavelet multiscale analysis framework, ridgelets and curvelets can effectively deal with such linear singularities in two-dimensional signals (Candès, 1998; Do and Vetterli, 2003, b). Therefore, many image enhancement algorithms used in ridgelet or curvelet domains develop rapidly and achieve better enhancement results (Nejad and Zakeri, 2011a, b, c; Shan et al.,

2005; Wang et al., 2007). Nejad and Zakeri (2011a, b, c) presented an automated imaging system for fast detection and isolation of distress in asphalt pavement images obtained by a pavement image acquisition system (PIAS). Classification of distress in a PMS using shape or gray-level information proves to be a particularly challenging task. This is due primarily to the changing shape of distress in a section of pavement images and the gray-level intensity overlap in soft distress. Even so, pavement without any distress is expected to have a consistent texture within distress across multiple slices. This research focused on using multiresolution texture analysis for the classification of distress from normal asphalt pavement. It offers a comprehensive analysis of three forms of multiresolution analysis. Texture features were computed from the following multiresolution transforms: the Haar wavelet (Stollnitz et al., 1995), Daubechies wavelet, Coiflet wavelet (Rajeev et al., 1997), ridgelet (Donoho, 2001), and curvelet transforms (Candès et al., 2006; Starck et al., 1999). Multiresolution analysis (Cheng et al., 1999a; Manjunath et al., 2000; Starck et al., 2003) have been successfully used in image processing with the recent emergence of applications to texture classification. Also, many methods have been proposed to detect and isolated pavement distress (Nejad and Zakeri, 2011a, b, c).

In the field of soft computing, Cheng et al. (2001) used a neural network (NN) to select a threshold for separating distress from the background. The main and standard deviation were used as parameters to train the NN for threshold selection (Cheng et al., 1999a, b; Nejad and Zakeri, 2011a). Cheng et al, 1999 also proposed a fuzzy set theory to detect and segment cracks, based on the fact that crack pixels are always darker than their surroundings. Many other methods have been proposed, such as the Otsu method, the regression method, the relaxation method, and the Kittler method. All these methods belong to two major categories: edge detection and thresholding. They all detect and segment distress in the space domain. Due to circumstances such as different type of pavements or distress, different lighting or weather conditions, and different external materials on pavement surfaces, pavement distress instances always have nonuniform backgrounds. It is difficult to find a certain threshold to detect and segment a distress according to traditional edge detection or threshold methods (Nejad and Zakeri, 2011b; Zhou, 2004). Several studies have investigated the discriminating power of wavelet-based features in various applications, including image compression (Mulcahy, 1997), image denoising (Li, 2004), and classification of natural textures (Wan and Shi, 2007). Instead of detecting distress in the space domain, wavelet transform (WT) is used to transform pavement images into the wavelet domain by Zhang et al. (2006). Zhou and Jian (2004) proposed several distress detection and isolation criteria based on wavelet coefficients in the high-frequency subbands of the wavelet domain (Zhou, 2004; Zhou et al., 2006).

Recently, the finite ridgelet and curvelet transforms have emerged as a new multiresolution analysis tool. In recent years, the ridgelet transform has been used in image contrast enhancement and image denoising (Do and Vetterli, 2003b). Automatic ridgelet image enhancement algorithm for road crack image, based on fuzzy entropy and fuzzy divergence, is done by Zhang et al. (2009a, b).

Experimental results from this research show that an image enhancement algorithm can effectively enhance the global and local contrast effects on road crack images. To the authors' knowledge, applications of ridgelets to texture classification have been investigated only in the context of natural images (Mulcahy, 1997), and curvelet-based applications have been investigated only in the image representation of astronomical images (Do and Vetterli, 2003b). Nejad and Zakeri, 2011a, b, c compared pavement distress detection and isolation using features derived from the wavelet, ridgelet, and curvelet transforms. This research shows clearly that curvelet-based features outperform all other multiresolution techniques in pothole distress and ridgelet-based features outperform all other multiresolution techniques in cracking distress, as well as other methods (Nejad and Zakeri, 2011a). The pavement distress detection and isolation algorithm proposed in this chapter consists of four main steps:

1. Segmentation of regions of interest from pavement images
2. Extraction of the most discriminative distress features
3. Creation of a classifier that automatically identifies the various distresses
4. Isolated distress are compressed and save to a distress database (DDB).

Finally, other images are thrown away.

Generally, in automatic pavement distress detection and isolation systems, the overall goal is to develop a fast and robust model for distress detection and classification toward smart management. In recent years, pavement image thresholding plays a crucial role in automatic pavement distress detection and classification, and researchers have paid more attention to this field.

Mainly, ordinary pavement image thresholding algorithms are the Ostu criteria, iterative threshold, histogram and fuzzy thresholding method, etc. (Albert and Nii, 2008; Wan and Shi, 2007; Zhang et al., 2006). All of these image preprocessing methods still suffer from the effects of noise when the images have lots of irregularities present (Zhou et al., 2006), as is the case for asphalt concrete surface. Chua and Xu, (1994) investigated the problem of pavement crack classification using moment invariant and NNs. Mohajeri and Manning (1991), as part of a fully automated PMS, described a rule-based system that incorporates knowledge about individual crack patterns and classifies by the process of elimination. In addition to classifying cracking by type, it is capable of quantifying the crack severity with parameters such as length, width, and area. Indeed, selecting an exact threshold with high accuracy is very important (Mohajeri and Manning, 1991; Nejad and Zakeri, 2011b, c).

Nallamothu and Wang (1996) proposed an artificial neural network (ANN) for pattern recognition for pavement distress classification. Cheng et al. (1999a) used the Hough transform to determine the type of the crack. Hsu et al. (2001) described a moment invariant technique for feature extraction and an NN for crack classification. The moment invariant technique reduces a two-dimensional image pattern into feature vectors that characterize the image, such as translation, scale, and rotation of an object in an image. After these features are extracted, the overall results of this study were satisfactory and the classification accuracy of the introduced system was 85% (Nejad and Zakeri, 2011b, c). (Zhou, 2004 and Zhou et al., 2006) used a two-step transformation method by wavelet and radon transform (RT) to determine the type of the crack; that method is similar to the method proposed in this research. Several statistical criteria are developed for distress detection and isolation, which include the high-amplitude wavelet coefficient percentage (HAWCP), the high-frequency energy percentage (HFEP), and the standard deviation (STD). These criteria are tested on hundreds of pavement images differing by type, severity, and extent of distress. Experimental results demonstrate that the proposed criteria are reliable for distress detection and isolation and that real-time distress detection and screening is feasible; however, the proposed method still suffers from the effects of noise, which are generated by an asphalt concrete surface (Nejad and Zakeri, 2011a, b, c; Zhou, 2004; Zhou et al., 2006).

Nejad and Zakeri, 2011a, b, c used a new method for optimum feature extraction based on the wavelet–radon (WR) transform and dynamic neural network (DNN) for pavement distress classification. This research demonstrated that the WR + DNN method can be used efficiently for fast automatic pavement distress detection and classification. DNN threshold selection played a key role in the high accuracy of the introduced method (Nejad and Zakeri, 2011b, c). The RT is applied on the M, Horizontal, Vertical and Diagonal (MHVD) results at the third level since the highest frequency components transformed into the wavelet coefficient in the high-frequency subband at the third level. From previous research, it can be seen that the high values of the wavelet modulus represent distress, while low values represent noise and background on the pavement surface (Zhou, 2004). When the RT is applied to wavelet modulus, a crack is transformed into a peak in the radon domain (Nejad and Zakeri, 2011b, c). Under some circumstances, the intensity of the background may be closed to the distress, and sometimes there can be some tiny, thin crack. This property has a tangible effect on threshold selection. To solve this problem, a new approach is proposed by Nejad and Zakeri (2011a, b, c). Image enhancement methods, the MHVD method, and then the RT are implemented. This approach makes it much easier to search for peaks in the radon domain. For the peak in the radon domain, we can use number, position, area, value, and volumes to describe them and call the patterns of the peaks (Nejad and Zakeri, 2011b, c).

The numbers of the peak correspond to the numbers of cracks. So the numbers can be used as one of the parameters of the NN to determine if there is a single crack or multiple cracks (Nejad and Zakeri, 2011a, b, c). Figures 19.4 and 19.5 show the reconstructed image from binary wavelet images and developing a three-dimensional radon transform (3DRT) from images by combination of the RT at any direction. Figure 19.5 shows the binarization of 3DRT and pattern extraction; the threshold of RT was modified by DTM (Nejad and Zakeri, 2011b, c). Despite of DTM consequences, the outcome is not still flexible.

Radon Transform

The RT uses a set of projections through an image at different angles (Nejad and Zakeri, 2011a, b, c). The RT can also be used for line detection. A technique for

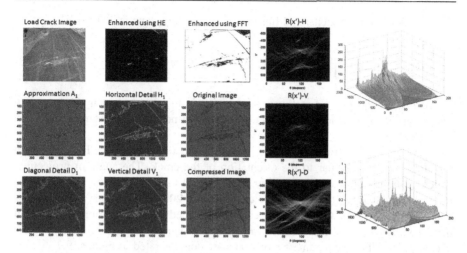

Figure 19.4 A reconstructed image from binary wavelet images and developing 3DRT.

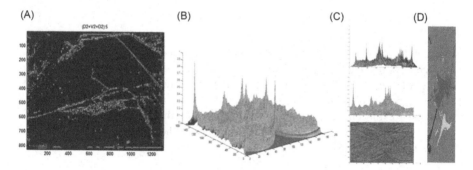

Figure 19.5 Binarization of 3DRT and pattern extraction. The threshold of RT was modified by Dynamic Threshold Method (DTM): (A) wavelet, (B) 3DRT, (C) DTM, and (D) pattern.

using RT to reconstruct a map of polar regions in a planet using a spacecraft in a polar orbit has also been devised (Nejad and Zakeri, 2011b, c). RT is based on the parameterization of straight lines and the evaluation of integrals of an image along these lines. Due to inherent properties of RT, it is a useful tool to capture the directional features of an image (Nejad and Zakeri, 2011b, c). The classical RT of a two-variable function u is Ru defined on a family of straight lines. The value of Ru on a given line is the integral of u along this line. Assuming that the line in the plane (t,q) is represented as $t = \tau + pq$, where p is the slope and τ is the offset of the line. The RT of the function over this line will be given as

$$\mathrm{Ru}(\tau, p) = \int_{-\infty}^{+\infty} u(\tau + pq, q)\,\mathrm{d}q \tag{19.1}$$

The RT of a two-dimensional function $f(x,y)$ in (r,θ) plane is defined as

$$R(r,\theta)[f(x,y)] = \int_{-\infty}^{+\infty} \int_{-\infty}^{+\infty} f(x,y)\delta(r - x\cos\theta - y\sin\theta)dx\,dy \qquad (19.2)$$

where $\delta(\ )$ is the Dirac function, $r\in[-\infty,\infty]$ is the perpendicular distance of a line from the origin, and $\theta\in[0,\pi]$ is the angle formed by the distance vector, as shown in Figure 19.6.

A crack will be projected into a valley in the radon domain. The projection direction is perpendicular to the orientation of the crack. The angle of projection, which is the coordinate of the valley, can be used to determine the orientation of the crack and classify the type of the crack (Zhou, 2004; Zhou et al., 2006). In this approach, the type of the distresses can be classified by thresholding the value in the radon domain, and the features from patterns can be extracted for training the classification NN. The parameters of peaks are demonstrated in Figure 19.7.

Moghadas Nejad and Zakeri, (2011b) defined the semi-flexible threshold as a DTM (Figure 19.7), which is defined as follows:

$$\text{if RTF} > 0 \text{ then } T = \text{RTF}_1 \text{ else } T + \text{RTF}_2 \text{ end} \qquad (19.3)$$

where, RTF_1 and RTF_2 are radon threshold function 1 and radon threshold function 2, respectively, and T denotes the final threshold (Figure 19.8). If did not come out any feature after using RTF_1, this means that threshold should be pulled down by using RTF_2. (Nejad and Zakeri, 2011b). Generally, researchers tried to select a unique threshold for the entire object, whereas the flexible threshold is more practical in many applications. For promoting the performance of thresholding, the fuzzy flexible threshold is introduced.

WT and New Production of NN

The wavelet neural network (WNN) is constructed based on the wavelet-transform theory and is an alternative to feed-forward NN. Wavelet decomposition (Avci et al., 2005b) is a powerful tool for nonstationary signal analysis. Let $x(t)$ be a piecewise continuous function. Wavelet decomposition allows one to decompose $x(t)$

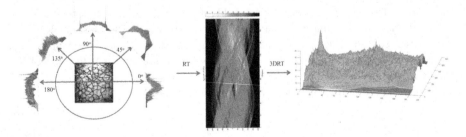

Figure 19.6 The principles of RT.

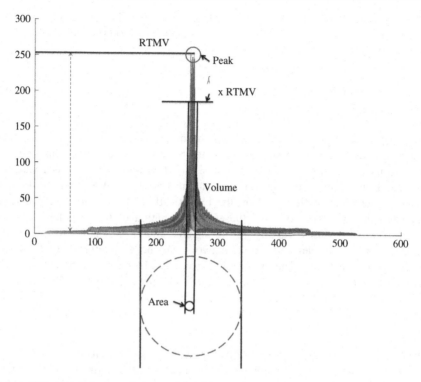

Figure 19.7 A peak in the radon domain and its parameters (Nejad and Zakeri, 2011a, b, c; Zhou et al., 2006).

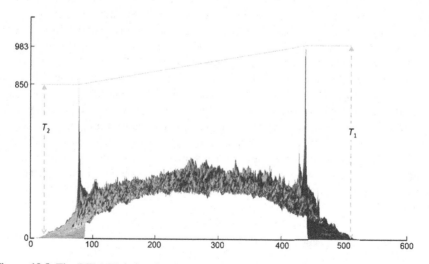

Figure 19.8 The DTM (Nejad and Zakeri, 2011b).

using a wavelet function $\Psi: R^n \to R$. Based on the wavelet decomposition, the wavelet network structure is defined by Avci and Turkoglu (2003), Avci et al. (2005a, b), and Avci (2007) as follows:

$$y(x) = \sum_{i=1}^{N} W_i \Psi \left[D_i(x - t_i) \right] + b \tag{19.4}$$

where D_i are dilation vectors specifying the diagonal dilation matrices, D_i, t_i are translation vectors, and the additional parameter b is introduced to help deal with nonzero mean functions on finite domains. An algorithm of the back-propagation type has been derived for adjusting the parameters of the WNN. Applications of WNN in the medical field include the classification of coronary artery disease, characteristics of heart valve prostheses, interpretation of the Doppler signals of the heart valve diseases, classifying bio signals, and ECG segment classification. Combination of wavelet and neural network can be used effectively to feature extraction and selection. The WRNN is constructed based on the wavelet and RT theory and is an alternative to feed-forward NNs. However, to date, WRNN pavement distress classification is a new approach (Avci, 2007; Avci and Turkoglu, 2003; Avci et al., 2005a).

Discrete WTs and Applications in Pavement Distress Evaluation

Nowadays, most popular methods of texture analysis are multiresolution or multichannel analyses such as wavelet decomposition and Gabor filters (Candès, 1998). Wavelet transform is superior to the Gabor transform, because its provides a true and framework for the processing of a signal and an image at variety scale. (Zhang et al., 2006; Zhou et al., 2006). Wavelet has several families, such as Daubechies 2 (D2), Haar (H), and Coiflet (C6; Rajeev et al., 1997; Stollnitz et al., 1995). The Daubechies, Haar, and Coiflet families were considered since they differ substantially in the number of adjacent pixels used to extract the wavelet coefficients, from which the pavement distress features are derived. These three families were investigated by Moghadas and Zakeri (2011) for their increasingly larger filters and smoother windows (Nejad and Zakeri, 2011a, b, c). A continuous wavelet transform (CWT) can decompose a signal or an image with a series of averaging and differencing calculations. Wavelets compute average intensity properties, as well as several detailed contrast levels distributed throughout the image. Wavelets can be calculated according to various levels of resolution or blurring depending on how many levels of averages are calculated. The general mother wavelet can be constructed from the following scaling $\varphi(x)$ and wavelet functions $\omega(x)$ (Dettori and Semlera, 2007):

$$\varphi(x) = 1.414 \sum g(k)\varphi(2x - k) \tag{19.5}$$

$$\omega(x) = 1.414 \sum h(k)\varphi(2x - k) \tag{19.6}$$

where $h(k) = -1^k g(N - 1 - k)$, and N is the number of scaling and wavelet coefficients. The sets of scaling $h(k)$ and wavelet $g(k)$ function coefficients vary depending on their corresponding wavelet bases. Wavelet representation of a signal is as follows (Zhang et al., 2006; Zhou et al., 2006):

$$
\begin{aligned}
f(x) = 2^{-k} \sum_{n=-\infty}^{\infty} [f(x), \varnothing(2^{-k}x - l)]\varnothing(2^{-k}x - l) + 2^{-k} \\
+ \sum_{n=-\infty}^{\infty} [f(x), \Psi(2^{-k}(x - l)]\Psi(2^{-k}x - l)
\end{aligned}
\tag{19.7}
$$

The approximation details are defined as follows:

$$
A_K = \langle f(x), 2^{-k/2}\varnothing(2^{-k}\varnothing(2^{-k}x - l)) = 2^{-k/2}\int f(x)\varnothing(2^{-k}x - l)\mathrm{d}x
\tag{19.8}
$$

$$
D_K = \langle f(x), 2^{-k/2}\Psi(2^{-k}\Psi(2^{-k}x - l)) = 2^{-k/2}\int f(x)\Psi(2^{-k}x - l)\mathrm{d}x
\tag{19.9}
$$

For image decomposition and reconstruction, an image can be decomposed by projecting the image in the space v_k, W_K^H, W_K^L, and W_K^D:

$$
\begin{aligned}
f_{k-1}(x, y) = 2^{-k} \sum_{m,n} \mathrm{LL}_k(m, n)\varnothing(2^{-k}x - m)\varnothing(2^{-k}y - n) \\
+ 2^{-k} \sum_{m,n} \mathrm{HL}_k(m, n)\varnothing(2^{-k}x - m)\Psi(2^{-k}y - n) \\
+ 2^{-k} \sum_{m,n} \mathrm{LH}_k(m, n)\Psi(2^{-k}x - m)\varnothing(2^{-k}y - n) \\
+ 2^{-k} \sum_{m,n} \mathrm{HH}_k(m, n)\Psi(2^{-k}x - m)\Psi(2^{-k}y - n)
\end{aligned}
\tag{19.10}
$$

where $\mathrm{LL}_k, \mathrm{HL}_k, \mathrm{LH}_k$, and HH_k are the orthogonal projections in the orthogonal space of V_K, W_K^H, W_K^L, and W_K^D, and they can be calculated by:

$$
\mathrm{LL}_k(i, j) = \sum_{m,n} l(m - 2i)l(n - 2j)\mathrm{LL}_{k-1}(m, n)
\tag{19.11}
$$

$$
\mathrm{HL}_k(i, j) = \sum_{m,n} h(m - 2i)l(n - 2j)\mathrm{LL}_{k-1}(m, n)
\tag{19.12}
$$

$$
\mathrm{LH}_k(i, j) = \sum_{m,n} l(m - 2i)h(n - 2j)\mathrm{LL}_{k-1}(m, n)
\tag{19.13}
$$

$$
\mathrm{HH}_k(i, j) = \sum_{m,n} h(m - 2i)h(n - 2j)\mathrm{LL}_{k-1}(m, n)
\tag{19.14}
$$

For more details on these wavelet families, see Kara and Watsuji (2003), Stollnitz et al. (1995), Rajeev et al. (1997), and Mulcahy (1997). The block diagram for image decomposition algorithm is shown in Figure 19.9, where L and H are the mirror filters.

There are several ways of generating a two-dimensional WT. The construction of the digital filters differs mainly in their scaling and wavelet coefficients. Scaling and wavelet function coefficients are characteristics of their particular families (Starck et al., 1999). The Haar wavelet uses only two scaling and wavelet function coefficients and calculates pairwise averages and differences. The Haar transform uses nonoverlapping windows and reflects changes between adjacent pixel pairs. The Daubechies wavelet with an order of two is used to decompose distress images into three levels. A weighted average is computed over four pixels, resulting in a smoother transform. These tests indicated that in general, the Daubechies wavelet was preferable. The Coiflet wavelet uses six scaling and wavelet function coefficients. The Coiflet wavelet presents the same problem with filter size; the mirror technique was also applied (Nejad and Zakeri, 2011b, c). Each of these wavelet filters could be applied multiple times to the averages (see C_2 in Figure 19.10), according to the desired level of resolution. In the following application, two levels of resolution were extracted for each wavelet. At each resolution level, the wavelet has three detailed coefficient matrices representing the vertical, horizontal, and diagonal structures of the image. HL_k, LH_k, and HH_k denote wavelet coefficients, which are also called the *details*, in the high-frequency horizontal, vertical, and diagonal subbands at level k. Distress, which is usually the high-frequency component, is most likely transformed into high-amplitude wavelet coefficients, and noise is transformed into low-amplitude wavelet coefficient in the high-frequency subbands of HL_k, LH_k, and HH_k (Zhou et al., 2006).

Generally, in good pavement images, their wavelet coefficients have a compact histogram, and bad pavement images have a widely spread histogram. There are many statistical parameters, such as the range, standard deviation, and moment to describe the spread of the histogram at any subband. Three criteria of HAWCP,

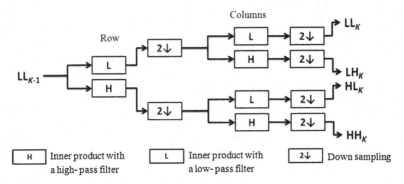

Figure 19.9 Image decomposition by WTs (Nejad and Zakeri, 2011a, b, c; Zhou et al., 2006).

HFEP, and STD are tested for pavement distress detection. The HAWCP threshold for distress images selects 45 as a final value. HAWCP has a value between 0% and 100%. The value 0% represents a good pavement surface and 100% represents the worst pavement surface. It can be used as a measure of the extent of distress, and therefore, it serves as a good criterion for distress detection and isolation (Nejad and Zakeri, 2011a, b, c; Zhou, 2004; Zhou et al., 2006).

One of the best ways to measure the severity of distress based on wavelet coefficients is to calculate the energy of coefficient. Distresses are transformed into a high-amplitude wavelet coefficient, which has higher energy than a low-amplitude coefficient. HFEP also has a value between 0% and 100%, where 0% means a perfect pavement surface and 100% means the worst pavement. The spread of the histogram can be used to characterize the type, severity, and extent of the distress. The STDs of the wavelet coefficients in the horizontal, vertical, and diagonal sub-bands at the second level are calculated. The largest STD of the horizontal STD_h, vertical STD_v, and diagonal STD_d subband is chosen to present the STD of the image. The groups of distress with close STD are similar to those grouped by HFEP. STD is a good criterion for distress detection and isolation. Previous research shows that HAWCP, HFEP, and STD are consistent for severe distress (see Table 19.1; Nejad and Zakeri, 2011a, b, c; Zhou, 2004; Zhou et al., 2006).

In order to detect all distresses, three criteria are combined for final distress detection and isolation. If any one of the three criteria detects the distress, the image is regarded as a distress image and is saved in a DDB. Experimental results demonstrated that the criteria used for distress detection and isolation are robust (Nejad and Zakeri, 2011a, b, c; Zhou, 2004; Zhou et al., 2006). The corresponding histograms of the horizontal, vertical, and diagonal wavelet coefficients at the first, second, and third levels have been plotted in Figure 19.11. For good pavement images, the wavelet coefficients have compact histograms, and for bad pavement images, the wavelet coefficients have widely separated histograms (Nejad and Zakeri, 2011b, c).

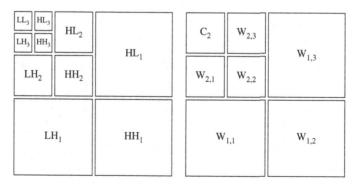

Figure 19.10 Wavelet decomposition: W (resolution, direction) (Brown, 2000; Nejad and Zakeri, 2011a, b, c; Zhou et al., 2006).

Table 19.1 Criteria for Distress Detection and Isolation by the Wavelet Method (Nejad and Zakeri, 2011a, b, c; Zhou, 2004; Zhou et al., 2006)

Criteria	Formula	What Is Measured?
HAWCP	$\text{HAWCP} = \sum\limits_{0}^{W/2} \sum\limits_{0}^{L/2} \dfrac{D_2(p,q) \times 4}{W \times L}$	W and L are the width and length of the image, and $D_2(p,q)$ is the binarized wavelet modulus at the second level.
HFEP	$\text{HFEP} = 1 - \sum\limits_{p=0}^{W/2^k} \sum\limits_{q=0}^{L/2^k} \text{LL}_k^2(p,q) / \sum\limits_{p=0}^{W} \sum\limits_{q=0}^{L} I^2(m,n)$	Here, $I(m,n)$ is the intensity at a pixel (m,n) of the original pavement image. The width and length of the kth level subband are $W/2^k$. HFEP can be obtained by subtracting the low-frequency energy percentage from 1 because of energy conservation. $\text{LL}_k(p,q)$ are the wavelet coefficients in the approximation subband.
STD	$\text{STD} = \sqrt{\sum\limits_{i=-\infty}^{+\infty} \left(c(i) - \sum\limits_{-\infty}^{+\infty} c(i)p(i) \right)^2 p(i)}$	$p(i)$ is the probability of the wavelet coefficient $c(i)$ of the standard deviation STD, which is a measure of the spread of a set data. The largest standard deviation in horizontal, vertical, and diagonal subband is chosen to present the STD of the image.
MWC	$M_q = \sum\limits_{-\infty}^{+\infty} |c(i) - \mu|^q p(i)$	M_q is the qth moment, The qth moment is centered at zero, and (i) is the probability of the wavelet coefficient $c(i)$.

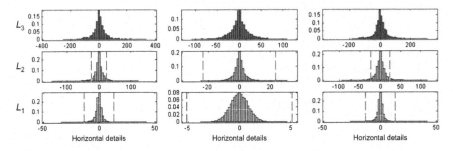

Figure 19.11 Histograms of wavelet coefficients for the pavement images at level k (Nejad and Zakeri, 2011a, b, c; Zhou, 2004; Zhou et al., 2006).

Ridgelet and Curvelet

In 2000, Donoho introduced the ridgelet transform (Candès and Donoho, 2000; Candès et al., 2006; Donoho, 2001). This section starts with briefly reviewing the continuous ridgelet transforms (CRTs) and draws its connection with the CWT (Starck et al., 1999, 2003). The CRT can be defined from a one-dimensional wavelet function oriented at constant lines and radial directions. The CRT in R^2 is defined by:

$$\text{CRT}_f(a, b, \theta) = \int_{R^2} \varphi_{a,b,\theta}(x)f(x)\mathrm{d}x \qquad (19.15)$$

where the ridgelets $\varphi_{a,b,\theta}(x)$ in two dimensions are defined using a wavelet function:

$$\varphi_{a,b,\theta}(x) = \frac{1}{\sqrt{2}}\varphi\left[\frac{x_1\cos\theta + x_1\sin\theta - b}{a}\right] \qquad (19.16)$$

Given an integrable bivariate function $f(\mathbf{x})$, we define its ridgelet coefficients by:

$$R_f(a, b, \theta) = \int \varphi_{a,b,\theta}(x)f(x)\mathrm{d}x \qquad (19.17)$$

The exact reconstruction formula can be presented as

$$f(x) = \int_0^{2\pi}\int_{-\infty}^{+\infty}\int_0^{\infty} R_f(a, b, \theta)\varphi_{a,b,\theta}(x)\frac{\mathrm{d}a}{a^3}\,\mathrm{d}b\,\frac{\mathrm{d}\theta}{4\pi} \qquad (19.18)$$

This is oriented at angles h, and constant along the lines of $x_1\cos h + x_2$ - $\sin h = \text{const}$. For details, see Do and Vetterli (2003a, b). The ridgelet transform is optimal for detecting cracks in the pavement surface image (Nejad and Zakeri, 2011a, b, c). Generally, wavelets detect objects with point singularities, while ridgelets are able to represent objects with line singularities. In 2000, Candès and Donoho introduced the curvelet transform (Candès and Donoho, 2000; Candès et al., 2006). The idea is to first decompose the image into a set of wavelet bands and to analyze each band with a local ridgelet transform. The block size can be changed at each scale level. Roughly, different levels of the multiscale ridgelet pyramid are used to represent different subbands of a filter bank output. The Continue Curvelet Transform (CCT) can be defined by a pair of windows $W(r)$ (a radial window) and $V(t)$ (an angular window), with W as a frequency-domain variable, and r and h as polar coordinates in the frequency domain:

$$\sum_{j=-\infty}^{\infty} W^2(2^j r) = 1, \quad r\varepsilon\left(\frac{3}{4}, \frac{3}{2}\right) \qquad (19.19)$$

$$\sum_{j=-\infty}^{\infty} V^2(t - 1) = 1, \quad r\varepsilon\left(-\frac{1}{2}, \frac{1}{2}\right) \qquad (19.20)$$

A polar "wedge," represented by U_j, is supported by W and V, the radial and angular windows. U_j is defined in the Fourier domain by:

$$U_j(r, \theta) = 2^{\frac{-3j}{4}} \left(W(2^{-j}r)V\left(\frac{2^{0.5j}\theta}{2\pi}\right) \right) \tag{19.21}$$

The curvelet transform can be defined as a function of $x = (x1, x2)$ at scale 2^{-j}, orientation θ, and position $x_k^{(j,l)}$ by:

$$\varphi_{i,j,k}(x) = \varphi_j(R_j(x - x_k^{(j,l)})) \tag{19.22}$$

where R_j is the rotation in radians. Further details are presented in Candès and Donoho (2000). Generally, curvelets will be superior to wavelets in optimally sparse representation of objects with edges (such as potholes and rutting), optimal image reconstruction in severely ill-posed problems and optimal sparse representation of wave propagators (Candès and Donoho, 2000; Candès et al., 2006; Donoho, 2001; Donoho and Duncan, 2000; Starck et al., 2003).

19.2.2　AI Methods and PMS

Artificial Neural Networks

ANN is a method that are constructed to make use of some organizational principles resembling those that mimic the human brain (Avci et al., 2005a, b). ANN can learn if provided with a range of examples and can produce valid answers from noisy data. They represent the promising new generation of information processing systems. ANNs are good at tasks such as pattern matching and classification, function approximation, optimization, and data clustering, while traditional computers, because of their architecture, are inefficient at these tasks (Turkoglu et al., 2003). These neurons are connected to each other through connecting links. Weights are assigned to these links, by which the signals transmitted in the network are multiplied. The output of each neuron is determined by using an activation function such as a sigmoid or step function. Nonlinear activation functions are usually used for this purpose. ANNs are trained by experience, and then they can generalize from past experiences and produce new results when an unknown input is applied to the networks (Bishop, 1996; Demuth and Beale, 2001; Hanbay et al., 2007; Karabatak and Cevdet Ince, 2009). The output of the neuron net is determined by Eq. (19.23). A simple artificial neuron model is shown in Figure 19.12.

$$y(t + 1) = a \left(\sum_{j=1}^{m} W_{ij}x_j(t) - \theta_i \right) \text{ and } f_i \triangleq \text{net}_i = \sum_{j=1}^{m} W_{ij}x_j - \theta_i \tag{19.23}$$

where $x = (x_1, x_2, \ldots, x_m)$ represents the m inputs applied to the neuron, W_i represents the weights for input x_i, h_i is a bias value, and a is the activation function. NN

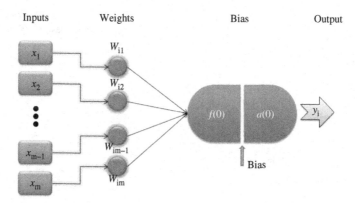

Figure 19.12 Artificial neuron model.

models have been used for pattern matching, nonlinear system modeling, communications, electrical and electronics industry, energy production, chemical industry, medical applications, data mining, and control because of their parallel processing capabilities. When designing an ANN model, a number of considerations must be taken into account (Bishop, 1996). First of all, the suitable structure of the ANN model must be chosen. After this, the activation function, the number of layers, and the number of units in each layer must be selected. Generally, a desirable model consists of a number of layers. The most general model assumes complete interconnections between all units. These connections can be bidirectional or unidirectional.

Today, ANNs can be trained to solve problems that are difficult for conventional computers or human beings (Demuth and Beale, 2001). The following section introduces our proposed method (Figures 19.12 and 19.13).

The ANN computes the net input, and uses a function to guess the outputs called *out*. The performance of the output is then compared to a threshold value. The net input to an ANN is generally computed as the weighted sum of all input signals; for a higher order of inputs, product units have been used. This method promoted information capacity:

$$\text{net} = \sum_{i=1}^{n} x_i \cdot W_{ij}, \quad \text{net} = \prod_{i=1}^{n} x_i^{W_{ij}} \tag{19.24}$$

The function $f_{(x)}$ used to production the output. Different types of activation functions are classified in Figures 19.14 and 19.15. Liner and several types of fuzzy functions can be considered as activation functions. Using a fuzzy function as an activation function, change a simple ANN to hybrid one.

Main types of learning ANNs are categorized in three ways. If a set consists of (input, output), errors between the real and the output's ANN must be minimized and defined as *supervised learning*. If discovering patterns in the input data without any assistance from an external source was the aim, this is categorized as

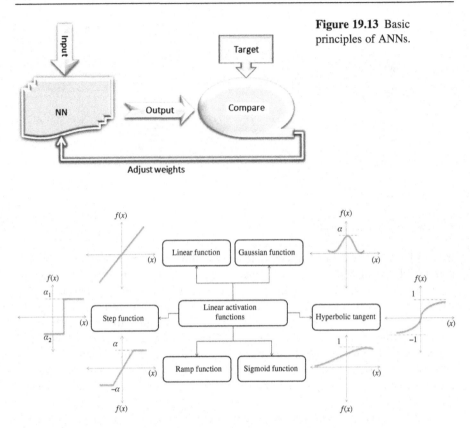

Figure 19.13 Basic principles of ANNs.

Figure 19.14 Linear activation functions.

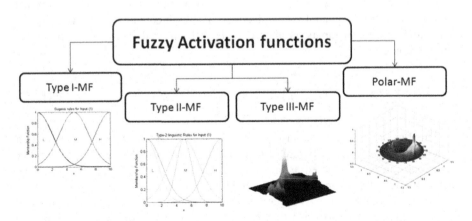

Figure 19.15 Fuzzy activation functions.

unsupervised learning. Finally, if reward and penalize the neuron applied, it is known *reinforcement learning*. Several learning rules have been developed for the different learning types that can be separated according to Figure 19.16.

Accuracy, complexity, and convergence are three indexes for performance evaluation of NNs. The network architecture, the training set size, and the complexity of the optimization method have important effects on the computational complexity of ANNs.

Hybrid Neural Networks

For one example of application NN and developing NN toward smart pavement management, hybrid neural networks (HNNs) are used for modeling the pavement condition index (PCI). One of these networks is the deduct-value neural network (DVNN), and the other is the correct deduct-value neural network (CDVNN). Various hidden layers with different neurons were examined to identify the "optimal architecture" for the analysis. The best test result has an accuracy of more than 97% for DVNN and near or more than 96% for CDVNN. A case study was conducted to clarify the capability of the proposed method. Experimental data obtained from Road Maintenance and Transportation Organization (RMTO) include type of distress, quantities, extent, deduct value, cumulative deduct value, and PCI. The developed HNN model used for the automated pavement condition index (APCI) has a higher regression value than manual PCI determination. Also, we evaluated the performance of the HNN based on its ability to correctly classify a section to the seven different options proposed. The HNN was able to reach a recognition rate (RR) of about 99.43% for the test data. APCI has the following advantages over the other model: it is swift and easy to operate, it has a simple structure, and one of

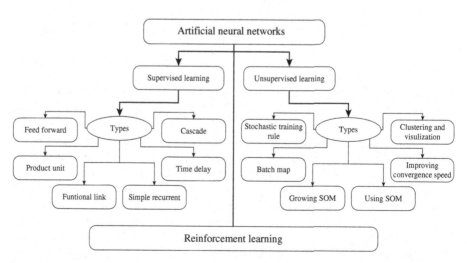

Figure 19.16 Learning methods.

its major benefits is that it does all the tiresome calculations. The proposed method involves the following five steps:

Step 1: Denote by DVNN the deduct values (DVs) obtained by the newly developed NN.
Step 2: Determine the maximum allowable number of deducts (m).
Step 3: Denote by CDVNN the maximum CDV obtained by the newly developed NN.
Step 4: Calculate APCI by subtracting the maximum CDVNN from 100.
Step 5: Calculate the APCI for the section.

The HNN developed for estimating PCI without any restrictive assumption by considering a combination of distress types as input variables and APCI panel data and primary causes (PRCs) of pavement deterioration as output variables. The PCI model developed based on NNs (i.e., APCI) can be mathematically expressed as

$$\text{APCI} = 100 - \left[\sum_{i=1}^{p}\sum_{j=1}^{m_i} \text{net}_{1ij}\right] \cdot \text{net}_2 \tag{19.25}$$

where the maximum value for APCI, on a 0–100 scale, net_{1ij} is the deduct value for distress type i and severity level j based on the HNN DV; i is the counter for the number of distress types; j is the counter for distress severity levels; p is the total number of distress types; m_i is the total number of distress severity levels for the ith distress type; net_2 is an adjustment function for multiple distresses that varies with TDV and q; TDV is the total deduct value, which is given by the NN developed (Figure 19.17).

We evaluate the performance of the HNN based on its ability to classify sections to their corresponding PCI class patterns correctly (Figure 19.18). The RR is defined as the ratio of the number of correctly classified samples to the total number of samples:

$$\text{Recognition rate} = \left(1 - \frac{\sum \text{Number of incorrectly classified samples}}{\sum \text{Total number of samples}}\right) \times 100\% \tag{19.26}$$

Through the experimentation of the proposed model, we tested APCI on the section for each class and with various rating scales (RSs). The results for testing the data set at seven classes are shown in Table 19.2.

The pavements in different conditions require different maintenance and repair operations. The pavement rehabilitation and maintenance methods are categorized into three groups: preventive maintenance (PM), major rehabilitation (MR), and reconstruction (RE). Figure 19.19 illustrates how the appropriate repair type varies with the APCI of a pavement section. According to the proposed algorithm and based on the selected thresholds, which can vary in each country or state depending on its policies and financial resources, different scenarios arise in selecting pavement maintenance and repair operations. Thus, evaluating the proficiency of the proposed method with regard to different scenarios is of great importance.

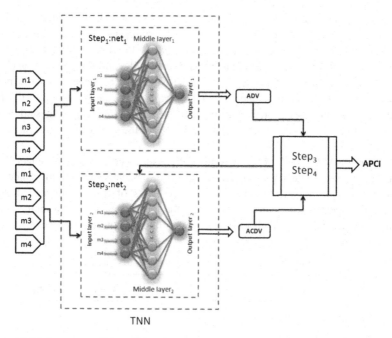

Figure 19.17 Summary of the APCI calculation for the proposed method based on HNN.

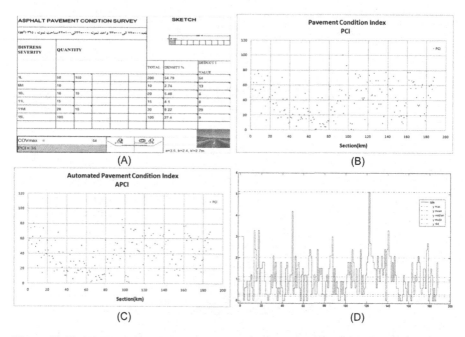

Figure 19.18 (A) Example sample unit condition survey sheet; (B) manually calculated indexes; (C) automated calculated indexes by HNN; (D) APCI versus PCI error.

Table 19.3 shows the maintenance and repair based on various RS patterns for the different conditions. The different options of the suggested maintenance consist of Insignificant Repair (IR), Current Maintenance (CM), Preventive Maintenance (PM), Suspension Activity (SA), Maintenance and Repair (MR) and Reconstruction (RE).

According to Table 19.3, RS_1 and RS_2, RS_5, and RS_6 have the same pattern options for suggested maintenance. Also, we evaluate the performance of the HNN based on its ability to classify sections into the seven different suggested options correctly. The results for the testing data set at the six classes are shown in Table 19.4.

The overall APCI method has a good performance and can be guaranteed in different maintenance and repair selection operations because the performance (RR) of the tested data is very good (98.9%). The types of distress identified during the APCI inspection provide insight into the cause of pavement deterioration. Examination of the pavement section's extrapolated distress types, severities, and

Table 19.2 Results Obtained with Various Thresholds

RS	Good	Satisfactory	Fair	Poor	Very Poor	Serious	Failed
RS_1	100	90	75	50	45	30	15
RS_2	100	89	74	51	44	29	14
RS_3	100	87	72	53	42	27	12
RS_4	100	85	70	55	40	25	10
RS_5	100	83	68	57	38	23	8
RS_6	100	81	66	59	36	21	6
RS_7	100	80	65	60	35	20	5

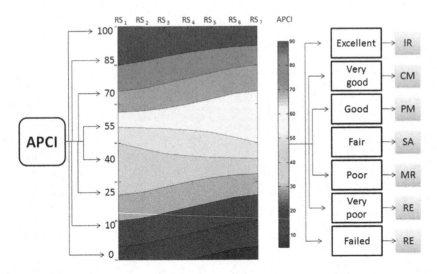

Figure 19.19 APCI for various repair types.

quantities provides valuable information used to determine the cause of pavement deterioration and eventually its maintenance and repair (M and R) needs. Table 19.5 classifies distress causes for pavements into load (CAS_1), climate (CAS_2), and other factors (CAS_3). Quantification of the relative effect of each cause can be determined corresponding to the automated deduct value (ADV) for the extrapolated section distresses.

Table 19.3 Appropriate Repair Type with the APCI of a Pavement Section

RS	100	85	70	57	40	25	0
RS_1	IR	CM	PM	SA	RE	RE	RE
RS_2	IR	CM	PM	MR	RE	RE	RE
RS_3	IR	CM	PM	MR	RE	RE	RE
RS_4	IR	CM	SA	MR	RE	RE	RE
RS_5	IR	PM	SA	MR	RE	RE	RE
RS_6	IR	PM	SA	RE	RE	RE	RE
RS_7	IR	PM	SA	RE	RE	RE	RE

Table 19.4 Results Obtained with Various Pavement Rehabilitation and Maintenance Methods

RS	Total Samples	Number	CM	PM	SA	MR	RE	Misclassified Samples	RR (%)
IR	11	11	0	0	0	0	0	0	100
CM	35	0	35	1	0	0	0	0	100
PM	42	0	0	41	0	0	0	1	97.5
SA	52	0	0	0	52	0	0	0	100
MR	32	0	0	0	0	32	0	0	100
RE	16	0	0	0	0	0	16	0	100
Total	188	11	34	42	52	32	16	1	99.4

Table 19.5 Case Distress Classification for Asphalt Surfaced and Portland Cement Concrete Roads

Case	Code of Distress	Code of Distress
	Asphalt Surfaced	*Portland Cement Concrete*
CAS_1	(01), (07), (13), (15), (16)	(22), (23), (24), (28), (34)
CAS_2	(03), (08), (10), (19)	(21), (24), (26), (37), (38), (39)
CAS_3	(02), (04), (05), (06), (09), (11), (12), (14), (17), (18)	(25), (27),(29), (30), (31), (32), (33), (35), (36)

The total ADV attributable to CAS_1 is determined, and those attributable to load, climate, and other factors are calculated. The percentage of ADVs attributable to load, climate, and other causes is computed as follows:

$$CAS_1 = \left(\frac{\sum_{i=0}^{n} (ADV_1)_i}{\left[\sum_{i=0}^{n} (ADV_1)_i + \sum_{j=0}^{m} (ADV_2)_j + \sum_{k=0}^{p} (ADV_3)_k \right]} \right) \times 100$$

(19.27)

$$CAS_2 = \left(\frac{\sum_{j=0}^{m} (ADV_2)_j}{\left[\sum_{i=0}^{n} (ADV_1)_i + \sum_{j=0}^{m} (ADV_2)_j + \sum_{k=0}^{p} (ADV_3)_k \right]} \right) \times 100$$

(19.28)

$$CAS_3 = 100 - [CAS_1 + CAS_2]$$

(19.29)

$$PRC = \max\{CAS_1, CAS_2, CAS_3\}$$

(19.30)

where ADV stands for the ADV for asphalt pavements i, j, and k are counters for the number of ADVs classified by cause, m is the maximum number of distress types considered, and ADV_1, ADV_2, and ADV_3 are the ADVs for distress CAS_1, CAS_2, and CAS_3, respectively. The percentage of deduct values attributed to each cause is an indication of the cause(s) of pavement deterioration. Primary pavement deterioration is determined by the maximum percentage of the CAS(s). We evaluate the performance of the HNN based on its ability to classify sections correctly into CAS_1, CAS_2, and CAS_3.The results for the testing data set are shown in Table 19.6.

The overall APCI method can be guaranteed to be efficient in section classification because the performance (RR) of the tested data is excellent (100%).

Fuzzy Logic Systems

Human reasoning, programming, and analyzing in the PMS, however, are almost always inexact. In the field of pavement distress detection and classification, fuzzy classification and thresholding theory require elements to be part of a class—threshold or not. Logic requires the values of parameters to be either 0 or 1, with

Table 19.6 Results of Classification by Causes Obtained with APCI Method

Case	Total Samples	CAS_1	CAS_2	CAS_3	Misclassified Samples	RR (%)
CAS_1	123	123	0	0	0	100
CAS_2	52	0	52	0	0	100
CAS_3	13	0	0	13	0	100
Total	188	123	52	13	0	100

similar constraints on the outcome of an inference process. Therefore, pavement assessment methods and reasoning usually include a measure of uncertainty. New facts by fuzzy logic (FL) can be inferred with a degree of certainty associated with each fact. Common sense can model by fuzzy sets and FL. The philosophy of FL is based on vagueness, imprecision, and/or ambiguity. An FL system is well known to a logical system that works by rule-based reasoning similar to ESs; however, it is more applicable and effective in ambiguous, vague, and not crisp applications (Hanbay et al., 2007; Karabatak and Cevdet Ince, 2009; Sandra et al., 2006; Yu, 2007).

FL methods combine data by following a logical decision-making process by incorporating descriptive linguistic rules (which may be extracted from other soft methods such as NN, ES, FL, or GA) through FL. Deeply knowledge of expert play a key role for membership function selection. Ambiguity in membership function assignment is the main problem with FL sets. Indeed different FL techniques may have various thresholds, class, and treatments in PMSs. To solve this problem, the Fuzzy Type II method is introduced.

Under title of Fuzzy Type I, Fuzzy Type II is developed. This method selects the upper and lower membership functions; however, the Footprints of Uncertainty (FOU) has a fixed value, 1, in all the upper and lower membership functions (Ioanniset et al., 2008; Tizhoosh, 2005). One of the applications of the Fuzzy Type II method is threshold selection based on Fuzzy Type II in automatic pavement distress detection and classification of an image-based algorithm. According to Ioanniset et al. (2008), "Type I fuzzy sets are not able to directly model such uncertainties because their membership functions are totally crisp. On the other hand, Type II fuzzy sets are able to model such uncertainties because their membership functions are themselves fuzzy. These fuzzy sets are characterized by their Footprints of Uncertainty (FOU), which in turn are characterized by their boundaries—upper and lower membership functions (MF)". They provided some solutions for nonsymmetrical Type II membership functions (Ioannis and Bob, 2002; Mendel, 2001; Mendel and Bob John, 2002; Mendel and Wu, 2007). The FOU implies that there is a distribution that sits on top of that shaded area. When they all equal 1, the resulting Fuzzy Type II sets are called *interval Type II Fuzzy sets*. Fuzzy sets of Type II, therefore, are fuzzy sets for which the membership function does not deliver a single value for every element but an interval. In an example of a Fuzzy Type II application, ambiguities in 3DRT assignment in automatic pavement cracking detection may have different thresholds. To solve this problem, Fuzzy Type II thresholding is introduced. The upper and lower membership functions are selected based on Fuzzy Type II theory (Figure 19.20). The main goal of using Fuzzy Type II is to create a flexible threshold for automatic thresholding. Building a high-accuracy relationship between the patterns of the peaks and the properties of the crack using automatic fuzzy threshold selection is the logical way of promoting the flexibility.

Like Type I, Type II has same definitions for membership functions. From Fuzzy Type I, one can derive the properties of Fuzzy Type II. The meaning of Fuzzy Type II sets is given by Tizhoosh (2005):

$$\tilde{A} = \{((x, u), \mu_{\tilde{A}}(x, u) | \forall x \in X, \forall u \in J_x \subseteq [0, 1]\} \tag{19.31}$$

A Type II membership function $\mu_{\tilde{A}}(x, u)$ is defined for a Type II Fuzzy set $\tilde{A}$, where $x \in X$ and $u \in J_x \subseteq [0,1]$, i.e.,

$$\tilde{A} = \{((x, u), \mu_{\tilde{A}}(x, u) | \forall x \in X, \forall u \in J_x \subseteq [0, 1]\}$$

in which $0 \le \mu_{\tilde{A}}(x, u) \le 1$. $\tilde{A}$ can also be expressed in the usual notation of fuzzy sets:

$$\tilde{A} = \int_{x \in X} \int_{u \in J_x} (\mu_{\tilde{A}}(x, u)/(x, u)), \quad J_x \subseteq [0, 1] \tag{19.32}$$

where the double integral denotes the union of all x and u. In order to define a Fuzzy Type II set, one can define a Fuzzy Type I set and assign upper and lower membership degrees to each element to reconstruct the FOU (Figure 19.21). A more practical definition for a Fuzzy Type II set can be given as follows:

$$\tilde{A} = \{((x, u), \mu_U(x), \mu_L(x) | \forall x \in X, \mu_L(x) \le \mu(x) \le \mu_U(x), \mu \in [0, 1]\} \tag{19.33}$$

When various rules are activated, the binary rules that define conventional ESs usually result in discontinuities at the exit of a system. This does not resemble human behavior, where a smooth relation usually exists between cause and consequence. Smooth relationships can be achieved by using fuzzy rules that include

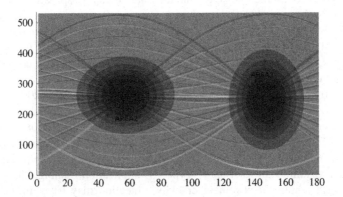

Figure 19.20 Ambiguities in RT assignment may have different thresholds.

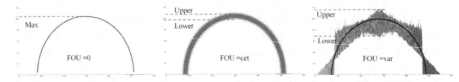

Figure 19.21 A possible way to construct Fuzzy Type II sets. The interval between lower and upper membership values (shaded region) should capture the FOU.

descriptive expressions such as poor, fair, or good (Figure 19.21) to categorize linguistic input and output variables. The FL was developed to provide soft algorithms for data processing that can both make inferences about imprecise data and use the data. It enables the variables to partially belong to a particular set and, at the same time, uses the generalizations of conventional Boolean logic operators in data processing. The main advantage of this approach is the possibility of introducing and using rules from experience, intuition, and heuristics, as well as the fact that a model of the process is not required.

A particular type of fuzzy system that holds great potential is the Fuzzy Type II system. The concept of a Fuzzy Type II set was introduced by Zadeh to account for some of the uncertainties in the FL systems (Zadeh, 1975a, b). Type II FL systems are those in which at least one of the antecedent or consequent sets is Type II. This type of fuzzy system allows one to consider uncertainty by more accuracy and a wider stretch.

Membership functions for fuzzy sets can be constructed by any heuristic or metaheuristic method in the domain. Two most important constraints must be considered for selecting a membership function. First, a membership function must be restricted between [0 1], and the next $\mu_A(x)$ must be unique. Four possible membership functions are Type I, II, III and polar. Type III and polar are new generation of fuzzy membership function that can be used in several application in the control and classification domains (see Figure 19.15). In the field of PMS, this new generation of membership functions forges a powerful link between several tools such as multiresolution methods (wavelet and beyond the wavelet), image processing, NN, and ES. A possible membership function can be defined for every category, with any tool. For example, using image processing techniques and RT, several membership functions are generated for pavement cracking distress (see Figure 19.15). Simpler and more complex functions can be used under the form of discrete and continuous. Generally, the ordinary functions are categorized as Triangular, Trapezoidal, Γ-membership, S-membership, Logistic, Exponential-like, and Gaussian functions. Additionally, several more advanced membership functions generated by an automatic generator is introduced. Cubic smoothing spline (CSS) is used to generate the membership functions because of nonuniform illumination of the 3DRTs. The estimation of the MF_U and MF_L is calculated from the fitting of a CSS (Mora et al., 2011) to the 3DRT(x,y). The select CSS is a special class of spline that can capture the low 3DRT value that limited the nonuniformity of the 3DRT (Culpin, 1986). The fitting objective is to minimize the equation

$$M = P \cdot \sum_{y=1}^{m} \sum_{x=1}^{n} (f(x,y) - s(x,y))^2 + (1-p) \int \int (D^2 S(x,y))^2 dxdy \qquad (19.34)$$

which includes two parts:

1. *Compactness:* Measures how close the spline is to the data that reflect to the summation term which is weighed by the smoothing factor p.
2. *Smoothness:* Measures the spline smoothness using its second derivative that reflect to the integral term weighed by $(1-p)$.

The smoothing factor p controls the balance between being an interpolating spline crossing all data points (with $p = 1$) and being a strictly smooth spline (with $p = 0$). The smoothing spline f minimizes when

$$\left[\left[P\sum_{j=1}^{n}(\omega(j)|y(:,j) - f(x(j))|^2)\right] + \left[(1-p)\int \lambda(t)|D^2f(t)|^2 \, dt\right]\right] \qquad (19.35)$$

where $|z|^2$ represents the sum of the squares of all the entries of n, in Eq. (19.35), n and m are the number of entries of x and y in Eq. (19.34) and the integral is over the smallest interval containing all the values of x and y. The default value for the weight vector w in the error measure is ones [size(x)]. The default value for the piecewise constant weight function λ in the roughness measure is the constant function 1. Further, D^2f denotes the second derivative of the function f. The default value for the smoothing parameter, p, is chosen in dependence on the given data sites x and y (Pal and Bezdek, 1994). The smoothing parameter is determines the relative weight to place on the contradictory demands of having f be smooth versus having f be close to the data. For $p = 0$, f is the least squares straight-line fit to the data, while, at the other extreme (i.e., for $p = 1$), f is the variation, or "natural" cubic spline interpolate. As p moves from 0 to 1, the smoothing spline changes from one extreme to the other (Figures 19.22 and 19.23).

The interesting range for p is often near $1/(1 + (\min(N,M))^3/600)$, with h the average spacing of the data sites; and it is in this range that the default value for p is chosen. For uniformly spaced data, one would expect a close following of the

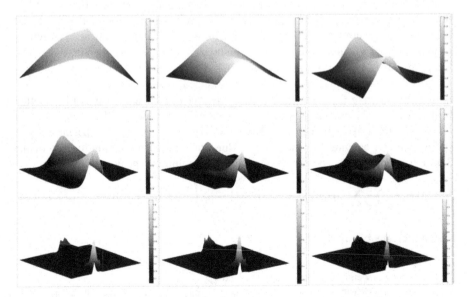

Figure 19.22 A sample membership function type III; as p moves from 0 to 1, the smoothing spline changes from one extreme to the other.

data for $p = 1/(1 + (\min(N,M))^3/6000)$ and some satisfactory smoothing for $p = 1/(1 + (\min(N,M))^3/60) \cdot p > 1$ can be input, but this leads to a smoothing spline even rougher than the variational cubic spline interpolate (Mora et al., 2011; Pal and Bezdek, 1994):

$$p = \left(\frac{1}{1 + ((\min(N,M))^3/600)} \right) \tag{19.36}$$

A reference smoothing factor $(p = 1 \times 10^{-4})$ was obtained empirically for constructed MF in upper bound and $(p = 0.93 \times 10^{-5})$ for constructed MF in the lower bound. After testing several thresholds, the general rule extracted from 3DRT thresholds for upper and lower bounds is

$$p_U = \left(\frac{1}{1 + ((\min(N,M))^3/583.2)} \right)_U \tag{19.37}$$

and

$$p_L = \left(\frac{1}{1 + ((\min(N,M))^3/542.4)} \right)_L \tag{19.38}$$

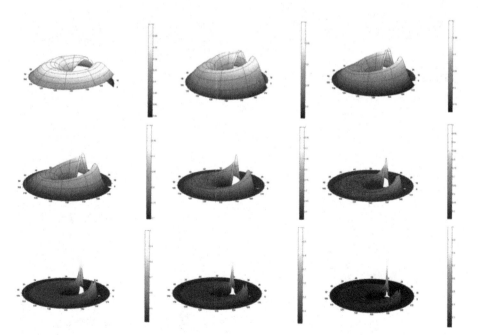

Figure 19.23 A sample of polar membership function; as p moves from 0 to 1, the smoothing spline changes from one extreme to the other.

Fuzzy Operators

The relations and operators in all four types are similar and defined next. For this purpose, for example, let ℓ, Θ be the polar domain, and P_1 and P_2 be fuzzy sets defined over the polar fuzzy domain.

Similarity of two polar fuzzy sets: Two fuzzy polar sets, P_1 and P_2, are similar if and only if the sets have the same ℓ, Θ polar domain, and $\mu P_1(x) = \mu P_2(x)$ for all $x \in X$. That is, $P_1 = P_2$ (Figure 19.24), all the membership functions are similar (Figure 19.25).

History of polar fuzzy sets: Polar fuzzy set P_1 is a history of polar fuzzy set P_2 if and only if $\mu P_1(x) \leq \mu P_2(x)$ for all $x \in X$. That is, $P_1 \subset P_2$. Figure 19.27 shows mode $P_1 \subset P_2$.

(NOT)—Complement of a polar fuzzy set: The complement of two polar fuzzy sets is simply the polar set containing the entire domain without the elements of that polar set. For polar fuzzy sets, the complement of the set O is $\tilde{O}$, the membership degrees differ. Let $\tilde{O}$ denote the complement of set O. Then, for all $x \in X$, $\mu \tilde{O}(x) = 1 - \mu O(x)$, $O \cap \tilde{O} = \varnothing$ and $\tilde{O} \cup O = \text{Domain} (2\pi\ell \times 1)$.

(AND)—t-norms of a polar fuzzy set: The roundabout or intersection is the minimum polar set in both sets at the intersection that is known by *t*-norms. Several

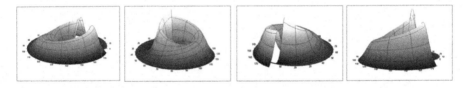

Figure 19.24 Similarity of polar fuzzy membership functions.

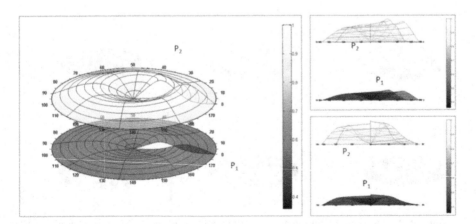

Figure 19.25 Illustration of the history of polar fuzzy membership functions.

methods are proposed for t-norm calculation in Type I and Type II. All t-norms can be developed for Type III and polar. If P_1 and P_2 are two fuzzy sets, then:

Polar min-operator: $\mu P_1 \ (\ell,\Theta) \cap \mu P_2 \ (\ell,\Theta) = \min \ \{\mu P_1 \ (\ell,\Theta), \mu P_2 \ (\ell,\Theta)\}, \ \forall \ (\ell,\Theta)\in O,$
Polar product operator: $\mu P_1(\ell,\Theta) \cap \mu P_2 \ (\ell,\Theta) = \mu P_2 \ (\ell,\Theta) \times \mu P_2 \ (\ell,\Theta), \ \forall \ (\ell,\Theta)\in O,$
$\mu P_1(\ell,\Theta) \cap \mu P_2 \ (\ell,\Theta) = \max\{0, \ (1+p)(\mu P_1(\ell,\Theta) + \mu P_2(\ell,\Theta) -1) - p\mu P_1(\ell,\Theta) \times \mu$
$P_2(\ell,\Theta)\},$ for $p \geq -1,$

$$\mu P_1(\rho,\vartheta) \cap \mu P_2(\rho,\vartheta) = \left[\frac{1}{1 + \sqrt[p]{[(1-\mu P_1(\rho,\vartheta)/P)^P + (1-\mu P_2(\rho,\vartheta)/P)^P]}}\right] \ \text{for } p>0$$

(19.39)

$$\mu P_1(\rho,\vartheta) \cap \mu P_2(\rho,\vartheta) = \left[\frac{\mu P_1(\rho,\vartheta) \times \mu P_2(\rho,\vartheta)}{p + [(1-p)(\mu P_1(\rho,\vartheta) + \mu P_2(\rho,\vartheta) - \mu P_1(\rho,\vartheta)\mu P_2(\rho,\vartheta))]}\right] \ \text{for } p>0$$

(19.40)

$$\mu P_1(\rho,\vartheta) \cap \mu P_2(\rho,\vartheta) = \left[\frac{\mu P_1(\rho,\vartheta) \times \mu P_2(\rho,\vartheta)}{\max\{\mu P_1(\rho,\vartheta) \times \mu P_2(\rho,\vartheta)\}}\right] \ \text{for } p[0,1]$$

(19.41)

(OR)—t-conorms of a polar fuzzy set: The roundabout or intersection is the maximum polar set in both sets at the intersection known as t-conorms. If P_1 and P_2 are two fuzzy sets, then:

Polar max-operator: $\mu P_1 \ (\ell,\Theta) \cup \mu P_2 \ (\ell,\Theta) = \min \ \{\mu P_1 \ (\ell,\Theta), \mu P_2 \ (\ell,\Theta)\}, \ \forall \ (\ell,\Theta)\in O,$
Polar summation operator: $\mu P_1 \ (\ell,\Theta) \cup \mu P_2 \ (\ell,\Theta) = [\mu P_1 \ (\ell,\Theta) + \mu P_2 \ (\ell,\Theta)] - (\mu P_1 \ (\ell,\Theta) \times \mu P_2 \ (\ell,\Theta), \ \forall \ (\ell,\Theta))\in O,$

$\mu P_1(\rho,\vartheta) \cup \mu P_2(\rho,\vartheta)$

$$= \left[\frac{1}{1 + \sqrt[p]{[((\mu P_1(\rho,\vartheta)/1 - \mu P_1(\rho,\vartheta)))^P + ((\mu P_2(\rho,\vartheta)/1 - \mu P_2(\rho,\vartheta)))^P]}}\right] \ \text{for } p\geq 0$$

(19.42)

$$\mu P_1(\rho,\vartheta) \cup \mu P_2(\rho,\vartheta) = 1 - \left[\frac{(1-\mu P_1(\rho,\vartheta)) \times (1-\mu P_2(\rho,\vartheta))}{\max[(1-\mu P_1(\rho,\vartheta)),(1-\mu P_2(\rho,\vartheta)),p]}\right] \ \text{for } p\in[0,1]$$

(19.43)

$\mu P_1(\rho,\vartheta) \cup \mu P_2(\rho,\vartheta)$

$$= \left[\frac{\mu P_1(\rho,\vartheta) + \mu P_2(\rho,\vartheta) - \mu P_1(\rho,\vartheta) \times \mu P_1(\rho,\vartheta) - (1-p)\mu P_1(\rho,\vartheta) \times \mu P_2(\rho,\vartheta)}{1 - (1-p)\mu P_1(\rho,\vartheta) \times \mu P_1(\rho,\vartheta)}\right] \ \text{for } p\geq 0$$

(19.44)

The effect of the operations polar membership function is illustrated in Figure 19.26. For the illustration in Figure 19.27, assume the polar set $\mu P_1(\rho,\vartheta)$, as an input polar set and $\mu P_2(\rho,\vartheta)$, as a rule polar set. The complement of $\mu P_1(\rho,\vartheta)$, the intersection and summation operations are illustrated in Figure 19.27B–D.

A Measure of Ultrafuzziness

Using a simple method, we change an ultra fuzzy to a 3DRT fuzzy set. According to a Type II membership function, MF must be in [0,1]. One can take out the non-dimensional form 3DRT by dividing every point by max 3DRT:

$$\text{RTMF}_{(i,j)} = \left(\left[v_{(3DRT)_{(i,j)}^h} \right] / \max \left[v_{(3DRT)_{(i,j)}^h} \right] \right)^{\frac{1}{h}} \tag{19.45}$$

$$\mu_{(i,j)} = \left(\left[v_{(3DRT)_{(i,j)}^h} \right] / \max \left[v_{(3DRT)_{(i,j)}^h} \right] \right)^{\frac{1}{h}} + H \tag{19.46}$$

and

$$\text{GR}_{(i,j)} = \frac{1}{(MN)^h} \sum_{j=1}^{N} \sum_{i=1}^{M} \left(\left[v_{(3DRT)_{(i,j)}^h} \right] / \max \left[v_{(3DRT)_{(i,j)}^h} \right] \right)^{\frac{1}{h}} \tag{19.47}$$

where M and N denote the size of the 3DRT platform, H is high-platform $H \in [0,1]$, $h \in (1, \infty)$, and $v_{(3DRT)_{(i,j)}^h}$ is the 3DRT value in position i and j. Select a bigger h is

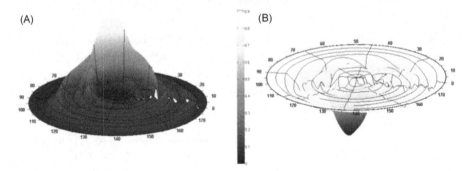

Figure 19.26 Illustration of the (A) polar fuzzy membership functions and (B) complement polar membership function $(\mu P_1(\rho, \vartheta))$.

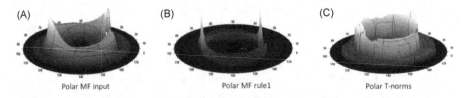

Figure 19.27 Illustration of the t-norm polar fuzzy membership functions: (A) polar membership functions input, (B) polar membership function rule 1, and (C) polar t-norms.

Select a bigger parameter is equivalent to more enhanced distress and smoother noisy background of image. (Figure 19.28). In order to define a Fuzzy Type II set, one can define a Fuzzy Type I set and assign upper and lower membership degrees to each element to reconstruct the FOU (Figure 19.29). In order to develop a Fuzzy Type III set, a Fuzzy Type II set must be published in 3D and assign upper and lower membership degrees to each element to reconstruct the FOU and for 3D polar, membership function Type III must be transformed in polar based on R and

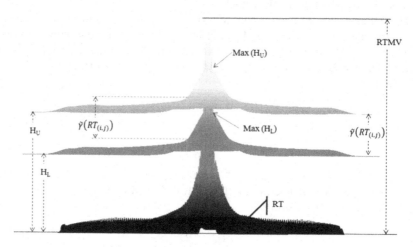

Figure 19.28 Basic rules for construction 3D Fuzzy Type III and the polar membership function.

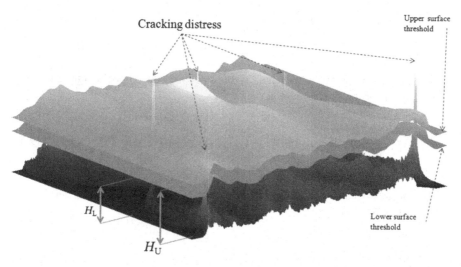

Figure 19.29 Three-dimensional domain of FOU for 3D fuzzy sets (3DFOU).

Θ. When RT (an effective tool in pavement cracking detection) is applied to a wavelet modulus, a distress (crack) is transformed into a peak in the radon domain. Originally, every distress reflects to RT and has different intensity in 3DRT histograms. For example, the mean of 3DRT has a wide range. According to the above, the max GR must equal 1. To extend fuzzy thresholding (in the control field) and thresholding in decision making (in pavement management) to Type III and polar fuzzy sets, ultrafuzziness should be zero, if the membership function can be selected without any ambiguity, such as Type I. The amount of ultrafuzziness will increase by rising the uncertainly bound.

The extreme case of maximal ultrafuzziness, equal to 1, is mean to completely vagueness. Pal and Bezdek (1994) extensively reviewed a well-known fuzziness index; two general classes proposed by them were the additive and multiplicative classes (Pal and Bezdek, 1994). Based on Kaufmanns index of fuzziness for a set $A \in r_n(x)$,

$$H_{ka}(A) = \left(\frac{2}{n^k}\right) d(A, A^{\text{near}}) \tag{19.48}$$

where $k \in R^+$, d is a metric, and A^{near} is the crisp set close to A. Generally, based on d, the weight of k is determined. The $d(A,A^{\text{near}})$ and linear or quadratic $H_{ka}(A)$ can be determined by q-norms:

$$d(A, A^{\text{near}}) = \left(\sum_{i=1}^{n} |\mu_i - \mu_{A^{\text{near}},i}|^q\right)^{1/q} \tag{19.49}$$

$$H_{ka}(q, A) = \left(\frac{2}{n^{1/q}}\right) \left(\sum_{i=1}^{n} |\mu_i - \mu_{A^{\text{near}},i}|^q\right)^{1/q} \tag{19.50}$$

where $q \in [1, \infty)$. On the other hand, Tizhoosh (2005) developed a simple ultrafuzziness index for the special case as follows:

$$\tilde{\gamma}(\tilde{A}) = \frac{1}{MN} \sum_{i=1}^{M-1} \sum_{j=1}^{N-1} [\mu_U(g_{ij}) - \mu_L(g_{ij})] \tag{19.51}$$

where $\mu_U(g) = [\mu_A(g)]^{1/\alpha}$ and $\mu_U(g) = [\mu_A(g)]^{1/\alpha}$, $\alpha \in (1,2]$. In general terms, it is presented as follows:

$$\tilde{\gamma}(\tilde{A}) = \frac{1}{MN} \sum_{g=0}^{L-1} [\mu_U(g_{ij}) - \mu_L(g_{ij})] \times h(g) \tag{19.52}$$

Based on this theory and with respect to Tizhoosh's method, for developing ultrafuzziness on two-dimensional data, we define a Type II fuzzy set, a measure of ultrafuzziness $\tilde{\gamma}$ for a platform 3DRT with $M \times N$ sets and surf 3DRT, and the membership function $\mu_{\tilde{A}(i,j)}$ can be developed as follows:

$$\tilde{\gamma}(A) = \left(\frac{1}{(MN)^{1/q}} \right) \left[\sum_{j=1}^{M-1} \sum_{i=1}^{N-1} |\mu_{U(i,j)} - \mu_{L(i,j)}|^q \right]^{1/q} \tag{19.53}$$

$$\frac{\partial}{\partial T} \tilde{\gamma}(RT_{(i,j)}) = \frac{\partial}{\partial T} \left(\frac{1}{(MN)^{1/q}} \right) \left[\sum_{j=1}^{M-1} \sum_{i=1}^{N-1} |\mu_{U((i,j),T)} - \mu_{L((i,j),T)}|^q \right]^{1/q} = 0$$

This basic definition relies on the assumption that the singletons sitting on the FOU are all equal in height. The variation in the space can be measured by this method; therefore, the new index is introduced in a three-dimensional domain of FOU for 3D fuzzy sets (3DFOU). By this novel method, the Damavand Hat cloud (DHC) resolves the problems with the ultrafuzziness index. According to Ioannis et al. (2008), "uncertainty (FOU) has a constant value, that equals one, in all the intervals of the universe of discourse" introducing a flexible membership function across the interval path (Figures 19.29 and 19.30).

Every membership function generator algorithm, such as Type I, Type II, Type III, and polar method can be evaluated based on four conditions *minimum ultrafuzziness*, *maximum ultrafuzziness*, *equal ultrafuzziness*, and *reduced ultrafuzziness* that every measure of fuzziness should satisfy, which is introduced by Kaufmann (1975).

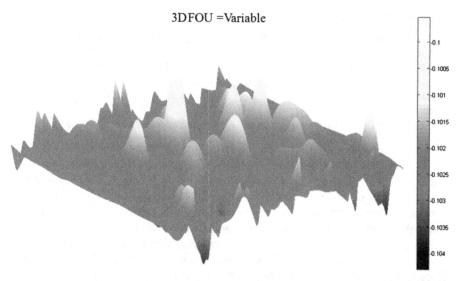

Figure 19.30 Example of 3DFOU using the DHC algorithm.

Finding the Optimum Thresholding Type III and Polar

The general approach for optimum defuzzification (e.g., in automatic pavement distress thresholding) based on the upper and lower surface is equal for Type III:

$$\xi = \left[1 - \frac{\min\tilde{\gamma}(i,j)}{\max\tilde{\gamma}(i,j)}\right], \text{SURF}(i,j) = \mu_{\text{L}}(g_{ij})[\xi + 1], \text{OR}\mu_{\text{U}}(g_{ij})[1 - \xi] \qquad (19.54)$$

Similarly, polar, based on the upper and lower surface, is equal:

$$\xi_p = \left[1 - \left(\frac{\min\tilde{\gamma}(\ell,\Theta)}{\max\tilde{\gamma}(\ell,\Theta)}\right)\right], \text{SURF polar}(\ell,\Theta) = \mu_{\text{L}}(g_{\ell,\Theta})\left[\xi_p + 1\right],$$
$$\text{OR } \mu_{\text{U}}(g_{\ell,\Theta})\left[1 - \xi_p\right] \qquad (19.55)$$

where (ξ) is the ultrafuzzy coefficient and $\tilde{\gamma}(i,j)$ is the ultrafuzzy value for $\mu_{\text{U}(i,j)}$ and $\mu_{\text{L}(i,j)}$, in upper and lower threshold in Type III, (ξ_p) is the ultrafuzzy coefficient, and $\tilde{\gamma}(\ell,\Theta)$ is the ultrafuzzy value for $\mu_{\text{U}(\ell,\Theta)}$ and $\mu_{\text{L}(\ell,\Theta)}$, in the upper and lower thresholds in the polar domain.

Expert Systems

Many kinds of ESs are used in management systems. The use of ES has a tangible ramification in several fields, such as control, design, diagnosis, monitoring, planning, perdition, selection, and simulation.

The main components that crystallize a simple ES are knowledge base, inference engine, and working memory.

1. The *knowledge base (KB)* is the main element of an ES, and it contains the domain knowledge.
2. The *inference engine (IE)* is an element that matched the facts contained in the working memory with the KB by the aim of drawing conclusions.
3. The *working memory (WM)* is the part of an ES that discovers the problem during firing the rules.
4. The *user interface (UI)* provides a friendly interface between the user and the system.

Antecedent and consequent are the two parts of every rule in the KB. Like humans, ESs can make mistakes. A chain of acquisition, organization, and studying knowledge is known as *knowledge acquisition*. In ES rules, the KB can represent uncertain or fuzzy inference, a large database, NNs, automatic knowledge generation, image processing, GA, or complex AI methods. These chains can spark the production of a new engine used to generate a HS in the ES domain. In the field of PMS, several ESs have been developed for use in pavement management and rehabilitation, such as Rose, Sceptre, Perserver, Expear, Pavement Expert, Pares, Paver and Micro Paver, Airpacs, Pmas, Pmdss, and AICPD.

Most of these systems were developed for flexible pavement and used the visual surface distress condition to obtain knowledge to draw conclusions and select recommendations.

For example, a complex ES (CES) can be supported with image processing for pavement distress detection and classification. This method is related to smart PMSs by CES for pavement distress classification. An RT ES is used to increase the effectiveness of the scale-invariant feature extraction algorithm. After preprocessing and binarization, the RT applied to the image modulus. This system first obtains distress detection information to form a hypothesis. The system then searches for supporting information to test the hypothesis. Finally, the knowledge extracted and the changing methods, forward and backward, are used to set CES in motion. Changing methods automatically shift from backward to forward, and vice versa.

In brief, CES has been demonstrated to be an effective tool for diagnosing. However, the proposed method is robustly semiautomatic for pavement distress detection. The most important aspect of the proposed CES model is the ability of self-organization in inference engines. These results point out the ability of the design of a new AI model for automatic acquisition systems. The test performances of this study show the advantages of proposed ES: it is more rapid than ordinary ES, easier to operate, and more robust than single chaining. The main steps for generating this system are (i) load the image of the pavement, (ii) ES interface for threshold selection, (iii) preprocessing on image and automatic RT, (iv) ES interface, (v) the forward chaining shifts to a backward chain, and vice versa, and (vi) sound and graphical message.

In the domain of computer sciences, AI dealing the components of intelligence such as the tuning, learning, and planning, whereas the in the agents world, AI referee to integrating these same components. Quantification of pavement crack data is one of the most important criteria in determining optimum pavement maintenance strategies. An intelligent agent can be developed for modeling the PCI. An intelligent agent is a self-healing intelligent agent based on fuzzy Takagi-Sugeno-Kang (TSK) and NN for pavement condition evaluation. The new agent has flexible autonomous action and verifies the general behavior of the system.

The degree of pavement deterioration is a function of distress type, distress severity, and the amount or density of distress. Producing an index that would take into account all the three factors was a considerable challenge. To overcome this problem, DVs were introduced as a weighing factor to indicate the degree of effect that each combination of distress type, severity level, and distress density has on pavement conditions. For another example, in a hybrid pavement diagnosing system, a fuzzy T.S.K is used for rule generation with 6 inputs and 10 rules, and NNs are used for training and self-healing as an AI HM. As we know in the agent's world, AI refers to integrating the same components with each other. Flexibility and intelligent agent (IA) reflects to (i) reactivity, (ii) proactivity, (iii) social ability, (iv) mobility, (v) veracity, and (vi) benevolence, rationality, and learning/adaptation.

19.3 Conclusions

19.3.1 Hybrid System

HS is a complex of AI techniques that uses, in parallel, a combination of AI methods or heuristics from various branches or the same one. This method handles uncertainty, ambiguity, and the huge and partial truth inherent as inputs in PMS to achieve the goal of smart pavement management. ES, FL, and NN can be integrated with image processing and multiresolution methods and born HS. This system can create an intelligent agent to handle tremendous calculations. Condition assessment, performance prediction, project selection, treatment selection priorities, and optimization are the main parts of PMS. Because of complexity in condition assessment, HS is an effective technique. Treatment selection is an application of HS.

19.3.2 Condition Assessment

Aggregating the condition of pavement distress can enhance the evaluation of a pavement section and reduces the various condition measurements into a single index. NNs (Pant et al., 1993) and FL methods (Type I, Type II, Polar, and T.S.K.) can be used to combine different pavement value index into a total condition index (Flintsch, 2003). These techniques can estimate functions from samples without requiring a mathematical formulation. Learning the ability of these methods from examples shows effective performance (Flintsch, 2003). Hybridization of NNs and FL systems, or neuro-fuzzy models, can merge the advantages and benefits of both techniques and incorporate them with multiresolution methods. Training, tuning, and learning this method can be easily achieved. HS in condition assessment has key properties of convergence and stability, as do NNs. Another area that has received significant contribution from the use of soft computing technologies is automatic identification distress. NNs, FL, and HS in the field of AI and image processing have been used for pavement distress detecting and classifying from detectors (Nejad and Zakeri, 2011b, c).

19.3.3 Performance Prediction

Pavement visual distress prediction, based on image processing, is a new field of PMS that can lead to a flourishing novel vision in smart pavement management system. This zone is another area where covers with uncertainty play a crucial role. Therefore, one potential area is to use HS as prediction performance. Uncertainly and vagueness in prediction application can be enhanced by the use of HS techniques. For a section, the crack propagation must be predicted over time based on past history of crack-print. The prediction models are based on general knowledge gained by training, learning, or even expert knowledge. Complex of NNs, FL, and ES with a multiresolution image system can build an effective tool to solve this problem. Neural models have been used for predicting cracking development pattern recognition, DV, CDV, and pavement index predicting (Flintsch, 2003).

A Fuzzy Type II method, especially a polar model, can work effectively for crack rotation prediction and rising distress index prediction. Good performance can be promoted by combining expert knowledge and ES with fuzzy and NN methods.

19.3.4   Need Analysis

The identification of pavement sections in need of maintenance, rehabilitation, reinforcement, or replacement, as well as appropriate strategies for the identified sections, involves a great deal of knowledge about the condition of the assets, effectiveness of the corrective strategies, and impact of the action on system performance (Flintsch, 2003). ESs, NNs, and FL systems are more effective techniques than traditional decision trees. It is highly recommended that performance of need analysis can be raised by designing and setting a multiagent combined HS for maintenance, rehabilitation, reinforcement, or replacement of pavement sections. FL Types I and II, can be used from a set of feasible treatments prepared using a HES for selection of maintenance and rehabilitation strategies. As shown in section 19.2.2. HNN, by name of TNN developed based on the distress types and extend present on the roadway section under evaluation for recommending appropriate pavement M and R treatments. Fuzzy Type II and III methods, combined with NNs, can be used for selecting optimum M and R strategies. From the previous discussion about smart pavement management, HS combined expert knowledge using AI techniques.

19.3.5   Ranking and Prioritization

Prioritization and ranking depend on core of thinking PMS and without the doubt to achieve a goal, uncertain, ambiguous, and numerical data. Thus, mixture of Type III, Polar, ES and ANN can be employed to handle all complex inputs (uncertain, ambiguous, and numerical data). This committee engine, designed in the center of complex systems, such as PMS. (Flintsch, 2003).

19.3.6   Optimization

One of the problems with strict mathematical optimization formulations is that the solution is not very stable. A small change in the budget allocation can result in significant changes in the optimized work program. For this reason, many decision makers have been reluctant to use this type of programming tools. PMS decisions are often based on imprecise or incomplete input parameters. There is often uncertainty surrounding specific annual costs and incomplete input parameters. HS, consisting of fuzzy techniques (such as Fuzzy Type II), can provide more flexibility with imprecise or incomplete input sets. When parameters in the optimization model have fuzzy behavior, HS can resolve dynamic programming problems and promote smart behavior of PMS by creating intelligence in the flexibility and stability of the programming process.

19.3.7 HS Implementation

HSs will improve their performance by tuning and learning over time. Self-healing in agents will improve the ability of the total performance and decrease the need for maintenance of the system. It simulates the behavior of a smart system.

19.3.8 General Conclusions

Uncertain, ambiguous, and incomplete information and data are inherent characters of PMS. This lack of data may be generated in all subsystems of PMS. These cases are hard to handle by traditional PMS methods. Maintenance and updates of such systems are neither easy nor cost-effective. Several architectures in the form of HS are proposed to solve these problems. The three main AI techniques as a base for developing HS elements were reviewed: ANNs, FL systems, and ES additionally, several examples of HS applications were presented. The following conclusions were drawn from this chapter:

- NNs, FL (Types I, II, and III), neuro-fuzzy, and multiresolution models are recommended for pavement distress detection, classification, flexible thresholding, and condition assessment.
- NNs, FL (type III and polar), and hybrid multiresolution models are recommended for pavement cracking prediction. An intelligent agent can be modified and updated intelligently.
- HSs can completely handle the computational role of combined numerical and subjective information. Complex AI methods such as NN−NN, FL−NN, FL−FL, ES−NN, ES−ES, and ES−FL have been used with acceptable outcomes.
- HSs could increase the flexibility and solution stability of the programming process. This ability makes HSs the best option to set in motion PMS toward smart PMS.

References

Alavi, A.H., Gandomi, A.H., 2011a. Robust data mining approach for formulation of geotechnical engineering systems. Eng. Comput. 28 (3), 242−274.

Alavi, A.H., Gandomi, A.H., 2011b. Prediction of principal ground-motion parameters using a HM coupling artificial neural networks and simulated annealing. Comput. Struct. 89 (23−24), 2176−2194.

Alavi, A.H., Ameri, M., Gandomi, A.H., Mirzahosseini, M.R., 2011. Formulation of flow number of asphalt mixes using a hybrid computational method. Constr. Build. Mater. 25 (3), 1338−1355.

Albert, A.P., Nii, A.O., 2008. Evaluating pavement cracks with bidimensional empirical mode decomposition. EURASIP J. Adv. Signal Process. Vol 28 doi: 10.1155/2008/861701, 7 pages.

Avci, E., 2007. An expert system based on wavelet neural network-adaptive norm entropy for scale invariant texture classification. Expert Syst. Appl. 32 (3), 919−926.

Avci, E., Turkoglu, I., 2003. Modeling of tunnel diode by adaptive-network-based fuzzy inference system. Int. J. Comput. Intell. 1 (1), 231−233.

Avci, E., Turkoglu, I., Mustafa, P., 2005a. Intelligent target recognition based on wavelet packet neural network. Expert Syst. Appl. 29 (1), 175−182.

Avci, E., Turkoglu, I., Poyraz, M., 2005b. Intelligent target recognition based on wavelet adaptive network based fuzzy inference system, Lecture Notes in Computer Science, Vol. 3522. Springer-Verlag, New York, NY, pp. 594−601.

Bishop, C.M., 1996. Neural Networks for Pattern Recognition. Clarendon Press, Oxford.

Brown, T.J., 2000. An adaptive strategy for wavelet based image enhancement. In: Proceeding of IMVIP2000 - Irish Machine Vision and Image Processing Conference Dublin, Ireland, pp. 67−81.

Candès, E.J., 1998. Ridgelets: Theory and Applications. Ph.D. thesis, Department of Statistics, Stanford University.

Candès, E.J., Donoho, D.L., 2000. Curvelets—A surprisingly effective nonadaptive representation for objects with edges. In: Cohen, A., Rabut, C., Schumaker, L., (Eds.), Curves and Surface Fitting: Saint-Malo 1999, Vanderbilt University Press, Nashville, pp. 105−120.

Candès, E., Demanet, L., Donoho, D., Ying, L., 2006. Fast discrete curvelet transforms. Applied and Computational Mathematics, pp. 1−43.

Chambon, S., Moliard, J.M., 2011. Automatic road pavement assessment with image processing review and comparison. Int. J. Geophy. 2011, 20

Chen, C., 2007. Soft Computing-based Life-Cycle Cost Analysis Tools for Transportation Infrastructure Management. Ph. D. thesis, Virginia Polytechnic Institute and State University, Blacksburg, VA, 24060.

Cheng, H.D., Chen, J.R., Glazier, C., Hu, Y.G., 1999. Novel approach to pavement distress detection based on fuzzy set theory. J. Comput. Civil Eng. 13 (4), 270−280.

Cheng, H.D., Glazier, C., Hu, Y.G., 1999a. Novel approach to pavement cracking detection based on fuzzy set theory. J. Comput. Civ. Eng. 13 (3), 270−280.

Cheng, H.D., Jiang, X., Li, J., Glazier, C., 1999b. Automated real time pavement distress analysis. Transport. Res. Rec. 1655, 55−64.

Cheng, H.D., Wang, J.L., Hu, Y.G., Glazier, C., Shi, X.J., Chen, X.W., 2001. Novel approach to pavement cracking detection based on neural network. Transport. Res. Rec. 1764, 119−127.

Chua, K.M., Xu, L., 1994. Simple procedure for identifying pavement distresses from video images. J. Transport. Eng. 120 (3), 412−431.

Claudio, R.J., Jacob, S., 2003. Adaptive image denoising and edge enhancement in scale-space using the wavelet transform. Pattern Recogn. Lett. 24, 965−971.

Culpin, D., 1986. Calculation of cubic smoothing splines for equally spaced data. Numer. Math. 48, 627−638.

Demuth, H., Beale, M., 2001. Neural Network Toolbox, User Guide. Version 4. The MathWorks Inc., Natick, MA, 4.

Dettori, L., Semlera, L., 2007. A comparison of wavelet, ridgelet, and curvelet based texture classification algorithms in computed tomography. Comput. Biol. Med. 37, 486−489.

Do, M.N., Vetterli, M., 2003a. The finite ridgelet transform for image representation. IEEE Trans. Image Process. 12 (1), 16−28.

Do, M., Vetterli, M., 2003b. Image denoising using orthonormal finite ridgelet transform. Proc. SPIE. 4119, 831−842.

Donoho, D., 2001. Ridge functions and orthonormal ridgelets. J. Approx. Theor. 111 (2), 143−179.

Donoho, D.L., Duncan, M.R., 2000. Digital curvelet transform: strategy, implementation and experiments. Proc. SPIE. 4056, 12−29.

Farook, S., Gao, X.T., 2003. Image enhancement based on a nonlinear multiscale method using dual-tree complex wavelet transform. In: IEEE Pacific RIM Conference on Communications, Computers and Signal Processing (2). pp. 716−719.

Flintsch, G.W., 2003. Soft computing applications in pavement and infrastructure management: state-of-the-art. Paper 03-3767, 82nd Annual Meeting of the Transportation Research Board, Washington, DC, Jan 11−14, 2003 (in a CD).

Gandomi, A.H., Alavi, A.H., 2011. Applications of computational intelligence in behavior simulation of concrete materials. In: Yang, X.S., Koziel, S. (Eds.), Computational Optimization and Applications in Engineering and Industry, 359. Springer, SCI, pp. 221−243.

Gandomi, A.H., Alavi, A.H., Mirzahosseini, M.R., Moqhadas Nejad, F., 2011a. Nonlinear genetic-based models for prediction of flow number of asphalt mixtures. J. Mater. Civ. Eng., ASCE. 23 (3), 248−263.

Gandomi, A.H., Alavi, A.H., Mousavi, M., Tabatabaei, S.M., 2011b. A hybrid computational approach to derive new ground-motion attenuation models. Eng. Appl. Artif. Intell. 24 (4), 717−732.

Hanbay, D., Turkoglu, I., Demir, Y., 2007. An expert system based on wavelet decomposition and neural network for modeling Chua's circuit. Expert Syst. Appl. 34 (4), 2278−2283.

Hatzilygeroudis, I., Prentzas, J., 2001. HYMES: a hybrid modular expert system with efficient inference and explanation. In: Eighth Panhellenic Conference on Informatics, Nicosia, Cyprus (1). pp. 422−431.

Hatzilygeroudis, I., Prentzas, J., 2004. Integrating (rules, neural networks) and cases for knowledge representation and reasoning in expert systems. Expert Syst. Appl. 27, 63−75.

Hsu, C.J., Chen, C.F., Lee, C., Huang, S.M., 2001. Airport Pavement Distress Image Classification Using Moment Invariant Neural Network. 22nd Asian Conference on Remote Sensing and processing (CRISP), Singapore, 2001.

Ioannis, K., Vlachos, G., Sergiadis, D., 2008. Image thresholding using type II fuzzy sets. Pattern Recognition. 41 (5), 1810−1811.

Ismail, N., Ismail, A., Atiq, R., 2009. An overview of expert systems in pavement management. Eur. J. Sci. Res. 30 (1), 99−111, ISSN 1450-216X.

Kara, B., Watsuji, N., 2003. "Using wavelets for texture classification," in IJCI Proceedings of International Conference on Signal Processing, September, pp. 920−924.

Prentzas, J., Hatzilygeroudis, I., Koutsojannis, C., 2001. A web-based ITS controlled by a hybrid expert system. In: Proceedings of the IEEE International Conference on Advanced Learning Techniques (ICALT'01).

Karabatak, M., Cevdet Ince, M., 2009. An expert system for detection of breast cancer based on association rules and neural network. Expert Syst. Appl. 33, 3465−3469.

Kaufmann, A., 1975. Introduction to the Theory of Fuzzy Subsets—Fundamental Theoretical Elements. Academic Press, New York, NY.

Laine, A.F., Fan, J., Schuler, S.A., 1994. Framework for contrast enhancement by dyadic wavelet analysis. Digital Mammography: Proceedings of the Second International Workshop on Digital Mammography. Elsevier, Oxford, UK, pp. 91−100.

Li, J., 2004. Pavement crack diseases detecting by image processing algorithm. J. Chang Univ. (Natural Science Edition). 24 (5), 24−29.

Liao, S., 2003. Knowledge management technologies and applications—literature review from 1995 to 2002. Expert Syst. Appl. 25, 155−164.

Liao, S., 2005. Expert system methodologies and applications—a decade review from 1995 to 2004. Expert Syst. Appl. 28, 93−103.

Luger, G., Stubbelfield, W., 1993. AI: Structures and Strategies for Complex Problem Solving. The Benjamin/Cummings Publishing Company Inc, University of New Mexico, Albuquerque.

Manjunath, B.S., Wu, P., Newsam, S., Shin, H.D., 2000. A texture descriptor for browsing and similarity retrieval. Signal Process. Image Commun. 16 (1), 33−43.

Mendel, J.M., 2001. Uncertain Rule-Based Fuzzy Logic Systems. Prentice-Hall, Englewood Cliffs, NJ.

Mendel, J.M., Bob John, R.I., 2002. Type-2 fuzzy sets made simple. IEEE Trans. Fuzzy Syst. 10 (2), 117−127.

Mendel, J.M., Wu, H., 2007. Type-2 fuzzistics for nonsymmetric interval type-2 fuzzy sets: forward problems. IEEE Trans. Fuzzy Syst. 15 (5), 916−930.

Mohajeri, M., 1991. An operating system of pavement distress diagnosis by image processing. Transport. Res. Rec. 1311, 120−130, TRB, National Research Council, Washington, DC.

Mora, A., Vieira, P.M., Manivannan, A., Fonseca, J.M., 2011. Automated Drusen detection in retinal images using analytical modelling algorithms. Biomed. Eng. Online. 10 (1), 59.

Mulcahy, C., 1997. Image compression using the Haar wavelet transform. Spelman Sci. Math J. 1, 22−31.

Nallamothu, S., Wang, K.C.P., 1996. Experimenting with recognition accelerator for pavement distress identification. Transport. Res. Rec. 1536, 130−135.

Nejad, F.M., Zakeri, H., 2011a. An expert system based on wavelet transform and radon neural network for pavement distress classification. Expert Syst. Appl. 38, 7088−7101.

Nejad, F.M., Zakeri, H., 2011b. An optimum feature extraction method based on wavelet radon transform and dynamic neural network for pavement distress classification. Expert Syst. Appl. 38, 9442−9460.

Nejad, F.M., Zakeri, H., 2011c. A comparison of multiresolution methods for detection and isolation of pavement distress. Expert Syst. Appl. 38, 2857−2872.

Pal, N.R., Bezdek, J.C., 1994. Measures of fuzziness: a review and several new classes. Fuzzy Sets, Neural Networks and Soft Computing. Van Nostrand Reinhold, New York, NY, pp. 194−212.

Pant, D.P., Zhou, X., Arudi, R.S., Bodocsi, A., Aktan, A.E., 1993. Neural-network-based procedure for condition assessment of utility cuts in flexible pavements. Transport. Res. Rec. 1399, TRB, National Research Council, Washington, DC.

Prentzas, J., Hatzilygeroudis, I., Koutsojannis, C., 2001. A web-based ITS controlled by a hybrid expert system. In: Proceedings of the IEEE International Conference on Advanced Learning Techniques (ICALT'01) pp. 239.

Rajeev, S., Vasquez, R.E., Singh, R., 1997. Comparison of Daubechies, Coiflet, and Symlet for edge detection. Proc. SPIE. 3074.

Sahina, S., Tolunb, M.R., Hassanpour, R., 2012. Hybrid expert systems: a survey of current approaches and applications. Expert Syst. Appl. 39 (4), 4609−4617.

Sandra, A.K., Vinayaka Rao, V.R., Raju, K.S., Sarkar, A.K., 2006. Prioritization of pavement stretches using fuzzy MCDM approach—a case study. In: 11th Online World Conference on Soft Computing in Industrial Applications, September 18-October 6, 2006. http://www.cs.armstrong.edu/wsc11/accepted.html.

Sattar, F., Floreby, L., Salomonsson, G., Lrivstriim, B., 1997. Image enhancement based on a nonlinear multiscale method. IEEE Trans. Image Process. 6, 88−95.

Shan, T., Wang, S., Zhang, X.R., Jiao, L.C., 2005. Automatic image enhancement driven by evolution based on ridgelet frame in the presence of noise. In: Proceedings of Applications of Evolutionary Computing (3449). pp. 304−313.

Starck, J.L., Donoho, D.L., Candès, E.J., 1999. Astronomical image representation by the curvelet transform. Astron. Astrophys. 398, 785−800.

Starck, J.L., Candès, E.J., Donoho, D.L., 2003. Gray and color image contrast enhancement by the curvelet transform. IEEE Trans. Image Process. 12, 706−716.

Stollnitz, E., DeRose, T., Salesin, D., 1995. Wavelets for computer graphics: a primer part 1. IEEE Comput. Graph. Appl. 15 (3), 76−84.

Tizhoosh, H.R., 2005. Previous image thresholding next using type-2 fuzzy sets. Pattern Recogn. 38, 2363−2372.

Turkoglu, I., Arslan, A., Ilkay, E., 2003. An intelligent system for diagnosis of the heart valve diseases with wave let packet neural networks. Comput. Biol. Med. 33, 319−331.

Vlachos, I.K., Sergiadis, G.D., 2008. Commenton: image thresholding using type II fuzzy sets. Pattern Recogn. 41 (5), 1810−1811.

Wan, Y., Shi, D., 2007. Joint exact histogram specification and image enhancement through the wavelet transform. IEEE Trans. Image Process. 2007, 2245−2250.

Wang, G., Xiao, L., He, A.Z., 2007. Algorithm research of adaptive fuzzy image enhancement in ridgelet transform domain. Acta Opt. Sin. 27, 1183−1190.

Winston, P.H., 1992. Artificial Intelligence. third ed. Addison-Wesley, Reading, MA.

Yu, W.-D., 2007. Hybrid soft computing approach for mining of complex construction databases. J. Comput. Civ. Eng. ASCE. 21 (5), 343−352.

Zadeh, L.A., 1975a. The concept of a linguistic variable and its application to approximate reasoning: I. Inform. Sci. 8, 199−249.

Zadeh, L.A., 1975b. The concept of a linguistic variable and its application to approximat reasoning: II. Inform. Sci. 8, 307−357.

Zhang, Y., Han, X., Ma, S.L., 2006. Feature extraction of hand-vein patterns based on ridgelet transform and local interconnection structure neural network. In: ICIC2006. LNCIS, Vol. 345, pp. 870−875.

Zhang, D., Shiru, Q., Li, H., Shuang, S., 2009a. Automatic Ridgelet image enhancement algorithm for road crack image based on fuzzy entropy and fuzzy divergence. Optic Laser. Eng. 47 (11), 1216−1225.

Zhang, D., Qu, S., He, L., Shi, S., 2009b. Automatic ridgelet image enhancement algorithm for road crack image based on fuzzy entropy and fuzzy divergence. Optic. Laser. Eng. 47 (11), 1216−1225.

Zhou, J., 2004. Automated Pavement Inspection Based on Wavelet Analysis. Ph. D. thesis, State University of New York, Stony Brook, NY.

Zhou, J., Huang, P.S., Chiang, F.-P., 2006. Wavelet-based pavement distress detection and evaluation. Opt. Eng. 45 (1), 10.

Zong, X., Laine, A.F., Geiser, E.A., Wilson, D.C., 1996. Denoising and contrast enhancement via wavelet shrinkage and nonlinear adaptive gain. Proc. SPIE. 2762, 566−574.

Printed in the United States
By Bookmasters